SCHAUM'S OUTLINE OF

THEORY AND PROBLEMS

OF

MICROBIOLOGY

•

I. EDWARD ALCAMO, Ph.D.
Professor of Biology
State University of New York at Farmingdale

•

SCHAUM'S OUTLINE SERIES

McGRAW-HILL
New York St. Louis San Francisco Auckland Bogotá Caracas
Lisbon London Madrid Mexico City Milan Montreal New Delhi
San Juan Singapore Sydney Tokyo Toronto

I. EDWARD ALCAMO is Professor of Biology at the State University of New York at Farmingdale. He has taught at the college level for thirty-plus years, specializing in microbiology for health science students. Dr. Alcamo has authored several books including *Fundamentals of Microbiology, AIDS: The Biological Basis, The Microbiology Coloring Book, DNA Technology: The Awesome Skill, Anatomy and Physiology The Easy Way,* and *The Anatomy Coloring Workbook.* He has also been honored for excellence in teaching by the State University of New York and the National Association of Biology Teachers. Dr. Alcamo is a member of numerous scientific organizations and a regular contributor to various scientific journals

Schaum's Outline and Theory and Problems of

MICROBIOLOGY

8 9 10 11 12 13 14 15 16 17 18 19 20 CUS CUS 9 8 7 6

ISBN 0-07-000967-8

Sponsoring Editor: Barbara Gilson
Production Supervisor: Pamela Pelton
Editing Supervisor: Maureen Walker

Library of Congress Cataloging-in-Publication Data

Alcamo, I. Edward
Schaum's outline of theory and problems of microbiology / I. Edward Alcamo
p. Cm. -- (Schaum's outline series)
Includes index.
ISBN 0-07-000967-8 (pbk.)
1. Microbiology--Outlines, syllabi, etc. 2. Microbiology -- Examinations, questions, etc. I. Title II. Series.
QR62.A43 1998 97-12020
579--dc21 CIP

McGraw-Hill
A Division of The ***McGraw-Hill*** *Companies*

Preface

Learning microbiology can be an exciting experience. You encounter new insights to infectious disease, new approaches to the systems of life, and new uses of technology. Microbiology has become one of the premier sciences of our day.

But microbiology can also be a daunting subject. New terms abound, new processes seem complicated, and a whole new encyclopedia of information waits to be learned. To some people, "microbiology" equates to "root canal." Even the term microbiology is imposing.

In *Schaum's Outline of Microbiology*, we've tried to reduce the stress of learning by using a theory and problem format, a type of question-answer approach. Instead of paragraph after paragraph of interminable information, we present microbiology in short bytes that can be learned piece by piece. We pose theories and problems in a logical sequence, then answer the problems accordingly. After a while it seems like you are holding a friendly conversation with your professor.

One of the beauties of this learning style is that you can stop at any time and take a breather. Each theory-problem-answer is a self-contained unit presented briefly and succinctly. You do not need to read page after page to get the concept. You take the problems one-by-one at your own pace. Soon you develop an overall sense of the topic. You are now ready to become the professor and hold a conversation with yourself.

Then it is time to prepare for an exam. You read the question, and answer it from memory, using a full essay that guides you through the information. (Educational psychologists agree that essay answers are far superior to simple words or phrases.) Since your essay is filled with information, you can expect to do well on short answer questions. As a bonus, you also learn to think and reason so you can guess intelligently. This will help your thought processes when you are confronted with critical thinking questions prominent in today's educational programs. This book teaches you how to learn conceptually and develop critical thinking skills.

Schaum's Outline of Microbiology also encourages collaborative learning because you can study with colleagues and friends. One individual poses the problem, then the group brainstorms back and forth using the answer as a guide. And when you review for an exam, all you need do is review the questions. You have a built-in study partner.

We've done our best to include all the essentials of microbiology in this book. *Schaum's Outline of Microbiology* conforms to the standard textbooks used in biology and health-related curricula. Even the sequence of topics follows the best-selling microbiology texts. We've ensured flexibility because the chapters stand alone, that is, we do not anticipate your using them in sequence.

The material in this book can be read easily and understood immediately. We've assumed no past experience in microbiology, and only a limited knowledge of biology and chemistry is expected. Essential vocabulary has been included to make you confident in using the terminology of microbiology, but the jargon has been omitted. To help you track essential processes and to provide a glimpse of microorganisms we have included over 100 pieces of art as well as numerous electron micrographs. So you can measure your progress, each chapter contains 60 review questions of various types (a total of 1500). All the answers are included in the back of the book for your reference. And do please write your notes all over this book--it is meant to be a welcome friend.

Students in various curricula will find this book valuable. For example, microbiology is part of the core curriculum in such fields as nursing, dental hygiene, pharmacy, and medical technology. It is part of premedical and predental programs, and it is included in the standard sequence for biology majors and those studying environmental and soil science, food and nutrition, biotechnology, health administration, and agriculture. Microbiology has even worked its way into the high school curriculum. And, graduate students will find the book a helpful tool for review.

But perhaps you are learning by yourself. You may be preparing for a board exam or participating in a self-directed learning program. Maybe you are studying for a college admissions exam, licensing exam, or other independent-study program such as the Regents College Exam. If any of these sound familiar, you stand to benefit from this book. The problem-answer approach is perfectly suited to your needs because it is an independent way of learning.

Here are a few other things to watch for: We've kept the chapters to approximately the same length so you can plan your study time accordingly. The key terms are bold-faced to draw your attention to them. The figures and photographs are evenly spaced so you can refer to them as you read. And each question-answer is about the same length to ensure uniformity and rhythm.

Before concluding, I would like to thank several individuals who contributed to this project. My editors at McGraw-Hill were Jeanne Flagg, Arthur Biderman, and Maureen Walker. Each brought a substantial measure of expertise to the book. Another key individual was Barbara Dunleavy. Barbara prepared the manuscript, then handled the production of the camera-ready copy you are reading. Barbara and I have worked together several years, and I continue to marvel at her prowess and dedication.

I'd also like to acknowledge the lights of my life, my children Michael Christopher, Elizabeth Ann, and Patricia Joy. Over the years, we have developed a love and friendship that has endured and deepened. Each has carved out a substantial place in society, a place that was beyond my wildest hopes or expectations for them. No father could be prouder of his brood.

And finally, a special word of love and thanks to Charlene Alice--you bring joy to my world; you put a song in my heart; you knock my socks off.

Now it's your turn. I sincerely hope you do well in microbiology, and I'll be very pleased when you get an A in your course. I would love to hear from you. Please write and let me know how well the book works for you, and how we can improve it. I can be reached at the Biology Department; State University of New York; Farmingdale, NY 11735. If you wish to call, my number is 516-420-2423. And if you subscribe to e-mail you can try me at alcamoie@farmingdale.edu.

In closing, I would like to welcome you to the wild and wonderful world of microorganisms. It is a fascinating world that elicits more than an occasional "Gee whiz." I hope you enjoy your experience.

I. EDWARD ALCAMO

Contents

Chapter 1

Introduction to Microbiology

OBJECTIVES

The opening chapter to this book introduces the science of microbiology and explores some of the basic concepts of this science. The objectives of the chapter are to:

1. Develop a broad perspective of microorganisms and recognize their place in the spectrum of living things.

2. Highlight some of the important events and personalities in the history and development of microbiology.

3. Briefly discuss the theory of spontaneous generation and the germ theory of disease as they relate to general and medical microbiology.

4. Describe the cellular characteristics of microorganisms and their designations as prokaryotes or eukaryotes.

5. Delineate how the names of microorganisms are derived.

6. Outline the various groups of microorganisms that make up the microbial world and briefly note some of their characteristics.

THEORY AND PROBLEMS

1.1 What is the subject matter of microbiology?

Microbiology is the study of microorganisms, a collection of organisms visible only with a microscope. The organisms in this group are very diverse and include the bacteria, cyanobacteria, rickettsiae, chlamydiae, fungi, microscopic algae, protozoa, and viruses, as displayed in Figure 1-1.

1.2 Are microorganisms the same as germs?

Microorganisms are the scientific name for what most people refer to as "germs." The term microorganism has a neutral connotation, whereas "germ" has a negative connotation and generally refers to something capable of causing disease. Medical microbiology is that branch of microbiology concerned with pathogenic microorganisms. General microbiology is involved with the total spectrum of microorganisms.

1.3 Are all microorganisms involved in infectious disease?

By far the largest majority of microorganisms have nothing whatever to do with infectious disease. Indeed, well over 99 percent of microorganisms contribute to the quality of human life. For example, microorganisms help maintain the balance of chemical elements in the natural environment by recycling carbon, nitrogen, sulfur, phosphorus, and other elements. In addition, microorganisms form the foundations of many food chains in the world; and microorganisms of the soil help break down the remains of all that dies.

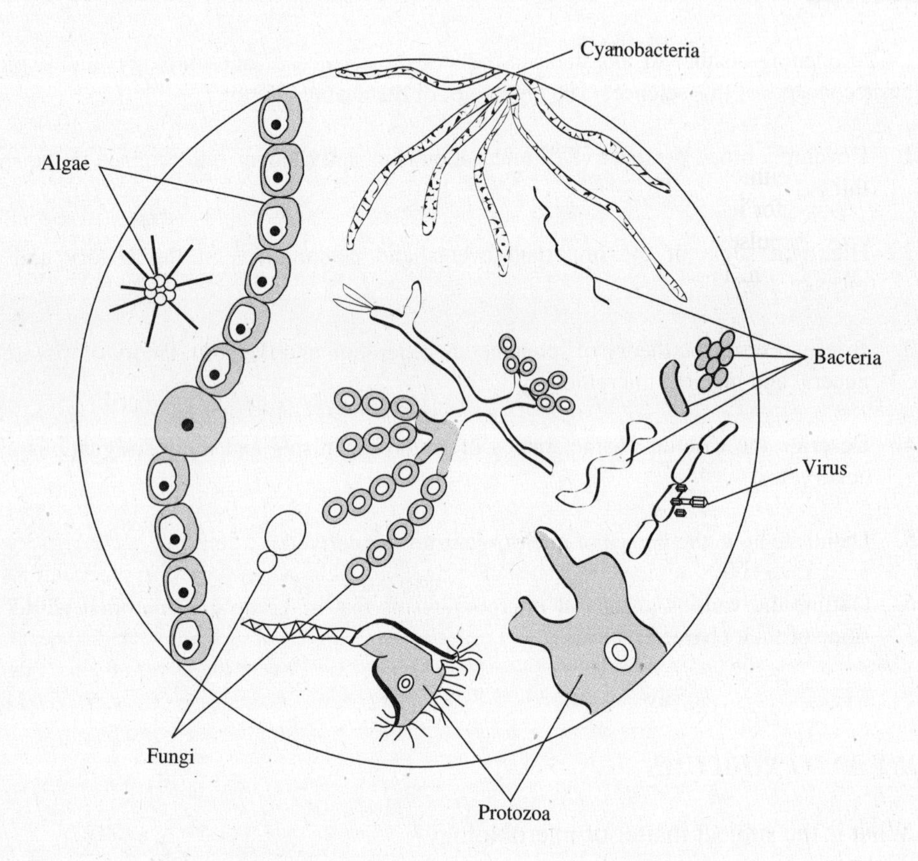

Fig. 1-1 A survey of the various types of microorganisms studied in the discipline of microbiology. Note the various shapes, sizes, and forms of the organisms.

1.4 Do any microorganisms perform photosynthesis?

Photosynthesis is the chemical process in which energy from the sun is used in the synthesis of carbon-containing compounds such as carbohydrates, which maintain the energy as chemical energy. We generally consider photosynthesis to be the domain of green plants, but certain species of microorganisms perform photosynthesis. Such organisms as **cyanobacteria** (formerly called blue-green algae) have the enzyme systems for photosynthesis. As a result of the process they contribute much of oxygen to the environment. Single-celled (unicellular) algae also perform photosynthesis and manufacture the carbohydrates used as energy sources by other organisms. In this way, microorganisms benefit all living things.

1.5 Are there any unique ways that humans derive benefits from microorganisms?

Humans derive substantial benefits from the activities of microorganisms. For example, many microbial forms live in various parts of the body and retard pathogenic bacteria from gaining a foothold. The mouth, skin, and small and large intestine are such areas. Microorganisms produce many of the foods we eat, including fermented dairy products (sour cream, yogurt, and buttermilk), as well as fermented foods as pickles, sauerkraut, breads, and alcoholic beverages. In industrial corporations, scientists cultivate microorganisms in huge quantities and use them to produce vitamins, enzymes, organic acids, and other essential growth factors.

1.6 In which way are microorganisms detrimental to health?

A minority of microorganisms cause disease in the human body. These organisms overwhelm human systems either by sheer force of numbers, or they produce powerful toxins that interfere with body systems (for instance, the bacteria that cause botulism and tetanus produce toxins that affect the flow of nerve impulses). Viruses inflict damage by replicating within tissue cells, thereby causing tissue degeneration. These conditions and many others may lead to infectious disease.

DEVELOPMENT OF MICROBIOLOGY

1.7 Who was the first to observe microorganisms?

No one is sure who made the first observations of microorganisms, but the microscope was available by the mid-1600s, and an English scientist named **Robert Hooke** made observations of cells in slices of cork tissue. He also observed strands of fungi among the specimens he viewed. Beginning in the 1670s, a Dutch merchant named **Anton van Leeuwenhoek** made careful observations of microscopic organisms which he called **animalcules**. Among his descriptions were those of protozoa, fungi, and various kinds of bacteria. Van Leeuwenhoek revealed the microscopic world to scientists of the day and is regarded as one of the first to provide accurate descriptions of the world of microorganisms.

1.8 Did the study of microbiology progress after van Leeuwenhoek's death?

After van Leeuwenhoek died, the study of microbiology did not progress rapidly because microscopes were rare and the interest in microorganisms was not high. In those years, scientists debated the theory of **spontaneous generation**, the doctrine stating that living things including that microorganisms arise from lifeless matter such as beef broth. Earlier, in the 1600s, an investigator named Francesco Redi showed that maggots would not arise from decaying meat (as others believed) if the meat were covered to prevent the entry of flies, as shown in Figure 1-2. Years later, an English cleric named John Needham advanced spontaneous generation by showing that microorganisms appear spontaneously in beef broth, but a scientist named Lazarro Spallanzani disputed the theory by showing that boiled broth would not give rise to microscopic forms of life.

1.9 What is the importance of Louis Pasteur's work in the development of microbiology?

Louis Pasteur (pictured in Figure 1-3) lived in the late-1800s. He performed numerous experiments to discover why wine and dairy products became sour, and he found that bacteria were causing the souring. In doing so, he called attention to the importance of microorganisms in everyday life and stirred scientists to think that if bacteria could make the wine "sick," then perhaps they could make humans sick.

1.10 Did Pasteur become involved in the spontaneous generation controversy?

Pasteur believed that microorganisms were in the air. If so, they might be involved in disease. If this was the case, then it was possible to become sick by inhaling microorganisms. However, many scientists continued to believe that microorganisms arose spontaneously when illness was present. Therefore, Pasteur had to disprove spontaneous generation to maintain his own theory. He devised a series of **swan-necked flasks** filled with broth. He left the flasks of broth open to the air, but the flasks had a curve in the neck so that microorganisms would fall into the neck, not the broth. Pasteur left the flasks open to the air and showed that spontaneous generation would not occur, and the flasks would not become contaminated. Pasteur's experiments, shown in Figure 1-4, seriously disputed the notion of spontaneous generation and encouraged the belief that microorganisms were in the air and could cause disease.

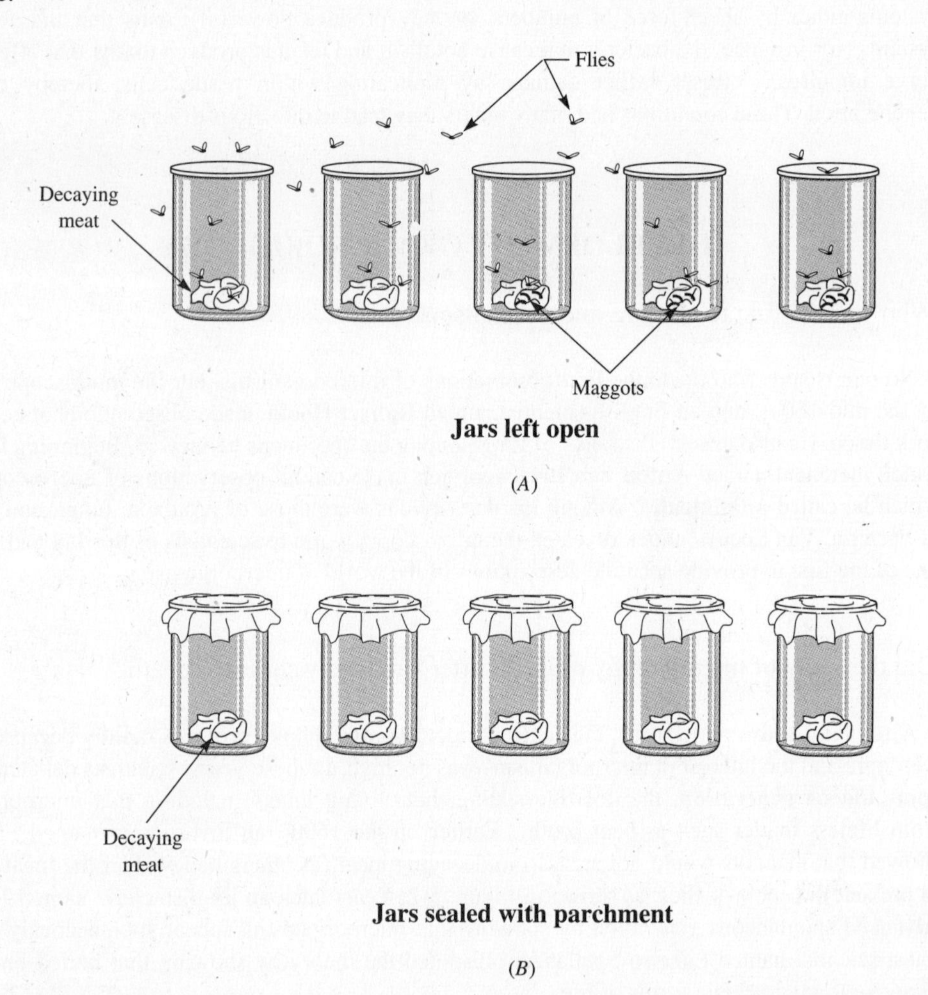

Fig. 1-2 Redi's experiments to disprove spontaneous generation. (*A*) When jars of decaying meat are left open to the air, they are exposed to flies; the flies lay their eggs on the meat, and the eggs hatch to maggots. Supporters of spontaneous generation believed that the decaying meat gives rise to the maggots. (*B*) Redi covered the jars with parchment and sealed them so the flies could not reach the decaying meat. No maggots appeared on the meat, and Redi used this evidence to indicate that the maggots did not arise from the meat but from flies in the air.

1.11 What is the germ theory of disease and what is Pasteur's role in proposing this theory?

During his work, Pasteur came to believe that microorganisms transmitted by the air could be the agents of human disease. He therefore postulated the **germ theory** of **disease**, which embodies the principle that infectious diseases are due to the activities of microorganisms in the body.

Fig. 1-3 Louis Pasteur, regarded by many to be one of the most influential scientists in the founding of microbiology.

1.12 Which scientist proved the truth of the germ theory of disease?

Although Pasteur performed numerous experiments, his attempts to prove the germ theory were unsuccessful. A German scientist named **Robert Koch** provided the proof by cultivating the bacteria that cause anthrax apart from any other type of organism. He then injected pure cultures of the anthrax bacilli into mice and showed that the bacilli invariably caused anthrax. These experiments proved the germ theory of disease. The procedures used by Koch came to be known as **Koch's postulates**. They are illustrated in Figure 1-5. They provided a set of principles whereby the cause of a particular disease could be identified.

1.13 Did other scientists develop the work of Pasteur and Koch?

In the late 1880s and during the first decade of the 1900s, scientists throughout the world seized the opportunity to further develop the germ theory of disease as enunciated by Pasteur and proved by Koch. There emerged a Golden Age of Microbiology during which many agents of different infectious diseases were identified. Many of the etiologic agents of microbial disease trace their discovery to that period of time.

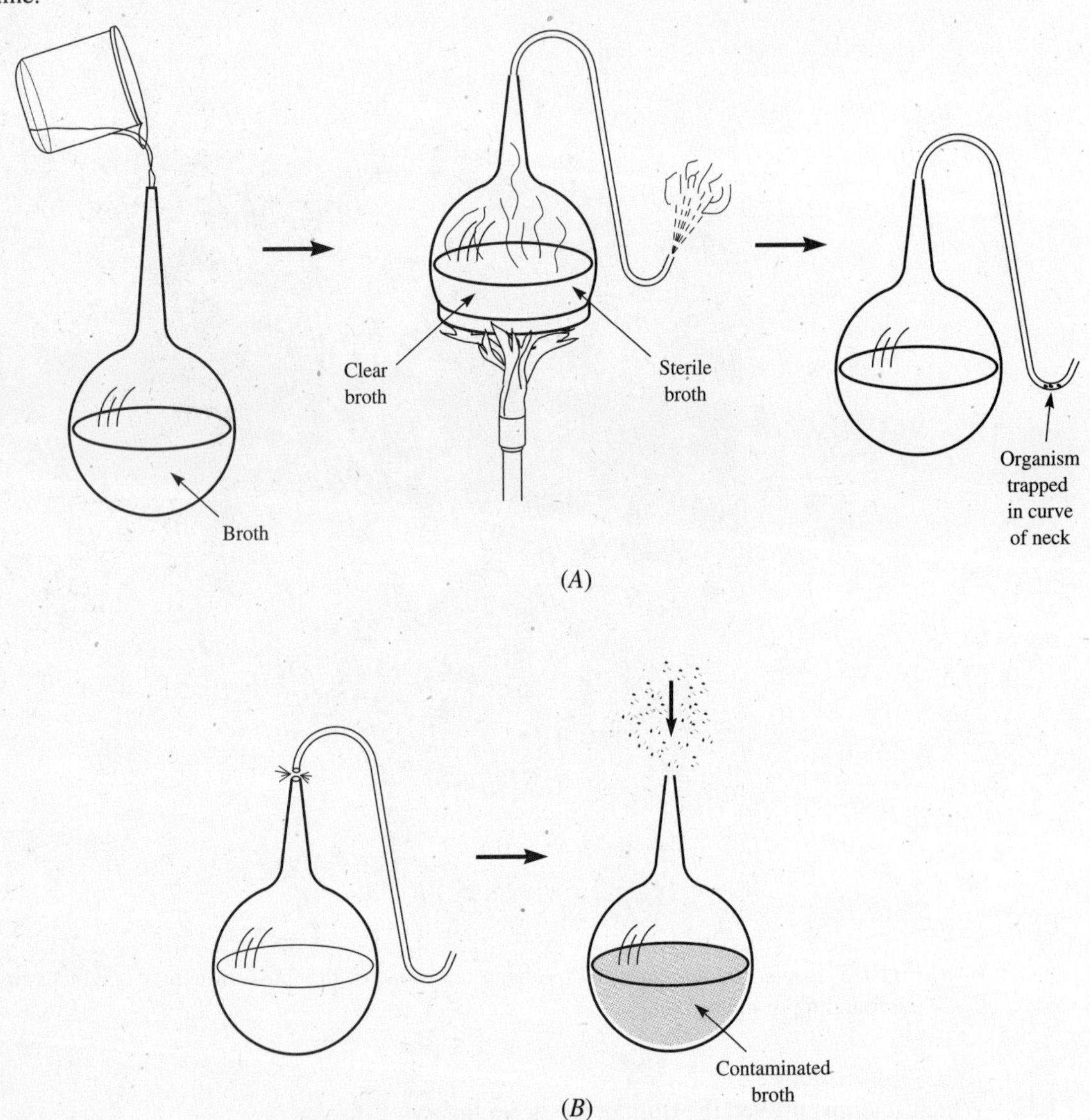

Fig. 1-4 The swan-necked experiment of Louis Pasteur, which he used to disprove spontaneous generation. (*A*) Nutrient-rich broth is placed in a flask and the neck is drawn out in the shape of a swan's neck. When the flask of broth is heated, the broth becomes sterile as air and organisms are driven away by the heat. The broth remains sterile because organisms entering the open-necked flask are trapped in the curve of the neck. Pasteur used this experiment to show that microorganisms come from the air rather than from the broth. (*B*) When the neck of the flask is removed, microorganisms enter the neck and the flask soon becomes filled with microorganisms and is contaminated. Pasteur used this evidence to further show that microorganisms exist in the air and that they originate from the air rather than from lifeless matter.

1.14 What was the practical effect of believing in the germ theory of disease?

Believing in the germ theory of disease implied that epidemics could be halted by interrupting the spread of microorganisms. Therefore, public health officials began a concerted effort to purify water, ensure that food was prepared carefully, pasteurize milk, isolate infected patients, employ insect control programs, and institute other methods to interrupt the spread of disease. Epidemics soon declined with these new methods of infection control.

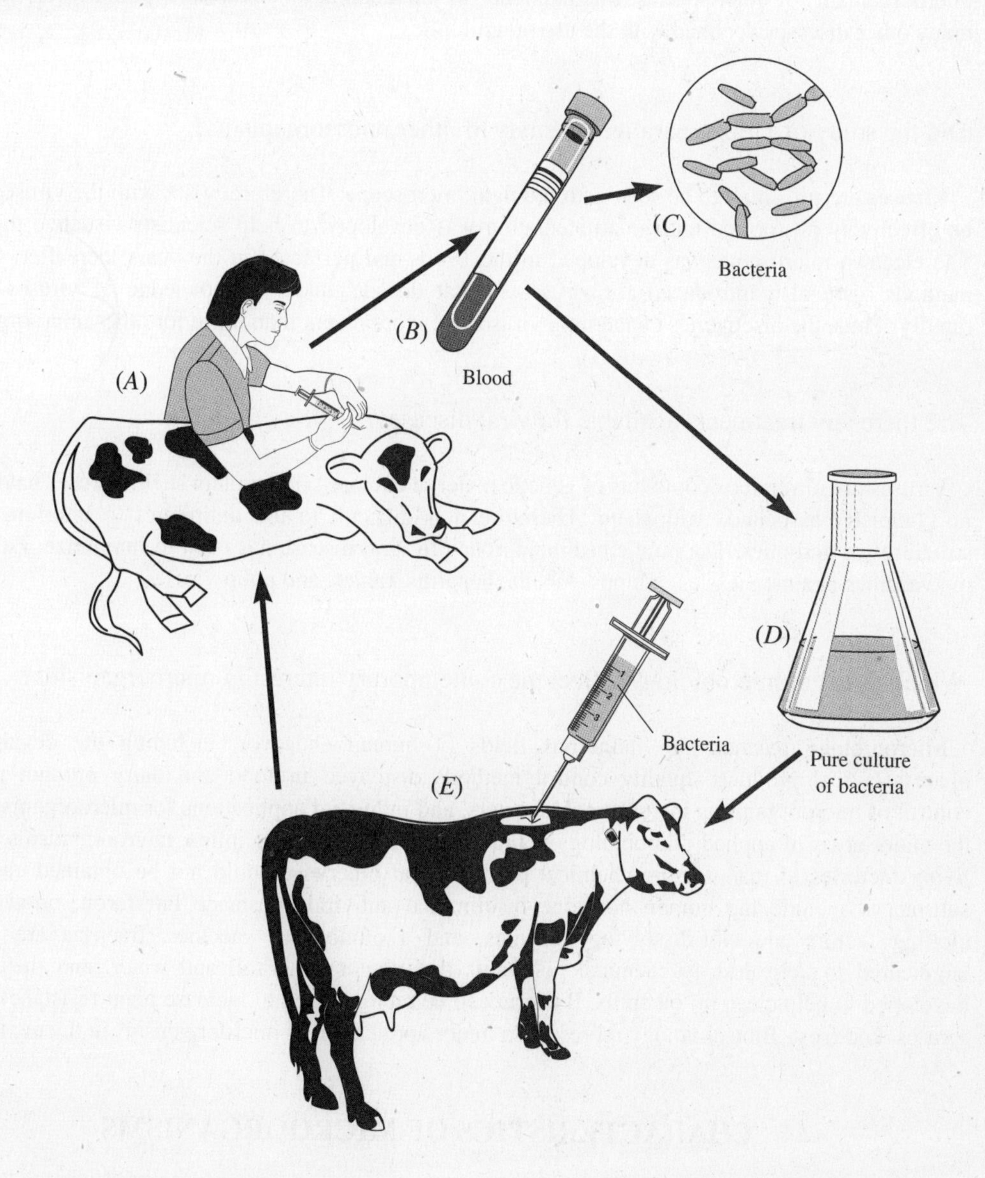

Fig. 1-5 Koch's postulates. (*A*) Blood is drawn from a sick animal and (*B*) brought to the laboratory. (*C*) A sample of the blood reveal bacteria. (*D*) The bacteria from the blood are cultivated in a pure culture in the laboratory. (*E*) A sample of the pure culture containing only one kind of bacteria is injected into a healthy animal. If the animal becomes sick and displays the same symptoms as the original animal, then evidence exists that this particular disease is caused by this particular organism.

1.15 When were cures for established diseases introduced to microbiology?

Through the early part of the 1900s it became possible to prevent epidemics, but it was rarely possible to render any life-saving therapy on an infected patient. Then, after World War II, **antibiotics** were introduced to medicine, and cures for infectious disease could be effected. Antibiotics are chemotherapeutic agents derived from microorganisms and having a substantial killing power on microorganisms of other species. The incidence of pneumonia, tuberculosis, typhoid fever, syphilis, and many other diseases declined with the use of antibiotics.

1.16 Did the study of viruses parallel the study of other microorganisms?

Viruses are too small to be seen with the light microscope. Therefore, work with the viruses could not be effectively performed until instrumentation was developed to help scientists visualize these agents. The electron microscope was developed in the 1940s and perfected in the years thereafter. Cultivation methods were also introduced for viruses in that decade, and the knowledge of viruses developed rapidly. Thus, the discoveries concerning viruses are more recent than those for other microorganisms.

1.17 Are there any treatments available for viral diseases?

Viruses are ultramicroscopic bits of genetic material enclosed in a protein shell. Viruses having little or no chemistry associated with them. Therefore it is difficult to use antibiotics to interfere with viral structures or activities. The public health approach to viral disease has been to immunize. Examples are the vaccines against measles, mumps, rubella, hepatitis, rabies, and polio viruses.

1.18 Which fields of microbiology reflect the contemporary interest in microorganisms?

Microbiology reaches into numerous fields of human endeavor, including the development of pharmaceutical products, quality control methods displayed in food and dairy product production, control of microorganisms in consumable waters, and industrial applications for microorganisms. One of the major areas of applied microbiology is **biotechnology**. In this discipline, microorganisms are used as living factories to manufacture chemical products that otherwise could not be obtained easily. These substances include the human hormone insulin, the antiviral substance interferon, numerous blood clotting factors and clot-dissolving enzymes, and a number of vaccines. Bacteria are also being engineered to help destroy chemical pollutants that contaminate soil and water, and they are being developed to help clean up oil spills. Bacteria can be reengineered to increase plant resistance to insects, viruses, and frost. Biotechnology represents a major application of microorganisms in the next century.

CHARACTERISTICS OF MICROORGANISMS

1.19 Do microorganisms share characteristics with other kinds of organisms?

With the exception of viruses, microorganisms share a cellular basis with all other organisms. Both microbial and other cells contain cytoplasm, in which enzymes are used to catalyze the chemical reactions of life. The hereditary substance of microbial and other living cells is deoxyribonucleic acid (DNA); and a major share of the energy is stored in adenosine triphosphate (ATP). In addition, microorganisms undergo a form of reproduction in which the DNA duplicates and is segregated to the new daughter cells. In all these instances, microorganisms are similar to other organisms.

1.20 Where are microorganisms classified with respect to other organisms?

Microorganisms have a set of characteristics that place them in either of the two major groups of organisms: prokaryotes and eukaryotes. Certain microorganisms such as bacteria are **prokaryotes** because of their cellular properties, while other microorganisms such as fungi, protozoa, and unicellular algae are **eukaryotes**. The specific differences between these groups are discussed in Chapter 4. Because of their simplicity and unique characteristics, viruses are neither prokaryotes nor eukaryotes.

1.21 Who devised the method for naming the microorganisms?

The system of nomenclature used for all living things is applied to microbial forms. This system was established in the mid-1700s by the Swedish botanist Carolus Linnaeus. All organisms are given a binomial name.

1.22 How is the binomial name for a microorganism developed?

The **binomial name** consists of two names: the genus to which the organism belongs and a modifying adjective called the species modifier. The first letter of the genus name is capitalized and the remainder of the genus name and the species modifier are written in lowercase letters. The entire binomial name is either italicized or underlined. It can be abbreviated by using the first letter of the genus name and the full species modifier. An example of a microbial name is *Escherichia coli*, the bacterial rod found in the human intestine. The name is abbreviated *E. coli.*

1.23 Are microorganisms placed in any one kingdom?

Microorganisms are distributed through various kingdoms of living things. The generally accepted classification of living things is that devised by Robert Whittaker of Cornell University in 1969. Whittaker suggested a five-kingdom classification of living things. Microorganisms are found in three of the five kingdoms.

1.24 What are the five kingdoms in the Whittaker classification?

The first of the five kingdoms in the Whittaker classification is **Monera**, as shown in Figure 1-6. Prokaryotes, such as bacteria and cyanobacteria (formerly blue-green algae) are in this kingdom; the second kingdom, **Protista,** includes protozoa, unicellular algae, and slime molds, all of which are eukaryotes and single-celled; in the third kingdom, **Fungi**, are the molds, mushrooms, and yeasts. These organisms are eukaryotes that absorb simple nutrients from the soil. The remaining two kingdoms are **Plantae** (plants) and **Animalia** (animals). Plants are multicellular eukaryotes that synthesize their foods by photosynthesis, while animals are multicellular eukaryotes that digest large food molecules into smaller ones for absorption.

1.25 What are some characteristics of the bacteria?

Bacteria are microscopic, relatively simple, prokaryotic organisms whose cells lack a nucleus or nuclear membrane. The bacteria may appear as rods (bacilli), spheres (cocci), or spirals (spirilla or spirochetes). Bacteria reproduce by binary fission, have unique cell walls, and exist in most environments on earth. They live at temperatures ranging from 0 to over 100°C and in conditions that are oxygen-rich or oxygen-free.

1.26 What are some of the important characteristics of fungi?

Fungi include unicellular yeasts and filamentous molds. The **yeasts** are single-celled organisms slightly larger than bacteria and are used in industrial fermentations and bread making. **Molds** are branched chains of cells that generally form spores for use in reproduction. The fungi prefer acidic environments, and most live at room temperature under conditions rich in oxygen. The common mushroom is a fungus.

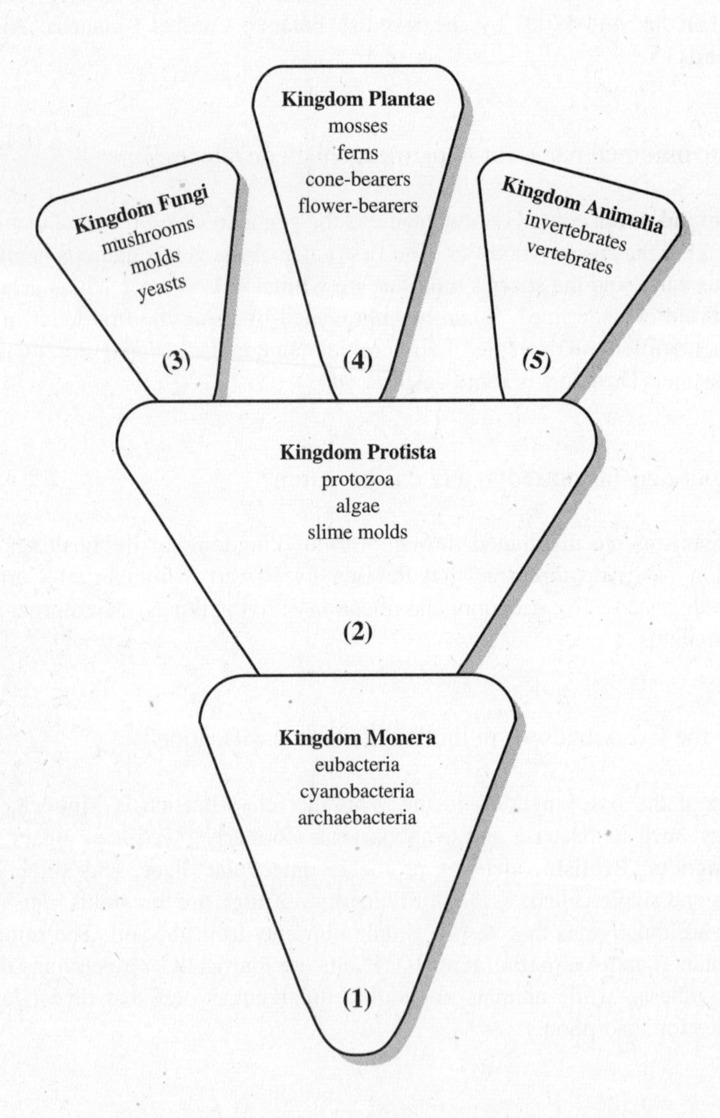

Fig. 1-6 The five kingdoms of microorganisms according to the classification system established by Robert Whittaker. Microorganisms occupy a prominent position in three of the five kingdoms.

1.27 Which major characteristics distinguish the protozoa?

Protozoa are eukaryotic single-celled organisms. They are classified according to how they move: some protozoa use flagella, others use cilia, and still others use pseudopodia. Protozoa exist in an infinite variety of shapes because they have no cell walls. Many are important causes of human diseases such as malaria, sleeping sickness, dysentery, and toxoplasmosis .

1.28 Which characteristics distinguish the algae?

The term **algae** implies a variety of plantlike organisms. In microbiology, there are several important types of single-celled algae that are important. Examples are the diatoms and dinoflagellates that inhabit the oceans and exist at the bases of the food chain. Most algae capture sunlight and transform it in photosynthesis to the chemical energy in carbohydrates.

1.29 Why are viruses not considered organisms in the strict sense?

Organisms are distinguished by their ability to grow, experience the chemical reactions of metabolism, reproduce independently, evolve in their environments, and display a cellular level of organization. Viruses do not have any of these characteristics. They consist of fragments of DNA or RNA enclosed in protein. Sometimes the protein is surrounded by a membranelike envelope. Viruses reproduce, but only within living host cells. They are acellular (noncellular) particles that display one characteristic of living things: replication. Replication happens only when living cells are available to assist the viruses and provide the chemical components, structure, and energy required.

Review Questions

Multiple Choice. Select the letter of the item that best completes each of the following statements.

1. The majority of microorganisms (*a*) are protozoa (*b*) contribute to the quality of life (*c*) live at the bottom of the ocean (*d*) are found in outer space.

2. The characteristic feature that applies to all microorganisms is (*a*) they are multicellular (*b*) their cells have distinct nuclei (*c*) they are visible only with a microscope (*d*) they perform photosynthesis.

3. Among the foods produced for human consumption by microorganisms is (*a*) milk (*b*) ham (*c*) yogurt (*d*) cucumbers.

4. Among the first scientists to see microorganisms was (*a*) Robert Hooke (*b*) Louis Pasteur (*c*) Joseph Lister (*d*) James T. Watson.

5. The theory of spontaneous generation states that (*a*) microorganisms arise from lifeless matter (*b*) evolution has taken place in large animals (*c*) humans have generated from apes (*d*) viruses are degenerative forms of bacteria.

6. Extensive studies on the microorganisms were performed in the 1670s by the Dutch merchant (*a*) van Gogh (*b*) van Hoogenstyne (*c*) van Dyck (*d*) van Leeuwenhoek.

7. Louis Pasteur's contribution to microbiology was that he (*a*) discovered viruses (*b*) supported the theory of spontaneous generation (*c*) attacked the doctrine of evolution (*d*) called attention to the importance of microorganisms in everyday life.

8. The concept stating that microorganisms are the causes of infectious disease is known as the (*a*) theory of evolution (*b*) germ theory of disease (*c*) cell theory of biology (*d*) theory of diminishing returns.

9. The swan-necked flasks were used by Pasteur to (*a*) disprove the theory of spontaneous generation (*b*) prove that microorganisms cause disease (*c*) learn the chemical structure of microorganisms (*d*) devise a classification scheme for bacteria and protozoa.

10. Cures for established cases of disease were introduced to microbiology with the (*a*) work of Hooke (*b*) discovery of antibiotics (*c*) description of the structure of DNA (*d*) developments of genetic engineering.

11. Effective work with the viruses depended upon the development of the (*a*) light microscope (*b*) dark-field microscope (*c*) ultraviolet light microscope (*d*) electron microscope.

12. All the following characteristics are associated with viruses except (*a*) they have little or no chemistry (*b*) antibiotics are used to interfere with their activities (*c*) they cause measles, mumps, and rubella (*d*) they are not types of bacteria.

13. The hereditary substance in all microbial cells is (*a*) ATP (*b*) DPN (*c*) DNA (*d*) AMP.

14. A packet of nucleic acid enclosed in protein best describes a (*a*) alga (*b*) RNA molecule (*c*) virus (*d*) bacterium.

15. The two components of the binomial name of a microorganism are the (*a*) order and family (*b*) family and genus (*c*) genus and species modifier (*d*) genus and variety.

16. Bacteria and cyanobacteria are classified together in the kingdom (*a*) Protista (*b*) Halophyta (*c*) Chlorophyta (*d*) Monera.

17. The two groups of organisms found in the kingdom Fungi are (*a*) viruses and yeasts (*b*) yeasts and molds (*c*) molds and bacteria (*d*) bacteria and protozoa.

18. Robert Koch is remembered in microbiology because he (*a*) proved the germ theory of disease (*b*) successfully cultivated viruses in the laboratory (*c*) developed a widely accepted classification scheme (*d*) devised the term prokaryote.

19. Among the single-celled algae of importance in microbiology are the (*a*) amoebas and ciliates (*b*) rods and cocci (*c*) RNA and DNA viruses (*d*) dinoflagellates and diatoms.

20. The cyanobacteria are notable for their ability to perform (*a*) binary fission (*b*) heterotrophic nutrition (*c*) photosynthesis (*d*) movement.

Matching. Use one of the following five choices for each of the characteristics noted below by placing the correct letter in the space.

____ 1. Observed only with electron microscope
____ 2. Classified as Protista

(*a*) Bacteria
(*b*) Fungi

____ 3. Perform photosynthesis
____ 4. Composed of nucleic acid and protein shell
____ 6. Rods, spheres, and spirals
____ 7. Filamentous, branched organisms
____ 8. Formerly called blue-green algae
____ 9. Cause hepatitis and polio
___ 10. Used in genetic engineering

(*c*) Viruses
(*d*) Protozoa
(*e*) Cyanobacteria

True/False. For each of the following statements, mark the letter "T" next to the statement if the statement is true. If the statement is false, change the underlined word to make the statement true.

___ **1.** Although photosynthesis is generally considered to be the domain of plants, certain microorganisms such as viruses and dinoflagellates also perform this process.

___ **2.** One way in which microorganisms cause disease is by producing powerful toxins that interfere with body systems.

___ **3.** Disease due to the activity of microorganisms is correctly known as physiological disease.

___ **4.** The English scientist Robert Hooke made observations of strands of bacteria in the specimens he viewed.

___ **5.** Microbiology failed to progress rapidly after the death of van Leeuwenhoek because microscopes were rare.

___ **6.** Both Redi and Spallanzani disputed the theory of evolution.

___ **7.** The experiments performed early in the career of Louis Pasteur demonstrated that meat became sour.

___ **8.** The germ theory of disease was initially postulated by a scientist named Robert Koch.

___ **9.** After the germ theory was established, it became possible to interrupt epidemics of disease.

__ **10.** Before the 1840s, the electron microscope was not available, and viruses could not be seen.

__ **11.** Among the favorite tools of genetic engineering are the fungi.

__ **12.** Among the characteristics shared by microorganisms and other kinds of organisms is the presence of cellular organelles.

__ **13.** Fungi, protozoa, and unicellular algae are classified together as prokaryotes.

__ **14.** Viruses consist of fragments of nucleic acids enclosed in a shell of carbohydrate.

__ **15.** The genus and species modifier together make up the binomial name for a microorganism.

__ **16.** The mushrooms, molds, and yeasts are classified together in the kingdom Protista.

__ **17.** Bacteria are prokaryotic organisms whose cell lacks a nucleus.

__ **18.** For the cultivation of fungi, an environment that is basic is preferred.

__ **19.** Among the organs of motion present in protozoa are cilia, flagella, and pseudopodia.

__ **20.** The algae are a group of microorganisms somewhat like the animals.

__ **21.** Among the diseases caused by viruses are measles, mumps, and plague.

__ **22.** Hooke used the term animalcules to refer to the microorganisms he observed.

__ **23.** Microorganisms form the foundations of many food chains in the world.

__ **24.** Viruses inflict their damage and cause tissue degeneration by replicating within living cells.

__ **25.** Antibiotics were introduced to medicine shortly after the Civil War.

__ **26.** The public health approach against viral disease has been to use antibiotics.

__ **27.** The system of nomenclature used for all organisms was established by Robert Koch.

__ **28.** When expressing the binomial name of a microorganisms, the name is either italicized or boldfaced.

__ **29.** The classification system suggested by Robert Whittaker includes six kingdoms.

__ **30.** Bacteria appear in numerous variations of rods, spheres, or triangles.

Chapter 2

The Chemical Basis of Microbiology

OBJECTIVES

All microorganisms have a chemical basis in their growth, metabolism, and pathogenic and environmental activities. This chapter explores the chemistry of microorganisms with an emphasis on the organic molecules that make up their structure and enzymes. The objectives of the chapter are to:

1. Differentiate organic molecules from inorganic molecules.
2. Review some fundamental elements of atomic structure and chemical bonding.
3. Define some essential concepts of chemical reactions, water, and acids and bases.
4. Identify the important characteristics and uses of carbohydrates, lipids, and proteins.
5. Understand how the nucleic acids interrelate with proteins to specify an amino acid sequence in the protein.
6. Specify how DNA replicates in the semiconservative mechanism.

THEORY AND PROBLEMS

2.1 When did scientists first realize that the chemical components of living things could be synthesized?

During the 1800s, scientists discovered that the compounds of living things such as microorganisms could be formulated in the laboratory. Friedrich Wohler's production of urea in 1828 was one of the first such syntheses. Wohler synthesized the organic substance urea, which is a component of human urine. After that time it became apparent that a study of chemistry is intimately linked with the study of biology. When work in microbiology developed in the mid-1800s, the chemistry of microorganisms came under close scrutiny. Louis Pasteur, one of the founders of microbiology, began his career as a chemist and performed seminal experiments on the chemistry of yeast fermentations (Chapter 1).

2.2 What are organic compounds, and how do they differ from inorganic compounds?

Chemical substances associated with living things such as microorganisms are called **organic compounds**; all other compounds in the universe are termed **inorganic compounds**. The four major organic substances found in all microorganisms and other living things are carbohydrates, lipids, proteins, and nucleic acids. They are the main subject matter of this chapter. The discipline of organic chemistry is essentially the chemistry of carbon.

CHEMICAL PRINCIPLES

2.3 Which are the fundamental substances of which all chemical compounds are composed?

All matter in the universe is composed of one or more fundamental substances known as **elements**. Ninety-two elements are known to exist naturally, and certain others have been synthesized by scientists. An element cannot be decomposed to a more basic substance by chemical means. Examples of elements are oxygen, iron, calcium, sodium, hydrogen, carbon, and nitrogen.

2.4 How are the elements designated?

Elements are designated by symbols often derived from Latin . For example, sodium (from the Latin word *natrium*) is abbreviated as Na; potassium (from *kalium*) is expressed as K; and iron (from *ferrum*) is expressed as Fe. Other symbols are derived from English names: H stands for hydrogen; O for oxygen; N for nitrogen; and C for carbon. Carbon, oxygen, hydrogen, and nitrogen make up over 90 percent of the weight of a typical microorganism such as a bacterium.

2.5 What are the fundamental units of elements and what are these units composed of?

Elements are composed of **atoms,** the smallest part of an element entering into combinations with other atoms (Figure 2-1). An atom cannot be broken down further without losing the properties of the element. Atoms consist of positively charged particles called **protons** surrounded by negatively charged particles called **electrons**. A proton is about 1835 times the weight of an electron. A third particle, the **neutron**, has no electrical charge; it has the same weight as a proton. Protons and neutrons adhere tightly to form the dense, positively charged nucleus of the atom; electrons spin around the nucleus. The **atomic number** is the number of protons found in an atom, while the **mass number** is the total number of protons and neutrons in an atom.

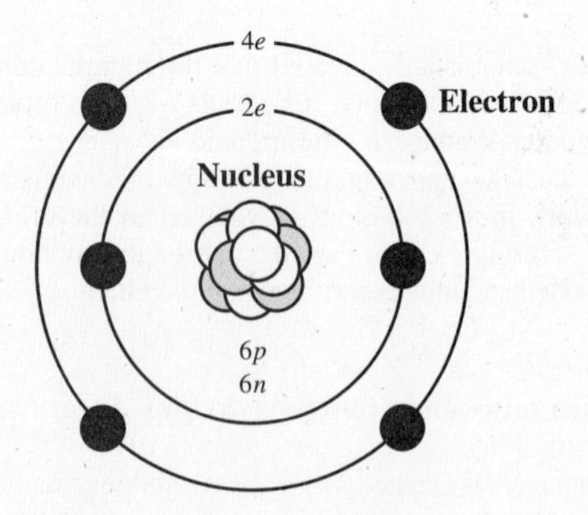

Fig. 2-1 A representation of an atom with illustrations of some of its important substructures.

2.6 Why is the arrangement of electrons in an atom important to its chemistry?

The arrangement of electrons in an atom is important to its chemistry because atoms are most stable when their outer shell of electrons has a full quota. For hydrogen this quota is two electrons, while for other elements it is eight electrons. Chemical reactions occur because electrons tend to gain or lose electrons until their outer shells are full and the atom is stable. An element whose atoms have a full outer shell is an **inert element** because its atoms do not enter into reactions with other atoms. Helium and neon are inert elements.

2.7 What is the difference between oxidation and reduction reactions?

When a chemical reaction results in a loss of electrons, it is called an **oxidation**. The molecule losing the electron is oxidized. When a reaction results in a gain of electrons, it is called a **reduction**. The molecule gaining the electron is reduced. Reactions as these usually occur together and are called oxidation-reduction reactions. They are important aspects of the energy metabolism occurring in the cytoplasm of microorganisms and are discussed later in this book.

2.8 Explain the difference between an atom, an ion, and an isotope.

Atoms are uncharged when they contain the same number of protons and electrons. When they lose or gain electrons, however, they acquire a charge and become **ions**. An ion may have a positive charge if it has an extra proton, or a negative charge if it possesses an extra electron. Sodium ions, calcium ions, potassium ions, and numerous other types of ions are important in microbial physiology. Although the number of protons is the same for all atoms of an element, the number of neutrons may vary. Variants such as these are called **isotopes**. Isotopes have the same atomic numbers, but different mass numbers. They are used as tracers in microbial research.

2.9 What are molecules and how do they relate to compounds?

Molecules are precise arrangements of atoms derived from different elements. An accumulation of molecules is a **compound**. A molecule may also be defined as the smallest part of the compound that retains the properties of the compound. For example, water is a compound composed of water molecules (H_2O), while glucose is composed of carbon, oxygen, and hydrogen ($C_6H_{12}O_6$) . In these situations, there are different kinds of atoms in the molecule. In other situations such as in hydrogen gas (H_2) or oxygen gas (O_2), the compound is composed of a single type of atom. Compounds make up the majority of the constituents of microbial cytoplasm.

2.10 How is the molecular weight of a compound determined and how is it expressed?

The **molecular weight** of a compound is equal to the atomic weights of the atoms in the molecule. For example, the molecular weight of water (H_2O) is 18 since the atomic weight of oxygen is 16 and hydrogen is 1. Molecular weights are expressed in **daltons** (a dalton is the weight of a hydrogen atom). They give a relative idea of a molecule's size. The molecular weight of an antibody molecule, for example, is measured in hundreds of thousands.

2.11 In what form are atoms linked to one another?

Atoms are linked to one another in molecules by associations called **chemical bonds**. In order for a chemical bond to form, the atoms must come close enough for their electron shells to overlap. Then an electron exchange or an electron sharing will occur.

2.12 What is an ionic bond and how does it form?

An **ionic bond** forms when the electrons of one atom transfer to a second atom. This transfer results in electrically charged atoms, or ions (Figure 2-2). The electrical charges are opposite to one another (i.e., positive and negative), and the oppositely charged ions attract one another. The attraction results in the ionic bond. Sodium chloride is formed from sodium and chloride ions drawn together by ionic bonding. Sodium and chloride ions often exist in the cytoplasm of a microorganism.

2.13 What is the basis for the formation of the covalent bond?

The second type of chemical bond is the **covalent bond**. A covalent bond forms when two atoms share one or more electrons. For example, carbon shares its electrons with four hydrogen atoms in methane molecules (CH_4), the gas formed by many bacteria when they grow in the absence of oxygen. Oxygen and hydrogen atoms share electrons in water molecules (H_2O). When a single pair of electrons is shared, the bond is a **single bond**; when two pairs are shared, then the bond is a **double bond**. Carbon is well-known for its ability to enter into numerous covalent bonds because it has four electrons in its outer shell. Thus it can combine with four other atoms or groups of atoms. So diverse are the possible carbon compounds that the chemistry of microorganisms is basically the chemistry of carbon.

2.14 What are the characteristics of a hydrogen bond?

A third type of linkage is the **hydrogen bond**. This is a weak bond. It forms between protons and free pairs of electrons on adjacent molecules, and is so-named because it exists in water molecules. The hydrogen bond is also known as an **electrostatic bond**, alternately called Van der Waal forces. It is also found between the components of nucleic acids in microorganisms, and it helps hold the strands of DNA together in the double helix.

2.15 What are the various types of chemical reactions occurring in microorganisms?

When molecules interact with one another and form new bonds, the process is called a **chemical reaction**. The **reactants** in a chemical reaction may enter interactions to form various **products**. For example, there may be a switch of parts among reactant molecules; or water may be introduced in a reaction known as a hydrolysis; or an oxidation-reduction reaction involving an exchange of electrons may occur.

2.16 Why is water important in the chemistry of microorganisms?

Water is an important aspect of many chemical reactions either as a reactant of the reaction or a molecule resulting from the reaction. Water is the universal solvent in microorganisms, and virtually all the chemical reactions of microbiology occur in water. Over 75 percent of the weight of a microorganism is water. Water participates in most functions of microorganisms and is the medium in which they live in the world.

2.17 What differences distinguish the acids and bases?

An **acid** is a chemical compound that releases hydrogen atoms when placed in water. Hydrochloric acid releases hydrogen atoms when placed in water. An acid can be a strong acid (such as hydrochloric, sulfuric, and nitric acids) if it releases many hydrogen ions, or a weak acid (for example, carbonic acid) if it releases few hydrogen ions. Certain chemical compounds attract hydrogen atoms when they are placed in water. These substances are **bases**. Typical bases include sodium hydroxide (NaOH) and potassium hydroxide (KOH). When these compounds are placed in water, they attract hydrogen ions from the water molecules, leaving behind the hydroxyl (—OH) ions. A basic (or alkaline) solution results. Both NaOH and KOH are strong bases, while substances such as guanine and adenine are weak bases. Acids and bases are generally destructive to microbial cytoplasm because they interfere with the chemistry taking place there.

2.18 How is pH defined?

The measure of acidity or alkalinity of a substance is the **pH**. The term pH refers to the hydrogen ion concentration of a substance. When the number of hydrogen ions and hydroxyl ions is equal, the pH of the substance is 7.0. (Pure water has a pH of 7.) Decreasing pH numbers represent more acidic substances, and the most acidic substance has a pH of 1.0. Alkaline substances have pH numbers higher than seven, and the most alkaline substance has a pH of 14.0. Most microorganisms prefer to live in a pH environment close to 7.0. The notable exception is fungi, which prefer a lower pH, close to 5.0.

ORGANIC COMPOUNDS OF MICROORGANISMS

2.19 What general characteristics apply to the carbohydrates?

Carbohydrates serve as structural materials and energy sources for microorganisms. They are composed of carbon, hydrogen, and oxygen; the ratio of hydrogen atoms to oxygen atoms is 2:1. Carbohydrates are produced during the process of photosynthesis in cyanobacteria, and they are broken down to release energy during the process of respiration taking place in all bacteria and other microorganisms.

2.20 Name some simple carbohydrates and describes their properties.

The simple carbohydrates are commonly referred to as **sugars**. Sugars are also known as **monosaccharides** if they are composed of single molecules. The most widely encountered monosaccharide in microorganisms is **glucose**, which has the molecular formula ($C_6H_{12}O_6$). Glucose is the basic form of fuel for microbial life and is metabolized to release its energy. Other monosaccharides are **fructose** and **galactose**. All have the same molecular formula ($C_6H_{12}O_6$), but the atoms are arranged differently. Molecules as these are called **isomers**. Glucose provides much of the energy used by microorganisms during their life activities such as movement, toxin formation, and ingestion of nutrients. The energy is released and used to form ATP, which is an immediate energy source. Chapter 6 treats this topic in more detail.

2.21 Discuss the characteristics of some disaccharides.

Disaccharides are composed of two monosaccharide molecules covalently bonded to one another. Three important **disaccharides** are associated with microorganisms: **maltose** is a combination of two glucose units broken down by yeasts during beer fermentations; **sucrose** (table sugar) is formed by linking glucose to fructose (Figure 2-3) and is digested by bacteria involved in tooth decay; and **lactose** is composed of glucose and galactose molecules and is broken down by bacteria during the souring of milk, which contains much lactose.

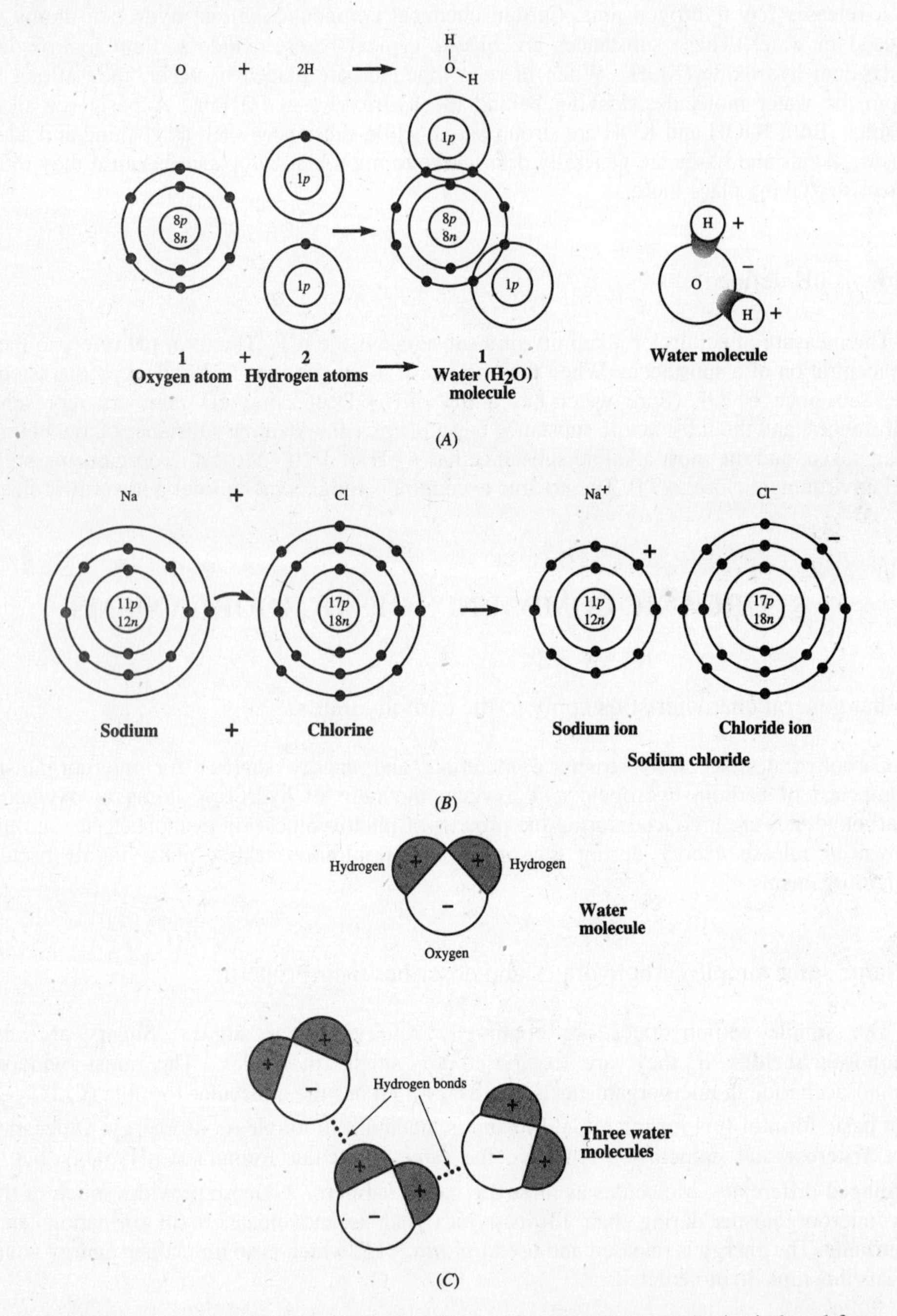

Fig. 2-2 Three types of bonds found in the atoms of molecules. (*A*) An ionic bond forms when an electron or electrons transfers from one atom to the next, thereby forming ions. The ions then attract one another, establishing the ionic bond. (*B*) A covalent bond forms when two or more atoms share electrons so as to complete the outer shell with eight electrons (two for hydrogen). (*C*) A hydrogen bond is a weak bond that forms between free pairs of electrons and nearby protons. Water molecules are held together by hydrogen bonds.

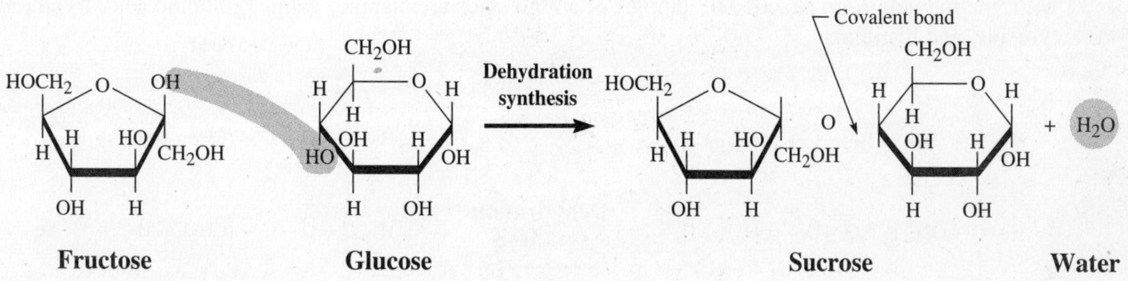

$$\text{Fructose} + \text{Glucose} \xrightarrow{\text{Dehydration synthesis}} \text{Sucrose} + H_2O$$

Fig. 2-3 Formation of the disaccharide sucrose.

2.22 What are some examples of polysaccharides and why are they important?

Complex carbohydrates are known as **polysaccharides**. Polysaccharides are formed by combining innumerable numbers of monosaccharides. Among the most important polysaccharides is **starch,** which is composed of thousands of glucose units. Starch serves as a storage form for carbohydrates and is used as a microbial energy source by those fungi and bacteria able to digest it. Another important polysaccharide is **cellulose**. Cellulose is also composed of glucose units, but the covalent linkages cannot be broken except by a few species of microorganisms. Cellulose is found in the cell walls of algae and of fungi, where another polysaccharide called chitin is also located.

2.23 What general characteristics apply to lipids?

Lipids are organic molecules composed of carbon, hydrogen, and oxygen atoms. The ratio of hydrogen atoms to oxygen atoms is much higher in lipids than in carbohydrates. Lipids include steroids, waxes, and fats. Other lipids include the phospholipids, which contain phosphorus and are found in the membranes of microbial cells. Lipids are also used by microorganisms as energy sources.

2.24 Describe the chemical composition of fat molecules.

Fat molecules are composed of a glycerol molecule and one, two, or three molecules of fatty acid (thus forming mono-, di-, and triglycerides). A fatty acid is a long chain of carbon atoms with associated hydroxyl (—OH) groups and an organic acid (—COOH) group. The fatty acids in a fat may be identical, or they may all be different. They are bound to the glycerol molecule in the process of **dehydration synthesis**, a process involving the removal of water components during covalent bond formation (Figure 2-4). The number of carbon atoms in a fatty acid may be as few as 4 or as many as 24. Certain fatty acids have one or more double bonds in the molecule where hydrogen atoms are missing. Fats that include these molecules are called **unsaturated** fats. Other fatty acids have no double bonds and are called **saturated** fats.

2.25 Which properties distinguish the proteins?

Proteins have immense size and complexity, but all proteins are composed of units called **amino acids.** Amino acids contain carbon, hydrogen, oxygen, and nitrogen atoms; sulfur or phosphorus atoms are sometimes present. There are 20 different kinds of amino acids, each having an amino ($—NH_2$) group and an organic acid (—COOH) group and usually an attached radical (—R) group. Amino acids vary with the radical group attached. Examples of amino acids are alanine, valine, glutamic acid, tryptophan, tyrosine, and histidine.

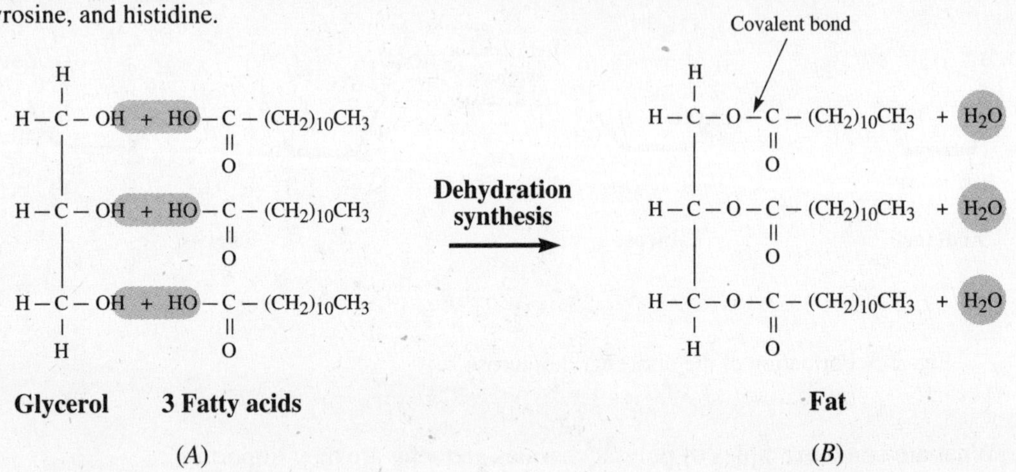

Covalent bond

$H-C-OH + HO-C(=O)-(CH_2)_{10}CH_3$ → $H-C-O-C(=O)-(CH_2)_{10}CH_3 + H_2O$

$H-C-OH + HO-C(=O)-(CH_2)_{10}CH_3$ **Dehydration synthesis** → $H-C-O-C(=O)-(CH_2)_{10}CH_3 + H_2O$

$H-C-OH + HO-C(=O)-(CH_2)_{10}CH_3$ → $H-C-O-C(=O)-(CH_2)_{10}CH_3 + H_2O$

Glycerol **3 Fatty acids** **Fat**

(*A*) (*B*)

Fig. 2-4 The chemistry of fats. (*A*) The components of a fat molecule illustrating the glycerol and fatty acid molecules. (*B*) The synthesis of a fat molecule by the process of dehydration synthesis. Note that the components of a water molecule are removed during the synthesis.

2.26 How are proteins formed in microbial cells?

To form a protein, amino acids are linked to one another by removing the hydrogen atom from the amino group of one amino acid and the hydroxyl group from the acid group of the second amino acid. The amino and acid groups then link up . Since the components of water are removed in the process, the reaction is a dehydration synthesis. The linkage forged between the amino acids is called a **peptide bond**; and small proteins are often called **peptides** (Figure 2-5).

2.27 In what places do proteins function in microorganisms?

Microbes depend upon proteins for the construction of cellular parts (e.g., flagella, capsules, cytoplasm, membranes) and for the synthesis of enzymes. **Enzymes** are chemical substances that catalyze most of the chemical reactions taking place within microbial cells. Enzymes are not used up in the reaction, but they remain available to catalyze succeeding reactions. Without enzymes, the chemistry of the organism could not take place. Proteins are also the component of toxins produced by many microorganisms, and they are the substance of antibodies produced by the body in defensive mechanisms. The amino acids of proteins also can serve as a reserve source of energy for microorganisms. When the need arises, enzymes remove the amino group from an amino acid and use the resulting compound for energy.

2.28 How is the sequence of amino acids determined in a protein?

Various microorganisms manufacture proteins unique to themselves. The information for synthesizing these proteins is found in the chromosomal material of the cell, where a genetic code specifies the sequence of amino acids occurring in the final protein. The chromosomes of the cell contain the genetic code in functional units of activity called genes. A single bacterial cell has a single chromosome, which has about 4000 genes.

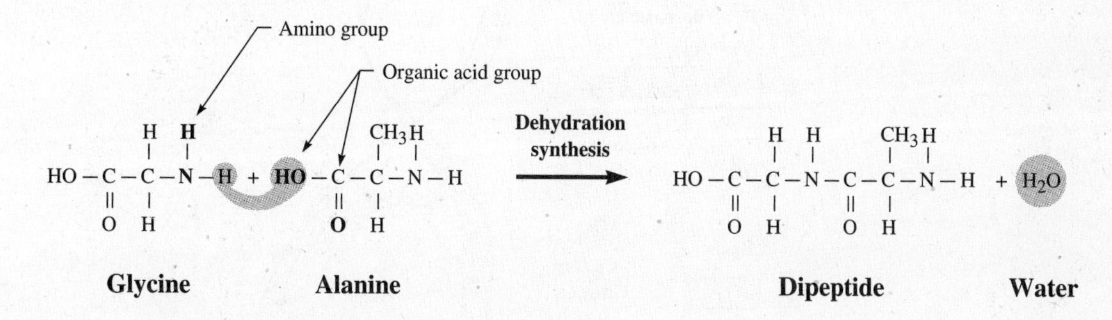

Fig. 2-5 Formation of a peptide bond between two amino acids.

2.29 Which general features are associated with nucleic acids?

Like proteins, nucleic acids are very large molecules, and like proteins, they are composed of building blocks. However, the units of nucleic acids are called **nucleotides**. Each nucleotide contains a carbohydrate molecule bonded to a phosphate group. It is also bonded to a nitrogen-containing molecule called a **nitrogenous base** so-named because it has basic properties.

2.30 Which are the two important nucleic acids in microorganisms?

Two important kinds of nucleic acids are found in microbial cells. One type is **deoxyribonucleic acid**, or **DNA**; the other is **ribonucleic acid**, or **RNA**. DNA is found primarily in the chromosome of the cell; it is the material of which the genes are composed. RNA is found in the cytoplasm of the cell. It participates with DNA in the synthesis of protein.

2.31 What differences distinguish DNA and RNA?

DNA and RNA differ slightly in the components of their nucleotides. DNA contains the five-carbon carbohydrate **deoxyribose**, while RNA has the five-carbon carbohydrate **ribose**. Both DNA and RNA have **phosphate groups** derived from a molecule of phosphoric acid. The phosphate groups connect the deoxyribose or ribose molecules to one another in the nucleotide chain. Both compounds contain the nitrogenous bases **adenine**, **guanine**, and **cytosine**, but DNA contains the base **thymine**, while RNA has **uracil**. Adenine and guanine are **purine** molecules, while cytosine, thymine, and uracil are **pyrimidine** molecules.

2.32 Describe the structure of the DNA molecule as it occurs in microorganisms.

In 1953, the biochemists James D. Watson and Francis H. C. Crick proposed a model for the structure of DNA that is now universally accepted. In the Watson-Crick model, DNA consists of two long chains of nucleotides standing opposite one another. In the nucleotide chains, guanine and cytosine molecules line up opposite one another, and adenine and thymine molecules oppose each other. Adenine and thymine are said to be **complementary**, as are guanine and cytosine. This is the principle of complementary base pairing. The two nucleotide chains then twist to form a **double helix** resembling a spiral staircase. Weak hydrogen bonds existing between the chains hold the chains together (Figure 2-6).

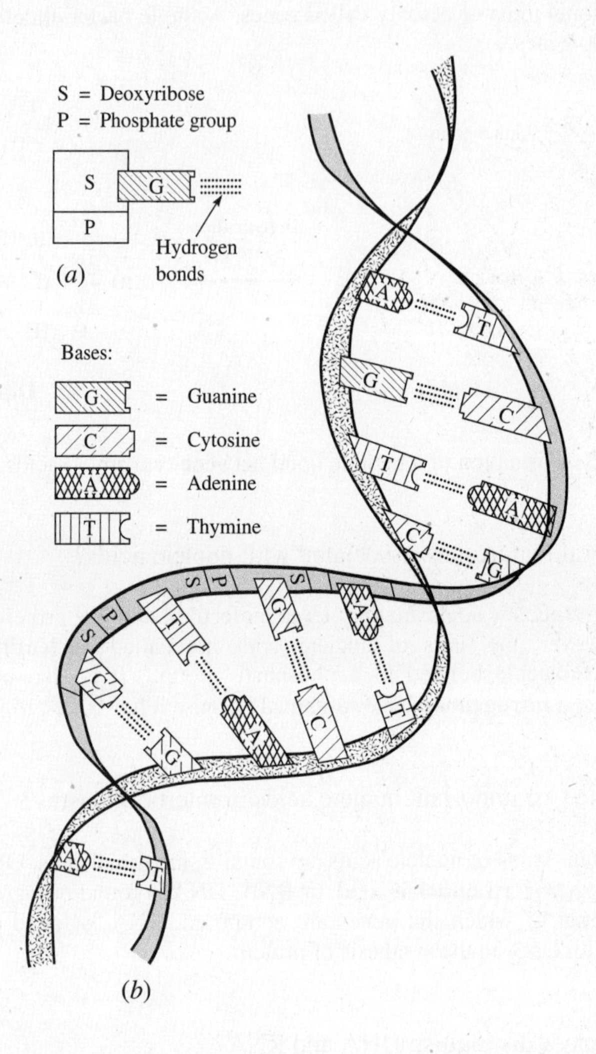

Fig. 2-6 Structure of a DNA molecule.

2.33 What is the significance of the sequence of nitrogenous bases in the DNA molecule?

The sequence of bases in the DNA molecule is the essential element in the proper placement of amino acids in proteins. How these bases are arranged is the essence of the genetic code. In microorganisms, the genetic code is transcribed to a molecule of RNA, then the code is translated to an amino acid sequence in a protein molecule.

2.34 What is the mechanism by which DNA replicates itself during microbial cell division?

Before a microbial cell splits in two, the DNA replicates itself. In bacteria, the single chromosome (or molecule of DNA) replicates to form two chromosomes. The replicated chromosomes then separate, and one chromosome passes into each new bacterial cell. The process of DNA replication begins when specialized enzymes separate or "unzip" the DNA double helix. As the two strands separate, the purine and pyrimidine bases on each strand are exposed. The bases then attract their complementary bases causing them to stand opposite. Deoxyribose molecules and phosphate groups are brought into the molecule, and the enzyme DNA polymerase unites all the nucleotide components to form a long strand of nucleotides.

2.35 Why is DNA replication described as "semiconservative?"

During the replication process, the old strand of DNA directs the synthesis of a new strand of DNA through complementary base pairing. The old strand then unites with the new strand to reform a double helix. This process is called **semiconservative replication** because one of the old strands is conserved in the new DNA double helix.

Review Questions

Completion. Add the word or words that best complete each of the following statements.

1. All matter in the universe is composed of one or more fundamental substances called ____________ .

2. Over 90% of the weight of a bacterium consists of carbon, oxygen, hydrogen, and ____________ .

3. Those compounds associated with living things such as microorganisms are referred to as ____________ .

4. Precise arrangements of atoms derived from different elements are found in ____________ .

5. The molecular weight in a compound is derived by adding together the ____________ .

6. The DNA strands in the double helix of a microorganism are held together by bonds referred to as ____________ .

7. An acid is a chemical compound that releases hydrogen atoms when it is placed in ____________ .

8. Cyanobacteria produce their carbohydrates during the process of ____________ .

9. The basic fuel for all microbial life and the carbohydrate metabolized to yield its energy is ____________ .

10. When they sour milk, bacteria break down the disaccharide____________ .

11. The cell walls of fungi and algae contain a polysaccharide composed of glucose units and ____________ .

12. Fats, waxes, and steroids are grouped together as a type of organic compound called ____________ .

13. In the formation of a fat molecule, the components of water are removed and fatty acids are bonded to glycerol molecules during the process of ____________.

14. All proteins of microbial cells are composed of units known as ____________ .

15. When amino acids are linked together to form larger organic compounds, the bond that forms is the ____________ .

16. Most of the chemical reactions taking place within microbial cells are catalyzed by a group of chemical substances called ____________ .

17. The chromosomal material of a microorganism is composed of the nucleic acid ____________ .

18. All nucleic acids are composed of nitrogenous bases, phosphate molecules, and a carbohydrate molecule which can be deoxyribose or ____________ .

19. RNA differs from DNA in that DNA contains thymine while RNA contains ____________ .

20. The replication of DNA results in two DNA molecules, each of which has an original DNA strand, and for this reason the replication is said to be ____________ .

True/False. Enter the letter "T" if the statement is true in its entirety. Enter the letter "F" if the statement or any part of the statement is false.

___ **1.** A single bacterial cell has numerous chromosomes in its cytoplasm and a set of genes that numbers about 40,000.

___ **2.** When a single pair of electrons is shared in a molecule, the bond is referred to as a single bond.

___ **3.** The type of chemical bond found between sodium and chloride ions in sodium chloride is the hydrogen bond.

___ **4.** Most microorganisms prefer to live in a pH environment close to 3.0.

___ **5.** Monosaccharides such as galactose and fructose have the same molecular formula with different arrangements of atoms, and these molecules are called isotopes.

___ **6.** Both cellulose and chitin are polysaccharides found in the cell walls of certain microorganisms.

___ **7.** Protein and carbohydrate molecules contain nitrogen, but nucleic acids and lipids have no nitrogen in their molecules.

___ **8.** One of the uses of proteins in microorganisms is for the synthesis of enzymes.

___ **9.** The chemical information for the synthesis of proteins is found in the chromosomal material of the cell.

___ **10.** Acids are those chemical substances that attract hydrogen atoms when they are placed in water.

___ **11.** An exchange of electrons occurs during an oxidation-reduction reaction occurring in microorganisms.

___ **12.** When an atom acquires a charge such as by acquiring an extra electron it becomes an ion.

___ **13.** Carbon is renowned for its ability to enter into numerous covalent bonds because it has eight electrons in its outer shell.

___ **14.** Over 75% of the weight of a microorganism is carbon.

___ **15.** Examples of nitrogenous bases found in nucleic acid are alanine, valine, tryptophan, and tyrosine.

___ **16.** Both maltose and sucrose are types of disaccharides used by microorganisms in their metabolism.

___ **17.** Both purine and pyrimidine molecules are widely found in proteins.

___ **18.** The arrangement of nitrogenous bases in the DNA molecule is the essence of the genetic code.

___ **19.** Molecular weights are expressed in daltons, and they give a relative idea of the size of a molecule.

___ **20.** The universal solvent in all microorganisms and the substance in which all chemical reactions occur is water.

Multiple Choice. Select the letter of the item that correctly completes each of the following statements.

1. When electrons are lost or gained by an atom, the atom acquires a charge and becomes (*a*) an isotope (*b*) a protein (*c*) an ion (*d*) a neutron.

2. All the following are types of chemical bonds existing in molecules except (*a*) covalent bonds (*b*) hydrogen bonds (*c*) acidic bonds (*d*) ionic bonds.

3. An inert element is one whose atoms (*a*) have no electrons (*b*) have an outer shell full of protons (*c*) have acidic qualities (*d*) have an outer shell full of electrons.

4. An example of a chemical substance having ionic bonds is (*a*) a protein (*b*) a triglyceride (*c*) sodium chloride (*d*) alanine.

5. A hydrogen bond is a weak bond forming between (*a*) three pairs of electrons (*b*) protons (*c*) protons and neutrons (*d*) protons and three pairs of electrons.

6. The ratio of hydrogen atoms to oxygen atoms is 2:1 in (*a*) nucleic acids such as DNA (*b*) proteins such as insulin (*c*) carbohydrates such as glucose (*d*) fats such as in fatty tissue.

7. A substance whose pH is 9.3 is said to be (*a*) alkaline (*b*) acidic (*c*) neutral (*d*) inert.

8. Glucose, galactose, and fructose all have the same molecular formula but different arrangements of atoms and are therefore said to be (*a*) ions (*b*) isotopes (*c*) isomers (*d*) polymers.

9. The carbohydrate maltose (*a*) is an example of a polysaccharide (*b*) contains numerous amino acids (*c*) is composed of two glucose units (*d*) is found in table sugar.

10. In all proteins there are (*a*) two or more monosaccharide units (*b*) two or more amino acids (*c*) either RNA or DNA molecules (*d*) numerous basic groups.

11. Proteins are used by microorganisms for the production of (*a*) energy compounds (*b*) enzymes and cellular parts (*c*) DNA molecules and chromosomes (*d*) ions.

12. The sequence of amino acids in a protein is determined by (*a*) the amount of energy in the cell (*b*) the presence of RNA in the nucleus of the cell (*c*) chemical information in the chromosomal material of the cell (*d*) amount of oxygen present.

13. Adenine, guanine, and thymine are examples of microbial (*a*) nucleotides (*b*) nitrogenous bases (*c*) amino acids (*d*) monosaccharides.

14. James D. Watson and Francis H. C. Crick are renowned for their determination of the (*a*) structure of DNA (*b*) the pH of bacterial cytoplasm (*c*) the acidity of potassium hydroxide (*d*) chemical composition of cellulose.

15. In fat molecules, fatty acids are bound to molecules of (*a*) chitin (*b*) cytosine (*c*) glycerol (*d*) amino acids.

16. A pH of 7.0 is found in (*a*) solutions of amino acids (*b*) all protein molecules (*c*) pure water (*d*) enzyme molecules.

17. The energy from glucose is released in metabolism and used to form the immediate energy source known as (*a*) ATP (*b*) DPN (*c*) CHO (*d*) RNA.

18. All of the following are found in an amino acid except (*a*) a radical group (*b*) an organic acid group (*c*) an amino group (*d*) a phosphate group.

19. The process of dehydration synthesis occurs in the production of (*a*) glucose molecules (*b*) ATP molecules (*c*) proteins and fats (*d*) isotopes.

20. Proteins appear in all the following places in microbiology except (*a*) they are a component of many microbial toxins (*b*) they are the substance of antibodies (*c*) they are used to synthesize enzymes (*d*) they are the hereditary material of the organism.

Chapter 3

Microbial Size and Microscopy

OBJECTIVES

Because microorganisms are invisible to the unaided eye, some form of microscope must be used to enlarge them and bring them into view. The following pages will survey various technological devices used for microscopy and point out some concepts of microbial size. The objectives of the chapter are to:

1. Identify the units used for measuring microorganisms and compare different microorganisms with respect to size.
2. Understand the theory underlying the use of the compound microscope and survey methods for increasing the resolution of this instrument.
3. Identify the important parts of the compound microscope and learn the functions of each part.
4. Summarize the special features that mark the dark-field microscope, the phase-contrast microscope, and the fluorescence microscope.
5. Recognize the principle underlying electron microscopy and explain the salient differences between transmission and scanning electron microscope.
6. Describe the principles for simple staining, negative staining, and Gram staining.
7. Review the procedures for the Gram stain, acid-fast stain, and other stain techniques.

THEORY AND PROBLEMS

3.1 What magnifications are achievable with contemporary microscopes?

The modern **bright-field microscope** (also called the light microscope) is used in most teaching laboratories. It will usually magnify an object up to 1000 diameters, although with special condensers and apparatus, this figure can be raised to 2000 diameters. The fluorescence, phase-contrast, and dark-field microscopes achieve similar magnifications. Electron microscopes, by comparison, utilize electron beams having considerably shorter wavelengths and achieve magnification of up to several million diameters.

SIZE RELATIONSHIPS

3.2 In what units are most microorganisms measured?

The metric system used in scientific research and writing is also used to measure microorganisms. However, the familiar meter and millimeter are not used. Instead, microorganisms are measured in units called micrometers. A **micrometer (μm)** is a millionth of a meter. Another way of expressing this is that a million micrometers is equivalent to a meter. A thousand micrometers equals a millimeter, and a thousandth of a millimeter is a micrometer. Bacteria, fungi, and protozoa are measured in micrometers.

3.3 Is the micrometer unit also used for viruses?

Viruses are among the smallest known objects that cause human disease. All viruses are measured in units called nanometers. A **nanometer (nm)** is a billionth of a meter. Some members of the Picornaviridae family of viruses (such as polio viruses) measure only about 20 nm in diameter. A billion nanometers is equivalent to a meter.

3.4 Are there any larger viruses?

Not all viruses are the size of the Picornaviridae. Certain viruses, such as members of the Poxviridae (the poxviruses), are considerably larger, measuring about 200 nm in diameter. This size equates to that of the smallest bacteria, the chlamydiae. An example of the latter is *Chlamydia trachomatis*, the etiologic agent of chlamydia, a sexually transmitted disease.

3.5 How do bacteria and viruses compare to one another in size?

The viruses are the smallest microorganisms, measuring about 20 to 200 nm in diameter. Bacteria generally range from about 0.5 μm to about 20 μm in length. There are three forms of "small" bacteria: rickettsiae, including *Rickettsia rickettsii*, the agent of Rocky Mountain spotted fever, are about 0.45 μm in diameter; chlamydiae, including *Chlamydia psittaci*, the agent of psittacosis, are about 0.25 μm in diameter; and mycoplasmas, including *Mycoplasma pneumoniae*, the etiologic agent of mycoplasmal pneumonia, are about 0.15 μm. Figure 3-1 shows these comparisons.

3.6 What are the size comparisons of fungi and protozoa?

Fungal cells comprise the hypha and mycelium of the fungus and range in size up to 25 μm in length, Single-celled fungi such as the yeast *Saccharomyces cerevisiae* are about 8 μm in length. Protozoal cells vary considerably in size. The etiologic agent of giardiasis, *Giardia lamblia,* measures about 20 μm in diameter, while the agent of balantidiasis, *Balantidium coli,* can be as large as 200 μm in length.

MICROSCOPY

3.7 What is the difference between a simple microscope and a compound microscope?

The **simple microscope** is a single lens system that magnifies all diameters of the object being viewed. The **compound microscope** consists of two lenses. The object produced by the first of the two lens is remagnified by the second lens. The magnification is thus compounded. An example of a compound microscope, the bright-field microscope, is shown in Figure 3-2.

3.8 What is the essential theory that underlies use of the compound microscope?

A compound light microscope has two magnifying lenses called **ocular** and objective lenses. The ocular lens is nearer the eye. The **objective lens** is nearer the object and magnifies the microorganism to create an image within the tube of the microscope. The image is then used as an object by the ocular lens, which remagnifies it to create the final virtual image. For example, a 40X objective forms an image that is 40 diameters greater than the object's normal size. A 10X ocular then remagnifies each of the 40 diameters another 10 times to form an image 400 diameters larger. This is the final magnification of the object.

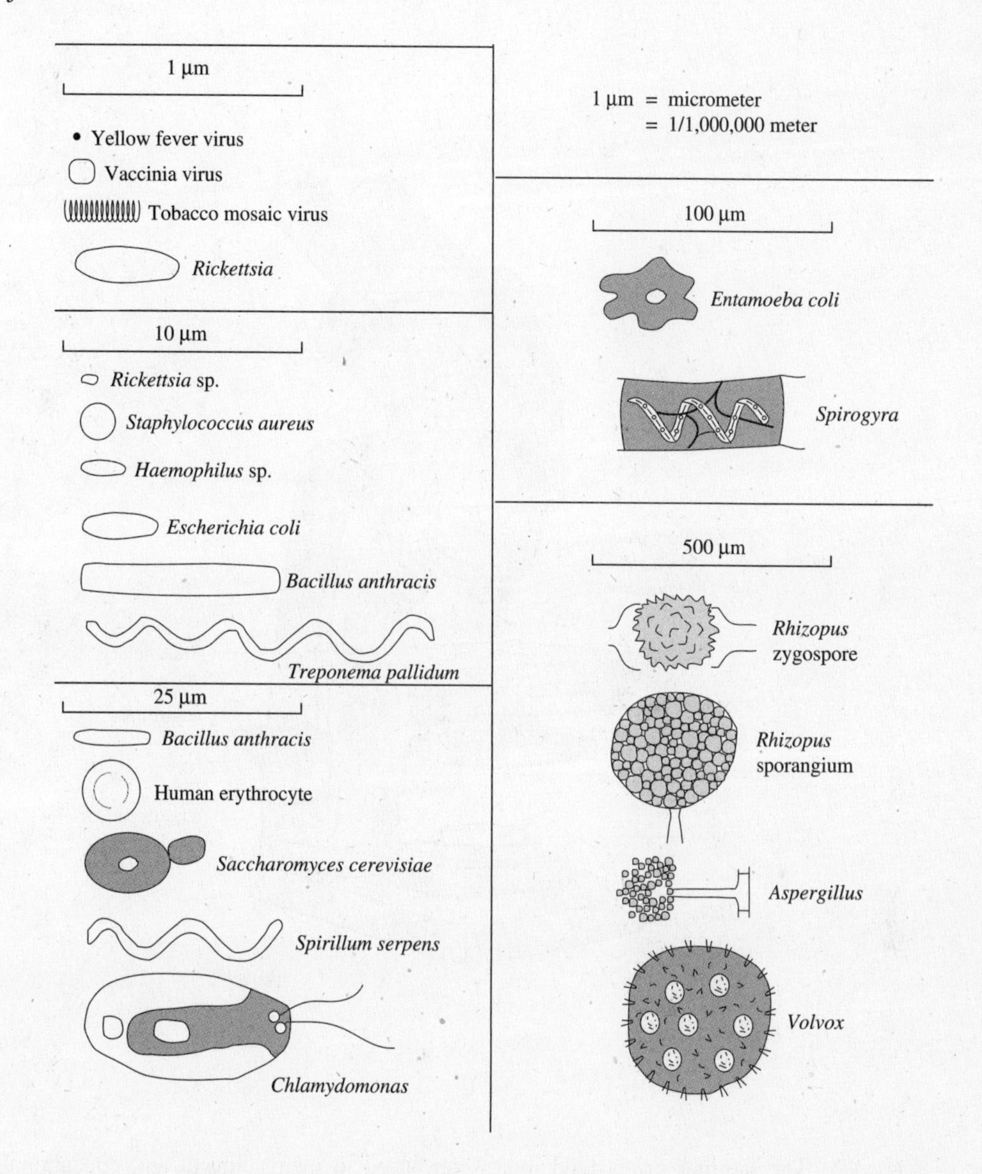

Fig. 3-1 A comparison of size relationships among various microorganisms. The shapes have been stylized for illustrative purposes.

3.9 What is meant by resolution as it is applied to the microscopy utilized in the science of microbiology?

Resolution is the ability of the microscope to distinguish two closely related points. Also known as resolving power, resolution gives an indication of the size of the smallest object that can be distinguished by a particular lens system. For example, if the resolution of the lens system is 0.5 μm, anything larger than 0.5 μm can be seen clearly, but anything smaller cannot be observed with clarity. For a microorganism to be viewed under the microscope its size must fall within the resolution of the lens system used.

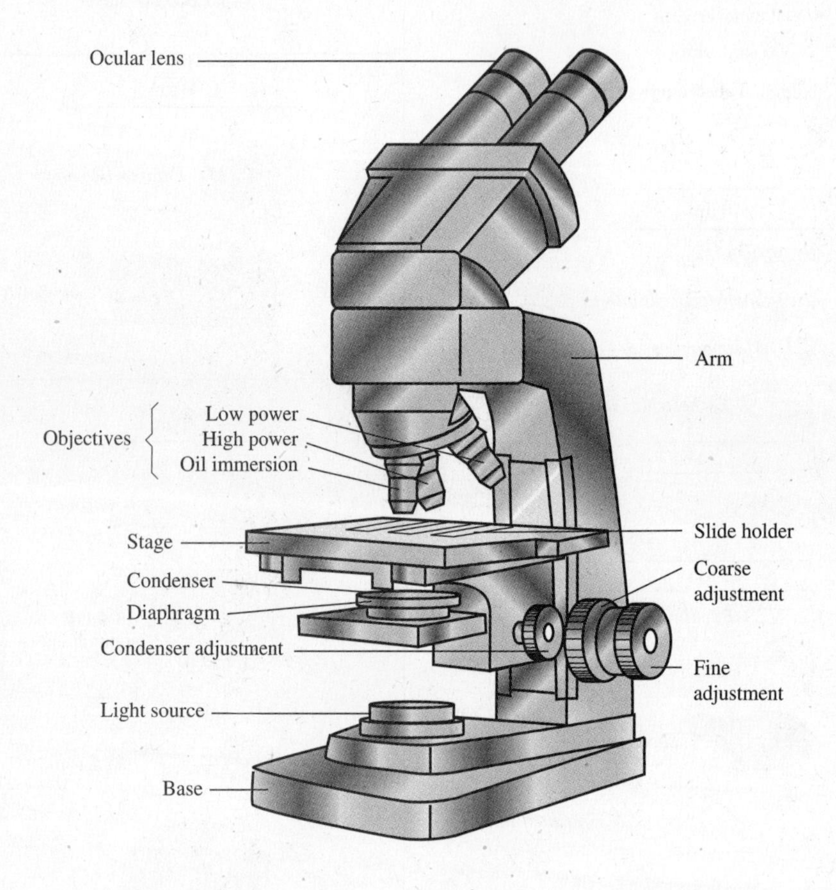

Fig. 3-2 The familiar bright-field microscope used in many clinical and educational laboratories. The important features of the microscope are noted in this view.

3.10 How is resolution determined?

To determine resolution, one must know the **numerical aperture (NA)** of the lens system. This denotes the size of the cone of light entering the aperture of the lens. The NA, typically etched into the lens, is multiplied by two. The product is then divided into the wavelength of the visible light, typically 550 nm. (If another form of light such as ultraviolet light were used, the wavelength of that light would be used in the formula.) The result is the resolution of the lens system expressed in nm. Conversion to micrometers is usually the final step.

3.11 Can the resolution of a microscope be increased?

The resolution of a microscope can be increased by shortening the wavelength of the energy employed in the magnification. This is the principle on which the electron microscope is based. The shorter wavelength of the election beam brings about a better resolution of the object. Another way to increase resolution is to increase the size of the cone of light (the numerical aperture). This is accomplished by using immersion oil.

3.12 What is meant by refractive index?

Refractive index refers to the light-bending ability of a medium such as air or glass. When the objective lens of the microscope is used, light bends as it passes from the glass slide into the air. This bending is due to the fact that the glass and the air have different refractive indices owing to their different densities.

3.13 How can the problem posed by the refractive index be solved in the light microscope?

To resolve this problem presented by different refractive indices, a drop of **immersion oil** is placed in the space between the glass slide and the glass lens. Immersion oil has the same refractive index as the glass. Because of this similarity, light travels in a straight line from the glass slide to the oil to the glass of the lens. The effect is to increase the amount of light entering the lens and to widen the cone of light entering. As the resolution of the lens system increases, viewing with the oil immersion lens is made possible.

3.14 What are some other important parts of the light microscope beside the lenses?

In addition to the lenses, there are several features of most light microscopes that enhance their usefulness. One part, called the **condenser**, consists of a series of lenses mounted under the stage. The condenser condenses light into a strong beam and sends it through the opening of the stage, as Figure 3-3 displays. On the condenser is mounted a shutterlike apparatus called the **diaphragm**. The diaphragm opens and closes to permit more or less light into the viewing area. Most microscopes have both coarse and fine adjustments. The **coarse adjustment** is used with the scanning or low power lenses, while the **fine adjustment** is used with the high power and oil immersion lenses.

3.15 How does a dark-field microscope compare to a bright-field microscope?

A **dark-field microscope** is a compound microscope that contains a special condenser. The condenser directs the light at the object to be viewed at an angle. Only the light bouncing off the object enters the objective lens. The effect is to create a light image on a dark background, as illustrated in Figure 3-4.

3.16 Under what circumstances might a dark-field microscope be useful?

Certain bacteria such as spirochetes are not easily viewed with the bright-field microscope, but can be easily seen moving about unstained using the dark-field microscope. *Treponema pallidum*, the etiological agent of syphilis and a very thin bacterium, is an example of such a spirochete. The disease can be diagnosed by securing scrapings from the skin of an infected patient and placing the specimen under the dark-field microscope, where spirochetes are seen moving about live and unstained.

3.17 What special features mark the phase contrast microscope?

A **phase contrast microscope** contains special condensers that scatter light and cause it to hit the object from various angles. The light rays illuminate parts of cells not otherwise visible with the ordinary light microscope.

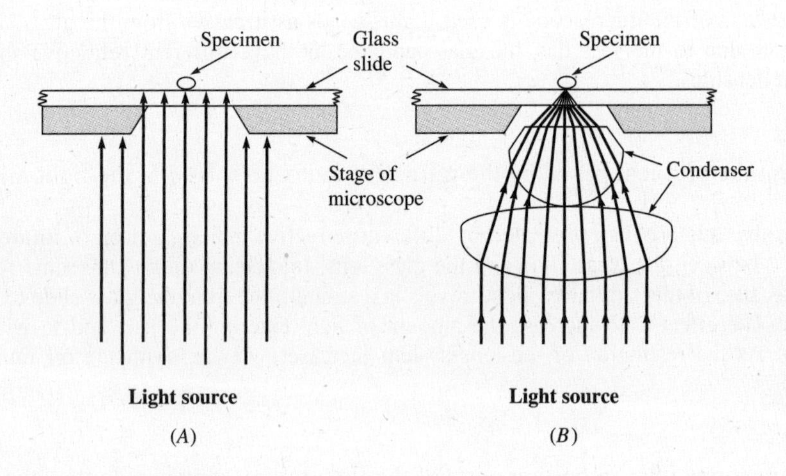

Fig. 3-3 An illustration of the effects a substage condenser has on the illumination of an object under the light microscope. Without a condenser (*A*), the light is dispersed and the specimen is seen without resolution. With the condenser (*B*), the light is focused on the specimen and the resolution is increased. The specimen is fully illuminated for viewing.

3.18 Of what value is the phase contrast microscope in microbiology?

In yeast cells such as *Saccharomyces cerevisiae*, the phase contrast microscope permits one to see internal details such as the nucleus, cytoplasm, and granules of these unicellular organisms. Moreover, unstained specimens can be seen with the phase contrast microscope in more detail than with other microscopes. *Saccharomyces cerevisiae* is widely used in genetic engineering studies as a vector organism to carry altered genes, and the phase contrast microscope is important to understanding the morphology and physiology of the yeast.

3.19 What is the principle behind use of the fluorescence microscope?

The **fluorescence microscope** is one that works with ultraviolet light, fluorescence dyes, and, on occasion, antibody molecules. The specimen is coated with a fluorescence dye and ultraviolet light is directed toward the specimen. The UV light raises electrons in the dye to a higher energy level. Then the electrons fall back to their original energy levels and give off the excess energy as light. The microbial specimen being observed appears to glow in the dark.

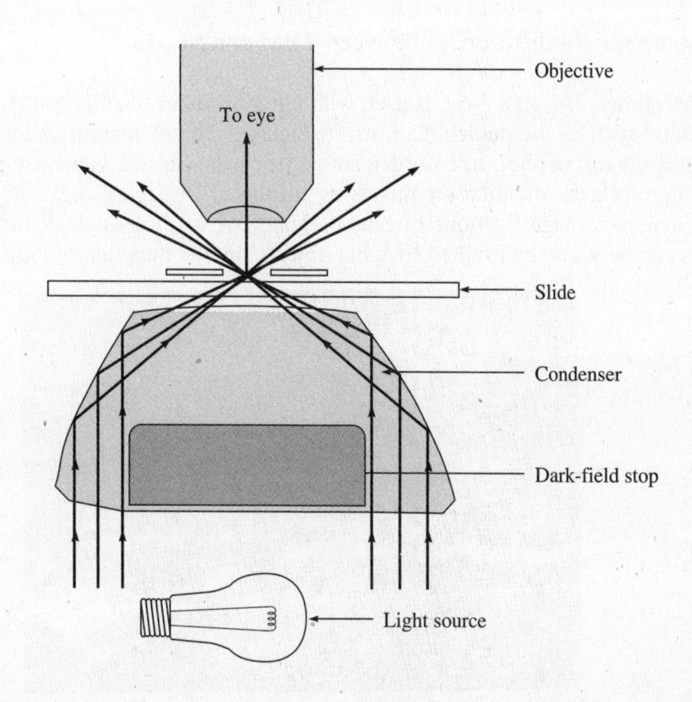

Fig. 3-4 Dark-field microscopy. In dark-field microscopy, a dark-field "stop" is placed in the condenser and light is permitted to reach the slide only from a number of angles. Very little light enters the objective lens, the exception being the few light rays that strike the specimen. The specimen is therefore seen on a dark background.

3.20 How can the fluorescence microscope be of value in the diagnostic laboratory?

The fluorescence microscope can be used in the diagnostic laboratory in the **fluorescence antibody test**. The object of this test is to identify an unknown organism. Antibodies that unite with a known species are used. The antibodies are chemically united with a fluorescent dye to form a dye-labeled antibody. Now the antibody is mixed with the microorganisms and a sample is placed on the slide. If the dye gathers on the microbial surface, then the name of the microorganism can be determined because the name of the antibody is known. If there is no fluorescence, then no accumulation took place and a different antibody will have to be tested.

3.21 What principle does the electron microscope use to achieve ultrahigh magnifications?

The **electron microscope** uses a beam of electrons as its source of energy. The electron beam has an exceedingly short wavelength and is able to detect very tiny objects such as viruses or molecules. Electrons are deflected by the object onto a viewing screen, and the object is observed. The minuscule wavelength of the electron beam contributes to the high resolution possible with the instrument.

3.22 What are the two types of electron microscopes currently in use in microbiology?

The two types of electron microscopes in use are the **transmission electron microscope** (**TEM**) and the **scanning electron microscope** (**SEM**). Each is used for its own unique different purpose. The major differences in use are the preparations of the sample and the cell parts examined.

3.23 Explain some salient differences between TEM and SEM.

The TEM, shown in Figure 3-5a, is used with ultrathin slices of cells and is valuable for seeing internal cellular details such as the nucleoid shown in Figure 3-5b. A magnification of several million times is possible, and details of the microorganism's cytoplasm can be viewed clearly. The SEM is used to illuminate the surfaces of microorganisms as displayed in Figure 3.5b. It provides three-dimensional views and achieves magnifications of tens or hundreds of thousands of times. Objects as small as 50 nanometers can be viewed with the SEM, but objects smaller than that require the TEM.

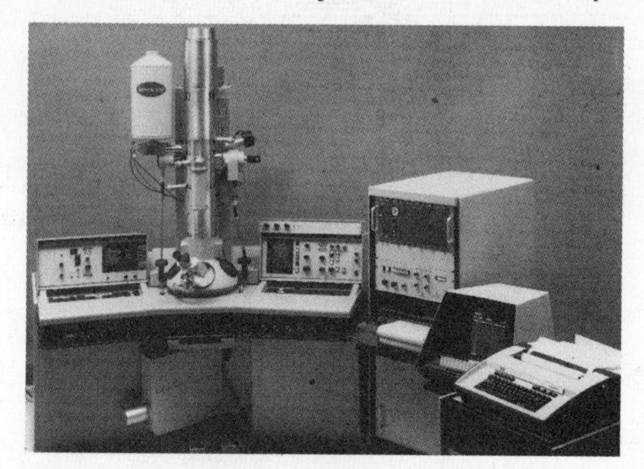

(A)

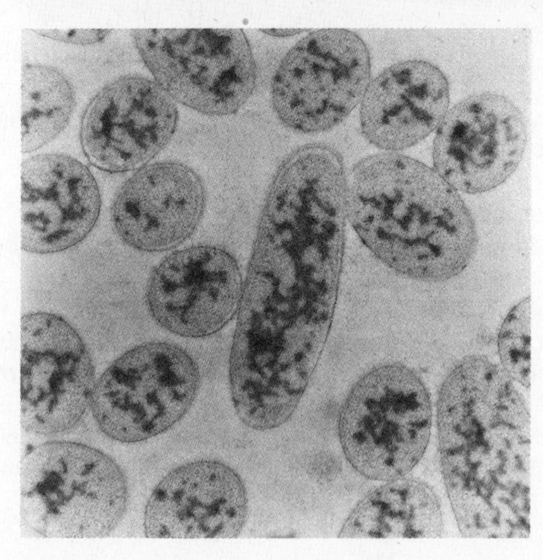

(B)

Fig. 3-5 (*A*) A transmission electron microscope operating at 100 kilovolts and capable of producing a magnification of several hundred thousand times. (*B*) A transmission electron micrograph of an ultrathin slice of a bacterium stained to show the nucleic acid in the cell. **(Permission granted by the American Society for Microbiology Journals Division.)**

3.24 Why must staining techniques be employed in microbiology?

The cytoplasm of bacteria, fungi, protozoa, and other microorganisms is transparent, and it would be very difficult to observe these organisms without the benefit of staining.

3.25 What are the different types of staining procedures used in microbiology?

Staining can be performed in a single step, such as the simple stain procedure or the negative stain procedure; or it may be a multiple-step procedure, such as the Gram technique or acid-fast technique. The multiple-step procedures usually yield some information about the microorganisms, while permitting one to see them. There are also procedures used to stain certain microbial structures such as flagella, capsules, and spores.

3.26 How are microorganisms prepared for staining procedures?

Prior to staining, organisms are generally placed on slides and subjected to **heat fixing**, a process by which heat is briefly applied to the slide to bind the microorganisms to the glass. The heat also kills any microorganisms that may be alive, and prepares them for staining by making the cell wall and membrane more permeable to the dye.

3.27 What is the biochemical basis for simple staining?

The cytoplasm of most microorganisms carries an overall negative charge. Stains used for the **simple stain procedure** carry positive charges and are known as basic dyes. Examples are methylene blue, crystal violet, and safranin. The dye is applied to the heat-fixed smear and the dye is attracted to the cytoplasm. Staining thus takes place.

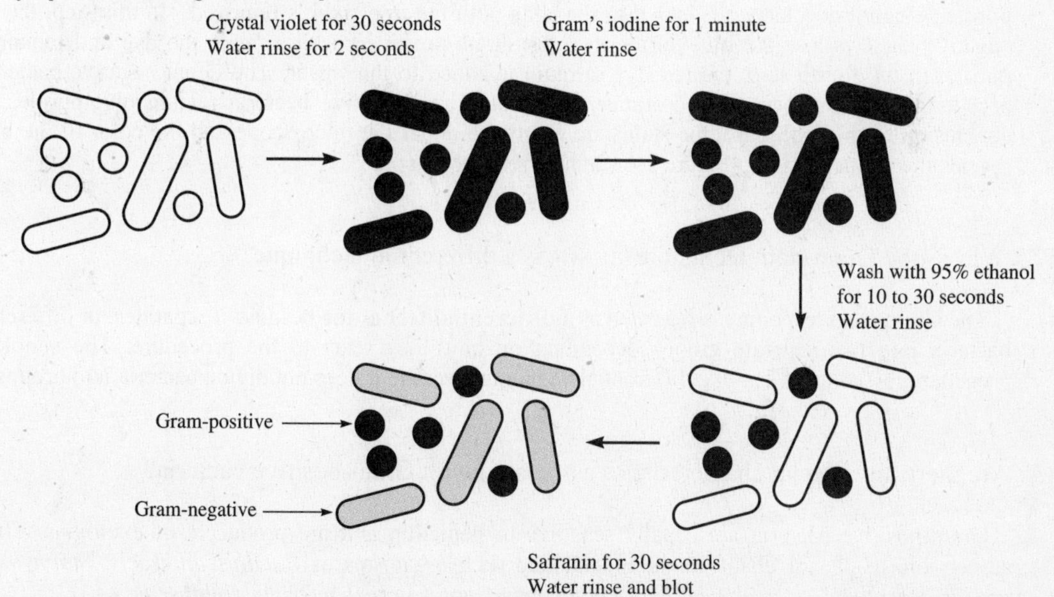

Fig. 3-6 The Gram stain technique. Crystal violet and iodine are used in the first two steps of the procedure, respectively. When 95% ethanol is used for washing, some of the bacteria are decolorized, but some retain their color. The decolorized bacteria are subsequently stained with safranin. These are the Gram-negative bacteria; the Gram-positive bacteria are those which resisted decolorization.

3.28 Can a dye that is repelled by microorganisms be used for staining?

In a procedure called the **negative stain procedure**, an acidic dye such as nigrosin or Congo red is used. This dye carries a negative charge. When mixed with microorganisms, the latter repels the dye and it gathers at the surface. Thus, the microorganisms are outlined in stain, but the organisms themselves remain clear and can be observed internally. Also, since they have not been subjected to chemical treatment, they remain intact.

3.29 What is the Gram stain technique and what is its value?

The **Gram stain technique** is named for Christian Gram who developed it in the 1880s. Gram found that virtually all bacteria may be subdivided into two major groups according to their reaction to this procedure. The groups are called Gram-positive bacteria and Gram-negative bacteria. Thus, the Gram stain procedure is an important tool for separating bacteria in groups as a first step to classifying them.

3.30 What is the basis for the Gram stain technique?

The composition of the bacterial cell wall is the basis for Gram staining. The cell wall of Gram-negative bacteria contains alcohol-soluble lipids, while the cell wall of Gram-positive bacteria lacks the lipids and therefore retains the crystal violet-iodine complex used.

3.31 How is the Gram stain procedure performed in the microbiology laboratory?

In the laboratory, a heat-fixed smear of bacteria is stained with crystal violet for one minute. Then, iodine is added to the smear for one minute, and the remainder is washed free. All bacteria are now blue-purple. Alcohol decolorizer is added to the slide until the free stain is removed. In this step, the Gram-negative bacteria lose the blue-purple dye, but Gram-positive bacteria retain the dye and remain blue-purple. In the fourth step, the red dye safranin is added to the smear. The Gram-negative bacteria will accept the dye and become red-orange, while the Gram-positive bacteria remain blue-purple. At the conclusion of the procedure, the stains are examined under the microscope and the color of the bacteria reveals the Gram reaction. Figure 3-6 summarizes these steps.

3.32 Why is the Gram stain technique known as a differential technique?

The Gram stain technique is regarded as a **differential technique** because it separates, or differentiates, bacteria into two separate groups depending on how they react to the procedure. The simple stain procedure, by contrast, is not a differential technique because it does not divide bacteria into groups.

3.33 Are there any special characteristics associated with Gram-positive bacteria?

Gram-positive bacteria are usually sensitive to penicillin and are producers of exotoxins. They are susceptible to phenol disinfectants and include such organisms as *Bacillus anthracis*, *Staphylococcus aureus*, *Streptococcus pyogenes*, *Clostridium tetani*, and *Corynebacterium diphtheriae*.

3.34 Which special characteristics are associated with Gram-negative bacteria?

Gram-negative bacteria are usually sensitive to the tetracycline antibiotics and to the aminoglycoside antibiotics such as gentamicin, neomycin, and kanamycin. Also, they produce endotoxins. They are susceptible to chlorine, iodine, and detergent disinfectants and include such organisms as *Salmonella typhi*, *Shigella sonnei*, *Bordetella pertussis*, and *Yersinia pestis*.

3.35 Are there any differential stain techniques beside the Gram stain technique?

Another important differential stain technique is the **acid-fast stain technique**. This technique separates species of *Mycobacterium* from other bacteria. All bacteria receive the first stain in the procedure, a red stain called carbolfuchsin. When acid-alcohol is added, however, all bacteria except *Mycobacterium* species lose the stain and become transparent. The following stain, methylene blue, stains these bacteria blue, while the *Mycobacterium* species remain red from the carbolfuchsin. Since species of *Mycobacterium* cause tuberculosis and leprosy, this stain technique is a valuable diagnostic tool.

3.36 Are there any procedures for the special staining of bacterial structures?

Certain special staining procedures exist for bacterial structures. **Bacterial flagella** are too small to be seen by simple staining. However, their presence and arrangement can be ascertained by treating them with an unstable colloidal suspension of tannic acid salts, which causes a heavy precipitate to form on the flagella. **Capsules** can be seen by the Welch method, a variation of the negative stain technique which involves a copper sulfate wash to retain the capsular integrity. The bacterial **nucleoid** can be enhanced with the Feulgen stain technique. **Bacterial spores** can be visualized by staining with malachite green in the presence of steam heat.

Review Questions

Completion. Add the word or words that correctly complete each of the following statements.

1. An alternative name for the bright-field microscope is the ______________ .

2. The unit most often used to measure the diameter of viruses is the ______________ .

3. Among the smallest bacteria are three groups known as rickettsiae, chlamydiae, and ______________ .

4. A meter is equivalent to micrometers that number ______________ .

5. Among the largest protozoa is the etiologic agent of balantidiasis, known as ______________ .

6. Whereas a simple microscope has a single lens, a compound microscope has ______________ .

7. The ability of a microscope to distinguish two closely related points is called ______________ .

8. The light bending ability of a medium such as air or glass is known as the ______________ .

9. The part of the microscope known as the condenser consists of a series of ______________ .

10. The shutterlike apparatus that controls the amount of light passing through the condenser is the ______________ .

11. Among the bacteria most often visualized with the dark-field microscope are the ______________ .

12. Among the microorganisms most often studied with the phase contrast microscope are the _______________ .

13. The type of energy used to illuminate specimens in the fluorescence microscope is _______________ .

14. The value of using an electron beam in the electron microscope is the beam's extremely small _______________ .

15. Two types of electron microscopes are the transmission electron microscope and the _______________ .

16. Bacteria and microorganisms must be stained before microscopy because their cytoplasm is usually _______________ .

17. Two multistep staining procedures used in microscopy are the Gram stain technique and the _______________ .

18. Those stains that carry a positive charge are referred to as _______________ .

19. The basis of Gram staining is believed to be centered in the bacterial structure called the _______________ .

20. The acid-fast stain technique is particularly useful for detecting species of the genus _______________ .

Microscope Matching.

___ 1. Used to visualize viruses
___ 2. Employs UV light
___ 3. Smallest wavelength energy
___ 4. Used with antibodies
___ 5. Gram stain results observed
___ 6. Light scattered by condenser
___ 7. Dark background
___ 8. Dye binds to microbial surface
___ 9. Spirochetes visible
___ 10. Yeast cells observed

(*a*) Dark-field microscope
(*b*) Phase-contrast microscope
(*c*) Bright-field microscope
(*d*) Fluorescence microscope
(*e*) Electron microscope

Multiple Choice. Select the letter of the item that correctly completes each of the following statements.

1. Among the smallest known agents that can cause human disease are the (*a*) protozoa (*b*) bacteria (*c*) viruses (*d*) rickettsiae.

2. To determine the resolution of a microscope system, one must be aware of the (*a*) ocular and objective magnifications (*b*) size of the condenser and diaphragm (*c*) numerical aperture and wavelength of light (*d*) working distance of the microscope.

3. Immersion oil has a refractive index identical to that of (*a*) air (*b*) glass (*c*) cytoplasm (*d*) water.

4. Syphilis can be most easily diagnosed in a patient by (*a*) observing the spirochete under the bright-field microscope (*b*) preparing an electron microscope preparation (*c*) using the Gram stain and bright-field microscope (*d*) placing scrapings under the dark-field microscope.

5. Which of the following represents a series of microorganisms of increasing size (*a*) fungi, viruses, bacteria (*b*) viruses, bacteria, fungi (*c*) protozoa, fungi, rickettsiae (*d*) rickettsiae, viruses, protozoa.

6. The Picornaviridae are among the (*a*) smallest viruses (*b*) largest chlamydiae (*c*) largest yeasts (*d*) smallest fungi.

7. Which of the following dyes is used in the Gram stain technique (*a*) methylene blue and Congo red (*b*) nigrosin and carbolfuchsin (*c*) safranin and crystal violet (*d*) Congo red and methylene blue.

8. A differential stain technique is one that (*a*) shows different structures in a microorganism (*b*) can be used under different circumstances (*c*) separates bacteria into different groups (*d*) uses different microscopes for observing bacteria.

9. The acid-fast stain technique is an important diagnostic tool in patients who have (*a*) pneumonia (*b*) diphtheria (*c*) tuberculosis (*d*) meningitis.

10. All the following apply to the negative stain procedure except (*a*) it utilizes a dye such as nigrosin (*b*) microorganisms repel the dye (*c*) microorganisms stain deeply (*d*) an acidic dye is used.

11. The process of heat fixing is performed (*a*) prior to staining (*b*) at the conclusion of the staining procedure (*c*) only during multistep procedures (*d*) only with electron microscopy.

12. The transmission electron microscope is more valuable than the scanning electron microscope because (*a*) whole specimens are used with transmission electron microscopy (*b*) the magnification possible with transmission electron microscopy is higher (*c*) bacteria can be seen with transmission electron microscopy but not with scanning electron microscopy (*d*) electrons are not used with transmission electron microscopy.

13. Phase contrast microscopy is valuable for visualizing (*a*) viruses (*b*) rickettsiae (*c*) chlamydiae (*d*) yeasts.

14. The dark-field microscope differs from the bright-field microscope because the dark-field microscope contains a special (*a*) ocular (*b*) fine adjustment (*c*) condenser (*d*) objective lens.

15. A nanometer is equivalent to a (*a*) thousandth of a millimeter (*b*) thousandth of a meter (*c*) millionth of a meter (*d*) billionth of a meter.

16. At the conclusion of the alcohol decolorizer step, Gram-negative bacteria (*a*) appear blue-purple (*b*) appear red (*c*) appear colorless (*d*) appear deep green.

17. The maximum magnification available to most bright-field microscopes is (*a*) one million times (*b*) one thousand times (*c*) one hundred thousand times (*d*) ten times.

18. The microscope condenser is responsible for (*a*) controlling the amount of light to the viewing area (*b*) reducing the refractive index of the glass (*c*) magnifying the object (*d*) bringing the light into a strong beam.

19. The fluorescent antibody technique is valuable for (*a*) detecting a particular microorganism (*b*) performing the Gram stain technique (*c*) avoiding the heat-fixing step for slides (*d*) use in electron microscopy.

20. Staining techniques are useful in microbiology because (*a*) most microscopes are of poor quality (*b*) the cytoplasm of microorganisms is transparent (*c*) stains deflect electrons propagated toward the slide (*d*) stains make the condenser and diaphragm of the microscope unnecessary.

Staining Technique Matching

___ 1. Dye repelled by bacteria
___ 2. Iodine used
___ 3. Developed in the 1880s
___ 4. Carbolfuchsin used
___ 5. Uses alcohol decolorizer
___ 6. Microorganisms outlined in stain
___ 7. Basic dye used
___ 8. Acid-alcohol decolorizer
___ 9. All bacteria divided to two groups
___ 10. *Mycobacterium* detected

(*a*) Gram stain technique
(*b*) Acid-fast technique
(*c*) Simple stain technique
(*d*) Negative stain technique

Chapter 4

Prokaryotes and Eukaryotes

OBJECTIVES

Certain microorganisms are classified as prokaryotes, and others as eukaryotes. The various structures found in prokaryotes and eukaryotes are also found in microorganisms, and knowledge of their structure gives insight to their physiology. In this chapter, the objectives are to:

1. Differentiate prokaryotes from eukaryotes and indicate why viruses fit into neither group.

2. Survey the sizes and shapes found in bacteria, with reference to any variations occurring in the three basic shapes.

3. Discuss how bacteria move and recognize the different forms of flagellation that occur in bacteria.

4. Learn the significance of bacterial pili, cell walls, and cell membranes.

5. Learn the components of the cytoplasm of the prokaryotic cell, while recognizing some aspects of the hereditary material of prokaryotes.

6. Outline the importance of bacterial spores in the life cycle of these microorganisms.

7. Summarize the key structures of eukaryotic cells and compare them to the structures of prokaryotic cells.

8. Summarize the methods for movement of molecules across the cell membrane in both prokaryotic and eukaryotic cells.

THEORY AND PROBLEMS

4.1 How are microorganisms characterized in relationship to other living things?

Like all other living organisms, microorganisms can be classified as prokaryotes or eukaryotes. These designations are based on the cellular and subcellular structures and ultrastructures observed with the electron microscope. Among the microorganisms, bacteria are **prokaryotes**, while other microorganisms, including fungi, protozoa, and simple algae, are **eukaryotes**. Complex plants and animals are also eukaryotes.

4.2 Which characteristics distinguish eukaryotes from prokaryotes?

Prokaryotes and eukaryotes differ in a number of physiological and structural ways. The cells of eukaryotes have a **nucleus** with a nuclear membrane, and **organelles** such as mitochondria, Golgi bodies, an endoplasmic reticulum, and lysosomes. By contrast, prokaryotes have no true nucleus (their genetic material is not bound by a membrane), nor do they have organelles. Their **ribosomes** are smaller than those of eukaryotic cells. In prokaryotic cells there is no evidence of mitotic structures (spindle, equatorial plate, and aster) and the hereditary material (DNA) is present in a single, closed loop chromosome. Eukaryotic cells have multiple chromosomes occurring in pairs and the characteristic structures involved in mitosis.

4.3 Are prokaryotes and eukaryotes similar in any respects?

Prokaryotes and eukaryotes share common features, among them the possession of nucleic acids and other organic substances such as proteins and carbohydrates. In addition, they utilize similar metabolic reactions such a glycolysis and chemiosmosis for the utilization of food and the production of energy and waste. Also, they exhibit many of the same physiological features such as motion and reproduction, although the mode of reproduction may be different, and different organs of motility may exist.

4.4 Are viruses considered either eukaryotes or prokaryotes?

No, viruses are acellular and do not fit either category. Viruses are composed of a fragment of nucleic acid, either RNA or DNA, enclosed in a protein shell (the capsid). They cannot reproduce themselves or carry out metabolic functions on their own.

PROKARYOTES

4.5 What are the major groups of prokaryotes considered in microbiology?

In the five-kingdom classification of organisms, prokaryotes constitute the Kingdom **Monera**. In this kingdom are the bacteria, the cyanobacteria, and many small forms of bacteria such as rickettsiae, chlamydiae, and mycoplasmas. These bacteria are referred to as Eubacteria. Also, placed with the Eubacteria are filamentous bacteria, gliding bacteria, sheathed bacteria, and many funguslike forms. Another broad category of prokaryotes are the Archaebacteria, which are believed to be among the earliest forms of life on Earth. These forms and numerous others are surveyed in Chapter 10.

4.6 What sizes and shapes do the bacteria have?

The smallest bacteria, the mycoplasmas, measure about 0.20 micrometer (μm) in diameter, while the larger bacteria range from 10 to 20 μm in length (some bacteria may be even larger). There are three basic shapes: the sphere, called the **coccus** (pl. cocci); the rod, or **bacillus** (pl. bacilli); and the spiral, called the **spirillum** (pl. spirilla) if the cell wall is rigid and the **spirochete** if the cell wall is flexible. The three shapes are illustrated in Figure 4-1. An example of a bacillus is *Bacillus subtilis*, a common Gram-positive sporeforming rod; an example of a coccus is *Streptococcus pyogenes*, the etiologic agent of scarlet fever; an example of a spirillum is *Spirillum minor*, the agent of rat bite fever; and an example of a spirochete is *Borrelia burgdorferi*, the cause of Lyme disease.

4.7 Are there any variations that occur within the three basic shapes of bacteria?

The rod forms of bacteria do not occur in any particular configurations, although some chains may be sometimes seen. However, the cocci occur in several variations. Pairs of cocci are called **diplococci** (e.g., *Neisseria meningitides*, the agent of spinal meningitis), while chains of cocci are called **streptococci** (e.g., *Streptococcus lactis*, commonly found in milk), and cubelike packets of four and eight cocci are called **sarcinae** (e.g., *Micrococcus luteus*, a common airborne contaminant) . An irregular cluster of cocci in a grapelike pattern is called a **staphylococcus**, often known as "staph." Among the spiral bacteria, there are the **spirilla** (singular, **spirillum**); the **spirochetes;** and those that have a single turn and appear as commas, the **vibrios** (e.g., *Vibrio cholerae*, the agent of cholera). In recent years, investigators have described star-shaped bacteria of the genus *Stella*, triangular bacteria, and square, flat cells of the genus *Haloarcula*.

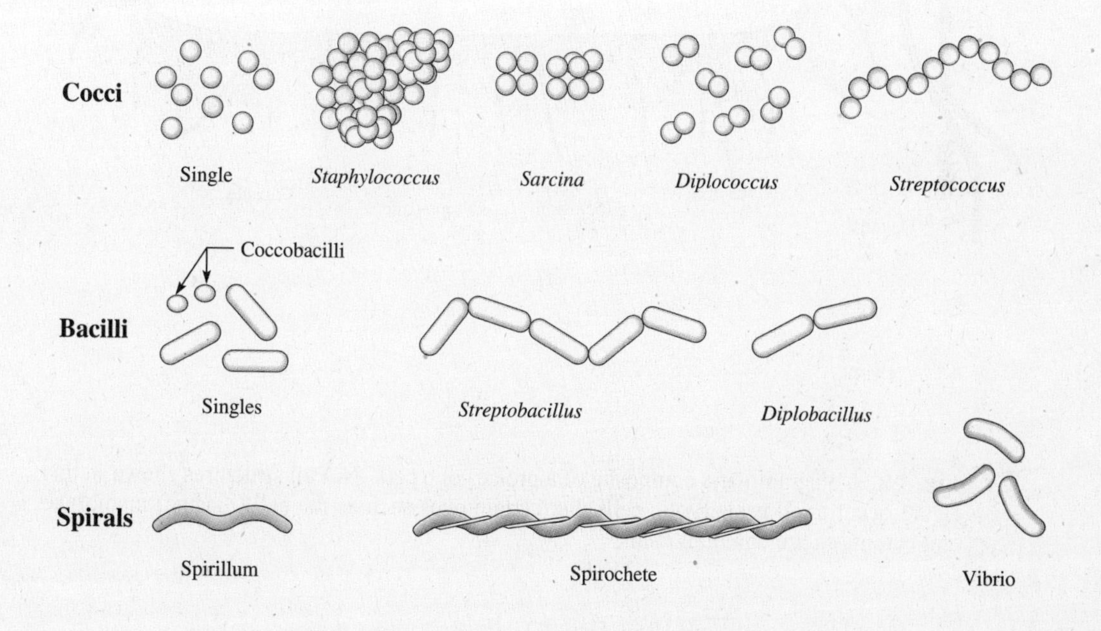

Fig. 4-1 The three shapes taken by the great majority of bacterial species. There are several arrangements within these basic shapes, as the figure displays.

4.8 How is the shape of a bacterium determined?

All bacterial shapes are determined by genetic information contained in the chromosome. Most bacteria maintain a single shape and are said to be **monomorphic.** Other bacteria alter their shapes and occur in numerous different forms. These bacteria are called **pleomorphic**.

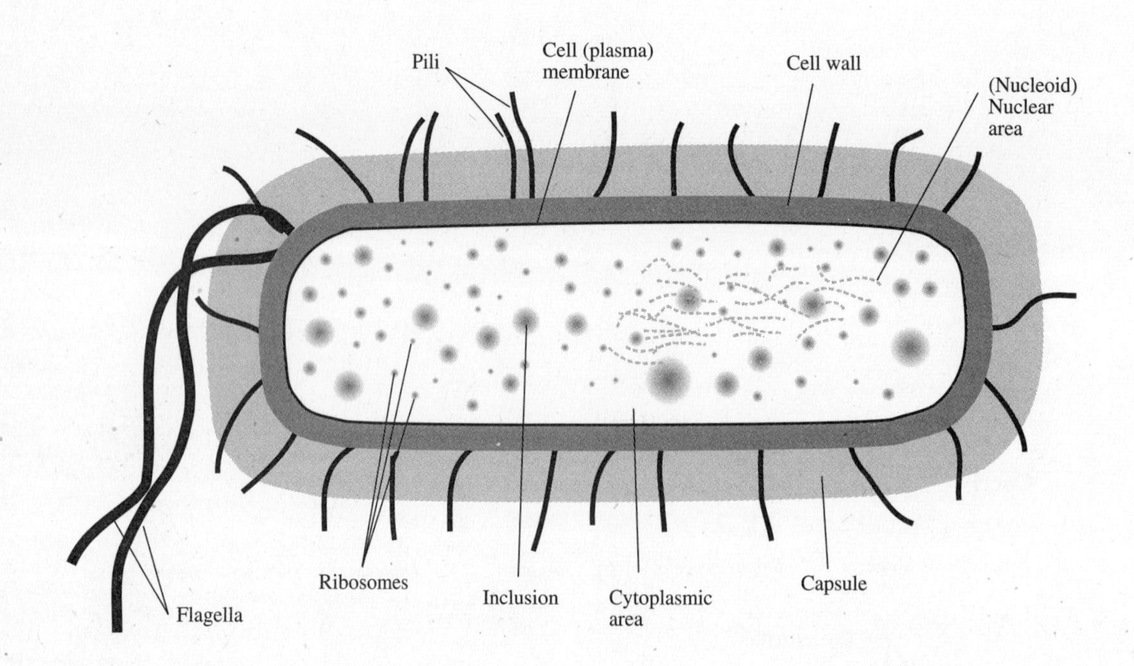

Fig. 4-2 A diagrammatic composite of a prokaryotic cell. Not all structures shown in this figure occur in all prokaryotic cells, but certain ones such as the cell (plasma) membrane and cytoplasm are common to all.

4.9 Are bacteria able to move, and if so, how?

Bacteria (and other prokaryotes) have the ability to move by means of appendages called **flagella** (Figure 4-2). Flagella are composed of the globular protein flagellin. They are extremely long and thin and cannot be seen by the light microscope unless specially stained. They are, however, readily visible under the electron microscope. Bacterial flagella propel the organism by a rotary motion. Each flagellum has three basic parts: the **filament**, which is the long, outermost region containing the flagellin; the **hook**, which is composed of a different protein and lies at the proximal end of the filament; and the **basal body**, which anchors the filament to the cell membrane and cell wall and is composed of a series of rings encircling a central, small rod (Gram-positive bacteria have only an inner pair of rings, while Gram-negative bacteria have both inner and outer pairs of rings).

4.10 Do all bacteria have the same form of flagellation?

No, the form of flagellation varies. Some bacteria, called **monotrichous** bacteria, have a single flagellum, while other bacteria, called **amphitrichous** bacteria, have a flagellum at each end of the cell. **Lophotrichous** bacteria have multiple flagella at the ends of the cells, and **peritrichous** bacteria have flagella distributed over the entire body of the cell.

4.11 Do all bacteria have flagella?

Flagella are not found on all bacteria. For example, cocci such as *Staphylococcus aureus* rarely have flagella and are unable to move independently. These bacteria cannot actively translocate to different regions of the body during infection, nor can they move to other areas of nutrients when the nutrients in their immediate area become scarce.

4.12 Do bacteria have any external cellular structures?

Certain species of bacteria are able to form a sticky, gelatinous layer of polysaccharides and proteins known as the **capsule**. The capsule protects the bacterium against environmental changes and creates a barrier against such things as drugs, antibodies, and other chemicals. It provides an advantage to the bacterium and is often found in pathogenic (disease-causing) species such as *Streptococcus pneumoniae*, the etiologic agent of bacterial pneumonia. When the capsule is less firm and more flowing it is called a **glycocalyx**, or slime layer.

4.13 What are the pili that many bacteria have?

Many species of bacteria, especially Gram-negative bacteria, have short, hairlike appendages called **pili**, also known as **fimbriae.** Pili are composed of protein. They enable the bacterial cell to attach to a surface and are found on many pathogenic species. Certain pili called **sex pili** function in genetic transfers, in which the pili connect adjacent bacteria to one another so that chromosomal or plasmid transfer can occur.

4.14 Do all bacteria have a cell wall?

One of the distinctive characteristics of most bacteria and other prokaryotes is the unique cell wall. This wall contains chemical components not found in the cell walls of eukaryotes. It is a rigid structure that gives bacteria their shape, as shown in Figure 4-3. The only bacteria that have no cell wall are the **mycoplasmas** such as *Mycoplasma pneumoniae*, the etiologic agent of primary atypical pneumonia.

4.15 What is unique about the bacterial cell wall?

The bacterial cell wall contains a large molecule called **peptidoglycan**. Peptidoglycan is a polysaccharide containing **N-acetylmuramic acid** and **N-acetylglucosamine**, with numerous chains of amino acids. In Gram-positive bacteria, the cell wall contains many layers of peptidoglycan together with another organic substance called **teichoic acid**. In Gram-negative bacteria, there is only a thin layer of peptidoglycan, surrounded by an outer membrane and an inner membrane. The peptidoglycan in Gram-negative bacteria is bonded to **lipoproteins**, which are not found in the Gram-positive cell wall.

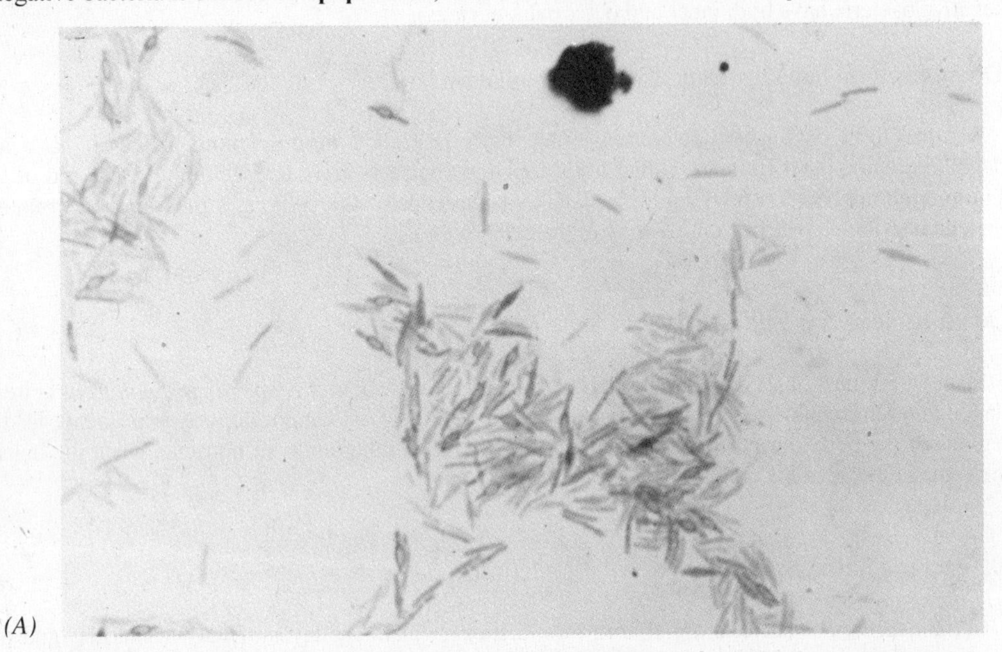

(*A*)

(*B*)

Fig. 4-3 Prokaryotic and eukaryotic cells. (*A*) A stained smear of bacteria as seen with the light microscope (X1000). The bacteria are relatively simple prokaryotic cells with little internal detail. Some of the cells are swollen with spores. (*B*) A view of the yeast *Candida albicans* under the light microscope (X1000). This eukaryotic organism is relatively large with significant internal details. Chains of cells give rise to circular and oval cells at the ends of the chains.

4.16 What is the structure of the cell membrane in prokaryotic cells?

The **cell membrane** is often referred to as the **plasma membrane** or the **cytoplasmic membrane**. In both prokaryotic and eukaryotic cells, the cell membrane is similar. It is a two-layered structure with space between the layers. Phospholipid molecules are arranged in two parallel rows forming a phospholipid bilayer. The outer and inner surfaces of the bilayer consist of polar heads, which are made up of phosphate groups and glycerol molecules. Nonpolar tails composed of hydrophobic fatty acids lie at the interior surfaces of the bilayer.

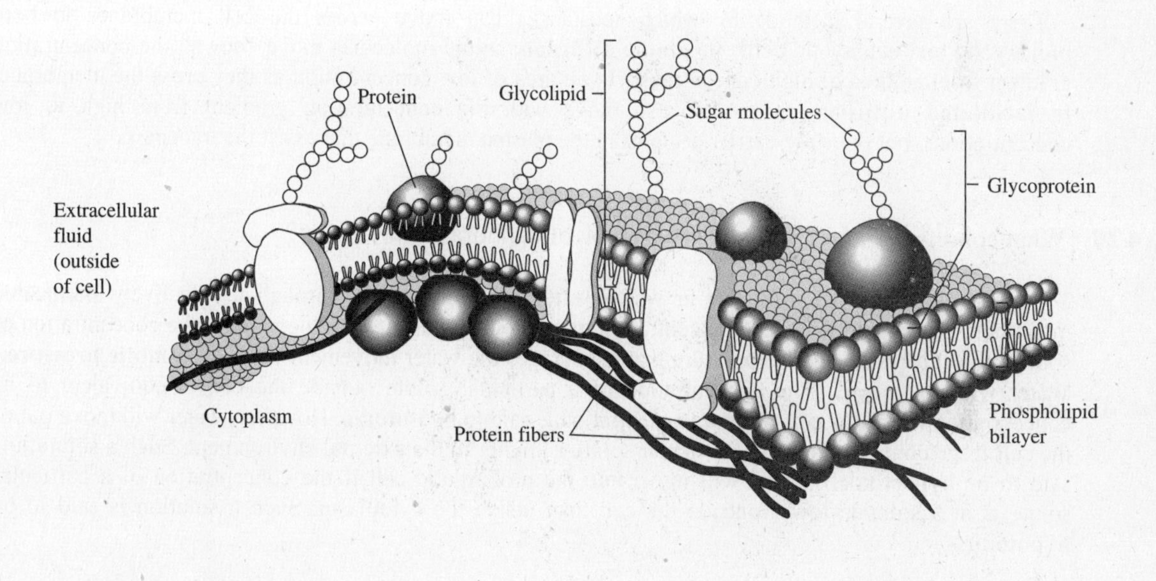

Fig. 4-4 The fluid mosaic model of the cell membrane. Two layers of phospholipids make up most of the cell membrane, and various types of protein molecules move freely within the bilayers.

4.17 Are there any proteins in the cell membrane?

The cell membrane contains protein molecules that may be arranged various ways. For example, certain protein molecules exist at the surface of the membrane, while others penetrate the membrane from outer to inner surface. Still other proteins function as enzymes in the membrane, while others serve as supportive proteins. Most have carbohydrate molecules attached and are glycoproteins. The protein molecules move freely within the phospholipid bilayer and give a constantly changing appearance to the cell membrane. For this reason, the cell membrane is described as a **fluid mosaic model**. This model is depicted in Figure 4-4.

4.18 What important roles are played by the cell membrane?

The cell membrane has several important functions, one of which is to provide a semipermeable barrier between the cytoplasm and the external environment. For this reason, certain substances pass through the membrane and others are held back. For example, substances that dissolve in lipids enter the cell more easily than ions, which do not dissolve in lipids. In addition, the cell membrane is the site of enzymes that participate in the respiration of prokaryotic cells. In bacteria, the enzymes and pigments of photosynthesis may be located in the cell membrane.

4.19 How do substances move across the cell membranes in prokaryotic and eukaryotic cells?

There are several methods by which substances can move across the cell membranes in both prokaryotic and eukaryotic cells. In **simple diffusion**, solute molecules move "down" the concentration gradient from an area of high concentration to an area of low concentration as they cross the membrane. In **facilitated diffusion**, molecules also move with the concentration gradient from high to low concentrations, but there are carrier proteins in the plasma membrane that assist the transport.

4.20 What conditions apply to osmosis as a type of membrane movement?

Osmosis is a type of diffusion in which water molecules move through a selectively permeable membrane from a region of low concentration of solute molecules to a region where the concentration of solute molecules is high. (The pressure that encourages the water movement is called **osmotic pressure**.) Water will not move if the concentration of a particular solute outside the cell is equivalent to its concentration inside the cell. Such a concentration is said to be **isotonic**. However, water will move out of the cell if the concentration of a particular solute is higher in the external environment. Such a solution is said to be **hypertonic**. Water will move into the prokaryotic cell if the concentration of a particular solute is in a solution lower outside the cell than inside the cytoplasm. Such a solution is said to be **hypotonic**.

4.21 Are there any cases in which molecules move against the concentration gradient when crossing the cell membrane?

In some cases, molecules will move against the concentration gradient (or "up the concentration gradient") into a region that already has a high concentration of a particular molecule. This movement is called **active transport,** as opposed to the passive transport of diffusion and osmosis. Active transport is an energy-utilizing process that depends on carrier proteins and ATP. In most cases, it brings substances into the cell from the external cellular environment.

4.22 Is the cytoplasm of prokaryotic cells considerably different from that of eukaryotic cells?

The **cytoplasm** of a prokaryotic cell is essentially similar to that in eukaryotic cells. The cytoplasm is about 80 percent water, with proteins, carbohydrates, lipids, ions, and many of the organic and inorganic compounds found in eukaryotic cells. There are, however, no organelles in the prokaryotic cytoplasm, and there are other differences with respect to the DNA, ribosomes, and other bodies. Moreover, there is no cytoskeleton of microtubules and microfilaments in the prokaryotic cytoplasm.

4.23 Do the cells of prokaryotes have nuclei?

The prokaryotic cell does not have a nucleus in the traditional sense of the term. It does, however, have a single, long **chromosome** in the form of a closed loop. The chromosome, which consists of double-stranded DNA, is many times the length of the cell. It is packed tightly into an area of the cytoplasm called the **nucleoid**. When the prokaryotic cell breaks open, such as through the activity of lysozyme, the DNA explodes out of the cell and is scattered about. Unlike eukaryotic cells, chromosome replication in prokaryotes is not coupled to cell division, so cells may have two or more copies of the chromosome.

4.24 In what form does the nucleoid appear under the light microscope when observed in the laboratory?

The nucleoid appears elongated, spherical, or shaped like a dumbbell, depending on the species of microorganism. It is located in one region of the cytoplasm and can be seen with the electron microscope.

4.25 Does DNA exist anywhere else in the prokaryotic cell beside the nucleoid?

Prokaryotic cells such as bacteria have small loops of double-stranded DNA called **plasmids** in their cytoplasm. The plasmids contain several genes believed by researchers to be nonessential to the cell's physiology and genetics. For example, the genes may function in the processes of antibiotic resistance or toxin production. In antibiotic resistance, the plasmids may encode enzymes such as penicillinase, which degrades penicillin molecules before they can interrupt the synthesis of penicillin by the bacterium. Plasmids are utilized in genetic engineering. They can be opened to receive new genetic information in the form of DNA fragments, then reclosed and inserted to fresh bacterial cells where the genes will express themselves in the synthesis of protein.

4.26 Where are the ribosomes located in prokaryotic cells?

In prokaryotic cells, the **ribosomes** are the sites of protein synthesis. They exist free in the cytoplasm, whereas in eukaryotic cells, ribosomes are both free and located along the endoplasmic reticulum.

4.27 Is there a size difference between prokaryotic and eukaryotic ribosomes?

The ribosomes of prokaryotic cells are smaller than eukaryotic ribosomes. Prokaryotic ribosomes measure 70 Svedberg units, while eukaryotic ribosomes measure 80 Svedberg units. (Svedberg units are determined by the sedimentation rate in an ultracentrifuge.) Both types of ribosomes have two subunits of protein and RNA.

4.28 What are the inclusions found in prokaryotic cells?

Inclusions are various kinds of ultramicroscopic bodies. In prokaryotic cells, inclusions may contain polysaccharides, lipids, sulfur, phosphate, or other important materials needed for the biochemistry of the cell.

4.29 Can inclusions be observed within microorganisms in the laboratory?

In certain instances, inclusions can be stained with certain dyes and observed in microscopic preparations. In the organism of diphtheria, for example, there are characteristic phosphate inclusions called **metachromatic granules**. These granules stain bright red with certain dyes and are often used as identifying markers for *Corynebacterium diphtheriae,* the diphtheria bacillus.

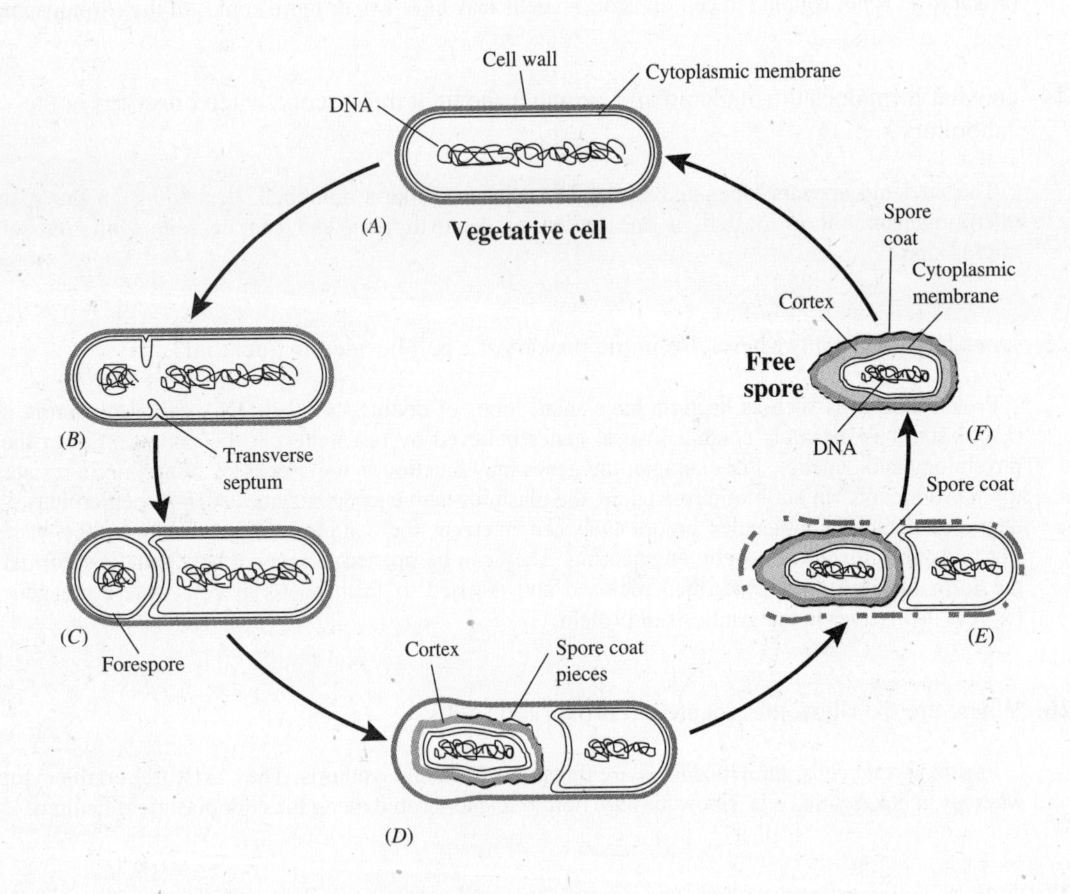

Fig. 4-5 The process of spore formation as it occurs in certain species of bacterial cells. (*A*) The DNA of the vegetative cell has duplicated and elongated. (*B*) One chromosome condenses at the end of the cell. The transverse septum begins to form. (*C*) The transverse septum has formed and the forespore is cut off from the remainder of the cell. (*D*) The cortex and pieces of the spore coat form around the forespore. (*E*) The walls of the spore are completed and the mature spore is formed. (*F*) The free spore has been released by the vegetative cell and is seen with an enclosing spore coat.

4.30 Do prokaryotic cells have any particular method for resisting changes in the environment?

Certain species of prokaryotic organisms are able to form highly resistant structures called **endospores**, also known as **spores**. Members of the genera *Bacillus and Clostridium* are notable for their ability to form spores. The organisms normally grow and multiply as vegetative cells, but when the environmental conditions become unfavorable, each cell reverts to a spore.

4.31 How does spore formation take place in bacterial cells?

In spore formation (sporulation), the DNA of the cell and a small amount of cytoplasm gather at one region of the cell as shown in Figure 4-5. In *Clostridium* species this region is usually at the end of the cell, a condition called **terminal sporulation**, while in *Bacillus* species it is usually at the center, a condition called **central sporulation**. Multiple layers of spore coatings form over the primordial spore and constitute several layers of protection against fluctuations in the environment.

4.32 How resistant to environmental conditions are bacterial spores?

Bacterial spores are among the most resistant forms of life known to science. They can resist boiling water temperatures, extensive drying, and the effects of disinfectants, antibiotics, and many other chemicals. To destroy spores, it is necessary to treat them in an autoclave with steam under pressure. In medical microbiology, spores can be extremely important because the etiologic agents of tetanus, botulism, gas gangrene, and anthrax are able to form spores that can transmit these diseases. Spores have recently been cultivated from samples of amber millions of years old.

EUKARYOTES

4.33 Which microorganisms are eukaryotic?

Among the microorganisms, the protozoa, fungi, and certain simple algae (e.g., diatoms and dinoflagellates) are eukaryotic. Plants and animals are also eukaryotic. The cells of eukaryotes are typically larger and more complex than those of prokaryotes. A notable feature of eukaryotic cells is the presence of a nucleus and a variety of organelles.

4.34 What is the structural makeup of eukaryotic flagella?

Like the prokaryotic cells, eukaryotic cells have appendages for motility called **flagella**. The flagella of eukaryotic cells are structurally and operationally different. In eukaryotic cells, the flagella consist of nine pairs of microtubules arranged in a ring, with two single microtubules in the center of the ring, a formation referred to as "9 + 2." The eukaryotic flagella have a whiplike rather than a rotary motion. Certain eukaryotes, especially many species of protozoa, possess shorter and more numerous structures called **cilia** that they use for locomotion. Cilia, like flagella, have the 9 + 2 microtubule arrangement.

4.35 What is the chemical composition of the eukaryotic cell walls?

Certain eukaryotic organisms such as fungi, algae, and plants have cells with cell walls. However, eukaryotic cell walls do not contain peptidoglycan, as do prokaryotic cell walls. The principal component of the fungal cell wall is **chitin**, a polymer of N-acetylglucosamine units. In the algae, the cell wall is composed of **cellulose**, the same polysaccharide as found in plant cell walls.

4.36 Do protozoal cells have cell walls?

Protozoa do not have cell walls. However, some species have an outer covering, called a **pellicle,** that encloses the plasma membrane.

4.37 How does the eukaryotic cell membrane compare to that of the prokaryotes?

The cell membrane (also called the plasma membrane) of eukaryotic and prokaryotic cells is quite similar in both structure and function. The fluid mosaic model applies to eukaryotic cells, but the proteins of the membrane are different. Eukaryotic cell membranes also contain **sterols**, a type of complex lipid, and the receptor sites on the membranes are composed of carbohydrates.

4.38 Does movement across the cell membrane in eukaryotes differ from that in prokaryotes?

Movement across the eukaryotic cell membrane occurs by the same mechanisms as in prokaryotes: diffusion, facilitated diffusion, osmosis, and active transport. In addition, material can be brought into the cell by **endocytosis**. In endocytosis, the cell membrane surrounds a particle or volume of fluid, then pinches off into the cytoplasm. When the material taken into the cell is of particulate nature, the endocytosis is called **phagocytosis**. When the material brought into the cell is dissolved in fluid, the endocytosis is referred to as **pinocytosis**. Endocytosis is also a feeding mechanism in such organisms as amoebas.

4.39 The cytoplasm of eukaryotic cells has a cytoskeleton. What is the structure of this organelle?

In eukaryotic cells, there is a complex internal structure composed of ultramicroscopic rods called **microfilaments** and ultramicroscopic cylinders known as **microtubules**. Together, the microfilaments and microtubules compose a three-dimensional structure, the **cytoskeleton**. The cytoskeleton provides shape and support for the eukaryotic cells and assists the transport of substances through the cytoplasm.

4.40 How is the nucleus organized in eukaryotic cells?

The **nucleus** of the eukaryotic cells is usually an oval or circular body and is frequently the largest cellular structure, as Figure 4.6 illustrates. It is composed of DNA and protein, and it is separated from the cytoplasm by a double membrane known as the **nuclear envelope**. Pores in the envelope permit communication between the nucleus and the cytoplasm. The nucleus contains one or more microscopic bodies called **nucleoli** (singular, nucleolus). These are the sites for the synthesis of ribosomal RNA. The nuclear fluid in which the nucleoli are suspended is known as **nucleoplasm**.

4.41 What is the major substance of the eukaryotic nucleus and how is it organized?

The major substance of the eukaryotic cell nucleus is the amorphous mass called **chromatin**. The chromatin consists of DNA and histone proteins organized into threadlike structures called chromosomes. The DNA is wound with histone proteins to form **nucleosomes**. The nucleosomes are linked together to form the chromosomes.

4.42 Which characteristics apply to the endoplasmic reticulum of eukaryotic cells?

Eukaryotic cells have in their cytoplasm a system of parallel membranes known as the **endoplasmic**

reticulum. Continuous with the cell membrane and nuclear envelope, the endoplasmic reticulum is an area where chemical reactions take place. It provides a pathway for the transport of molecules in the cell. In some cases, it serves as a storage area for synthesized molecules.

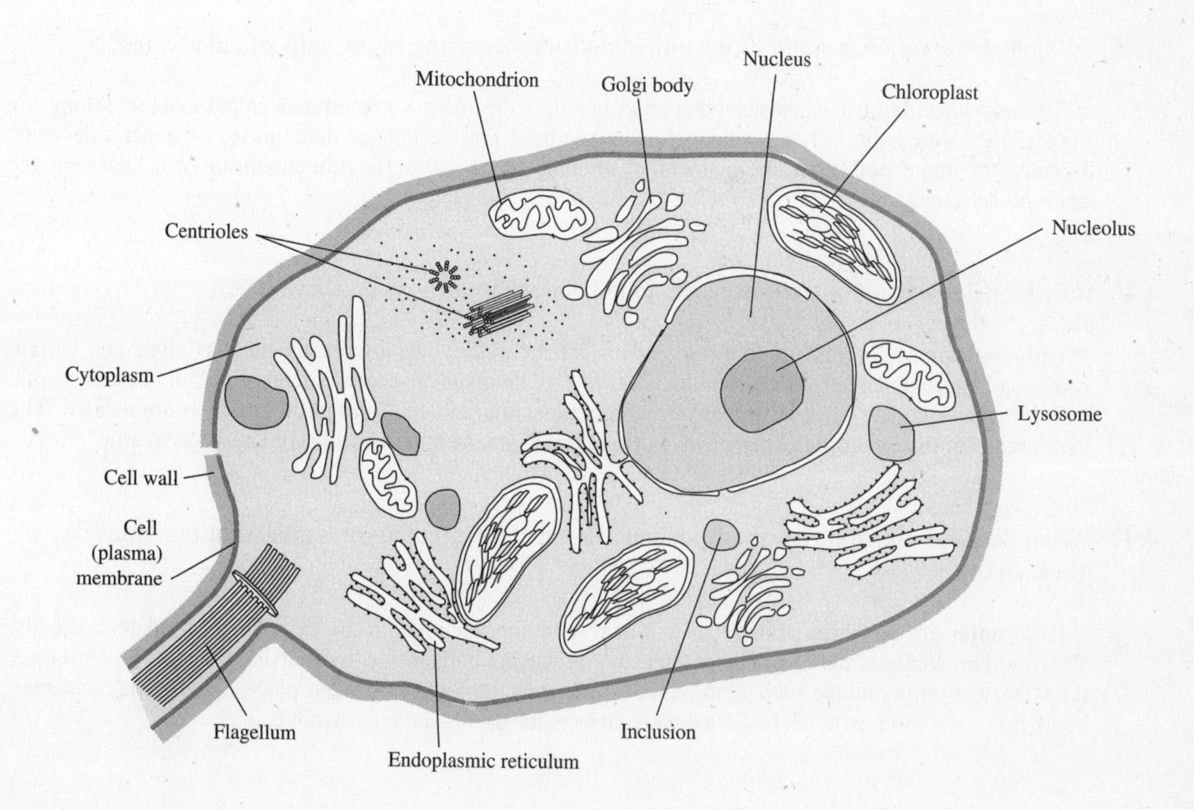

Fig. 4-6 A schematic eukaryotic cell showing the major organelles and cellular structures. Note the relative sizes and shapes of the organelles. Each organelle has a function consistent with its structure.

4.43 Where are the ribosomes located in eukaryotic cells?

In the eukaryotic cell, many of the ribosomes exist free in the cytoplasm, but most are attached to the surfaces of the endoplasmic reticulum. Eukaryotic ribosomes are somewhat larger and denser than prokaryotic ribosomes and they serve as the sites at which amino acids are bonded to one another to compose proteins.

4.44 What functions are served by the Golgi body in the eukaryotic cell?

The **Golgi body** of eukaryotic cells is a system of flattened membranes resembling sacs. Believed to be derived from the endoplasmic reticulum, the Golgi body is the site for packaging and releasing many cellular proteins and lipids. It is also the site where glycoproteins (combinations of protein and carbohydrate molecules) are formed.

4.45 Which characteristics apply to the mitochondrion occurring in the cells of eukaryotes?

The **mitochondrion** is a sausage-shaped organelle with internal membranes called **cristae**. Along the cristae are located the enzymes and cofactors required for the energy metabolism of eukaryotic cells. Because the major portion of energy metabolism takes place within the mitochondrion, it is known as the cell's powerhouse.

4.46 In which eukaryotes are chloroplasts found and what purpose do they serve?

Chloroplasts are organelles found in photosynthetic eukaryotic organisms, such as algae and certain protozoa. The chloroplast, like the mitochondrion, is involved in energy metabolism for the eukaryotic cell. It is the organelle where chlorophyll and other essential molecules of photosynthesis are located. The membranes of the chloroplast are known as **thylakoids**; stacks of thylakoids are known as **grana**.

4.47 What description applies to the lysosomes in the cells of eukaryotes and what functions do they serve?

Lysosomes are membrane-bound structures resembling vacuoles in the cytoplasm of eukaryotic cells. They contain enzymes that function in digestive processes both inside and outside the cell. For instance, the lysosome will combine with food particles after phagocytosis has taken place. Its enzymes will then break down the food particles to release the nutrients for use by the eukaryotic cell.

Review Questions

Multiple Choice. For each of the following statements, select the letter of the item that correctly completes the statement.

1. All the following microorganisms are considered prokaryotes except (*a*) bacteria (*b*) viruses (*c*) chlamydiae (*d*) mycoplasmas.

2. Eukaryotes differ from prokaryotes in that (*a*) eukaryotes do not have organelles (*b*) eukaryotes have a single chromosome (*c*) eukaryotes have a nucleus and organelles (*d*) eukaryotes do not divide by mitosis.

3. Among the common features shared by prokaryotes and eukaryotes are (*a*) the same shapes and sizes (*b*) the same types of movement (*c*) common organic substances such as proteins and carbohydrates (*d*) ribosomes of the same weight.

4. The three basic shapes found in most common bacteria are (*a*) triangles, squares, and rectangles, and

hexagons, icosahedrons, and helices (*c*) spheres, spirals, and rods (*d*) cubes, filaments, and rhomboids.

5. The shape of a bacterium is determined by (*a*) the genetic material contained in its chromosome (*b*) the structural composition of its cell wall (*c*) whether or not it forms spores (*d*) the cytoskeleton located in its cytoplasm.

6. The flagella found in bacteria (*a*) number the same in all bacteria (*b*) are composed of carbohydrate (*c*) are found only at one end of the cell (*d*) are composed of protein.

7. The substance peptidoglycan is found in the (*a*) ribosomes of eukaryotes (*b*) cell wall of bacteria (*c*) chromosomes of eukaryotes (*d*) cell membrane of bacteria.

8. Substances move across the membranes of prokaryotic and eukaryotic cells by all the following methods except (*a*) active transport (*b*) facilitated diffusion (*c*) osmosis (*d*) translation.

9. The ribosomes of prokaryotic cells (*a*) are found in the nucleus (*b*) exist free in the cytoplasm (*c*) are located along the endoplasmic reticulum (*d*) can be seen with the light microscope.

10. Mitochondria are found (*a*) in all bacteria (*b*) in all prokaryotic cells (*c*) within the nucleus of a prokaryote (*d*) in eukaryotic cells.

11. The Golgi body of eukaryotic cells (*a*) is considered the powerhouse of the cell (*b*) is used to package proteins and lipids before they are released (*c*) functions in cell motility (*d*) functions in cell reproduction.

12. In eukaryotic cells, the process of endocytosis permits (*a*) cell motion (*b*) the breakdown of carbohydrates and the release of energy (*c*) materials to pass into a cell (*d*) mitosis to take place.

13. In the prokaryotic cell, DNA is found in the (*a*) cell membrane and Golgi body (*b*) chromosome and plasmid (*c*) flagellum and cilium (*d*) cell wall and cell membrane.

14. The fluid mosaic model describes the (*a*) chromosomal material of prokaryotes (*b*) structure of the flagellum in eukaryotes (*c*) structure of the capsule in prokaryotes (*d*) structure of the cell membrane in prokaryotes.

15. Two components of the cell membrane in prokaryotes are (*a*) nucleic acids and carbohydrates (*b*) ATP and peptidoglycan (*c*) protein and lipid (*d*) DNA and RNA.

16. Chains of bacteria cocci are known as (*a*) streptococci (*b*) micrococci (*c*) sarcinae (*d*) staphylococci.

17. The cell walls of Gram-positive and Gram-negative bacteria (*a*) are the sites of spore formation (*b*) both contain peptidoglycan (*c*) are connected to the cell nucleus by the endoplasmic reticulum (*d*) are composed exclusively of proteins.

18. In the process of active transport, molecules move (*a*) down the concentration gradient (*b*) with the concentration gradient (*c*) against the concentration gradient (*d*) toward a region of low molecular concentration.

19. The form taken by the chromosome in a prokaryotic cell is (*a*) a long, linear structure (*b*) the structure of an X (*c*) a C-shaped structure (*d*) a closed loop.

20. The lysosomes of eukaryotic cells contain (*a*) enzymes that function in digestion (*b*) chlorophyll molecules for photosynthesis (*c*) storehouses of ATP molecules (*d*) the chromosomes of the organism.

True/False. For each of the following, specify whether the statement in its entirety is *true* or *false*.

___ **1.** All bacteria occur in variations of five basic shapes.

___ **2.** The structure of the cell membrane is identical in both prokaryotic and eukaryotic cells.

___ **3.** About 80% of the cytoplasm of prokaryotes and eukaryotes consists of ATP molecules.

___ **4.** In the process of diffusion, molecules move across the cell membrane from a region of high solute concentration to a region of low solute concentration.

___ **5.** In eukaryotic cells, the ribosomes are located primarily along the endoplasmic reticulum.

___ **6.** In prokaryotic cells, DNA may be found in the chromosome and plasmids.

___ **7.** Bacterial flagella are composed of a carbohydrate known as peptidoglycan.

___ **8.** Viruses are considered prokaryotes because they have all the characteristics of prokaryotes.

___ **9.** The cell membrane is the site of enzymes that participate in the respiration of prokaryotic cells.

___ **10.** All bacteria have a cell wall.

___ **11.** If the fluid outside a prokaryotic cell contains a higher concentration of a particular solute than is present within the cell, the outside environment is said to be hypertonic.

___ **12.** All eukaryotic cells have a nucleus.

___ **13.** Members of the genera *Bacillus* and *Clostridium* are noted for their ability to produce capsules.

___ **14.** Among the microorganisms, the protozoa, fungi, and bacteria are considered eukaryotic.

___ **15.** Eukaryotic flagella consist of nine pairs of microtubules arranged in a ring with two single microtubules at the center of the ring.

___ **16.** A principal component of the cell wall in fungi is peptidoglycan.

___ **17.** The pellicle is an outer covering found on the cells of protozoa.

___ **18.** The structure of cilia in eukaryotic cells is similar to the structure of flagella in prokaryotic cells.

___ **19.** Bacterial spores are among the most resistant forms of life known to science.

___ **20.** There is no evidence of mitotic structures present in prokaryotic cells.

Completion. For each of the following, add the term that completes the statement best.

1. The cell membrane of prokaryotic and eukaryotic cells conform to a model known as the ___________.

2. Osmosis is a type of diffusion that is concerned with the movement of molecules of ___________.

3. One of the most resistant forms of life known to science is the bacterial ___________.

4. The principal component of the cell wall of fungi is ___________.

5. The name given to an irregular cluster of bacterial cocci is the ____________ .

6. When the bacterial capsule is less firm and more flowing, it is referred as a slime layer, or ____________ .

7. The cell membrane of prokaryotic and eukaryotic cells contains two parallel rows of ____________ .

8. One of the functions of the cell membrane in prokaryotes is to serve as the site of ____________ .

9. In prokaryotic cells, the ribosomes used in metabolism are suspended in the ____________ .

10. Among the diseases that can be caused by sporeforming bacteria are anthrax, gas gangrene, botulism, and ____________ .

11. The "9 + 2" arrangement of microtubules is commonly found in microbial flagella and ____________ .

12. In eukaryotic cells, the ribosomes are attached to the surfaces of the ____________ .

13. In eukaryotes, one or more bodies called nucleoli exist within the nucleus and are the sites for synthesis of ____________ .

14. The form of endocytosis in which particles are taken into the cell is known as ____________ .

15. Among the eukaryotic microorganisms are classified certain algae, fungi, and ____________ .

16. Those bacteria having multiple flagella at the ends of the cell are described as ____________ .

17. An alternative name for pili is ____________ .

18. Disease-causing entities that are considered neither eukaryotes nor prokaryotes are ____________ .

19. The rod form of a bacterium is known as a ____________ .

20. With the exception of mycoplasmas, all bacteria possess a ____________ .

Chapter 5

Microbial Growth and Cultivation

OBJECTIVES

In microorganisms, growth refers to the accumulation of organic matter and the conversion of the organic matter into more of the organism. Growth results in an increase in size, followed by division of the organism into two cells. This chapter is concerned with the characteristics associated with microbial growth and the conditions that encourage growth. The chapter's objectives are to:

1. Review the patterns for cell division in eukaryotic microorganisms and prokaryotic microorganisms, with an emphasis on the stages of mitosis.

2. Explain the process of binary fission as it occurs in bacteria, and note the consequences if bacteria do not break apart following binary fission.

3. Discuss the phases that a population of microorganisms pass through during a normal growth curve.

4. Distinguish the format taken by colonies of various types of microorganisms, and show how the numbers in a microbial colony can be counted.

5. Identify the various environments in which microorganisms can live, with reference to the temperature, gaseous, and pH requirements for microbial growth.

6. Discuss the importance of water and moisture in the growth of microorganisms.

7. Describe how pure cultures of microorganisms can be obtained and specify the various types of media in which microorganisms can be cultivated in the laboratory.

8. Specify how media can be selective, differential, and enriched for the cultivation of different kinds of microorganisms.

THEORY AND PROBLEMS

5.1 Do all microorganisms experience growth?

Bacteria, fungi, protozoa, and algae all display the characteristic of growth. Although they are considered microorganisms, viruses do not display growth patterns, and so the discussions in this chapter do not apply to viruses. Many microbiologists consider viruses to be particles with the ability to multiply in living cells. The expression "viral growth" is a misnomer.

CELL DUPLICATION AND POPULATION GROWTH

5.2 What are the patterns for cell division in microorganisms?

There are two general patterns for cell division in microorganisms. The eukaryotic protozoa, fungi, and unicellular algae undergo cell division by the process of mitosis as illustrated in Figure 5-1. The prokaryotes such as bacteria, by contrast, do not display the structures associated with mitosis; these microorganisms reproduce by the process of binary fission.

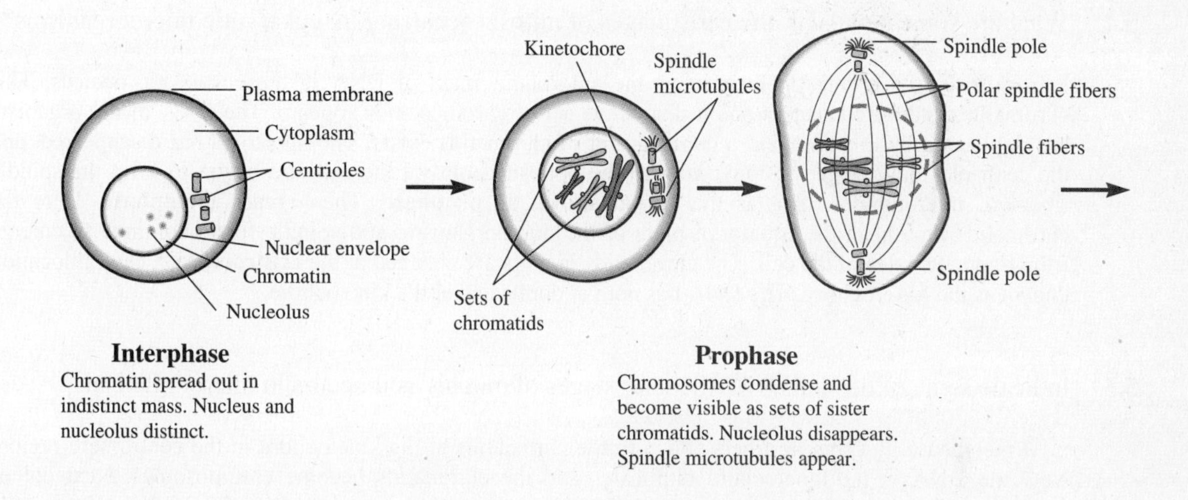

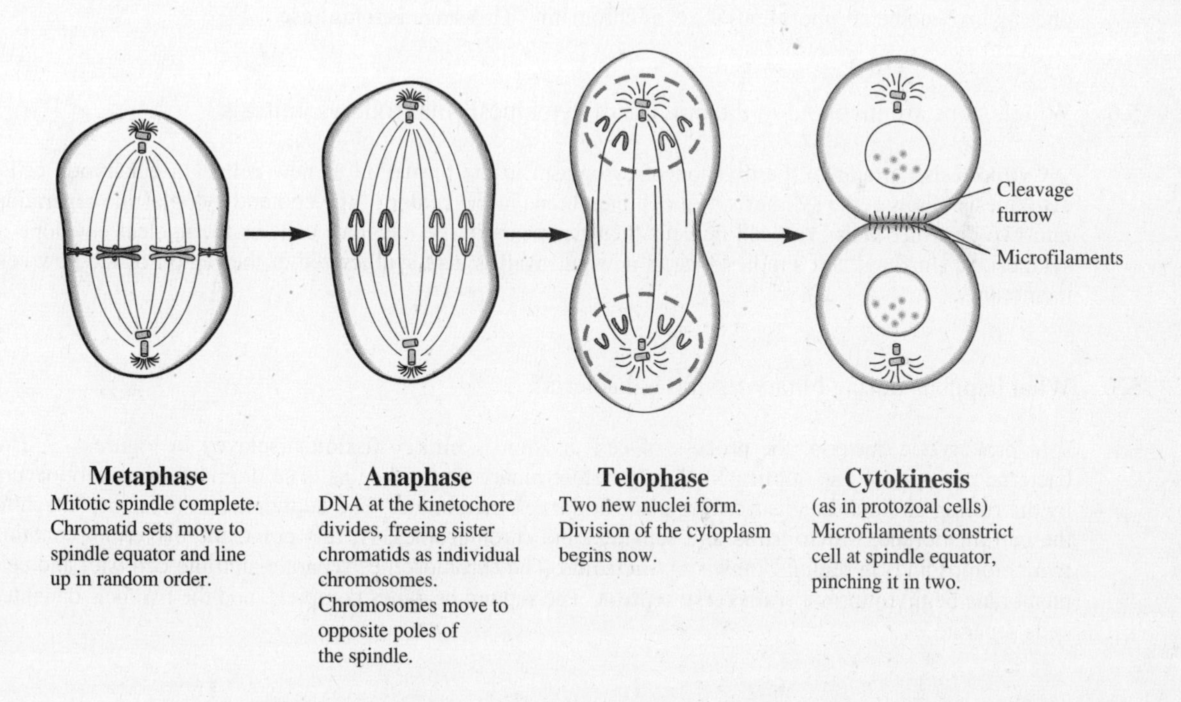

Fig. 5-1 The stages of mitosis as they occur in eukaryotic microorganisms such as simple algae, fungi, and protozoa.

5.3 Describe a general overview of mitosis as it takes place in eukaryotic microorganisms?

In the process of mitosis, the DNA contained in the nuclear chromatin of the organism undergoes a replicative process in which semiconservative replication accounts for the duplication of the nuclear material. Then the cell divides. The process is similar to that taking place in binary fission, but there is more complexity in mitosis.

5.4 What are some details of the early stages of mitosis occurring in eukaryotic microorganisms?

In **mitosis**, the chromatin material is the amorphous mass of DNA in the eukaryotic nucleus. This chromatin material condenses and a distinctive set of chromosomes appears. The DNA of the organism has duplicated by this time and a duplicate set of chromatids exists. The nucleolus has disappeared, and the centrioles have begun to move toward the opposite poles of the cell and begun to form the spindle fibers of microtubules. This is the period known as **prophase**. Then comes **metaphase**. Here the chromatids line up in the equatorial plate of the microorganism, and spindle fibers continue to emerge from opposite poles of the cell. The chromatids in pairs are attached at the **centromere**, a central location known as the **kinetochore**. The DNA has not yet duplicated at the kinetochore.

5.5 Indicate some of the details of the later stages of mitosis as it occurs in microorganisms?

At metaphase, the spindle fibers attach to the chromatids at the kinetochore in the centromere region. Now the DNA in the kinetochore duplicates and the chromatids become chromosomes. Next comes **anaphase**. During this phase, one set of chromosomes is drawn to either side of the cell by the spindle fibers. The fibers, composed of microtubules, are believed to be disassembled in sequential fashion as they draw toward the chromosomes toward the poles. When the chromosomes arrive at the poles, they once again become an amorphous mass of chromatin. This stage is **telophase**.

5.6 Which steps are involved in the process of cytokinesis that follows mitosis?

Cytokinesis pertains to the division of cytoplasm in the formation of new cells. In eukaryotic cells, cytokinesis involves the synthesis of a cell membrane at the center of the cell and cytoplasmic separation into two daughter cells. The cell nucleus then reappears in the daughter cells as the nuclear envelope is synthesized. In fungi and simple algae, a new cell wall is also synthesized at the region of the new cell membrane.

5.7 What happens during binary fission in bacteria?

In prokaryotic bacteria, the process of cell division is **binary fission** displayed in Figure 5-2. The bacterial chromosome is duplicated shortly before binary fission begins. The duplication usually occurs by the rolling circle mechanism taking place at the cell membrane. The duplicate chromosomes attach to the cell membrane, which grows and separates the chromosomes. At this point, the bacterium contains two chromosomes in regions known as **nucleoids**. The chromosomes separate, and the cell wall and cell membrane begin to form a **transverse septum**. The septum becomes complete, and the two new daughter cells separate.

5.8 Do the bacteria always break apart after cell division?

Depending on the planes of division, bacterial cells do not necessarily break apart. In some cases, the bacteria divide in the same plane and remain linked as a **streptococcus**; or they divide at right angles and form cuboidal packets of four or eight cocci called **sarcinae**; or they divide in random planes and form grapelike clusters called **staphylococci**.

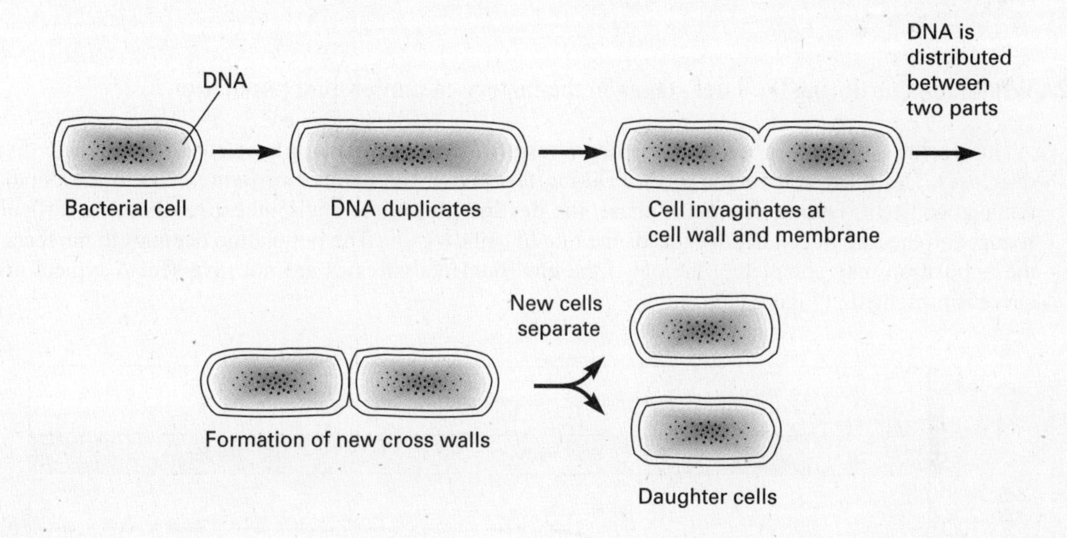

Fig. 5-2 Binary fission, the process of reproduction taking place in prokaryotic microorganisms such as bacteria.

5.9 Are there any other forms of cell division besides mitosis and binary fission in microorganisms?

In yeasts and in a few species of bacteria, reproduction takes a different form than binary fission or mitosis. DNA duplication occurs and a new cell develops at the surface of the existing cell. This small cell gathers cytoplasm and organelles (if a eukaryotic cell) and develops and eventually breaks free from the parent cell. The reproductive process is called **budding**.

5.10 What are some different patterns that microorganisms undergo during cell division?

Different microorganisms undergo cell division at different time intervals. For example, in the fungi, cell division occurs at the tips of the hyphae, and the organism elongates with each new cell produced. In protozoa, cell division may occur every several hours or days. In certain bacteria, cell division occurs very often. In one species called *Escherichia coli*, the cells undergo binary fission every 20 to 30 minutes. The time passing in between cell divisions is known as the **generation time.**

5.11 Which early phases do a population of microorganisms pass through after they has been introduced into a fresh culture medium?

On introduction to a fresh growth environment (such as a culture medium), a population of microorganisms will pass through four distinct phases. The first phase, the **lag phase**, will encompass several hours. During this time the organisms grow in size, accumulate organic matter and store large quantities of chemical energy such as ATP for biosynthesis. During the next phase, the **logarithmic (log) phase**, the microorganisms undergo rapid cell division and fulfill their generation time. The population doubles during each generation time, and the population increases in size at a logarithmic, or exponential rate.

5.12 What happens during the later stages in the history of a microbial population?

During the third phase of a population's history, the **stationary phase,** the rate of cell division decreases, and older cells begin to die. During this phase, the number of living cells in the population remains constant. During the fourth phase, the **decline phase**, the environment has become difficult for living, and the rate of cell death exceeds the rate of cell division. The population declines in numbers, and the population may completely die out if the environmental stresses are not reversed. A typical growth curve is presented in Figure 5-3.

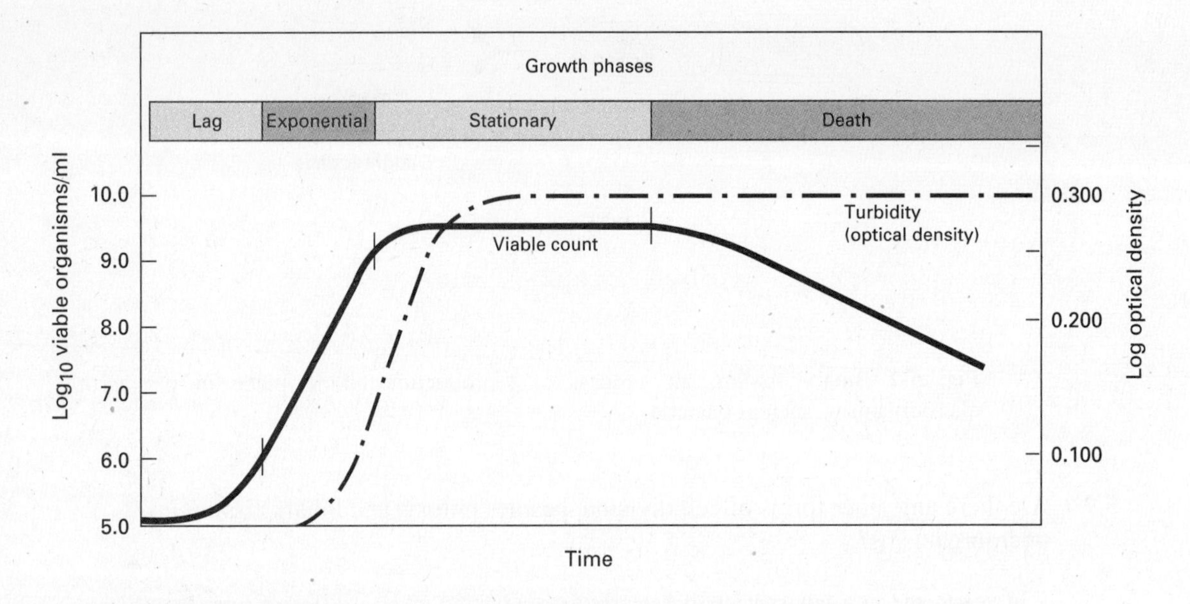

Fig. 5-3 The growth curve for a population of bacteria showing the four phases of the population. The turbidity of the growth medium is shown in terms of optical density. It remains constant even though the number of viable bacteria declines because dead cells contribute to the density.

5.13 What is a microbial colony?

When a microorganism such as a bacterium divides in a population, the result of those divisions is a visible mass of bacteria called a **colony**. The members of a colony such as shown in Figure 5-4 are all derived from a single ancestor, and the bacteria in the colony are genetically identical, unless mutations have taken place. The mass resulting from a single fungal spore is a **mycelium**; this represents a fungal colony.

5.14 Do protozoa form colonies?

With certain exceptions such as members of the genera *Naegleria* and *Acanthamoeba* protozoa cannot be cultivated on solid culture media, so they are unable to form colonies. However, they can be cultivated in liquid media where they form cultures consisting of only one protozoan species. This is called an axenic culture.

5.15 How can the numbers of bacteria be measured in a bacterial population or colony?

There are several methods for determining the number of bacteria in a colony, and most revolve around cultivation in a Petri dish containing a culture medium. Generally the medium contains **agar**, a polysaccharide from marine algae that solidify the medium without adding any nutrients. To determine the number of bacteria within a population, dilutions of a population are made in a sterile water or saline solution, and predetermined amounts are placed on plates of agar medium. The bacteria will form colonies during incubation, and by measuring the number of colonies in a particular dilution, one can calculate back to find the original number of bacteria in the original population. This method is referred to as the **plate count method**.

5.16 Are there any other ways of determining the number of bacteria in a population?

Bacterial growth can be calculated by direct **microscopic counts**. To perform this technique, a known volume of the sample is introduced into a specially calibrated counting chamber such as a Petroff-Hauser chamber. The number of organisms in the chamber are counted, and by utilizing the dilution factor, the original number can be determined. This method also works for fungal spores and for protozoa.

5.17 What is the most probable number test used for bacterial counting?

It is possible to determine the number of bacteria in population by the **most probable number test**. This is an estimate of the number of cells in the population determined as a statistical probability falling within a particular range. The most probable number is calculated by observing the number of tubes that display bacterial growth in broth media and bring about an established metabolic change such as lactose fermentation to gas.

5.18 How can filtering be used to determine a bacterial count?

A bacterial count can also be determined by filtering a known quantity of fluid, catching the bacteria on the filter and placing the filter on a plate of agar media where the bacteria will form colonies. This test is used in the **membrane filter technique** where a special medium such as eosin methylene blue agar (EMB) or Endo agar is used. Counting the bacterial colonies gives an estimate of the number of original bacteria. Turbidity measurements can also be used to determine numbers, since the cloudier the broth becomes, the higher is the microbial population number.

ENVIRONMENTAL GROWTH CONDITIONS

5.19 In what sorts of environments can microorganisms live?

Animals and plants tend to be restricted to the types of environments they grow in, but microorganisms grow in a much broader series of environments. For example, microorganisms can grow in ice or in hot springs, in fresh water or in salt water, with or without oxygen, or at highly acidic to highly alkaline environments. Microorganisms inhabit virtually all environments on Earth.

5.20 Are microorganisms restricted to the particular environment in which we find them?

Some microorganisms grow only under specified environmental conditions. Scientists describe those microorganisms as **obligate**. Other microorganisms exist in a number of different environments. Scientists call those microorganisms **facultative**.

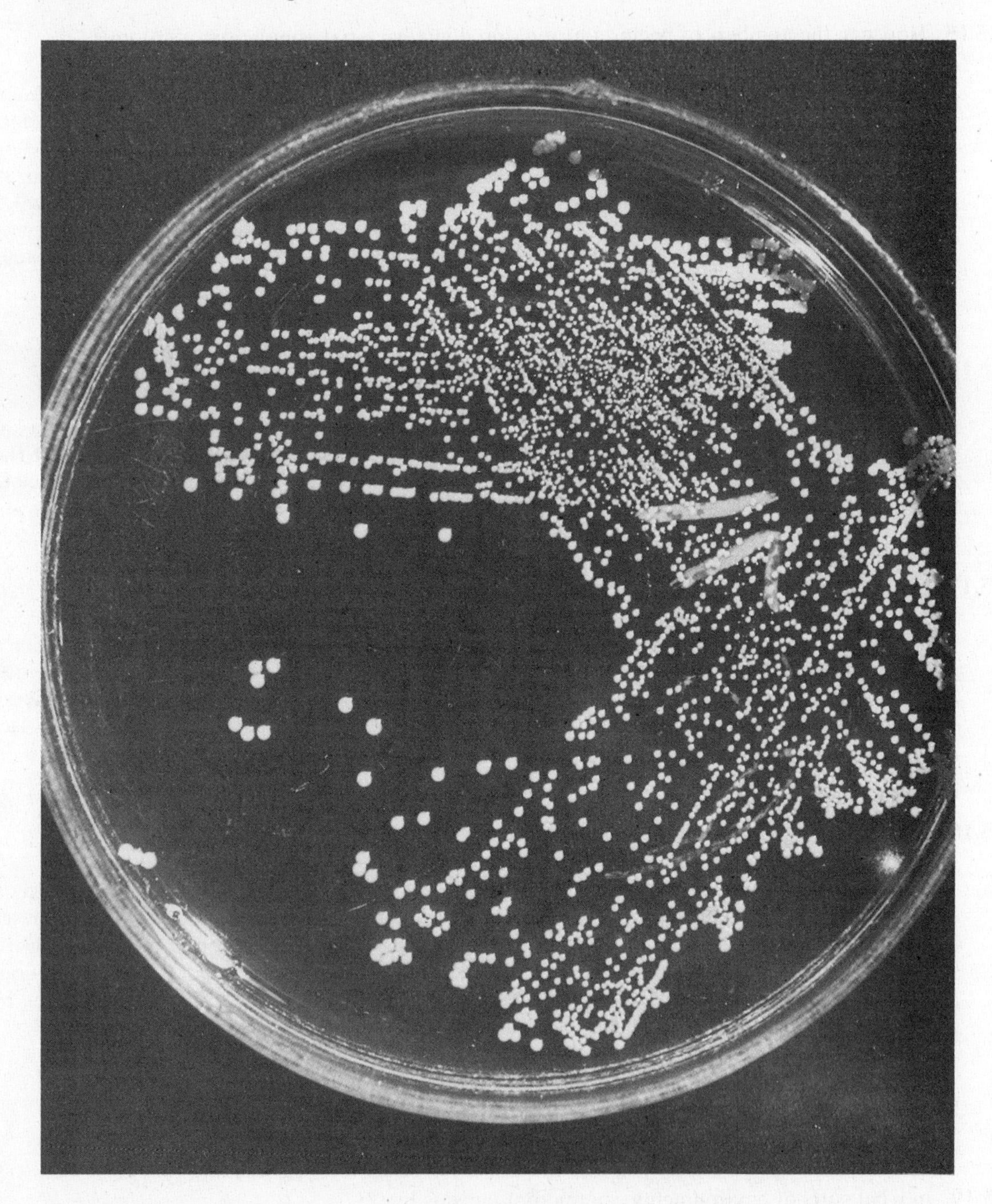

Fig. 5-4 A plate of nutrient agar containing bacterial colonies. This plate has been prepared by the streak plate method. Each dot represents a mass of visible growth derived from a single bacterium, which was deposited on the agar surface. Each colony represents a pure culture.

5.21 At which temperatures do protozoa and fungi survive best?

Protozoa and fungi and unicellular algae commonly exist under environmental temperatures that are fairly restricted. For example, fungi and protozoa live primarily at temperatures close to room temperature, about 25°C. There are many protozoa and fungi that live at the higher body temperature of 37°C. These microorganisms are generally involved in human disease. In addition, there are many fungi that live at colder temperatures such as the 5°C found in the refrigerator. The growth of a mold on a refrigerated piece of food is evidence of this ability to grow at this colder temperature.

5.22 At which temperatures do bacteria survive best?

Of all microorganisms, bacteria have the broadest ranges of temperatures for growth. They can grow at temperatures as low as 0°C or over 100°C.

5.23 Into which groups are bacteria classified according to their temperature requirements?

Bacteria are divided into three broad categories according to the temperature at which they grow best, as illustrated in Figure 5-5. Psychrophilic bacteria, or **psychrophiles**, grow best at temperatures between 5°C and 20°C, with some species growing at 0°C in ice. Psychrophiles are known to cause refrigerated foods to spoil (as in spoiled milk), but they generally do not grow at human body temperatures. The second category are the mesophilic bacteria, or **mesophiles**. These bacteria grow at temperatures between 20°C and 40°C, and they include the species that grow at body temperature (37°C). Many pathogenic bacteria are in this group. In the final group are the thermophilic bacteria, or **thermophiles**. These bacteria grow at temperatures between 40°C and 80°C, and some tolerate temperatures as high as 110°C. Many thermophiles grow at temperatures at which milk is pasteurized and they survive the pasteurization process. Thermophiles are found in hot springs, geysers, and thermal vents found at the bottoms of oceans. Many of the archaebacteria are thermophiles.

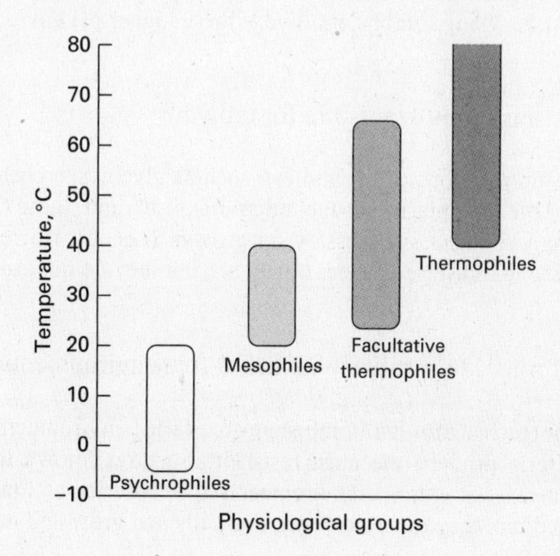

Fig. 5-5 The physiological groups of bacteria arranged according to their temperature requirements.

5.24 Do all bacteria require oxygen for growth?

Many bacteria are **aerobic**, that is, they require oxygen to grow and perform the biochemical reactions of their metabolism. There are, however, a large number of bacterial species that are **anaerobic** and are cultivated in a special apparatus where the oxygen has been removed. These bacteria (known as anaerobes) live in the absence of oxygen, and they are found in the muddy bottom of swamps, landfills, and garbage depots. Some anaerobes are **obligate anaerobes**, while others are **facultative anaerobes**, meaning that they live with or without oxygen.

5.25 Can any bacteria live in a reduced oxygen atmosphere?

Some bacteria are **microaerophilic**. These bacteria grow best when a small amount of oxygen is present. They are found in such environments as the urinary and digestive tracts in humans, and some species can cause infections in these systems. Many microaerophiles are dependent on carbon dioxide for their metabolism. These organisms are called **capnophiles**.

5.26 What are the pH requirements for best microbial growth?

The term **pH** refers to the acidity or alkalinity level of the solution in which microorganisms grow; it can range from the most acidic level of 0 to the neutral level of 7.0 to the most alkaline level of 14.0. Fungi grow best under conditions that are somewhat acidic. They often prefer pH levels of 5.0. Protozoa tend to grow best where the pH is near neutral (7.0). Bacteria grow over a wide range of pH levels. It is important to note that microorganisms have an optimum pH, the acidity of alkalinity level where they grow best. This level is not variable more than one unit if the microorganisms are to survive.

5.27 At what pH levels can various types of bacteria grow?

Different bacterial species can grow over a broad range of pH levels. Certain bacteria can live at pH levels as low as 0 and ranging up to 5.0. These bacteria are called **acidophiles**. Those bacteria that can exist from pH of 5 to 8.5 are called **neutrophiles**. Most of the human pathogens are neutrophiles. Those bacteria that live at alkaline pH levels are called **alkalinophiles**. These bacteria live at pH levels from about 7.0 to about 11.5. Many soil bacteria live at these higher pH environments.

5.28 Do all microorganisms require moisture for growth?

In order to grow, the reactions of metabolism such as glycolysis, chemiosmosis, and protein synthesis must take place in the cytoplasm of the microorganism and these reactions take place in **water**. Therefore, a water environment is necessary for growth to occur. However, many microorganisms have the capacity to survive in moisture-free environments, but they do not grow in these environments.

5.29 Which microbial forms can survive in moisture-free environments?

Fungi produce **spores** that survive in moisture-free (arid) environments, and certain protozoa produce resistant **cysts**. Bacteria produce the most resistant structures known to science, the **bacterial spores**. Bacterial spores can exist at extreme environments for centuries or longer. When they are placed on a nutritious culture medium, they revert to vegetative cells that grow and multiply as normal bacterial cells.

5.30 In what unusual environmental conditions can microorganisms be found?

Microorganisms can be found in numerous other environments on Earth. In the oceans, there exist thousands of species of marine microorganisms, which thrive in salt water. At the bottoms of the oceans, scientists locate barophilic microorganisms (or **barophiles**), which survive the intense pressures of the water above them. Microorganisms may also be found in environments that contain much radiation, in environments where there are high osmotic pressures, and in environments where there are exceptional quantities of salt. The latter microorganisms as well as the marine microorganisms are known as halophilic microorganisms or **halophiles**.

LABORATORY CULTIVATION METHODS

5.31 Why is it important to cultivate microorganisms in the laboratory?

There are many reasons for cultivating microorganisms in the laboratory. Microorganisms are extensively used in research in many of the metabolic processes. Many of the metabolic processes discussed in Chapter 6 were originally discovered in microorganisms, then applied to complex animals and plants. Also, cultivating microorganisms is essential for diagnostic purposes when infectious disease is involved. Furthermore, studying microorganisms in the laboratory gives an opportunity to develop methods for interrupting their spread and controlling their growth. Finally, many important applications of microbial growth can be developed by understanding patterns of microbial metabolism in the laboratory.

5.32 Why are pure cultures of microorganisms required in the laboratory?

In order to cultivate a species of microorganism, the latter must be obtained in a pure culture. A **pure culture** is one in which only one species of microorganism is growing. This is important because if there were two or more species being cultivated together it would be difficult to know which effect was being brought about by which species. Also, if a diagnosis for disease is desired, it would be difficult to know which microorganisms were responsible for the disease.

5.33 How can pure cultures be obtained?

A pure culture is obtained by isolating a single microorganism and permitting that microorganism to grow and form a colony. For a fungus, this usually means obtaining a single fungal spore and permitting that spore to germinate to produce a mycelium. For a bacterium, a single cell must be separated from the remainder and permitted to undergo binary fission. Viruses must be cultivated in fertilized eggs, as shown in Figure 5-6, because they require living tissue for multiplication.

5.34 How is the streak plate method performed for obtaining pure bacterial cultures?

For bacteria, there are two methods available for obtaining pure cultures. The first method, called the **streak plate method,** utilizes a plate of agar medium. Bacteria are obtained on a wire loop, and the loop is streaked across the agar surface, depositing bacteria with each streak. The loop moves around the plate, and as streaking continues, fewer and fewer organisms are deposited. Eventually, there will be few enough organisms to form well-isolated, discrete colonies. Each colony represents a pure culture.

5.35 Describe the pour plate methods used for obtaining pure cultures of bacteria?

The second method, called the **pour plate method**, involves diluting a sample of bacteria in several tubes of liquid agar medium, as Figure 5-7 illustrates. Each tube contains fewer and fewer bacteria. The tubes are then poured into empty Petri dishes, and where there are the fewest bacteria, the most isolated colonies will form. As before, each colony represents a pure culture, and a sample of the colony can be taken to obtain one species of bacteria.

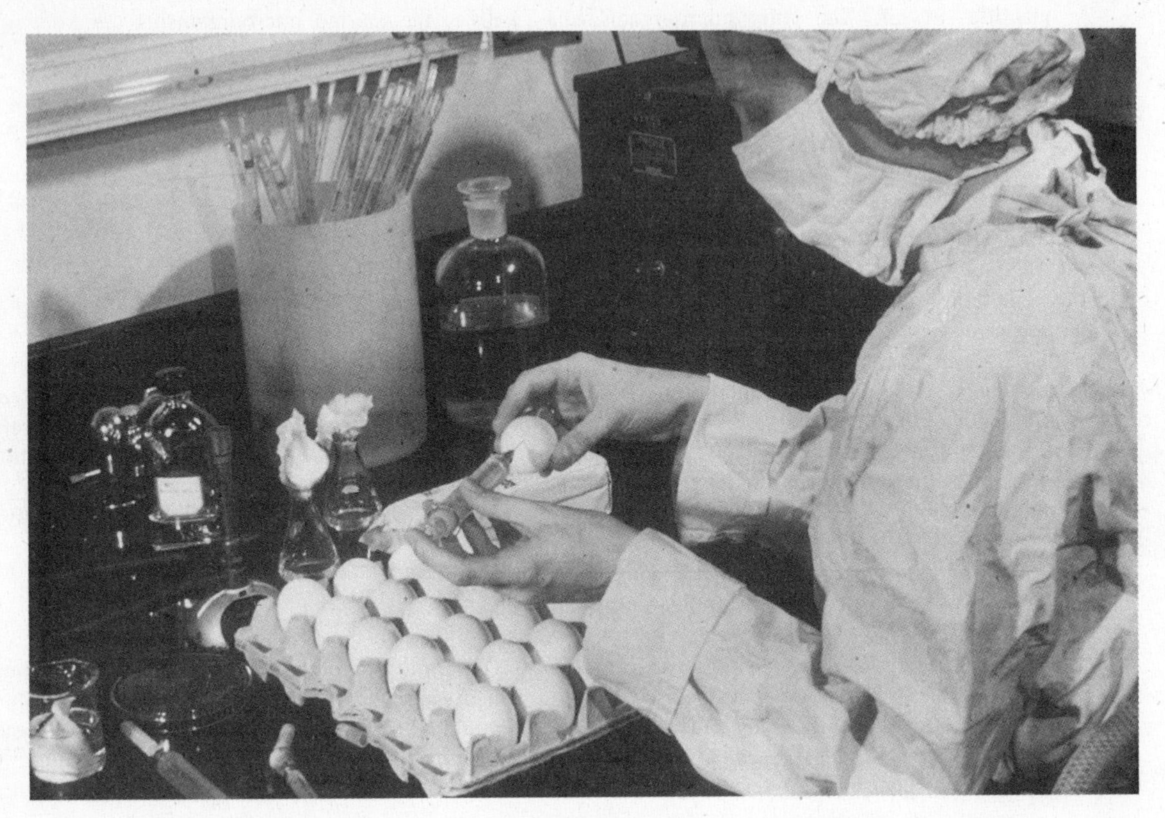

Fig. 5-6 Inoculation of fertilized eggs with viruses to encourage their multiplication. Viruses multiply only within living tissues cells.

5.36 In what sort of material do microorganisms grow in the laboratory?

Protozoa are generally cultivated in a liquid (broth) medium. Fungi and bacteria, by contrast, can be cultivated on solid media containing agar. There are exceptions, however, and certain organisms such as those of syphilis (*Treponema pallidum*) and leprosy (*Mycobacterium leprae*) have never been cultivated on agar media. These bacteria must be cultivated in live animals. The bacilli that cause leprosy are cultivated in the footpads of mice.

5.37 What two different types of media are available in the laboratory?

Two general types of media are available for cultivating microorganisms in the laboratory. The first type, called a **synthetic medium**, is prepared by combining precisely determined components such as sodium chloride, potassium phosphate, magnesium sulfate, amino acids, glucose, vitamins, and water. In this medium, the nature and quantity of each component is known. This medium is also called a **chemically defined media**. The second kind of medium, known as a **natural medium,** contains nutrient sources in which the nature and quantity of each component cannot be defined. This medium is also referred to as a **chemically nondefined medium**. It may contain such things as beef extract, soybean extract, a protein called peptone, and water. A medium such as **nutrient agar** is a natural medium.

5.38 Are there any media available for cultivating certain kinds of microorganisms?

Certain types of media are called **selective media**. In this medium, one species of microorganism will grow to the exclusion of other species. The medium will "select out" one species of microorganism. An example is **mannitol salt agar**. In this medium, staphylococci such as *Staphylococcus aureus* will grow well because the bacteria ferment the alcoholic carbohydrate mannitol in the presence of a high salt environment; other microorganisms will be inhibited by the high salt concentration.

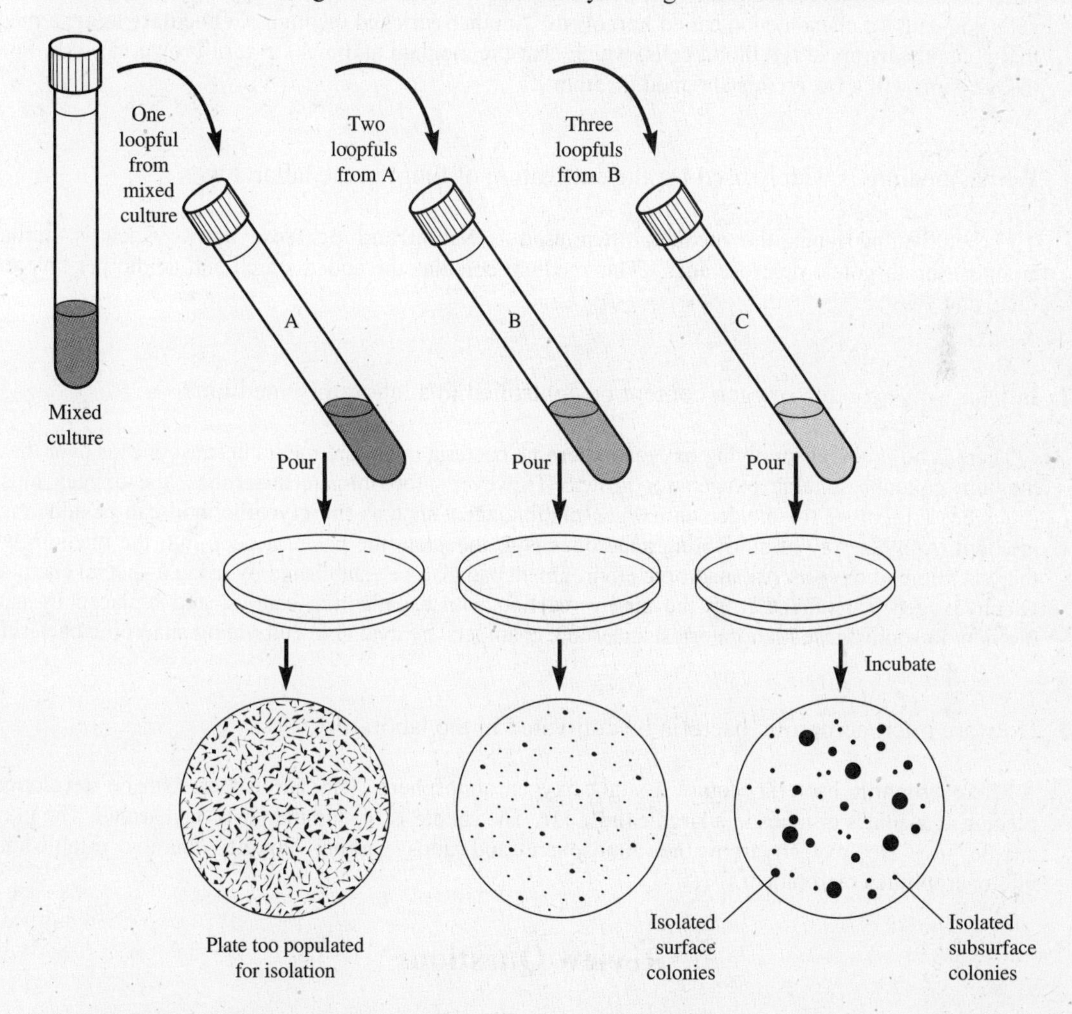

Fig. 5-7 The pour plate method used in the microbiology laboratory to isolate the bacteria in a mixed culture.

5.39 Is it possible to differentiate bacteria on a particular medium?

Bacteria may be differentiated from one another on a medium known as a **differential medium**. In this medium, different bacteria will exhibit different qualities. For example, in **MacConkey agar**, certain lactose-fermenting bacteria will appear red because they absorb the dye neutral red; nonfermenters of lactose fail to absorb the dye and will remain colorless. Another differential medium is **eosin methylene**

blue agar (EMB agar). *Escherichia coli* grows in this medium and absorbs the two dyes to produce green colonies with a metallic fluorescent sheen; other bacterial colonies are mucoid and purple.

5.40 How can a laboratory medium be enriched?

Certain bacteria will grow only if the medium has been enriched. An example of an **enriched medium** is **blood agar**, a medium in which blood has been added to a rich agar base such as trypticase soy agar in order to encourage streptococci to grow. Pathogenic species of streptococci will break down the blood cells and cause a phenomenon called **hemolysis**. Another enriched medium is **chocolate agar,** a medium that contains disrupted red blood cells, which char the medium and make it turn brown in color. Species of *Neisseria* will grow on this enriched medium.

5.41 Which medium is widely used for the cultivation of fungi in the laboratory?

For cultivating fungi, the medium often used is **Sabouraud dextrose agar**, which is similar in composition to potato dextrose agar. This medium contains the added starch and acidic pH favored by fungi and yeasts.

5.42 In what ways can the oxygen content be controlled in a laboratory medium?

There is no problem supplying oxygen to aerobic bacteria in culture media because the air over the agar medium contains suitable oxygen for growth. However, for obligate anaerobes, the oxygen must be removed. To remove the oxygen an oxygen-binding agent such as **thioglycollic acid** can be added to the medium. Another way of cultivating anaerobes is to inoculate the bacteria deep into the medium where there is minimal oxygen. An anaerobic environment can also be established by using a special chamber in which oxygen is removed from the air by reaction with a palladium catalyst and replaced by carbon dioxide. In sophisticated laboratories, anaerobic chambers are available cultivating anaerobic bacteria.

5.43 How are microaerophilic bacteria be cultivated in the laboratory?

Microaerophilic bacteria require a slight oxygen atmosphere. This atmosphere can be developed by placing agar plates or tubes in a large **candle jar**. The candle is lit, and the jar is then sealed. The burning candle uses up oxygen from the atmosphere and adds carbon dioxide, thereby establishing a microaerophilic environment.

Review Questions

Multiple Choice. Select the letter of the item that correctly completes each of the following statements.

1. Psychrophilic bacteria are those bacteria that grow best (*a*) in oxygen-free environments (*b*) at pH levels of 8 or above (*c*) at cold temperatures (*d*) only in the presence of viruses.

2. Microorganisms that survive in moisture-free environments do so because they (*a*) form spores (*b*) metabolize glucose molecules only (*c*) have no cell membranes (*d*) have no chromosomes.

3. Mannitol salt agar is an example of a medium that will support (*a*) viruses only (*b*) chlamydiae and rickettsiae (*c*) protozoa as well as fungi (*d*) only certain bacterial species.

4. A laboratory medium for bacteria can be enriched by (*a*) adding ATP (*b*) adding blood (*c*) including salt in the medium (*d*) increasing the level of potassium.

5. Those bacteria designated microaerophilic are distinguished by their ability to grow (*a*) in high concentrations of salt (*b*) in low concentrations of oxygen (*c*) without ATP or glucose (*d*) only in the presence of viruses.

6. The process of mitosis occurs (*a*) only in bacteria (*b*) in viruses as well as bacteria (*c*) in eukaryotic organisms (*d*) only in chemically defined media.

7. The generation time for all microorganisms (*a*) varies (*b*) is 30 minutes (*c*) is three hours (*d*) is 12 hours.

8. The process of budding takes place (*a*) primarily in viruses (*b*) in bacteria that form branches (*c*) only within the protozoa (*d*) in the yeasts.

9. During the stationary phase of growth of microorganisms (*a*) the rate of cell division increases (*b*) the rate of cell division decreases (*c*) the population is at its most vigorous state (*d*) the population is at its least vigorous state.

10. Fungi and protozoa live primarily at temperatures (*a*) close to 100°C (*b*) about 50 to 60°C (*c*) close to room temperature (*d*) between 5 and 10°C.

11. Bacteria that are pathogenic in the body grow at temperatures (*a*) of over 100°C (*b*) at body temperature (*c*) at the same temperature as viruses (*d*) at thermophilic temperatures.

12. A pure culture is a culture in which (*a*) only one species of microorganism is present (*b*) only one nutrient is required by the bacterium for growth (*c*) only one organism other than the main organism is present (*d*) there are no waste products in the culture.

13. All the following are required to perform the streak plate isolation method except (*a*) a wire loop (*b*) a plate of agar medium (*c*) a culture of bacteria (*d*) an electrophoresis machine.

14. A differential medium is one in which (*a*) fungi and viruses grow differently (*b*) two different bacteria can be distinguished (*c*) a particular nutrient is used differently by two different bacteria (*d*) two different temperatures are utilized in the incubation period.

15. Inoculating bacteria deep into a tube of solid medium provides (*a*) conditions for anaerobic growth (*b*) an opportunity to remove waste products (*c*) enhanced oxygen atmospheres (*d*) increased numbers of potassium and sodium ions.

16. Direct microscopic counts can be used to determine the number in a population of all of the following except (*a*) protozoa (*b*) fungal spores (*c*) bacteria (*d*) viruses.

17. Barophilic microorganisms are those microorganisms able to grow (*a*) at cold temperatures (*b*) at high pressures (*c*) at high temperatures (*d*) at high pH values.

18. Those bacteria that live at pH levels from 5 to 8.5 are known as (*a*) thermophiles (*b*) mesophiles (*c*) neutrophiles (*d*) capnophiles.

19. A filter may be used in microbiology for all the following purposes except (*a*) to separate viruses from bacteria (*b*) to determine a bacterial count (*c*) to separate glucose molecules from ATP molecules (*d*) to separate viruses from protozoa.

20. The process of binary fission proceeds most frequently in (*a*) viruses (*b*) protozoa (*c*) bacteria (*d*) fungi.

True/False. For each of the following statements, mark the letter "T" next to the statement if the statement is true. If the statement is false, change the underlined word to make the statement true.

___ **1.** Microorganisms undergo rapid cell division during the lag phase of their growth.

___ **2.** Among the microorganisms, the ones that do not display the characteristic of growth are viruses.

___ **3.** The time passing between cell divisions of a microorganism is known as the mitosis time.

___ **4.** The members of a microbial colony are all derived from a single ancestor.

___ **5.** In order to resist moisture-free environments, certain species of protozoa produce spores.

___ **6.** Halophilic microorganisms are those that are able to live in high sugar environments.

___ **7.** In the laboratory, bacteria are unable to form colonies.

___ **8.** The polysaccharide used to solidify bacterial media is agar.

___ **9.** In order for the reactions of metabolism to take place, there must be present a certain amount of water.

___ **10.** A medium in which precise amounts of certain components are combined and the name of all components are known is called a natural medium.

___ **11.** Chocolate agar is a bacteriological medium that contains disrupted kidney cells.

___ **12.** Those bacteria that require oxygen for growth are said to be aerobic.

___ **13.** In fungi, protozoa, and unicellular algae, the primary method for reproduction is binary fission.

___ **14.** The reproductive process of budding is widely seen in viruses.

___ **15.** Facultative microorganisms are microorganisms capable of living in a number of environments.

___ **16.** Those bacteria designated alkalinophiles are able to grow at pH levels above 7.0.

___ **17.** Most protozoa are generally cultivated in a solid medium.

__ **18.** An example of a natural medium used in the laboratory is <u>nutrient</u> agar.

__ **19.** Fungi grow best under conditions that are somewhat <u>basic</u>.

__ **20.** During the decline phase of microbial growth, the rate of cell death <u>exceeds</u> the rate of cell division.

Matching

___ 1. Found at the ocean bottom
___ 2. Prefer high temperature
___ 3. Grow in sour cream
___ 4. Survive pasteurization
___ 5. Live in high salt environments
___ 6. Thrive in high pressure environments
___ 7. Live at pH 7.0
___ 8. Prefer acidic environments
___ 9. Use carbon dioxide
__ 10. Live in slight oxygen environments

(*a*) Barophiles
(*b*) Neutrophiles
(*c*) Halophiles
(*d*) Thermophiles
(*e*) Acidophiles
(*g*) Microaerophiles
(*h*) Capnophiles

Selection. Select one of the four types of microorganisms for each of the characteristics listed below.

___ 1. Grow best where pH is neutral
___ 2. Can be microaerophilic
___ 3. Multiply by binary fission
___ 4. Do not form any resistant structures
___ 5. Certain species grow on mannitol salt agar
___ 6. Cell division occurs at the tips of hyphae
___ 7. Grow primarily in liquid environments in the laboratory
___ 8. Some important species grow anaerobically
___ 9. Do not undergo any sort of cell division
__ 10. Multiply only in living tissues

(*a*) Viruses
(*b*) Bacteria
(*c*) Fungi
(*d*) Protozoa

Chapter 6

Metabolism of Microorganisms

OBJECTIVES

Metabolism is one of the essential aspects of microbial life. Together with growth, reproduction, adaptability, and other factors, metabolism helps define the microorganism, while providing insights to its chemical activities. The objectives of this chapter are to:

1. Develop an understanding of the broad aspects of catabolism and anabolism and how they underlie the life of a microorganism.

2. Highlight the characteristics of enzymes and the important place they have in metabolism.

3. Make note of the importance of adenosine triphosphate (ATP) in the energy metabolism of a microorganism.

4. Study the metabolic pathways of glycolysis, Krebs cycle, electron transport, and chemiosmosis, and understand how a microorganism obtains the energy present in a carbohydrate molecule for its use.

5. Recognize the importance of oxygen as a final electron acceptor in the energy metabolism of microorganisms.

6. Understand how proteins and fats can be used as energy sources in microbial life.

7. Identify the mechanisms of microbial photosynthesis for synthesizing energy compounds in anabolism.

8. Study the importance of water as an electron donor in photosynthesis and note how oxygen is produced this process.

THEORY AND PROBLEMS

6.1 What is metabolism?

Metabolism is the general term used for the thousands and thousands of chemical reactions occurring in all living things. Except for viruses, metabolism is a characteristic of all microorganisms, plants, and animals.

6.2 Into which broad areas can metabolism be subdivided ?

Metabolism is generally subdivided into anabolism and catabolism. **Anabolism** is the overall process by which cells synthesize molecules and structures. It is also referred to as biosynthesis. Anabolism is a building process that generally requires an input of energy. The second broad category is catabolism. **Catabolism** is the overall chemical process in which cells break down large molecules into smaller ones. Catabolism is a process that generally results in the liberation of energy. Taken together, the chemistry of anabolism and catabolism together make up metabolism.

6.3 Why is metabolism necessary for a cell?

Metabolism contributes to the stability of a living cell while providing a dynamic pool of building blocks for synthesis reactions. In addition, metabolism permits cells to extract energy for their life processes and to continue to grow and multiply.

ENZYMES

6.4 How are the reactions of metabolism brought about?

The chemical reactions of metabolism are catalyzed through the activity of enzymes. **Enzymes** are biological catalysts that influence chemical reactions while themselves remaining unchanged. The reactions would probably occur, but they would take an enormously long period of time. Enzymes reduce the time for a reaction to occur to milliseconds.

6.5 How do enzymes perform their functions?

For any chemical reaction to take place, a certain amount of energy must be put into the reaction. This energy is called the energy of activation, or **activation energy**. For example, heat energy might be the activation energy introduced to a chemical reaction to encourage the reaction to occur. Enzymes reduce the amount of activation energy necessary by bringing the components of a chemical reaction together or by forcing them to separate.

6.6 What terms are given to the reactants in a metabolic reaction?

In any metabolic reaction catalyzed by enzymes, the chemical components taking part are known as **substrates**. Once the enzyme has acted on the substrate or substrates, the products of the reaction are known as **end products**.

6.7 Do enzymes act in different chemical reactions?

Enzymes are highly specific. Each enzyme participates in one reaction and one reaction only. Therefore, if thousands of different chemical reactions occur in a cell, the cell must posses thousands of different enzymes. After the reaction has occurred, the enzyme is set free to activate another chemical reaction like it as Figure 6-1 illustrates.

6.8 What are enzymes composed of?

Enzymes are composed of proteins and, in some cases, they contain nonprotein parts. **Simple enzymes** contain proteins alone. **Conjugated enzymes** contain a nonprotein portion. The conjugated enzyme is sometimes referred to as a **holoenzyme**; the protein portion is called an **apoenzyme**, while the nonprotein portion is called a **cofactor**.

6.9 Are there different kinds of cofactors in enzyme molecules?

The cofactors of an enzyme molecule can either be **organic molecules** or **inorganic molecules**. Possible organic molecules functioning as cofactors include NAD (nicotinamide adenine dinucleotide) and FAD (flavin adenine dinucleotide). Another organic molecule serving as a cofactor is coenzyme A. Inorganic molecules sometimes found as cofactors include metals such as iron, magnesium, copper, manganese, zinc, and cobalt. In many cases, the cofactor is the active portion of the enzyme molecule.

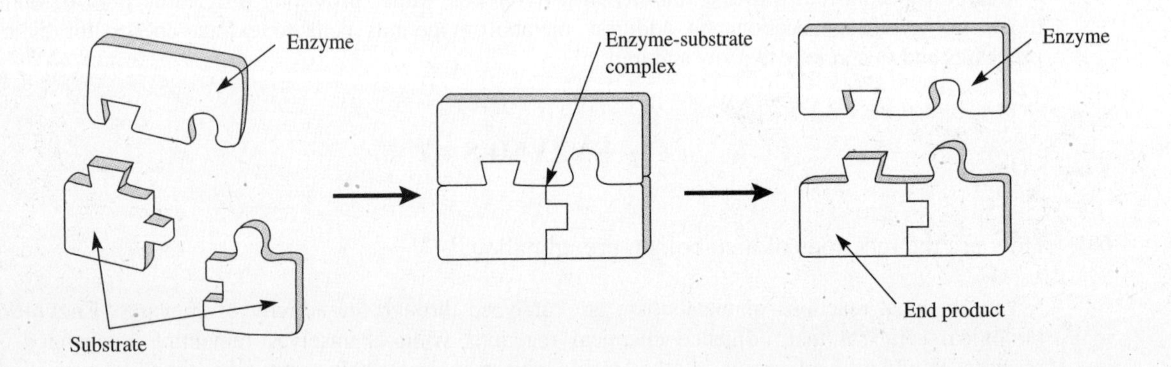

Fig. 6-1 The action of enzymes. An enzyme is a protein molecule that acts on substrate molecules and brings about a reaction resulting in end products. In this example, an enzyme molecule is combining two substrate molecules to produce a single molecule.

6.10 How large are enzyme molecules and what form do they take?

Enzyme molecules have molecular weights that range from a few thousand to over a million daltons. As protein molecules, they have primary, secondary, and tertiary structures, and they are often folded into three-dimensional molecules. In this huge molecule, there is generally one or more areas called the **active site**, where the essential chemical reaction occurs.

6.11 How do enzyme molecules work to bring about a chemical reaction?

An enzyme molecule and its substrate or substrates combine at the active site of the enzyme molecule. Generally, the substrate and active site fit closely together, such as a lock fitting together with a key. Once the union has occurred, the complex is called the **enzyme-substrate complex**. The enzyme then undergoes an alteration in its structure, and the chemical reaction occurs. This is the **induced fit hypothesis**. As a result of the reaction, two chemical molecules may be joined together to produce a single molecule, or a single substrate molecule may be broken apart to produce two or more molecules. Thus, an enzyme can form chemical bonds between substrate molecules or it may break chemical bonds within a substrate molecule.

6.12 Do enzymes have any special names?

By international agreement, contemporary enzyme names end in the suffix -ase. What precedes the suffix may be the name of the substrate; or it may have to do with the reaction the enzyme causes; or it may refer to the group to which the enzyme belongs. For example, the enzyme lactase breaks down the substrate lactose; the enzyme hexokinase breaks down the six-carbon sugar glucose; and the enzyme ligase catalyzes the union of two chemical molecules to form a single chemical molecule. It should be noted that some enzymes have traditional names. For example, pepsin is the gastric enzyme that breaks down protein.

6.13 Are enzymes susceptible to chemical and physical environmental changes?

Enzymes are proteins, and because proteins are susceptible to chemical and physical changes in the environment, enzymes are equally susceptible. For example, excessive heat causes enzyme proteins to lose their three-dimensional structure, a process called denaturation. Enzyme denaturation is one reason why high heat kills cells. Similarly, chemical substances such as phenol and many disinfectants destroy microorganisms because they destroy their enzymes.

ENERGY AND ATP

6.14 Is energy required for the reactions of metabolism to take place?

Many of the chemical reactions of metabolism are types of anabolism, and anabolism generally requires an input of energy. In addition, an input of energy is generally required to begin the reactions of catabolism. Moreover, microorganisms require energy for such things as motion, for responses to environmental stimuli, and for reproductive processes. Energy is one of the most essential requirements for all life processes in all organisms.

6.15 Can energy be stored in microorganisms?

The storage form for microbial energy is the chemical compound **adenosine triphosphate (ATP)**. ATP is somewhat like a portable battery. It can be taken to any part of the cell and used there to energize a chemical reaction or participate in a cellular process. When the energy is released from an ATP molecule, the molecule breaks down into **adenosine diphosphate (ADP)** and a phosphate group as Figure 6-2 displays. The breakdown of a single ATP molecule provides 10,000 calories of energy for cellular work.

6.16 Can microorganisms reconstitute an ATP molecule from its breakdown products in the microbial cell?

The energy of ATP is stored in the chemical bonds that unite the phosphate group to the ADP molecule. The ATP molecule can be reconstituted, but to do so the chemistry requires an input of energy. Much of this energy for ATP synthesis is supplied in the reactions of catabolism.

6.17 For how long can ATP be stored in the microorganism?

ATP molecules cannot be stored for more than a few minutes in any living cell, nor can ATP molecules be transported among cells. As it is used up in microorganisms, ATP is resynthesized using the energy present in energy compounds such as carbohydrates. The ATP molecules so-produced are then utilized almost immediately, and the resulting ADP molecule and phosphate group are made available for resynthesis of the ATP molecule. Thus, there is a constant turnover.

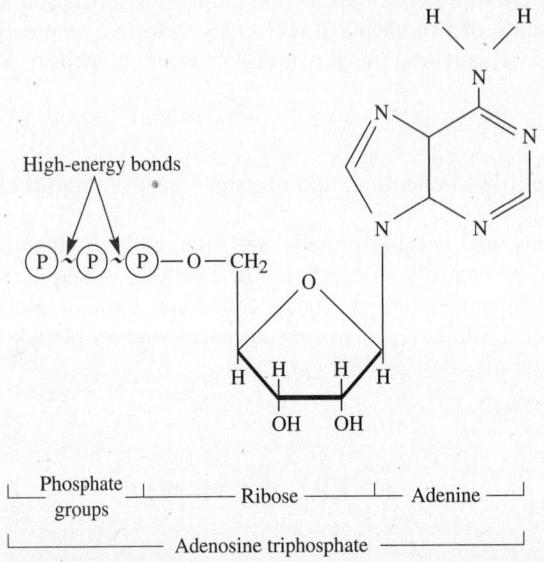

High-energy bonds

Phosphate groups — Ribose — Adenine

Adenosine triphosphate

Fig. 6-2 The structure of adenosine triphosphate (ATP). ATP is a principal storage form for microbial energy. It is composed of an adenine molecule, a ribose molecule, and three phosphate groups.

6.18 How is energy made available to microorganisms for synthesizing ATP molecules?

In the chemical reactions of metabolism, energy is released from molecules during various oxidation and reduction reactions. **Oxidation reactions** are energy-yielding reactions in which electrons are lost from a substrate molecule. The molecule is said to be oxidized. In a **reduction reaction,** electrons are added to a substrate molecule and the molecule is said to be reduced. Hydrogen atoms may also be added to the molecule in the reduction process. Oxidation and reduction reactions usually happen together, and the energy released is then captured in ATP.

6.19 Where does the energy for microbial metabolism come from?

The ultimate source of all energy on the Earth is the sun. The sun's energy is captured during the process of **photosynthesis** and is incorporated into carbohydrate molecules. The carbohydrate molecules, such as glucose, sucrose, lactose, maltose, and starch, are then utilized by living things as energy sources. The energy is released from the carbohydrates during metabolism and used to form ATP molecules.

6.20 How are carbohydrates used by microorganisms to obtain energy?

Carbohydrates are used as energy sources through the processes of a metabolic pathway. A **metabolic pathway** is a series of metabolic reactions, each catalyzed by an enzyme. The product of one reaction in a metabolic pathway is usually the substrate for the next reaction. Many metabolic pathways have branches that permit the processing of nutrients or the synthesis of materials. In other cases, the metabolic pathways have a cyclic form where the starting molecule is generated at the end of the pathway to initiate another turn of the cycle. The reactions of metabolism are usually interconnected and often merge at many points. The effect of metabolism is to integrate many reactions for the benefit of the cell.

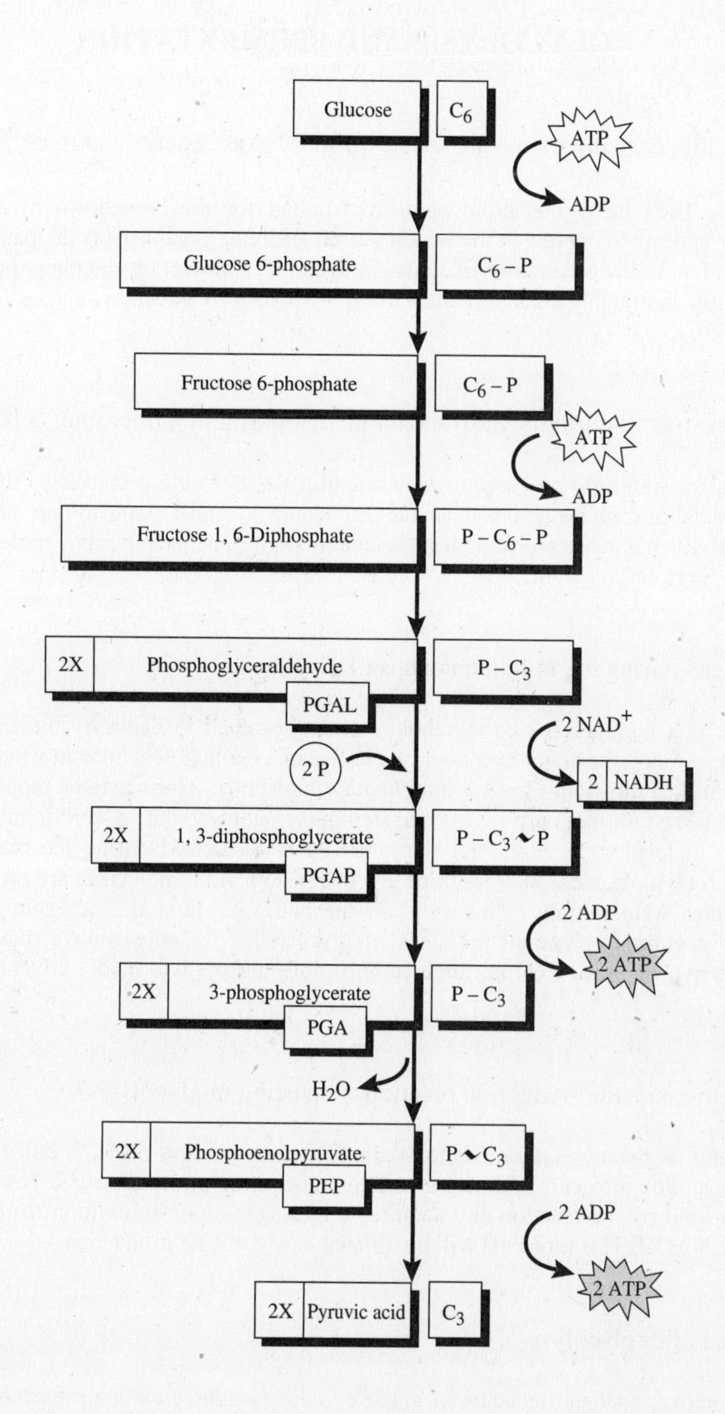

Fig. 6-3 A brief outline of glycolysis showing the individual steps. Two molecules of ATP are "invested" in the early reactions of the pathway and four ATP molecules are produced toward the end of the pathway, for a net gain of two ATP molecules. Two molecules of NADH are also produced for later use in electron transport. The product of glycolysis is pyruvic acid.

GLYCOLYSIS AND FERMENTATION

6.21 Which of the carbohydrates are used primarily as energy sources by microorganisms?

There have been many metabolic pathways studied for the breakdown of carbohydrates, but the pathway that appears to operate in the widest variety of living organisms is the pathway called glycolysis. **Glycolysis** refers to the breakdown of sugar, in this case, glucose. Of the many pathways for glycolysis, one of the most thoroughly studied is the **Embden-Meyerhoff pathway,** named for two scientists of the 1930s.

6.22 What occurs overall during glycolysis in the cytoplasm of a microbial cell?

Glycolysis is a multistep process in which one **glucose** molecule is converted to numerous compounds through a metabolic pathway (such as the Embden-Meyerhoff pathway) to yield two molecules of **pyruvic acid**. Each glucose molecule has six carbon atoms, and each pyruvic molecule resulting from the pathway has three carbon atoms.

6.23 What happens during the individual steps of glycolysis?

Glycolysis is a highly involved metabolic pathway studied in depth by biochemists and students of biochemistry. Among the major events of glycolysis are chemical reactions in which a six-carbon glucose molecule is broken down into several intermediary molecules. These actions require the input of energy, and ATP is utilized for that purpose. At one step in the pathway, the six-carbon molecule is split into two three-carbon molecules. In additional steps, the energy liberated during the reactions is used for the synthesis of ATP molecules. As a result of glycolysis, four ATP molecules are produced. However, two ATP molecules were initially "invested" in the pathway, and the net gain is therefore two ATP molecules. Enzymes catalyze all the steps of glycolysis. As noted above, the final products of the pathway are two molecules of the three-carbon molecule pyruvic acid. Glycolysis is summarized in Figure 6-3.

6.24 Are there any oxidation-reduction reactions occurring in glycolysis?

At one point in glycolysis, an oxidation-reduction reaction takes place. A pair of electrons is removed from a three-carbon molecule and transferred to a coenzyme molecule called NAD. The NAD molecule becomes reduced by the reaction and acquires a hydrogen atom in addition to the two electrons. It is converted to NADH. This molecule will be utilized later for ATP production.

6.25 What occurs after glycolysis?

As noted above, a single molecule of glucose yields two three-carbon molecules of pyruvic acid as a result of glycolysis. These pyruvic acid molecules will then be utilized for ATP production through the reactions of the Krebs cycle and electron transport, or they may be utilized in fermentation.

6.26 How can pyruvic acid be utilized in the process of fermentation?

Fermentation is a metabolic pathway that does not involve oxygen. In fermentation specific enzymes take the pyruvic acid molecules and convert them to acids, alcohols, and other end products. Yeast cells, for example, have the enzyme yeast alcohol dehydrogenase that converts the pyruvic acid to acetaldehyde while liberating a carbon dioxide molecule. The acetaldehyde molecule then converts to an ethyl alcohol molecule, and the alcohol accumulates. This process is the basis for the fermentation industry that produces beer, wine, spirits, and other alcohols. The fermentation can also yield various acids such as acetic acid, butyric acid, and lactic acid, depending upon which enzyme is present and which organism is producing the enzyme.

6.27 Must the environment be free of oxygen for fermentation to occur?

Fermentation is a process that occurs only in the absence of oxygen. It is a way for the yeast or other microorganism to obtain the two molecules of ATP per molecule of glucose metabolized through glycolysis. If oxygen were present, fermentation would probably not occur and the pyruvic acid would be utilized in the Krebs cycle. The Krebs cycle yields a large number of molecules that can be utilized for ATP synthesis.

THE KREBS CYCLE

6.28 What is the nature of the Krebs cycle used by microorganisms in energy metabolism?

The **Krebs cycle** is a cyclic series of chemical reactions in which the end product of the cycle is the starting point for a new cycle of events, as Figure 6-4 demonstrates. The objective of the Krebs cycle is to produce a large number of reduced coenzymes that can be utilized in energy metabolism leading to the synthesis of ATP molecules.

6.29 Is pyruvic acid utilized directly in the Krebs cycle?

Pyruvic acid is not used directly in the Krebs cycle. Instead, pyruvic acid undergoes an enzyme-catalyzed conversion. The enzyme removes one of the carbon atoms from the pyruvic acid molecule and liberates this as carbon dioxide. The remaining part of the molecule unites with a coenzyme known as **coenzyme A**. The result of this reaction is a compound called **acetyl CoA.** Acetyl CoA will then be utilized in the Krebs cycle. It is important to note that the reaction also liberates electrons, which are taken up by NAD. This acquisition of electrons results in reduced NAD, which takes up a hydrogen atom to become NADH. A molecule like this was formed earlier in glycolysis.

6.30 How does acetyl CoA function in the Krebs cycle?

The Krebs cycle is somewhat like a whirlpool of chemical events. To enter the Krebs cycle, the acetyl CoA molecule combines with a four-carbon molecule called **oxaloacetic acid.** This combination forms the six-carbon molecule **citric acid.** Citric acid is then converted to a five-carbon compound, and during the reaction the sixth carbon atom is released as CO_2. The five-carbon molecule is converted by an enzyme to a four-carbon molecule and the fifth carbon is given off as carbon dioxide. These reactions account for the release of the carbon atoms originally present in the glucose molecule. They also demonstrate how the carbon atoms of glucose are released as the waste gas, carbon dioxide.

6.31 How does the Krebs cycle come to an end?

The reactions of the Krebs cycle continue as a metabolic pathway catalyzed by a series of enzymes. At the end of the pathway a molecule of oxaloacetic acid forms. This molecule now unites with a new molecule of acetyl CoA, and the Krebs cycle proceeds to another turn. During that turn, additional carbon atoms will be released as CO_2 molecules.

6.32 Why is the Krebs cycle important in energy metabolism?

The Krebs cycle is essential for energy metabolism because electrons are given off at several places during the reactions of the cycle. At three different points of each cycle, pairs of electrons are assumed by NAD molecules to produce three molecules of reduced NAD, or NADH. In addition, another reaction yields a pair of electrons taken up by the coenzyme FAD. This capture of electrons yields reduced FAD, or $FADH_2$. Also during the Krebs cycle reactions, a reaction occurs in which enough energy is liberated to synthesize a molecule of ATP. Since two turns of the Krebs cycle occur for every glucose molecule, two molecules of ATP are produced. These two ATP molecules are in addition to the two ATP molecules resulting from the reactions of glycolysis.

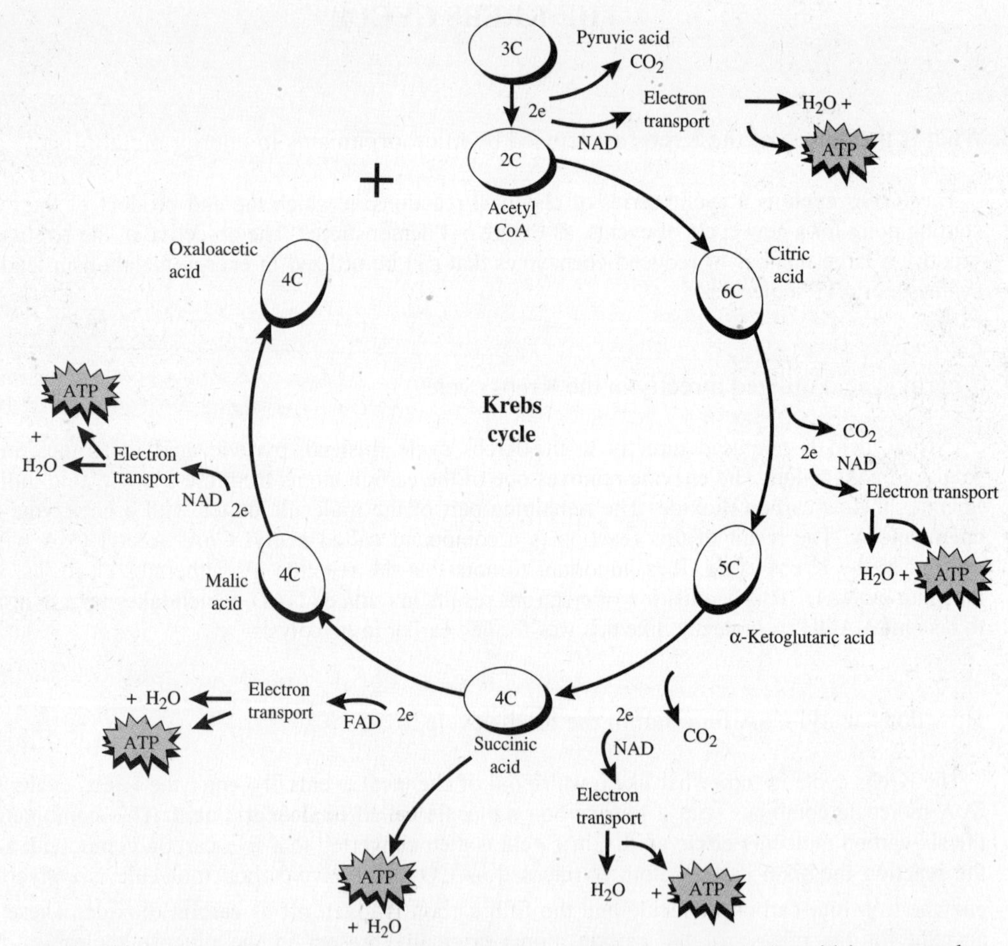

Fig. 6-4 A brief overview of the reactions of the Krebs cycle. Pyruvic acid is prepared for the Krebs cycle by conversion to acetyl-CoA. This molecule then combines with oxaloacetic acid to form citric acid. During a series of succeeding conversions, electrons are given off to NAD or FAD molecules for use in electron transport. Also in the process, three molecules of carbon dioxide evolve.

ELECTRON TRANSPORT AND CHEMIOSMOSIS

6.33 What happens to the NADH molecules formed in the Krebs cycle and in glycolysis, and what is the fate of the $FADH_2$ molecules?

The NADH and $FADH_2$ molecules are of supreme importance in energy metabolism. The NADH molecule is the first in a series of electron carriers in a chain commonly known as the **electron transport system.** Other participants in the electron transport system are molecules of FAD and molecules of chemical compounds known as **quinones.** Also involved are a series of pigmented cofactors known as **cytochromes.** The NAD, FAD, quinone, and cytochrome molecules participate in a series of electron transfers. In these transfers, electrons move from one molecule to the next , and each molecule becomes oxidized as it loses electrons or reduced as it gains electrons. Thus, the electron transport chain is a series of oxidation-reduction reactions in which electrons are passed along. Figure 6-5 summarizes the process.

6.34 What is the ultimate fate of the electrons in the electron transport chain?

The final acceptor for electrons in the electron transport chain is a molecule of **oxygen.** When the oxygen atom accepts electrons it also takes on two hydrogen atoms and becomes a water molecule. Because of the participation of oxygen, the overall process is often referred to as **cellular respiration,** and the electron transport chain is sometimes called the cellular respiration chain. Oxygen is the only acceptor that will take on the electrons; its participation in this metabolic process is indispensable.

6.35 How is energy trapped during the electron transport system?

The electrons of NADH and $FADH_2$ (which originated in Krebs cycle reactions) pass through the system of quinone and cytochrome molecules found within the cell membrane of prokaryotes and inner mitochondrial membrane in eukaryotic organisms. As the electrons are transported among the molecules, the energy released is used to pump hydrogen atoms from the inside to the outside of the membrane. The result is a high concentration of protons on the outside of the membrane. A force called the **proton motive force** soon develops, and this force suddenly drives the protons back across the membranes to equalize the concentrations on each side. As the protons flow back across the membrane, their energy is released and used to synthesize ATP molecules, using ADP molecules and phosphate groups as the building blocks. An enzyme in the membrane called ATP synthetaze accomplishes the synthesis. The process of proton movement and ATP synthesis is called **chemiosmosis.** Figure 6-6 illustrates the process.

6.36 Why is the term oxidative phosphorylation used for the electron transport and chemiosmosis processes?

The reactions in which electrons are passed among coenzymes, quinones, and cytochromes are known as oxidation reactions, and the addition of a phosphate group to an ADP molecule is known as phosphorylation. Therefore the term **oxidative phosphorylation** is used for the process in which electrons are passed along and their energy is released to synthesize ATP molecules.

6.37 What is the energy yield from oxidative phosphorylation processes going on in microorganisms?

There is a considerable yield of energy from the metabolic processes involving glycolysis, the Krebs cycle, and oxidative phosphorylation. For each NADH molecule metabolized and its pair of electrons transported through the electron transport chain, three ATP molecules are synthesized. For each $FADH_2$ molecule metabolized and its electrons passed along to oxygen, there are two ATP molecules synthesized. In microorganisms having this energy metabolism, a total of 34 molecules of ATP result from the reactions of oxidative phosphorylation. In addition to these 34 molecules, two molecules of ATP are produced during the two turns of the Krebs cycle and two molecules of ATP are the net result of glycolysis. The total number of ATP molecules is therefore 38 molecules of ATP per glucose molecule metabolized.

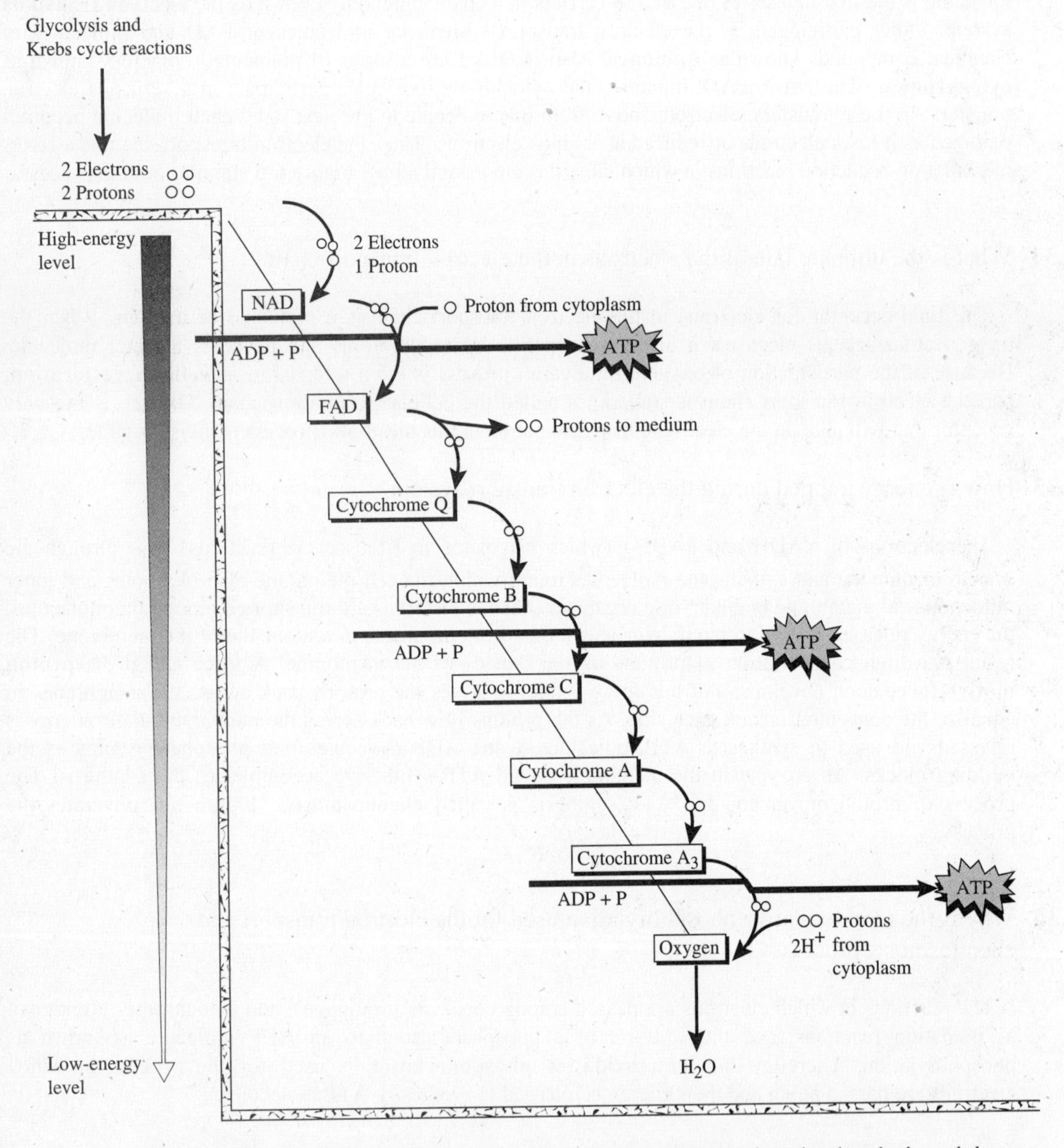

Fig. 6-5 The electron transport system. Electrons from the reactions in glycolysis and the Krebs cycle are taken up by molecules of NAD, then passed to FAD molecules, and a series of quinones and cytochromes. Ultimately, the electrons are passed to an oxygen atom, which acquires protons from the cytoplasm and becomes water. During the electron transport, the energy is released to drive the proton motive force in chemiosmosis. This energy is used to synthesize ATP.

6.38 Do alternative pathways for energy metabolism exist in microorganisms?

The process of glycolysis, Krebs cycle, and electron transport are not the only mechanisms available for energy release in microorganisms. Certain bacterial species follow a different pathway called the **hexose monophosphate (HMP) shunt** to break down carbohydrates and release the energy, and other species use a pathway called the **Entner-Douderoff (ED) pathway** in their energy metabolism. The HMP produces large numbers of five-carbon sugars that can be used for nucleic acid synthesis; the ED pathway occurs in *Pseudomonas* species.

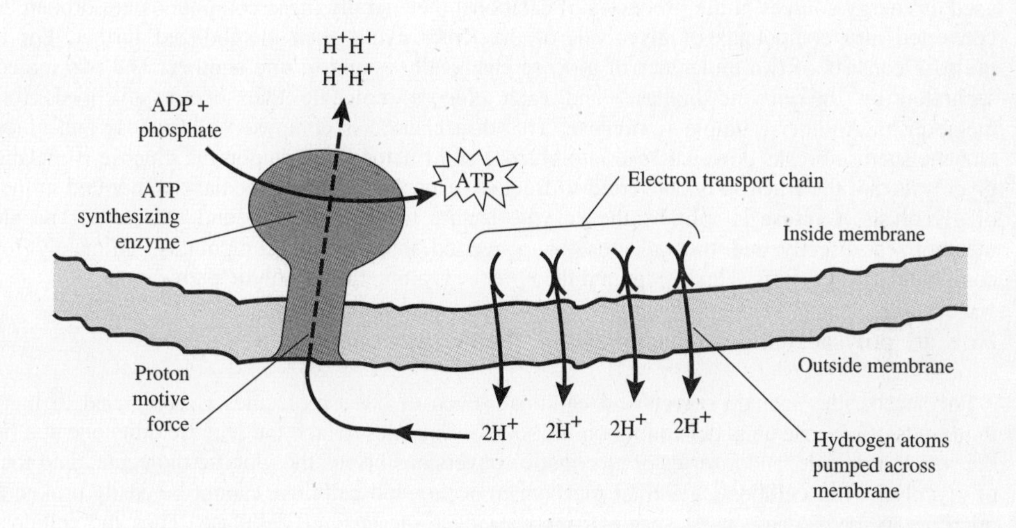

Fig. 6-6 Chemiosmosis. The transport of electrons among NAD, FAD, and cytochromes releases enough energy to pump hydrogen atoms from inside the microbial membrane to outside. When protons accumulate, a proton motive force develops and the protons are driven back across the membrane to equalize the concentrations. During the proton flow, ADP molecules unite with phosphate groups to form ATP molecules. The energy for this synthesis is derived from the proton motive force.

OTHER ASPECTS OF CATABOLISM

6.39 Is oxygen the only substance that can be used as an electron acceptor in energy metabolism?

Many species of microorganisms, including many species of bacteria, participate in **anaerobic respiration.** The process is somewhat similar to aerobic respiration (involving oxygen), and it utilizes an electron-transport system. However, it uses various kinds of molecules as electron acceptors for the energy metabolism.

6.40 What are some examples of molecules that can be used as electron acceptors in anaerobic metabolism?

Among the common anaerobic electron acceptors are nitrate and sulfate. When these molecules accept electrons they are reduced to nitrite and hydrogen sulfide, respectively. Both chemical transformations are important aspects of cycles that go on in soils and permit the recycling of nitrogen and sulfur. Carbon dioxide can also be used as an electron acceptor; it is reduced to methane gas, CH_4. This conversion is very important in anaerobic soils, since methane is an extremely pungent gas, known as natural gas.

6.41 How are carbohydrates other than glucose used in energy metabolism?

Numerous other carbohydrates, including monosaccharides, disaccharides, and polysaccharides, can be used as energy sources in the processes of catabolism. Generally these compounds are broken down and converted into compounds of glycolysis or the Krebs cycle, then metabolized further. For example, **maltose** consists of two molecules of glucose chemically bound to one another. The two molecules are separated by the enzyme maltase, and each glucose molecule then enters glycolysis for further metabolism. Another example is **sucrose.** This disaccharide is composed of fructose and glucose. The enzyme sucrase breaks down sucrose into glucose and fructose, whereupon the glucose is metabolized in glycolysis and the fructose is converted to fructose-phosphate, an intermediary compound in the process of glycolysis. **Lactose** is split by the enzyme lactase to yield glucose and galactose. The glucose is metabolized directly and the galactose is converted into several compounds, ultimately forming a compound of glycolysis. This compound then proceeds along the glycolytic pathway.

6.42 How are polysaccharides metabolized for their energy content?

Polysaccharides include **starch** and **cellulose.** Each of these molecules is composed of hundreds or thousands of glucose units depending on the source. Enzymes detach the glucose units one at a time from the starch molecule, and a series of enzymatic conversions brings the glucose molecules into the scheme of glycolysis. For **cellulose,** a similar mechanism occurs, but cellulose cannot be easily broken down by microorganisms because most species cannot produce the enzyme cellulase. Thus the cellulose of the plant cell wall remains undigested in the environment except in those few species of microorganisms that produce cellulase. The rumen of the cow contains cellulase producers.

6.43 Can proteins be utilized for energy metabolism?

Both proteins and fats can be used for their energy content. In the processes of catabolism, a protein is broken down into its constituent amino acids. Through the process of **deamination,** the amino group of the amino acid is removed and an oxygen atom is inserted. The usual result is a molecule normally found in either glycolysis or the Krebs cycle. For example, the amino acid alanine undergoes deamination and becomes pyruvic acid, which then can be metabolized for its energy content.

6.44 Is it possible to utilize fats as energy sources in microorganisms?

As for proteins, the fats can be utilized successfully as energy sources in metabolism. A fat generally consists of a glycerol molecule and one, two, or three fatty acids. Once a glycerol molecule is separated from the fat, it is converted to dihydroxy-acetone-phosphate (DHAP), which is an intermediary in the process of glycolysis. The DHAP is then metabolized further along the pathway. Fatty acids are long chains of carbon atoms, having 16, 18, or 20 carbons long. These long chains are broken into two-carbon units by appropriate enzymes. The two-carbon units are then converted into molecules of acetyl CoA by a process called **beta-oxidation**, or the fatty acid spiral. The latter molecules then are absorbed into the Krebs cycle and the metabolism continues, as Figure 6-7 shows.

6.45 Can synthesis processes occur in connection with breakdown processes?

The metabolic reactions of catabolism are intimately linked to the metabolic processes of anabolism, or biosynthesis. For example, many amino acids can be produced by reversal of the process of deamination. When alanine is synthesized from pyruvic acid, for instance, the oxygen portion of the pyruvic acid is replaced by an amine group in the process called **amination.** Also, a Krebs cycle compound called alpha-ketoglutarate can be utilized to form the amino acid glutamic acid.

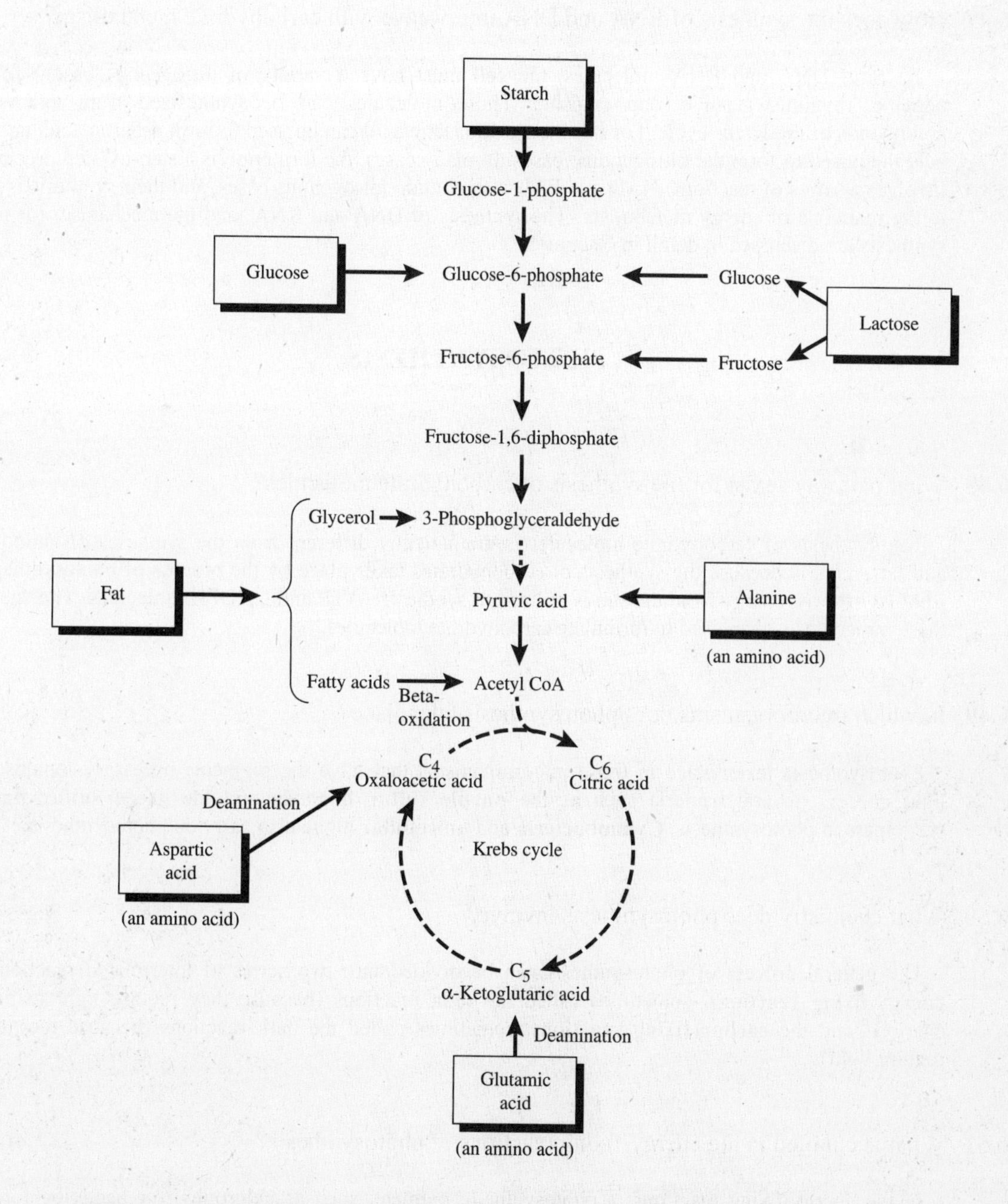

Fig. 6-7 An overview of metabolism in microorganisms showing how various carbohydrates, amino acids, and fats can be used for energy metabolism.

6.46 Can fats and lipids be produced by utilizing the reactions of energy metabolism?

Fatty acids and fats can be produced by reversal of certain reactions occurring in energy metabolism. For instance, acetyl CoA molecules can be utilized to form fatty acid molecules, and glycerol can be produced from the three-carbon molecule DHAP formed during glycolysis. When united to one another by a dehydration synthesis, the fatty acids and glycerol molecules form a fat.

6.47 How does the synthesis of RNA and DNA interweave with carbohydrate metabolism?

To form DNA and RNA molecules, the cell must have a variety of nitrogenous bases including adenine, thymine, guanine, and cytosine. These molecules can be synthesized from intermediary compounds of the Krebs cycle. For instance, oxaloacetic acid can be used to form aspartic acid, an amino acid then used to form the nitrogenous bases. In many cases, the formation is a step-by-step process and involves a series of reactions. NAD and FAD also contain nitrogenous bases, and their synthesis is traced to the reactions of energy metabolism. The synthesis of DNA and RNA and the mechanism for protein synthesis are discussed in detail in Chapter 7.

PHOTOSYNTHESIS

6.48 What pathways exist for the synthesis of carbohydrate molecules?

The synthesis of carbohydrate molecules is dramatically different from the syntheses of amino acids and fats. This is because the synthesis of carbohydrates takes place by the process of photosynthesis. In **photosynthesis,** energy from the sun is utilized to synthesize ATP and NADPH molecules. The energy in these molecules is then used to formulate carbohydrate molecules.

6.49 In which microorganisms does photosynthesis take place?

Photosynthesis takes place in those microorganisms that have the pigments necessary for absorbing light energy. Several bacteria such as the **purple sulfur bacteria** and the **green sulfur bacteria** participate in photosynthesis. **Cyanobacteria** and **unicellular algae** also carry on photosynthesis.

6.50 What chemistry does photosynthesis involve?

The general process of photosynthesis can be divided into two series of interrelated reactions: the **energy-fixing reactions,** sometimes called the light reactions (because they require light to provide energy) and the **carbon-fixing reactions,** sometimes called the dark reactions (because they do not require light).

6.51 What is entailed in the energy-fixing reactions of photosynthesis?

In the energy-fixing reactions, a photosynthetic pigment, such as chlorophyll or bacteriochlorophyll serves as an energy harvesting molecule. The pigment absorbs light energy, which is transferred to its electrons. The energized electrons then pass through an electron transport chain. During the various oxidation-reduction reactions that occur, a **proton motive force** is established as hydrogen ions pass through the membranes. In cyanobacteria, the chlorophyll molecules are located at the cell membrane, while in the unicellular algae, the chlorophyll molecules are located in chloroplasts.

6.52 What are the effects of the proton motive force in the energy-fixing reactions of photosynthesis?

The proton motive force establishes chemiosmosis and accounts for the synthesis of ATP molecules. The ATP is stored in the cell. Since this form of ATP generation depends on light, it is called **photophosphorylation.** The final resting place for the electrons after their transport through a series of cytochromes is NADP molecules. The NADP molecules are reduced, and they become NADPH molecules, as displayed in Figure 6-8.

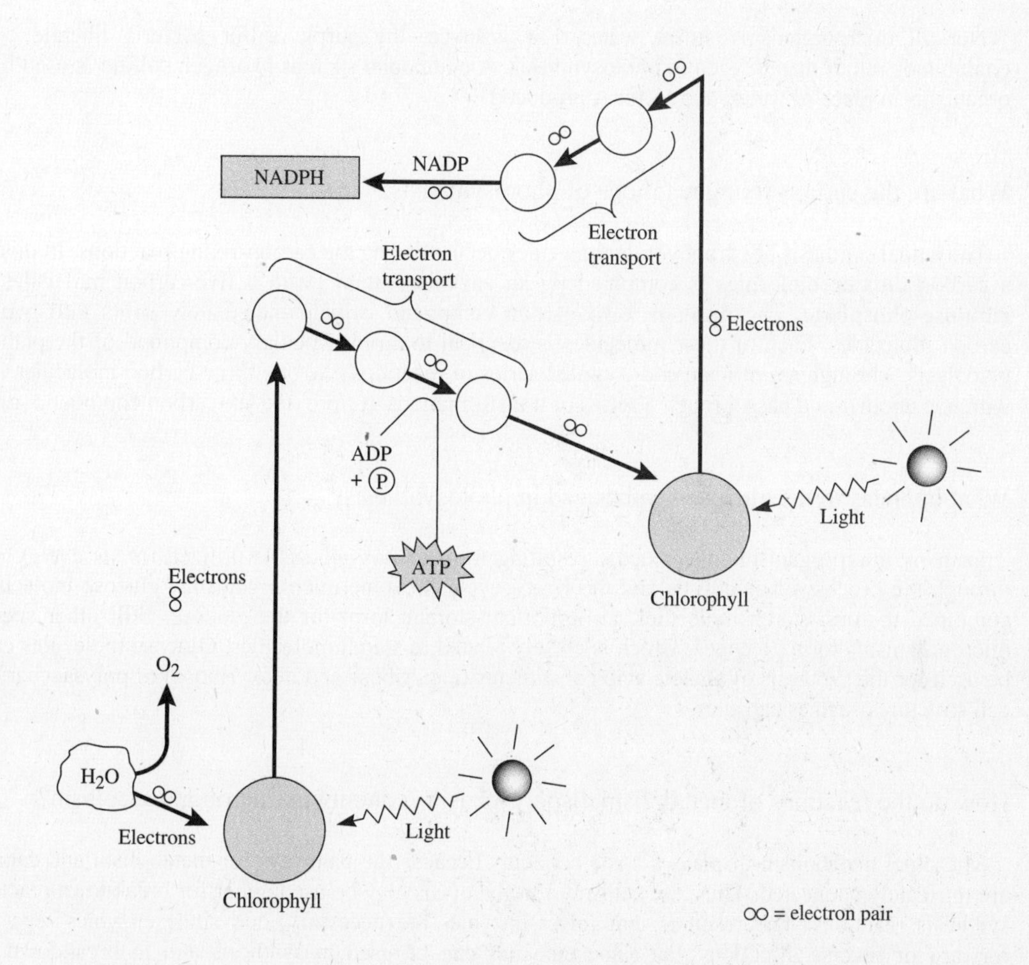

Fig. 6-8 An overview of the energy-fixing reactions of photosynthesis. This scheme occurs in unicellular algae. Sunlight strikes the chlorophyll molecule, causing electrons to be emitted. The electrons are transported among carriers, then enter another chlorophyll molecule, where they are again stimulated by light. The electrons eventually unite with an NADP molecule to form NADPH. During the electron transport, ATP is produced. This ATP and the NADPH will then be used in the carbon-fixing reactions. Water molecules are used to replace the electrons in the original chlorophyll molecule, and the leftover oxygen atoms form molecules of atmospheric oxygen.

6.53 How are electrons replaced in the energy-fixing reactions of photosynthesis?

The electrons lost from chlorophyll molecules in the energy-fixing reactions are replaced by electrons from water molecules, which are broken down by enzymes. The oxygen ions that remain reorganize themselves as molecular oxygen gas, which is given off to the environment. Approximately 20 percent of the atmosphere contains oxygen, which is largely a byproduct of the photosynthesis taking place in cyanobacteria. This is the same oxygen gas utilized in the energy metabolism taking place by cellular respiration.

6.54 Do all microorganisms use water as a source of replacement electrons?

Not all microorganisms utilize water. For instance, the purple sulfur bacteria liberate oxidized compounds rather than oxygen in photosynthesis. A compound such as hydrogen sulfide is used by these organisms in place of water, and sulfur is produced.

6.55 What are the carbon-fixing reactions of photosynthesis?

The actual synthesis of carbohydrate molecules occurs during the carbon-fixing reactions. In this series, a carbon dioxide molecules is combined by an enzyme system with a five-carbon molecules called **ribulose-phosphate.** The result is a six-carbon compound, which immediately splits into two three-carbon molecules. Each of these molecules is identical to an intermediary compound of the pathway of glycolysis. Through an intricate and complex series of reactions, the two three-carbon molecules interact with one another and pass through a series of transformations to form the six-carbon compound glucose.

6.56 What happens to the glucose synthesized in photosynthesis?

In many microorganisms, the glucose resulting from photosynthesis is utilized for its energy content through the process of glycolysis and the Krebs cycle. In some microorganisms, glucose molecules are combined to form starch molecules, an important storage form for the glucose. Still other species of microorganisms form glycogen, which is closely related to starch molecules. Glucose molecules can also be used for the synthesis of nucleic acid constituents (e.g., ribose and deoxyribose) or polysaccharides for cell structures such as capsules.

6.57 How do the reactions of metabolism display a basic economy in microbial cytoplasm?

Microbial metabolism displays a basic economy because the pathways for metabolism and catabolism are intricately connected. Thus, the pathway for glycolysis may be used either for breakdown reactions or synthesis reactions. The result is that fewer enzymes are necessary, and since enzymes react in the forward or reverse directions, the same enzymes can be used in synthesis and in breakdown. Many molecular intermediates can be diverted into numerous other pathways, and a given molecule may serve multiple purposes. Despite the complexity in this metabolism, there is an overall simplicity. The study of metabolism sheds light on many aspects of microbial growth.

Review Questions

Multiple Choice. Select the letter of the item that correctly completes each of the following statements.

1. All the following characteristics apply to enzymes except (*a*) they reduce the amount of activation energy necessary for a chemical reaction to take place (*b*) they act on substances known as substrates (*c*) they are composed solely of carbohydrate molecules (*d*) they operate only in the reactions of catabolism.

2. The process of anabolism is one in which microbial cells (*a*) synthesize molecules and structures (*b*) transport electrons among electron carriers (*c*) microbial cells break down larger molecules into smaller ones (*d*) glycolysis and the Krebs cycle are key intermediaries.

3. Among the organic molecules functioning as cofactors of enzymes are (*a*) iron ions (*b*) FAD and NAD (*c*) ATP molecules (*d*) pyruvic acid molecules and acetaldehyde molecules.

4. The induced fit hypothesis helps to explain (*a*) the alteration of an enzyme molecule during a chemical reaction (*b*) the transfer of electrons through organic carriers in photosynthesis (*c*) the conversion of proteins to carbohydrates in the Krebs cycle (*d*) the involvement of coenzyme A in the metabolic processes.

5. In order to produce ATP molecules during metabolism, all the following are necessary except (*a*) adenosine diphosphate molecules (*b*) energy (*c*) phosphate groups (*d*) DNA and RNA.

6. During a reduction reaction such as occurs in metabolism (*a*) electrons are lost from a substrate molecule (*b*) large amounts of energy are usually obtained (*c*) electrons are added to a substrate molecule (*d*) the substrate molecules is oxidized.

7. During the chemical reactions of glycolysis (*a*) carbohydrates are converted into proteins (*b*) enzymes do not play a role (*c*) carbohydrate molecules are produced from carbon dioxide molecules (*d*) two molecules of pyruvic acid result from a single molecule of glucose.

8. The net gain of ATP molecules resulting from glycolysis in microorganisms is (*a*) two (*b*) four (*c*) 36 (*d*) 38.

9. Which of the following statements applies to fermentation? (*a*) fermentation occurs in the absence of oxygen (*b*) DNA is needed for fermentation to occur (*c*) a product of fermentation is starch molecules (*d*) fermentation occurs in most microbial cells.

10. The chemical substance that enters the Krebs cycle for further metabolism is (*a*) ethyl alcohol (*b*) pyruvic acid (*c*) acetyl-CoA (*d*) adenosine triphosphate.

11. All the following characteristics apply to the Krebs cycle except (*a*) carbon dioxide molecules are released as waste products (*b*) citric acid is formed during the cycle (*c*) all the reactions are catalyzed by enzymes (*d*) the reactions result in the synthesis of glucose.

12. In the electron transport chain (*a*) oxygen is used as final acceptor (*b*) cytochrome molecules do not participate in the electron transfers (*c*) one possible result of the transfers is fermentation (*d*) the source of electrons for electron transport is DNA.

13. The process of chemiosmosis accounts for (*a*) the conversion of amino acids to carbohydrate molecules (*b*) the breakdown of starch molecules into glucose molecules for glycolysis (*c*) the trapping of energy in ATP molecules (*d*) the synthesis of glucose molecules using light as an energy source.

14. When an NADH molecule is metabolized and its electrons transported through the electron transport chain (*a*) six amino acid molecules are formulated (*b*) a single glucose molecules results (*c*) three ATP molecules are synthesized (*d*) one triglyceride and two diglycerides result.

15. The hexose monophosphate shunt and the Entner-Douderoff pathway are alternative mechanisms for (*a*) DNA synthesis in microorganisms (*b*) photosynthesis in photosynthetic organisms (*c*) energy metabolism in certain species of microorganisms (*d*) chemiosmosis.

16. In microbial metabolism, nitrate and sulfate may be used for accepting electrons in the (*a*) absence of enzymes (*b*) absence of ATP (*c*) absence of oxygen (*d*) presence of cytochromes.

17. In order for starch and cellulose to be metabolized for their energy content (*a*) they must first be changed into fat molecules (*b*) their glucose units must be released (*c*) an environment free of oxygen must be present (*d*) the genetic code must be favorable.

18. The process of deamination accounts for the conversion of (*a*) polysaccharides to disaccharides (*b*) disaccharides to polysaccharides (*c*) fats to carbohydrate molecules (*d*) amino acids to energy compounds.

19. In order for cells to utilized fatty acids for their energy content, the fatty acids are broken down and converted into molecules of (*a*) quinone and ribulose phosphate (*b*) acetyl-CoA (*c*) various amino acids (*d*) DHAP molecules.

20. Amino acids can be formed from intermediaries of catabolism by (*a*) substituting an acid group where there is a carbon atom on a carbohydrate molecule (*b*) binding an ATP molecule to a carbohydrate molecule (*c*) altering the active site on an enzyme molecule (*d*) replacing an oxygen atom on an intermediary compound with an amine group.

21. The energy used to drive the reactions of photosynthesis is obtained from (*a*) oxidation-reduction reactions (*b*) the sun (*c*) ATP molecules (*d*) acetyl-CoA molecules.

22. Among the bacteria participating in photosynthesis are the (*a*) purple bacteria and green sulfur bacteria (*b*) intestinal bacteria such as *E. coli* (*c*) soil bacteria such as the actinomycetes (*d*) rickettsiae and chlamydiae.

23. The replacement of electrons lost from chlorophyll molecules in photosynthesis in certain microorganisms depends on the presence of (*a*) ATP molecules (*b*) cytochrome molecules (*c*) water molecules (*d*) nitrogenous base molecules.

24. During the carbon-fixing reactions of photosynthesis (*a*) carbon dioxide molecules react with ribulose phosphate molecules (*b*) pyruvic acid molecules combine to form glucose molecules (*c*) starch molecules are converted to cellulose molecules (*d*) DNA molecules change into RNA molecules.

25. Among the immediate products of microbial photosynthesis are (*a*) amino acids and proteins (*b*) oxygen gas and glucose molecules (*c*) pyruvic acid molecules (*d*) viruses.

Completion. For each of the following, add the term that best completes the statement.

1. The broad category of metabolism in which cells break down large molecules into smaller ones is called ______________.

2. The organic component of which all enzymes are composed is ______________.

3. An enzyme molecule can be identified because its name generally ends in the suffix ______________.

4. The major storage form for microbial energy is the compound ______________.

5. A chemical reaction in which electrons are lost from a substrate molecules and from which energy is usually derived is an ______________.

6. Glycolysis is a multistep metabolic pathway in which two molecules of pyruvic acid result from the breakdown of one molecule of ______________.

7. Of the many processes for glycolysis studied in microorganisms, the one most thoroughly studied is the pathway known as the ______________ .

8. In the absence of oxygen, pyruvic acid molecules are converted to alcohol molecules by microorganisms known as ______________ .

9. Before pyruvic acid molecules are used in the Krebs cycle, an enzyme removes one of the carbon atoms and converts the two remaining carbon atoms to the compound ______________ .

10. An important cofactor that takes up electrons during the reactions of the Krebs cycle and glycolysis is ______________ .

11. The ultimate fate of electrons transported through the electron transport chain is to unite with a molecule of ______________ .

12. Pigmented coenzymes functioning in the electron transport system in microorganisms are called ______________ .

13. The proton motive force drives proteins across microbial membranes, and the energy is used to synthesize ATP molecules in the process known as ______________ .

14. When a single molecule of glucose is metabolized through cellular respiration in microorganisms, the total number of ATP molecules formed is generally ______________ .

15. Certain microorganisms use nitrite and hydrogen sulfide molecules in a form of metabolism that is ______________ .

16. In order to be used in energy metabolism, the disaccharide lactose is broken down into molecules of glucose and ______________ .

17. In the process of deamination, energy-yielding compounds are produced by the activity of enzymes on ______________ .

18. To be useful for energy purposes, fatty acids are broken down into two-carbon units through the process of ______________ .

19. Many of the bases of DNA and RNA can be formed from a Krebs cycle compound known as ______________ .

20. Photosynthesis occurs in those bacteria that have colored pigments used for the absorption of ______________ .

21. The energy-fixing reactions of photosynthesis result in the formation of molecules of NADH and ______________ .

22. An important byproduct of water utilization in the energy-fixing reactions of photosynthesis is the gas ______________ .

23. The compound ribulose phosphate is used in the carbon-fixing reactions of photosynthesis, in which the gas utilized is ______________ .

24. An important monosaccharide resulting from the carbon-fixing reactions of photosynthesis is ______________ .

25. The polysaccharides that may result from photosynthesis is microorganisms include glycogen and ______________.

Matching. Select one of the four metabolic processes from Column B for each of the characteristics listed in Column A.

Column A	Column B
___ 1. Occurs in cyanobacteria and green sulfur bacteria	(*a*) Glycolysis
___ 2. Two molecules "invested" to drive the process	(*b*) Krebs cycle
___ 3. Involves energy-fixing and carbon-fixing reactions	(*c*) Chemiosmosis
___ 4. Hydrogen atoms pumped across a membrane	(*d*) Photosynthesis
___ 5. Involves citric acid and oxaloacetic acid	
___ 6. An important byproduct is oxygen gas	
___ 7. Results in two molecules of pyruvic acid	
___ 8. Begins with the entry of acetyl-CoA molecules	
___ 9. Results in the synthesis of glucose molecules	
___ 10. Products of the process can be used for fermentation	

Chapter 7

DNA and Gene Expression

OBJECTIVES

The hereditary characteristics of microorganisms are expressed through the physiology of deoxyribonucleic acid (DNA). The biochemistry of DNA activity and gene expression are essential aspects of the metabolism of microorganisms. In this chapter, the objectives are to:

1. Review the composition of DNA with emphasis on the arrangement of the nucleotides to form the DNA molecule.
2. Learn how DNA replicates and how DNA controls the synthesis of proteins in cells.
3. Identify the key steps of transcription during the early stages of protein synthesis.
4. Recognize the importance of ribonucleic acid (RNA) in the process of protein synthesis occurring in the microbial cytoplasm.
5. Understand some of the key steps of translation as it takes place in protein synthesis.
6. Learn the stages that peptides pass through in order to form their final structure.
7. Understand the nature of the genetic code and how altering the genetic code can result in mutations.
8. Summarize the methods by which gene control can occur in microorganisms.

THEORY AND PROBLEMS

7.1 Where does DNA and gene expression fit in with the metabolism of microorganisms?

The structure and function of DNA and the expression of genes are intimately associated with microbial metabolism. As defined in Chapter 6, microbial metabolism is concerned with the anabolism and catabolism of organic compounds and the chemistry taking place in a microorganism. A significant aspect of anabolism is the synthesis of proteins. This synthesis relates to the structure and function of DNA and the expression of the genetic code in DNA as a protein molecule.

7.2 When did scientists first realize that DNA directs the synthesis of protein and is the molecule of heredity?

As early as 1928, experiments by Frederick Griffith indicated that the hereditary characteristics of bacteria could change, and experiments by Oswald Avery in 1944 indicated that DNA was the source of heredity. In 1952, Alfred Hershey and Martha Chase performed convincing experiments that DNA directs the synthesis of protein, and the work of James D. Watson and Francis H.C. Crick in 1952 resulted in a description of the structure for DNA and a sense of how it could bring about protein synthesis.

STRUCTURE AND PHYSIOLOGY OF DNA

7.3 What is DNA composed of?

DNA is one of two important nucleic acids found in microorganisms and all other living things. DNA stands for **deoxyribonucleic acid.** The other important nucleic acid is ribonucleic acid, or RNA. Both DNA and RNA are long chains (polymers) of chemical units known as nucleotides.

7.4 What are the components of nucleotides and how are they arranged to form a nucleic acid?

A **nucleotide** consists of three basic units: a molecule of carbohydrate (deoxyribose in DNA; ribose in RNA), a phosphate group (a group of phosphorus, hydrogen, and oxygen atoms derived from phosphoric acid), and a nitrogenous base. Nucleotides are joined to one another in a nucleic acid by forming chemical bonds between the carbohydrate molecule of one nucleotide and the phosphate group of the next. This combination results in a sugar-phosphate "backbone," a repeating pattern of sugar-phosphate-sugar-phosphate-sugar-phosphate molecules. The nitrogenous bases are arranged as appendages along this backbone and are connected to the carbohydrate molecule.

7.5 Which nucleotides are found in DNA and in RNA?

In DNA, there are only four different **nitrogenous bases:** adenine, cytosine, guanine, and thymine. Adenine and guanine belong to a group of organic molecules called **purines;** cytosine and thymine belong to a group of molecules called **pyrimidines.** In RNA, the four nitrogenous bases are adenine, cytosine, guanine, and uracil. Uracil, like cytosine and thymine, is a pyrimidine.

7.6 How does RNA differ from DNA?

As the previous discussions note, RNA differs from DNA in two respects: the carbohydrate in RNA is ribose (rather than deoxyribose in DNA); and one of the four bases in RNA is uracil (rather than thymine in DNA).

7.7 What is the importance of DNA in microorganisms?

In prokaryotes such as bacteria, DNA is the sole component of the chromosome, which is the structure of heredity. In eukaryotic microorganisms such as fungi, protozoa, and algae, the DNA is combined with histone proteins to form the cellular chromosomes. Prokaryotic (bacterial) chromosomes are single, closed loops of DNA, while eukaryotic chromosomes are strands of DNA and protein.

7.8 In what form does DNA exist in the chromosome?

There are two strands of DNA in the chromosome or chromosomes of all living things. The two strands are arranged as a double helix, an arrangement that resembles a spiral staircase. In the double helix, the base adenine always is aligned opposite thymine and vice versa as Figure 7-1 displays; the base guanine is always aligned opposite cytosine and vice versa; the A-T bond and the C-G bond is a weak bond. Adenine is said to be complementary to thymine and guanine is complementary to cytosine. This structure of DNA was first proposed in 1953 by Watson and Crick.

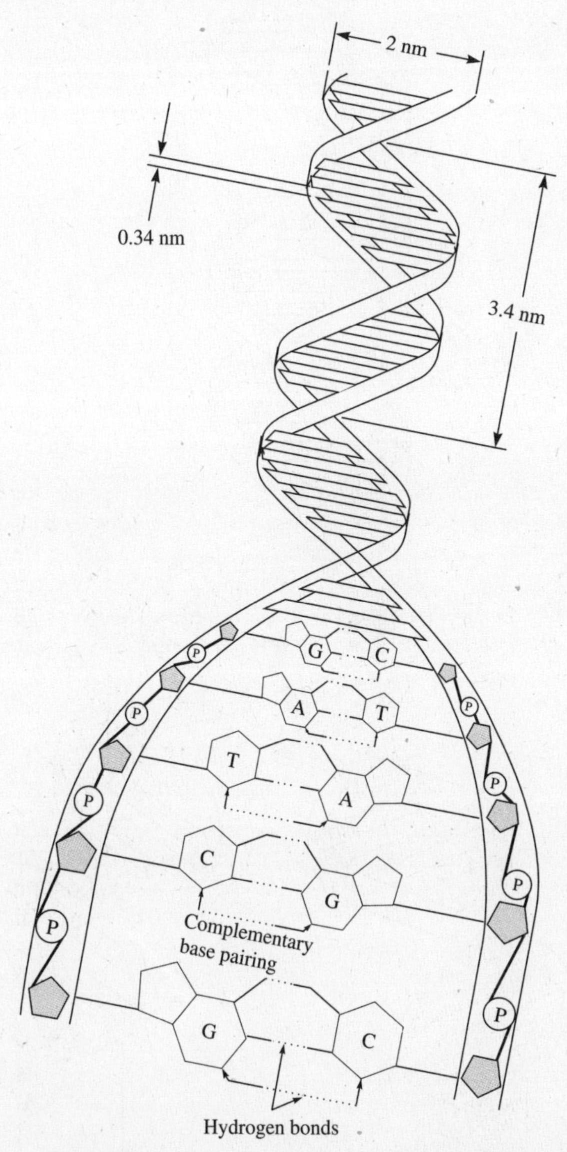

Fig. 7-1 The chemical composition of the DNA molecule. The molecule is arranged as a double helix of two intertwined chains consisting of deoxyribose and phosphate groups. Complementary nitrogenous bases extend from the deoxyribose molecules toward one another. The distances between various aspects of the molecule are shown.

7.9 Do all molecules of DNA have the same amounts of nitrogenous bases?

The number and sequence of nitrogenous bases in the DNA molecule are key elements in variations that are found in chromosomes. Different organisms have different numbers of the four bases, and the sequence varies in countless ways. In all cases, however, adenine and thymine always stand opposite each other, and cytosine and guanine oppose one another.

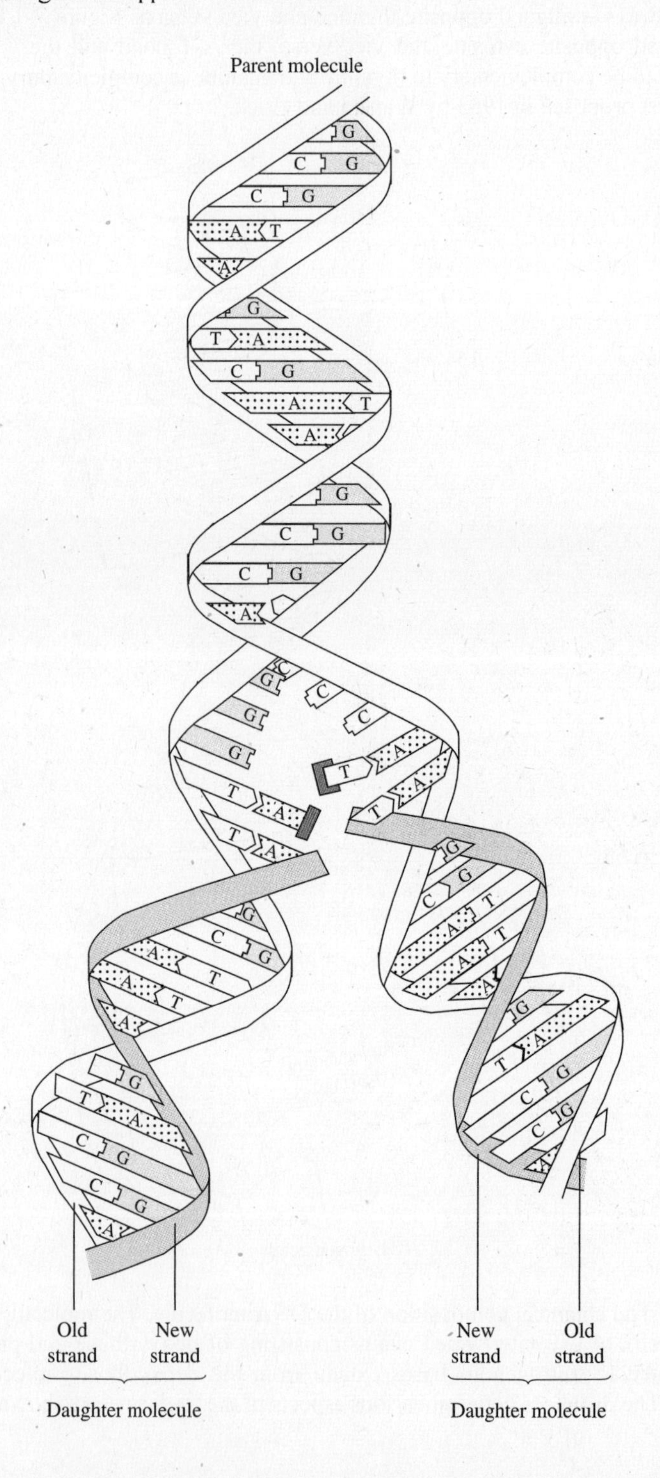

Fig. 7-2 The semiconservative replication of the DNA molecule. The double helix unwinds and each strand of DNA serves as a template for the synthesis of a new strand. The principle of complementary base pairing is observed during the synthesis. A parent strand of DNA then unites with a new strand of DNA to form a new double helix.

7.10 Is a gene the same as a DNA molecule?

Genes are the units of heredity. A gene is a segment or section of a DNA molecule. This segment of DNA provides a genetic code for the synthesis of proteins.

7.11 How are genes transferred between parent and offspring organisms?

In the duplication and replication of organisms such as bacteria, protozoa, and complex living things, the DNA of the parent organism is duplicated, and one of the duplicates passes into each of the offspring cells. In this way the heredity characteristics pass to the offspring.

7.12 How does the replication of the DNA molecule take place?

Watson and Crick first suggested the method of DNA replication that has now been verified. The double helix of DNA unwinds, and each strand of DNA serves as a template, or model, for the synthesis of a new strand, as demonstrated in Figure 7-2. The important factor in the new strand is the placement of bases. This placement is influenced by the bases in the parent molecule. Thus, A aligns T in the new DNA molecule and C aligns G. An enzyme called DNA polymerase then attaches the nucleotides together and two new strands of DNA are formed.

7.13 Do the two new strands of DNA unite with one another to form a double helix?

No. Experiments have shown conclusively that one parent strand of DNA unites with one new strand of DNA to form a double helix. The other parent strand unites with the other new strand to form another double helix. This method of replication is called **semiconservative replication.** It results in two double helices of DNA, where there was originally one double helix. The new DNA molecules then pass into their respective cells during cell division.

7.14 How does DNA control the synthesis of protein in cells?

DNA expresses itself in the synthesis of proteins by specifying the sequence of amino acids occurring in the new protein. A gene (segment of DNA) does not build a protein directly, but instead it dispatches instructions in the form of RNA. The RNA then programs protein synthesis. Biochemical information passes from DNA in the nucleus of the cell to RNA acting as a messenger, to protein synthesis taking place in the cytoplasm of the cell.

PROTEIN SYNTHESIS

7.15 What are the first stages of protein synthesis in which genes express themselves?

The first stages of protein synthesis are grouped together as the major concept of **transcription.** Transcription begins when the DNA double helix unwinds at a certain point, which denotes the gene. The nucleic acid language of the gene is written as a sequence of bases on the DNA molecule. Such a sequence might read G-C-T-T-A-C-C-G-A-T-T.... This is the molecular "language" that will ultimately specify an amino acid sequence in a protein.

7.16 Is the nucleotide language of DNA written in any particular format?

The nucleotide language in DNA is written in groups of three bases called **triplet codes**. Each group of three bases will eventually encode a single amino acid. These groups of three nucleotides (for example, G-G-A) are the smallest code that will specify an amino acid.

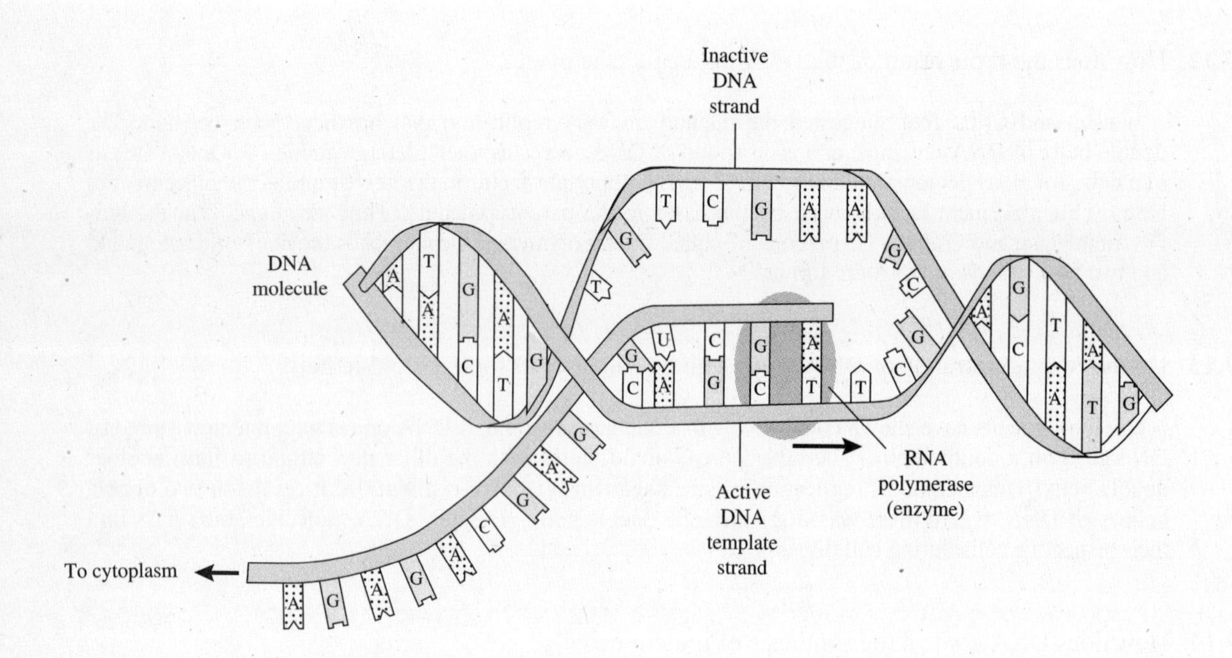

Fig. 7-3 Transfer of the triplet code of DNA to a molecule of messenger RNA (mRNA). A single strand of DNA (the template strand) serves as a model for producing a complementary strand of RNA. This process is catalyzed by RNA polymerase. The process is called transcription.

7.17 How are the triplet codes transferred to RNA?

The triplet codes of DNA are transferred to a molecule of RNA as a series of three-base codes known as **codons**. This transfer occurs in transcription and takes place in the nucleus of the cell. A molecule of RNA is transcribed from a DNA molecule by a process that resembles DNA synthesis. The active strand of DNA specifies the new RNA molecule by placing complementary bases in place. Note, however, that uracil (U) is positioned rather than thymine (T), since there is no thymine in RNA. Figure 7-3 shows this pattern. The RNA nucleotides are then linked together by the enzyme **RNA polymerase,** and the RNA molecule is synthesized.

7.18 What name is given to the RNA molecule so-produced in transcription?

The RNA molecule produced from the message in DNA is known as a **messenger RNA** molecule **(mRNA)** because it will carry the genetic message into the cytoplasm. In prokaryotic cells, there is no nucleus and mRNA is produced in the region known as the nucleoid. In eukaryotic organisms, mRNA synthesis occurs in the nucleus.

7.19 Is the synthesis of mRNA a random event?

The synthesis of RNA is a highly ordered event that begins with a "start" signal known as a **promoter.** The promoter region is a sequence of nitrogenous bases found on the DNA molecule next to the gene. Transcription begins when RNA polymerase attaches to the promoter region. After RNA elongation has occurred for a period of time, a region on the DNA molecule called the **terminator region** is reached. This region signals the end of the gene and the conclusion of mRNA synthesis.

7.20 How does mRNA leave the region of the DNA?

In prokaryotic organisms such as bacteria, mRNA synthesis occurs in the cytoplasm and the molecule moves away from the DNA molecule to another area where protein synthesis will take place. In eukaryotic molecules, the mRNA molecule leaves the nucleus through pores and carries the genetic message to the cytoplasm. The mRNA molecule is a collection of codons illustrated in Figure 7-4. Each codon will specify a single amino acid.

7.21 Is the mRNA molecule used exactly as specified by the DNA molecule?

In prokaryotic organisms, the mRNA molecule is used exactly as it is synthesized. In eukaryotic organisms, certain sections of the mRNA molecule are spliced out. These sections are called **introns**. The remaining sections called **exons** are stitched together to form the final mRNA molecule. Thus, in eukaryotic organisms, the mRNA molecule is a collection of useful segments or exons.

7.22 Where does the mRNA molecule move to for protein synthesis?

Protein synthesis occurs at ultramicroscopic bodies in the cytoplasm known as **ribosomes.** In prokaryotes, the ribosomes float free in the cytoplasm, while in eukaryotes, most ribosomes are associated with membranes of the endoplasmic reticulum. Ribosomes are composed of RNA and protein.

7.23 How does protein synthesis actually take place?

In protein synthesis, the message in mRNA molecules is translated to an amino acid sequence in a protein. The process, called **translation,** requires a number of enzymes, a supply of chemical energy stored in ATP and another type of RNA called transfer RNA.

7.24 What is transfer RNA?

Transfer RNA (tRNA) is a small molecule having roughly the shape of a cloverleaf. The molecule has a key section of three nitrogenous bases called an **anticodon.** The anticodon of a particular tRNA molecule is complementary to a codon somewhere on the mRNA molecule. In protein synthesis, a codon will come together with complementary anticodon on the tRNA molecule.

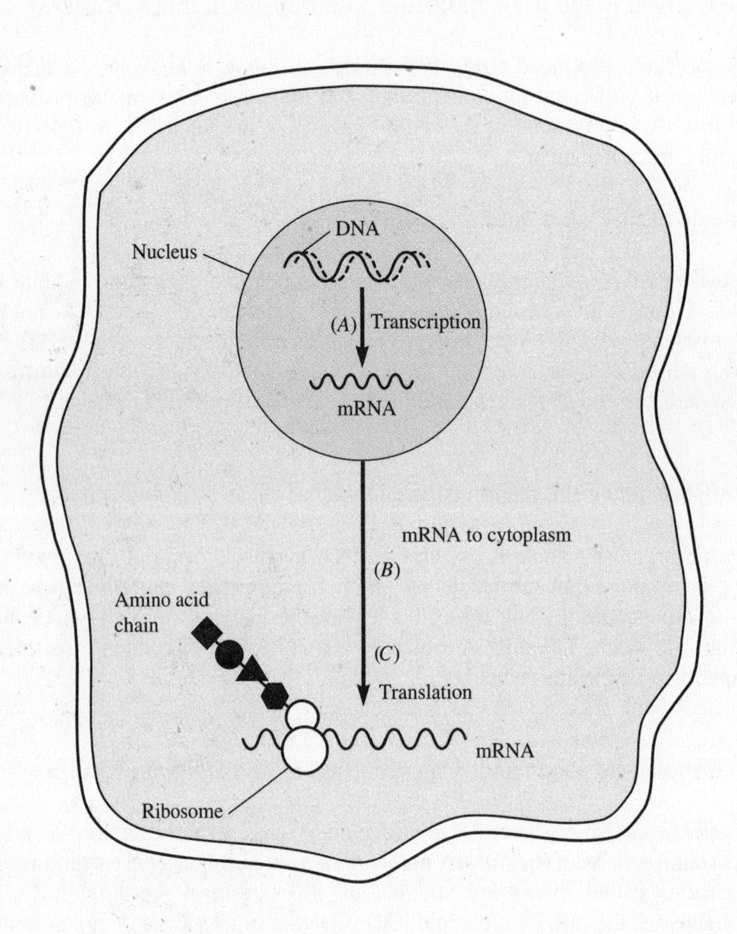

Fig. 7-4 An overall view of DNA activity and protein synthesis. (*A*) During transcription, the DNA molecule transcribes its message to a molecule of messenger RNA (mRNA). (*B*) In eukaryotic cells, the mRNA leaves the nucleus and moves to the cytoplasm carrying the message for protein synthesis. (*C*) At the ribosome, the genetic message of mRNA is used to position amino acids in a chain during the process of translation.

7.25 How do tRNA molecules accomplish their function?

For each of the 20 amino acids, there is a specific tRNA molecule. For example, the amino acid glycine has a tRNA molecule that will unite only with glycine. This tRNA molecule will not unite with any of the other 19 amino acids. Similarly, there is a specific tRNA molecule for the amino acids alanine, tyrosine, valine, trytophan, and all the other amino acids. These tRNA molecules and their associated amino acids function in translation.

7.26 By what process does translation occur in a living organism?

Translation occurs at the ribosomes of the cell. The mRNA molecule arrives at the ribosomes and meets the tRNA molecules here. Various tRNA molecules are attached to their various amino acids. A single codon of the mRNA molecule is exposed at the ribosome (for example, G-C-C). A tRNA molecule with the complementary anticodon (C-G-G-) approaches it and lines up next to it . In doing so, it brings its amino acid into position. Then a new codon is exposed (A-C-G). Another tRNA molecule with the complementary codon (U-G-C) will approach this codon. In doing so, it brings a second amino acid into position alongside the first. A covalent bond forms between the two amino acids and they join together to form a dipeptide. The first tRNA molecule breaks away and reenters the cytoplasm to unite with another molecule of its amino acid. Meanwhile, a third tRNA molecule brings a third amino acid into position next to the mRNA molecule and a bond forms between its amino acid and the first two amino acids. The result is a tripeptide consisting of three amino acids. Figure 7-5 summarizes this process.

7.27 How does the peptide chain continue to grow?

A peptide is a small protein formulated at the ribosome. With the approach of successive tRNA molecules and their amino acids, matches are made between the codons of the mRNA and anticodons of the tRNA. Thus, specific amino acids are brought into position and attached to the growing peptide chain. ATP energy is utilized throughout this process, and specific enzymes are responsible for uniting tRNA molecules with their specific amino acid.

7.28 By what process do ribosomes connect the amino acid to the growing peptide chain?

Each ribosome has a binding site for an mRNA molecule on the smaller of its two subunits. It also has binding sites for tRNA molecules. Another section of the ribosome holds the tRNA molecules in place while the amino acids attach to the growing peptide chain.

7.29 Does the synthesis of amino acids occur simultaneously throughout the mRNA molecule?

No. The synthesis of a peptide is a step by step sequence. The mRNA molecule contains a special codon called the **start codon** where translation begins. When a tRNA molecule binds to this codon the protein synthesis is initiated. Often, the first tRNA molecule carries the amino acid methionine. Amino acids are added to the chain one by one, and elongation continues until a **stop codon** occurs on the mRNA molecules. Stop codons are U-A-A, U-A-G, and U-G-A. These codons signal translation to stop.

7.30 What happens after translation comes to an end?

After translation has ended, the peptide undergoes a transformation to a final protein. The peptide may coil into a **secondary structure**, and the coils may fold over one another or form adjacent sheets of protein in a **tertiary structure.** In some cases, several polypeptides come together and intertwine to form a **quaternary structure.** This final molecule is the functional protein.

7.31 What is the significance of transcription and translation?

The processes of transcription and translation explain how protein is synthesized in the catabolism processes of the organism. They also display how DNA, functioning as genes, directs and encodes the synthesis of proteins. The final protein formed may be a cytoplasmic protein, flagellar protein, membrane protein, or protein toxin, or a protein of myriad other uses.

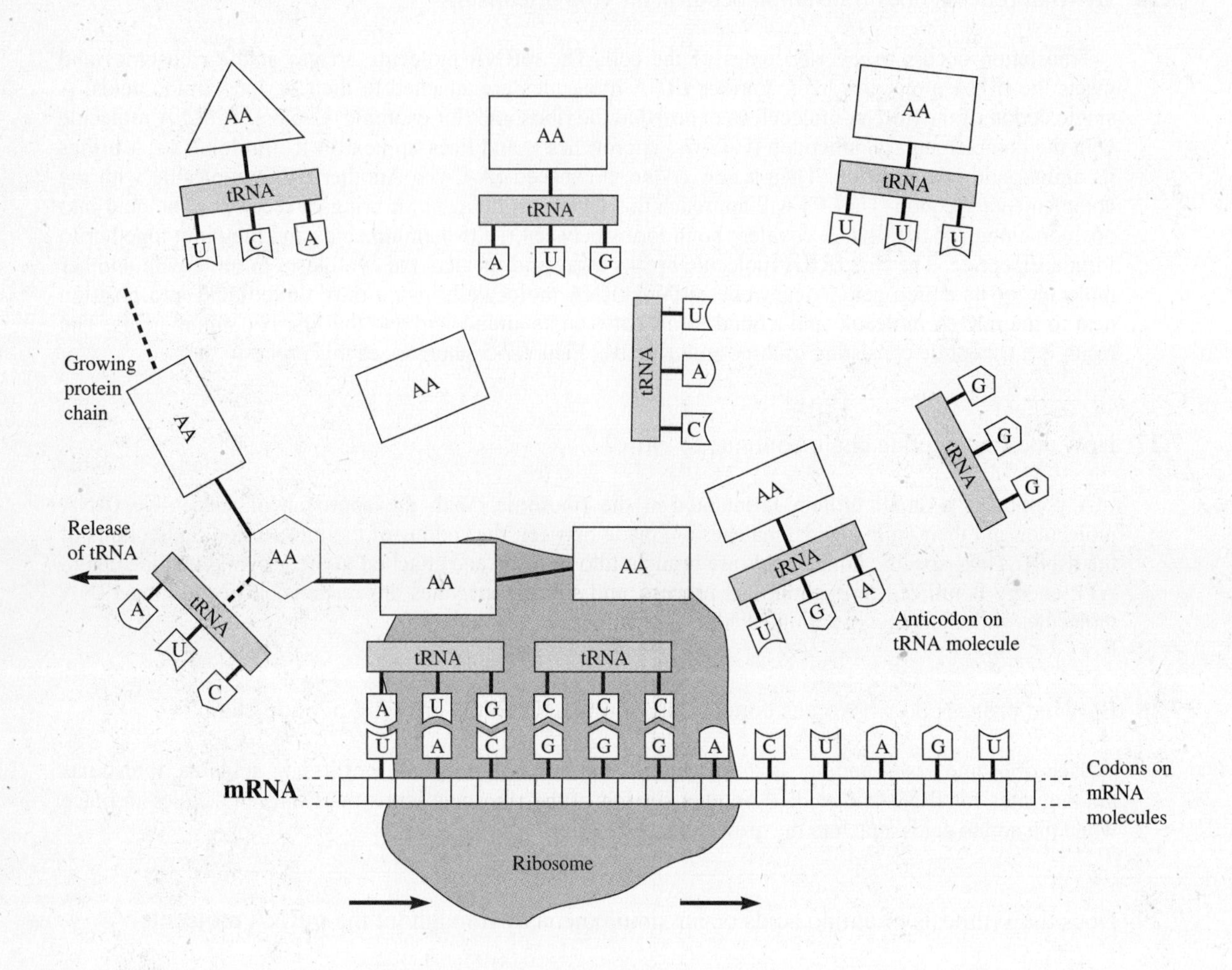

Fig. 7-5 Translation. During this process, transfer RNA molecules (tRNA) bring their amino acids to the ribosome where they meet the mRNA molecule. The codons of the mRNA molecule complement the anticodons of the tRNA molecules, and an amino acid is positioned next to the previous amino acid. A peptide bond forms, and the amino acid joins the growing peptide chain. The tRNA molecule then returns to the cytoplasm to bond with another molecule of its amino acid.

7.32 Explain the nature of the genetic code?

The **genetic code** is the series of three-base sequences that specify an amino acid in a protein chain as Figure 7-6 demonstrates. The amino acid code is nearly universal among all organisms. This universality implies that the code G-U-U specifies the amino acid valine in bacteria, fungi, protozoa, plants, animals, and humans. Studying the expression of genes and the genetic code in microorganisms helps us understand how it operates in more complex species.

7.33 Can the genetic code change?

If the sequence of nitrogenous bases in the DNA molecule were to change, then the genetic code itself would change. Such a change is known as a mutation. A mutation may involve a single base pair or multiple changes in base pairs.

7.34 How do mutations occur?

Mutations can occur by any of a number of means. For example, inserting an extra nucleotide in the DNA molecule during replication will disrupt the genetic code. Also, deleting a nucleotide will change the code. Any chemical that changes the structure of a nitrogenous base can change the genetic code. Many chemicals such as formaldehyde and urea are known to cause mutations in this way. Mutagens such as x-rays and ultraviolet light can also bring about mutations by changing the nature of the DNA.

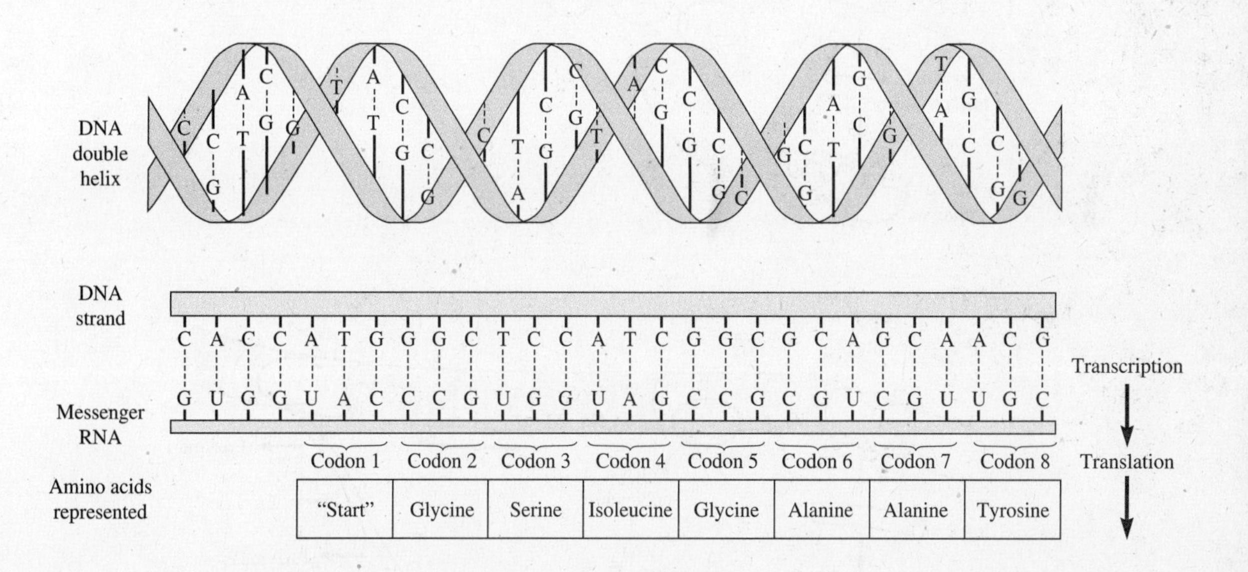

Fig. 7-6 The genetic code. The DNA molecule as a double helix contains the base sequences that constitute the genetic code. A single strand of DNA transcribes the genetic code to messenger RNA in transcription. The messenger RNA molecule carries the code to the ribosome, where the code is translated into a chain of amino acids. Each codon of the messenger RNA molecule specifies one amino acid of the protein chain.

GENE CONTROL

7.35 In a normal cell, is it possible to control the expression of genes?

It is not only possible, it is necessary that the expression of genes be controlled. It would be chemically unwise for all the genes to be expressing themselves all the time because this would rapidly deplete the resources of the cell. Instead, living cells have complex systems of control whereby they can switch genes on and off.

7.36 What is an example of gene control?

An elegant example of gene control is seen in the organism *Escherichia coli*. The cluster of genes that is responsible for synthesizing a particular protein is called an **operon.** This cluster of genes includes a promoter region, an operator gene, a repressor gene, a regulatory gene, and a number of structural genes that actually encode the protein, as Figure 7-7 shows.

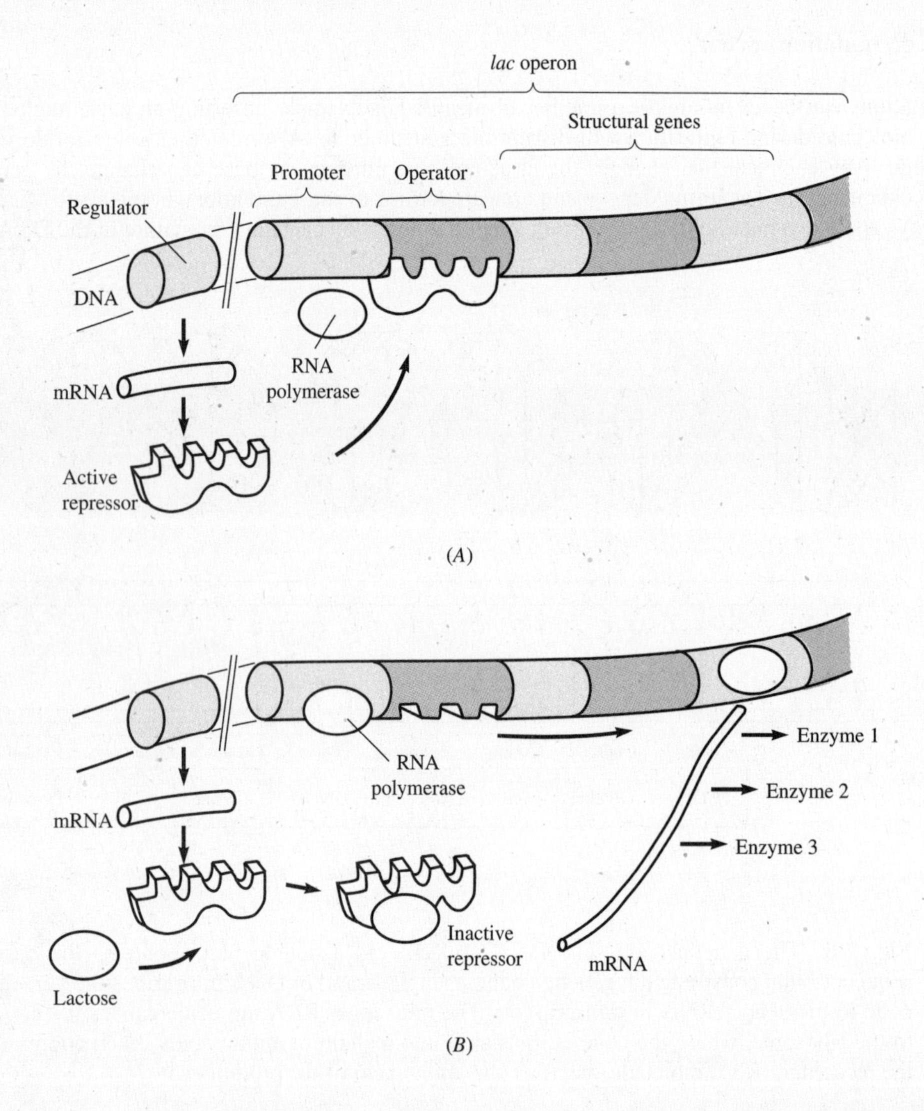

Fig. 7-7 Activity of the operon. (*A*) The operon is a cluster of genes essential in the synthesis of one or more enzymes. In the *lac* operon, when there is no lactose present, the repressor molecule binds to the operator and prevents the enzyme RNA polymerase from catalyzing enzyme formation. (*B*) When lactose is present in the environment, it binds to the repressor substance, and the enzyme is free to activate the promoter and operator genes, which leads to the formation of enzymes necessary for the breakdown of lactose.

7.37 How does the operon work in *E. coli.*

One of the most well-known operons is the operon that encodes the enzyme lactase. This enzyme breaks down lactose when lactose is present. When lactose is absent, the **repressor gene** produces a repressor protein that binds to the operator gene. This binding blocks the expression of the structural genes, and no lactase is produced. However, when lactose enters the environment, the lactose molecules bind to the repressor substance and prevent the repressor from binding to the operator. The operator gene is now freed and it will not block the expression of the structural genes. The structural genes encode the enzyme lactase and the lactase acts on the lactose in the environment and prepares it for energy metabolism.

7.38 Do operons exist in all microorganisms?

Operons are known to exist in bacteria, but they are much more complex in protozoa, fungi, and more complex organisms. In eukaryotes for example, many genes are controlled or encoded by regulatory proteins on different chromosomes. Indeed, as cells become more complex, the level of gene expression becomes equally complex. Studying gene expression is one of the major thrusts of modern biochemistry.

Review Questions

True/False. For each of the following statements, mark the letter "T" next to the statement if the statement is true. If the statement is false, change the underlined word to make the statement true.

___ **1.** The bases adenine, guanine, and cytosine are found in both RNA and DNA, but RNA also has thymine.

___ **2.** In the cellular chromosomes of eukaryotic microorganisms, the DNA is combined with histone proteins to form the chromosomes.

___ **3.** In the double helix form of DNA in the microbial chromosome, the base adenine is complementary to the base guanine.

___ **4.** A segment or section of a DNA molecule is generally referred to as a gene.

___ **5.** The genetic code in DNA expresses itself in the sequence of monosaccharides occurring in a protein.

___ **6.** In the first stage of protein synthesis, known as translation, the genetic message in DNA is passed to a base sequence in RNA.

___ **7.** When DNA undergoes replication, one new strand combines with an old strand in the method known as conservative replication.

___ **8.** The RNA molecule that receives the genetic code from DNA in protein synthesis is known as transfer RNA.

___ **9.** The chromosomal DNA of prokaryotic cells is contained in the region known as the nucleus of the cell.

__ **10.** A start signal known as the promoter initiates the synthesis of RNA using the genetic message in DNA.

__ **11.** Once the preliminary mRNA molecule has been produced in eukaryotic organisms, sections called exons are spliced out in the production of the final mRNA molecule.

__ **12.** The actual synthesis of protein molecules from amino acid units takes place at cellular bodies known as centrioles.

__ **13.** Amino acids are delivered to the sites of protein synthesis by fragments of RNA referred to as conductor RNA.

___ **14.** Amino acids are placed alongside one another as complementary molecules of RNA match the codon with the anticodon.

___ **15.** A messenger RNA molecule having the code GAC will match with a transfer RNA molecule having the code CTG.

___ **16.** Small proteins formulated from amino acids at the ribosome particles are referred to as glycerides.

___ **17.** In eukaryotic microorganisms, the ribosomes are associated with membranes of the mitochondrion.

___ **18.** In order to unite amino acids together to form proteins, a supply of chemical energy stored in ATP must be necessary.

___ **19.** A change in the sequence of nitrogenous bases of the DNA molecule results in a mutation.

___ **20.** A codon is a cluster of genes in bacteria that is responsible for synthesizing a particular protein.

Multiple Choice. Select the letter of the word or phrase that best completes each of the following.

1. The synthesis of proteins in a microorganism represents an aspect of metabolism known as (*a*) catabolism (*b*) photosynthesis (*c*) anabolism (*d*) chemiosmosis.

2. The chemical units of both DNA and RNA are (*a*) amino acids (*b*) nucleotides (*c*) enzyme molecules (*d*) NAD and FAD.

3. All of the following refer to both cytosine and thymine except (*a*) both are nitrogenous bases (*b*) both are pyrimidine molecules (*c*) both may be found in RNA (*d*) both may be found in DNA.

4. All of the following apply to the DNA of bacteria except (*a*) the DNA is concentrated in the nucleoid (*b*) the DNA contains histone proteins (*c*) the DNA is a single, closed loop (*d*) the DNA contains the genetic code.

5. In the double helix arrangement of DNA in the chromosome, the base cytosine stands opposite the base (*a*) adenine (*b*) thymine (*c*) cytosine (*d*) guanine.

6. During the replication of the DNA molecule, one of the original DNA strands (*a*) combines with the other original DNA strand (*b*) is broken down for nucleotide units (*c*) combines with one of the new strands of DNA (*d*) is converted into carbohydrate molecules for energy metabolism.

7. The instructions for building a protein from amino acids pass from the DNA to (*a*) a molecule of RNA (*b*) an enzyme molecule containing FAD (*c*) a series of protein enzymes (*d*) the cell membrane where the protein is synthesized.

8. A single amino acid is encoded by a group of (*a*) three enzymes (*b*) three nucleotides (*c*) three anticodons (*d*) three molecules of ATP.

9. The terminator and promoter regions functioning in protein synthesis exist on the (*a*) endoplasmic reticulum (*b*) DNA molecule (*c*) ribosome (*d*) nuclear membrane.

10. In eukaryotic microorganisms, the messenger RNA molecule consists of a collection of (*a*) introns (*b*) enzymes (*c*) protein fragments (*d*) exons.

11. Transfer RNA molecules exist in the metabolizing cell in the (*a*) nucleus (*b*) mitochondrion (*c*) cytoplasm (*d*) Golgi body.

12. Translation and transcription are subdivisions of the general process of (*a*) DNA replication (*b*) glycolysis (*c*) protein synthesis (*d*) the electron transport chain.

13. In order to form proteins in the metabolizing cell, all the following factors are required except (*a*) ATP molecules (*b*) enzymes (*c*) messenger RNA molecules (*d*) fatty acid molecules.

14. A mutation occurring in the cell is most often due to (*a*) disrupting the genetic code (*b*) eliminating the supply of ATP (*c*) breaking down enzyme molecules (*d*) depleting the supply of glucose molecules.

15. The operon that encodes an enzyme in *E. coli* is an example of (*a*) how enzymes work in the cell (*b*) how gene control exists in the cell (*c*) how proteins are converted to energy compounds in the cell (*d*) how eukaryotic organisms have evolved from prokaryotic organisms.

Completion. For each of the following, add the term that completes the statement best.

1. In eukaryotic microorganisms such as fungi and protozoa ____________ is combined with histone proteins to form chromosomes.

2. One difference between RNA and DNA is the presence of the carbohydrate ____________ in RNA molecules.

3. Parent organisms pass their ____________ to their offspring when the DNA of the parent organism duplicates and one of the duplicates moves to the offspring cell.

4. During the replication of DNA, the enzyme known as ____________ attaches nucleotides together to form the new DNA strands.

5. The genetic code is expressed in the sequence of ____________ that occurs in new proteins.

6. The nucleotide language of DNA is written in groups of ____________ bases called triplet codes.

7. The transfer of genetic information from the DNA molecule to the mRNA molecule occurs in the process known as ____________.

8. Prokaryotic cells lack a ____________ and mRNA is synthesized at the nucleoid.

9. In prokaryotic organisms, the ____________ molecule is used as it is synthesized, while in eukaryotic organisms, it is modified.

10. The transfer RNA molecule has roughly the shape of a cloverleaf and functions in the transport of ____________ to the ribosome.

11. There are approximately ____________ different amino acids that are used to make all the proteins of the cell.

12. Each ribosome has a ____________ for an mRNA molecule on its smaller subunit.

13. When a protein coils, it forms its ____________ structure.

14. When an alteration occurs in the sequence of ____________ of a DNA molecule, a mutation may result.

15. The operon has been studied in depth in the common bacterium ____________.

Chapter 8

Microbial Genetics

OBJECTIVES

Microbial genetics is concerned with the biochemical changes occurring in the DNA of microorganisms. These changes bring new characteristics to the microorganisms and influence the physiology of those organisms. The objectives of the chapter are to:

1. Explain mutation and recombination as the two major changes occurring in the genes of a microorganism.
2. Discuss the effects mutations can bring to a microorganism and how those mutations come about.
3. Explain the meaning of recombinant DNA and identify three general methods for recombination as it occurs in microorganisms.
4. Define the special features of transformation and note the importance of plasmids in this process.
5. Summarize the stages of conjugation that occur in bacteria, with reference to the various forms of DNA that may be involved.
6. Conceptualize the importance of viruses in the recombination process of transduction.
7. Explain the genetic engineering process, with reference to the structures and enzymes that participate in this mechanism.
8. Illustrate the practical significance of genetically engineered bacteria by identifying several instances in which the mechanism has medical use.

THEORY AND PROBLEMS

8.1 What is the general subject matter of microbial genetics ?

The study of microbial genetics is concerned with the deoxyribonucleic acid (DNA) contained in the chromosome and other genetic elements of microorganisms. The DNA is subdivided into functional units called genes; microbial genetics has to do with changes that occur in the genes of a microorganism. Experiments in this field are often conducted in atmospherically controlled environments as shown in Figure 8-1.

8.2 Which types of changes can occur in a microorganism's genes?

Two major types of changes may occur in the genes: first, a permanent change called a **mutation** may occur. A mutation is usually due to a chemical or physical agent affecting the gene and changing its nature; second, a change may occur when genetic material is acquired or lost by a microorganism. Such an alteration in the genetic material is called a **recombination**.

MUTATIONS

8.3 What are the different kinds of mutations occurring in a microorganism?

Various types of mutations may occur in the DNA of a microorganism. If a single nitrogenous base is lost and replaced by a different base, then the mutation is known as a **point mutation**. A point mutation may result in a single amino acid change in a protein (Chapter 7). Another kind of mutation is a **deletion mutation**, where an entire section of a microbial chromosome is lost. Still another type is an **inversion mutation**, one in which a piece of DNA is replaced by an identical piece of DNA in reverse order. Finally, an **insertion mutation** is one where an entirely new sequence of nucleotides is inserted in a microbial chromosome.

Fig. 8-1 A photograph of a scientist in a positive-pressure enclosed suit performing experiments in microbial genetics. Precautions as these are important to prevent mutated organisms from escaping the lab and for preventing infection of the scientist.

8.4 Describe the effects a mutation can have on microorganisms?

If a permanent mutational change occurs in the DNA, the effect will be seen on the proteins synthesized. For example, important enzymes may not be produced; or structural proteins may not be synthesized; or functional proteins such as those of the cilia or flagella may not be manufactured. The effect on the microorganism may vary from no effect to minor effect to microbial death.

8.5 Which physical and chemical agents are known to cause mutations.

Anything that causes a mutation in the DNA of a microorganism is called a **mutagenic agent** or **mutagen**. Such chemical agents as urea, phenol, and various acids can be mutagens. Such physical agents as ultraviolet light or x-rays may also be mutagens.

8.6 How many different kinds of mutations are there?

There are two general types of mutations: spontaneous mutations and induced mutations. **Spontaneous mutations** occur in nature without human intervention and may be due to UV light in sunlight or chemicals in the soil. **Induced mutations** are brought about in the laboratory when researchers apply mutagens to microorganisms.

8.7 How do mutagenic agents act in microorganisms?

Mutagenic agents may induce various types of point mutations. Ultraviolet light, for example, causes adjacent thymine bases in DNA to bind to one another as demonstrated in Figure 8-2. This binding prevents the bases from functioning in the formation of messenger RNA molecules, and the appropriate protein cannot be produced. Another kind of mutation, the **frameshift mutation,** is caused by chemicals such as nitrous acid, formaldehyde, and acridine dyes. These chemicals cause a base of two to be lost from the DNA molecule, and the triplet coding sequence is put out of phase (that is, the reading frame is shifted.)

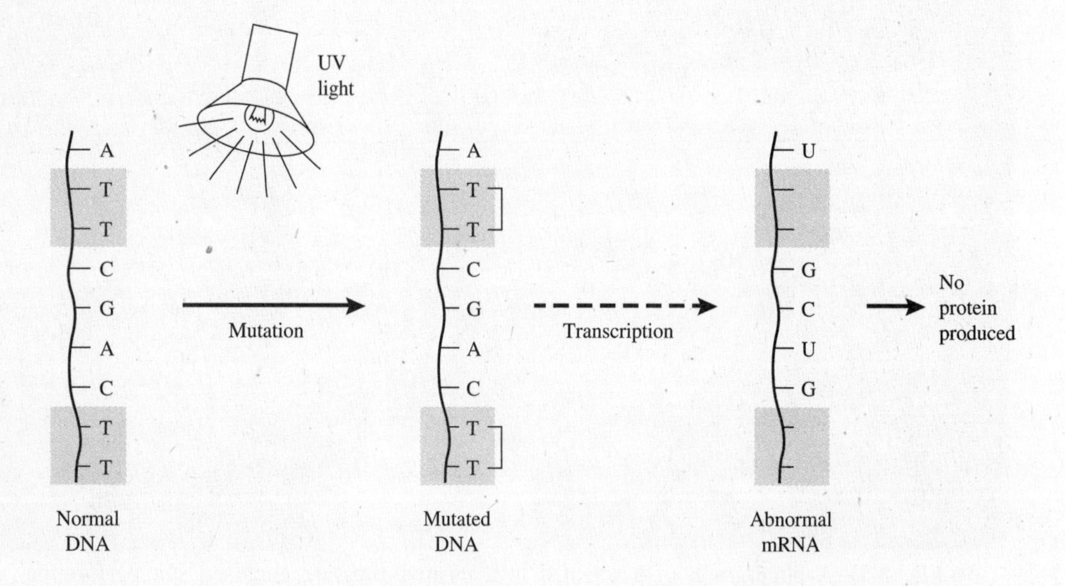

Fig. 8-2 Mutation by ultraviolet light. When ultraviolet light is directed at a bacterium, the radiation causes adjacent thymine molecules of normal DNA to bind to one another, forming dimers. In the mutated DNA, the bound thymine molecules are unable to function in the formation of a messenger RNA molecule, with the result that an abnormal mRNA molecule results. The defects in the mRNA molecule prevent a normal protein from being produced.

8.8 Can mutations in microorganisms be reversed?

In certain cases, mutations can be reversed. For example, dimers are created when adjacent thymine molecules bind together in a mutation. In the presence of visible light and a repair enzyme, bacteria can repair this defect by binding the enzyme to the dimer and permitting light energy to break the dimer apart. This action reconstitutes the thymine molecules in their original form and repairs the mutation.

8.9 What are some long-range effects of mutation in microorganisms?

Mutations are believed to be a driving force in the evolution of microorganisms. For example, the emergence of drug-resistant bacteria is probably due to mutations that occurred in bacteria centuries ago. However, the ability to resist drugs was not expressed until antibiotics were first used in this century. Those bacteria possessing the mutation survived when exposed to the antibiotics, and they multiplied to yield a population of drug-resistant bacteria. This emergence is a form of natural selection.

RECOMBINATION

8.10 How does mutation compare to recombination?

Where mutation is a permanent change in the DNA of a microorganism, recombination refers to the acquisition of new segments of DNA. That acquisition may consist of DNA fragments from the local environment or DNA donated by another microorganism.

8.11 What is recombinant DNA?

When a recombination takes place, new genes or gene combinations are added to those already present in a microorganism. The newly reconstituted DNA is known as **recombinant DNA,** since the DNA has been "recombined."

8.12 How does recombination occur in microorganisms?

There are three general methods for recombination in microorganisms: transformation, conjugation, and transduction. These methods have been thoroughly studied in bacteria but not in fungi, protozoa, and simple algae. Their occurrence in these microorganisms is uncertain.

8.13 What does the recombination process of transformation involve?

Transformation occurs when a recipient bacterium takes up a segment of DNA from its surrounding environment. The DNA could have been released from a donor bacterium after the cell died. Because the nucleic acid is tightly packed in the bacterial nucleoid, the DNA scatters about in the environment when the bacterium dies.

8.14 Can all bacteria take up DNA fragments from the environment?

The only bacteria that can take up DNA fragments are called **competent cells**. A bacterium has competence by virtue of its biochemical composition and because it is closely related to the bacterium donating the DNA fragments. In the laboratory, competence can be artificially stimulated in a bacterium by alternately heating and cooling the cells in the presence of calcium chloride.

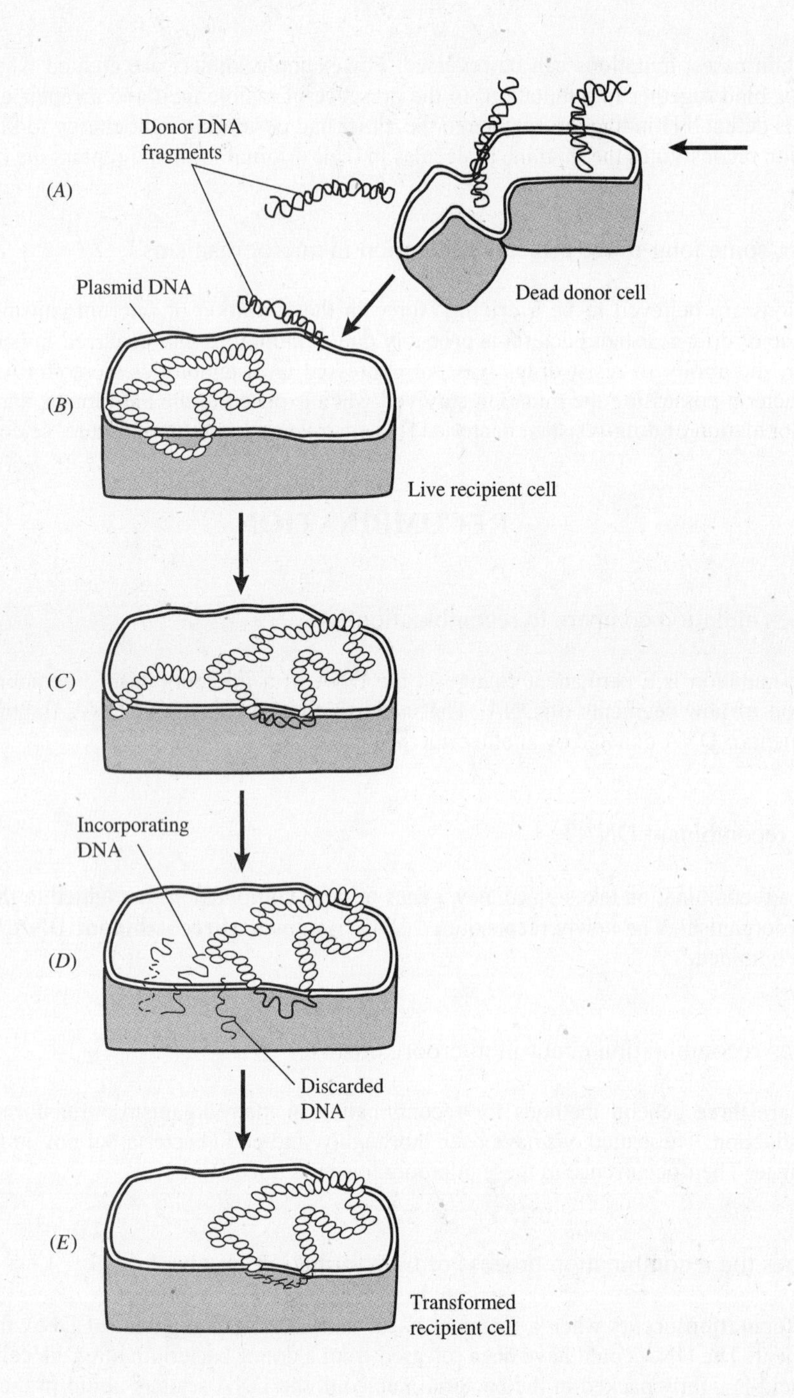

Fig. 8-3 Transformation in bacteria. (*A*) A bacterium disintegrates and its DNA is released into the surrounding environment. (*B*) A live, competent bacterium in the area serves as a recipient and takes up the donor's DNA fragments. Enzymes carry the fragments into the cell. (*C*) Within the recipient cell, the DNA fragment remains whole in the cytoplasm. (*D*) One strand of the fragment DNA disintegrates, and the remaining strand integrates into the DNA of the recipient bacterium. It therefore replaces a similar single-stranded molecule in the recipient cell. (*E*) The recipient cell is now considered transformed.

8.15 What are the specific features of transformation occurring in bacteria?

During transformation, a large segment of DNA is bound to a receptor site on the surface of the competent cell. The segment is then biochemically severed into smaller fragments, and enzymes move the fragments into the cell where they are prepared for recombination with the bacterial chromosome. A single-stranded molecule is used for transformation; it replaces a similar single-stranded molecule in the recipient cell. The process is summarized in Figure 8-3.

8.16 Are chromosomes the only parts of bacteria that can be transformed?

Bacteria posses extrachromosomal loops of DNA called **plasmids**. Plasmids are small, self-replicating loops of DNA with about 10 to 50 genes. They are not essential for the survival of the bacterium. New segments of DNA acquired in transformation may attach to the plasmids of a bacterium, or entire plasmids may be recovered from the local environment.

8.17 In which bacteria has transformation been identified?

The original experiments in transformation were performed in 1928 by Frederick Griffith on the bacterium *Streptococcus pneumoniae*. Griffith showed that these bacteria could be transformed from nonencapsulated to encapsulated forms as a result of transformation. Other organisms that have been researched include *Escherichia coli* and species of *Haemophilus*, *Bacillus*, and *Pseudomonas*.

8.18 How does conjugation differ from transformation?

Conjugation varies considerably from transformation. In conjugation two live bacteria come together, and a donor cell contributes its DNA to the recipient cell. Conjugation generally involves plasmids. The donor cells are referred to as F^+ cells, and recipient cells are called F^- cells. The DNA that is transmitted between donor and recipient is referred to as the fertility, or F factor. Commonly, the F factor is a plasmid. Possession of the F factor permits a bacterium to act as a donor while absence of the F factor indicates a recipient bacterium.

8.19 What are the stages of conjugation taking place in bacteria?

During conjugation, a donor cell and recipient cell come together and a filamentous structure known as the sex pilus attaches the two. A conjugation tube forms between the cells, and the F factor (the plasmid) replicates within the donor cell. The F factor then moves through the conjugation tube into the recipient cell. The recipient cell now becomes a donor cell because it now has the F factor. Figure 8-4 depicts the process.

8.20 Is the bacterial chromosome ever involved in conjugation?

In some cases, the F factor (plasmid) integrates itself into the bacterial chromosome. The bacterium so-produced is called an Hfr cell, or high-frequency-recombinant. During conjugation between an Hfr and F^- cell, the DNA undergoes replication, and one of the chromosomes breaks at the point of the F factor. With the F factor leading, the broken chromosome moves through the conjugation tube into the recipient cell. In rare instances, the entire chromosome moves through, but more commonly, only a portion of the DNA moves into the recipient. The recipient then becomes an F^+ cell because it has the F factor. It also contains several of the genes from the donor cell. The recipient bacterium has now been recombined.

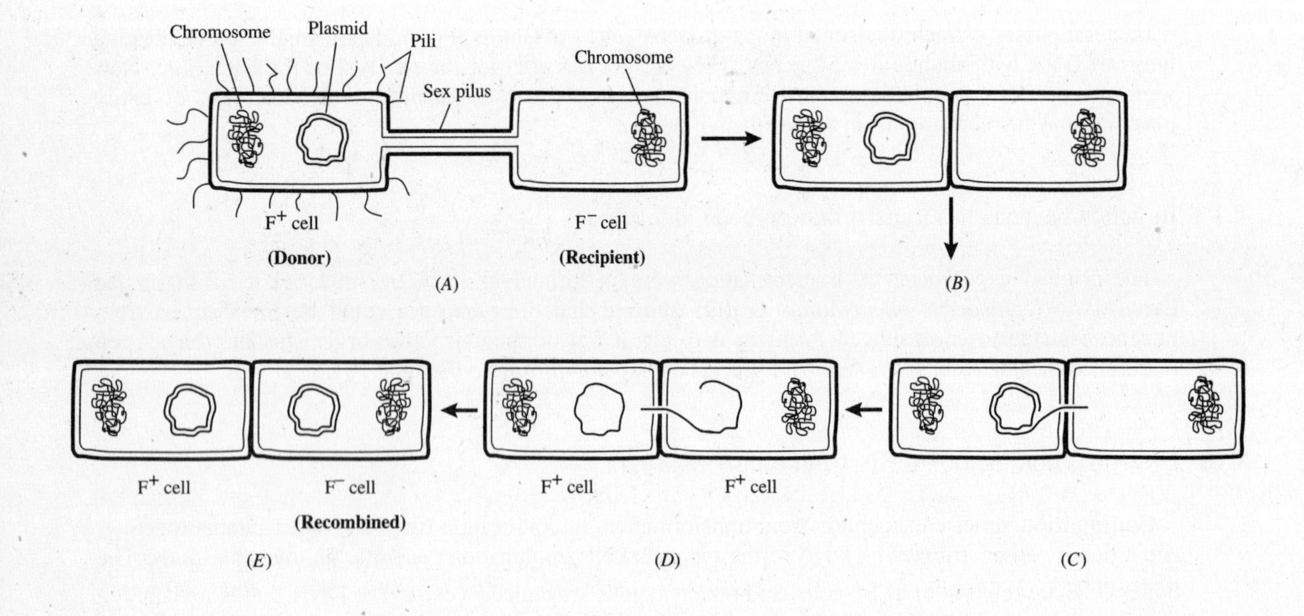

Fig. 8-4 Bacterial conjugation. (*A*) A donor cell designated f^+ is bound to a recipient cell designated f^- by a sex pilus. The donor cell has pili, the recipient cell has no pili. (*B*) The sex pilus retracts and the two cells come close to one another. (*C*) The plasmid of the donor cell replicates and a copy consisting of single-stranded DNA moves through a conjugation tube into the recipient cell. (*D*) When the copy enters the recipient, it serves as a template for producing a second single-stranded DNA molecule, which then unites with the template strand. (*E*) The recipient organism now contains the f factor and is now considered a donor cell.

8.21 Does the recombined bacterium have any advantage over other bacteria in the population?

When a recombined bacterium acquires gene combinations not found in other bacteria, it acquires new biochemical abilities. It will undergo reproduction by binary fission and it may be able to form new structures or synthesize new enzymes. The recombination may result in an organism better-adapted to its environment. For example, it may acquire drug resistance.

8.22 Are there any other forms of conjugation occurring in bacteria?

In a third form of conjugation called **sexduction**, the F factor first attaches to the chromosome, then detaches at some later time. In so doing, it takes along a segment of chromosomal DNA and now becomes an F' factor. Should conjugation take place, the new F' factor with its bacterial genes will pass into the recipient and transform that recipient. As before, a new gene combination results.

8.23 In which types of bacteria has conjugation been observed?

Conjugation has been observed in numerous species of Gram-negative bacteria, but few Gram-positive species. It has been researched thoroughly in *Escherichia coli* and is known to exist in species of *Salmonella, Shigella,* and other Gram-negative organisms. In addition, it is known to occur between cells of different species.

8.24 Can genes move between plasmids prior to conjugation?

Research studies indicate that genes move among and between plasmids in a microbial cell. These so-called "jumping genes" are able to migrate from plasmid to plasmid, from plasmid to chromosome, or from chromosome to chromosome. The phenomenon is important because it contributes to the development of multiply-drug-resistant bacteria. For instance, if drug resistance genes congregate to a single plasmid, it is possible that the single plasmid may be transferred to a recipient organism that is pathogenic. The result is a pathogenic bacterium, which is now resistant to many drugs.

8.25 How does transduction compare with transformation and conjugation?

Transduction varies considerably from the other two recombinant processes because it relies upon a virus to transfer DNA between donor and recipient bacteria.

8.26 How do viruses acquire segments of DNA for transferal in transduction.

Many bacterial viruses undergo a lysogenic, or latent, cycle. In this cycle, the viral DNA attaches to the bacterial chromosome and remains with the host chromosome for a long period of time. In this form, the viral DNA is known as a prophage or "provirus." The prophage (provirus) will continue to replicate with the host chromosome and will remain a part of that chromosome. At some time in the future, the provirus detaches from the bacterial chromosome and carries along a number of bacterial genes. Now it enters its lytic cycle. In this cycle, the virus replicates its DNA many times and in so doing, also replicates the bacterial DNA. All the proviruses that form during replication carry both viral and bacterial DNA. Should the virus leave the bacterium and enter a second bacterium, it will carry along the bacterial DNA and incorporate this DNA when it attaches to the chromosome. The recipient bacterium is now transduced, or changed, through its acquisition of new segments of DNA. The process described is referred to as **specialized transduction**.

8.27 Is specialized transduction the only type of transduction that occurs in bacteria?

There is a second type of transduction called **generalized transduction**. In this process a virus enters a bacterium and releases its DNA. Rather than attaching to the bacterial chromosome, however, the virus immediately enters its lytic cycle and begins to replicate. During the replication, the DNA of the bacterial chromosome is broken into small fragments, and certain viruses will incorporate some of these bacterial fragments into their genomes. Thus, a small percentage of the viruses have become carriers of bacterial DNA. Figure 8-5 presents a view of the process.

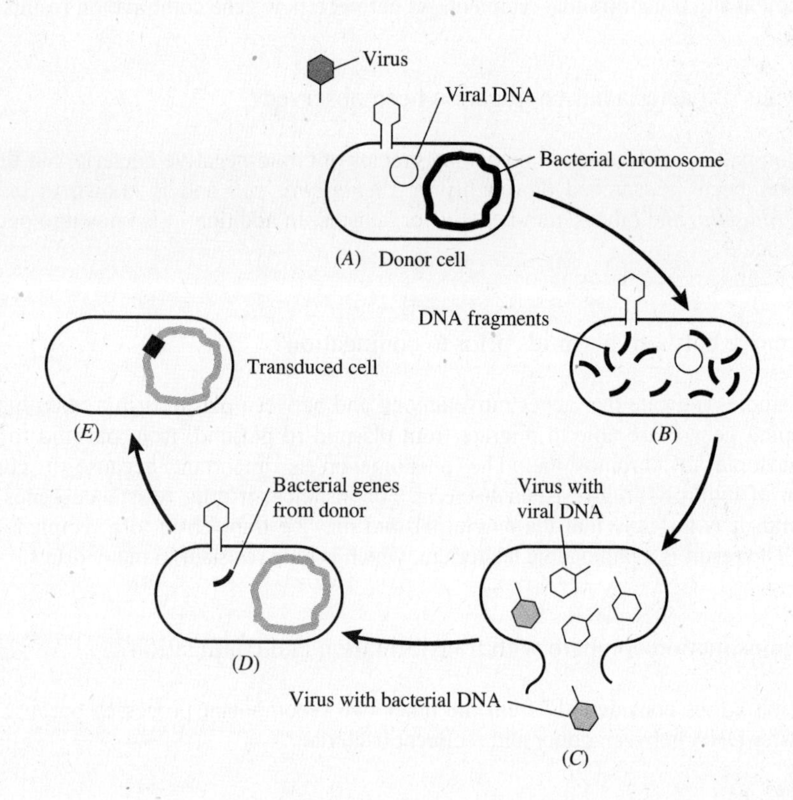

Fig. 8-5 Generalized transduction. (*A*) In generalized transduction a virus known as a bacteriophage infects a host cell. (*B*) The virus begins the replication cycle, also known as its lytic cycle. During the process, the host DNA is broken into small fragments. (*C*) During the replication process, some of the new viral particles envelope fragments of bacterial DNA instead of the usual viral DNA. (*D*) After these viruses are released from the bacterium, they may subsequently infect fresh bacteria and introduce their viral DNAs into the recipient. (*E*) If the viral DNA is integrated into the new host cell, the former bacterial genes are introduced, and the recipient bacterium is transduced (recombined).

8.28 How do the viral carriers transmit the bacterial DNA to a recipient bacteria?

When the viruses are released from the donor bacteria, they enter the cytoplasm of fresh recipient bacteria. Here, they release their DNA. For a small percentage of viruses this DNA is actually the DNA of the previous (donor) bacterium . The DNA from the donor bacterium then attaches to the DNA of the recipient bacterium, and the genes begin expressing themselves. The transduction is completed.

8.29 Is transduction widely observed in bacteria?

Transduction has been studied thoroughly in species of *Salmonella* and their viruses. Transduction has also been studied in *Escherichia coli*, but it does not appear to occur in numerous other bacterial species. However, the potential for transduction exists because many bacteria carry viruses attached to their chromosomes. For example, the diphtheria bacillus *Corynebacterium diphtheriae* possesses such a virus.

GENETIC ENGINEERING

8.30 What is genetic engineering and how does it relate to microbial genetics?

Since the 1970s, scientists have learned that they can manipulate the genes of bacteria for the benefit of humans. These manipulations form the basis for the processes of genetic engineering and recombinant DNA technology. From observations of bacterial transformation and transduction scientists learned they could isolate plasmids from bacteria and insert into those plasmids a section of foreign DNA. Then, utilizing the ability of bacteria to undergo transformation, they discovered that they could insert recombined plasmids into fresh bacteria where the genes would encode proteins unlike those normally formed by the bacteria.

8.31 What steps are necessary to form a recombined plasmid?

One of the prerequisites for producing recombined plasmids is a group of bacterial enzymes called **endonucleases**. These enzymes cleave the phosphate-deoxyribose bond in the DNA molecule and thereby open the bacterial plasmid or cleave the bacterial chromosome. Endonucleases are also known as **restriction enzymes** because they operate at a restricted site. Restriction enzymes will open a DNA molecule at the same site regardless of the source so long as a certain sequence of nitrogenous bases is present. Many hundreds of restriction enzymes are now known to exist. Figure 8-6 illustrates this process.

8.32 How are restriction enzymes used to form a new, recombined plasmid for use in genetic engineering?

Scientists begin with plasmid molecules derived from bacteria such as *E. coli.* Then they obtain DNA from a different species of organism and use the restriction enzyme to open both plasmid and foreign DNA. Once the DNA molecules are broken, the ends of the plasmid are joined with the ends of the foreign DNA, and an enlarged plasmid containing the foreign DNA is produced. For example, it is possible to place DNA from *Staphylococcus aureus* into the plasmids of *Escherichia coli.* An enzyme called **DNA ligase** is then used to stitch together the ends of the plasmid and form a recombined DNA molecule. Such a molecule is called a **chimera.** The chimeras are then introduced into fresh *E. coli* cells by alternate heating and cooling in calcium chloride. When the *E. coli* cells are cultivated on fresh medium they produce all normal proteins plus proteins normally found in *S. aureus*. The bacteria have been genetically engineered to produce foreign proteins.

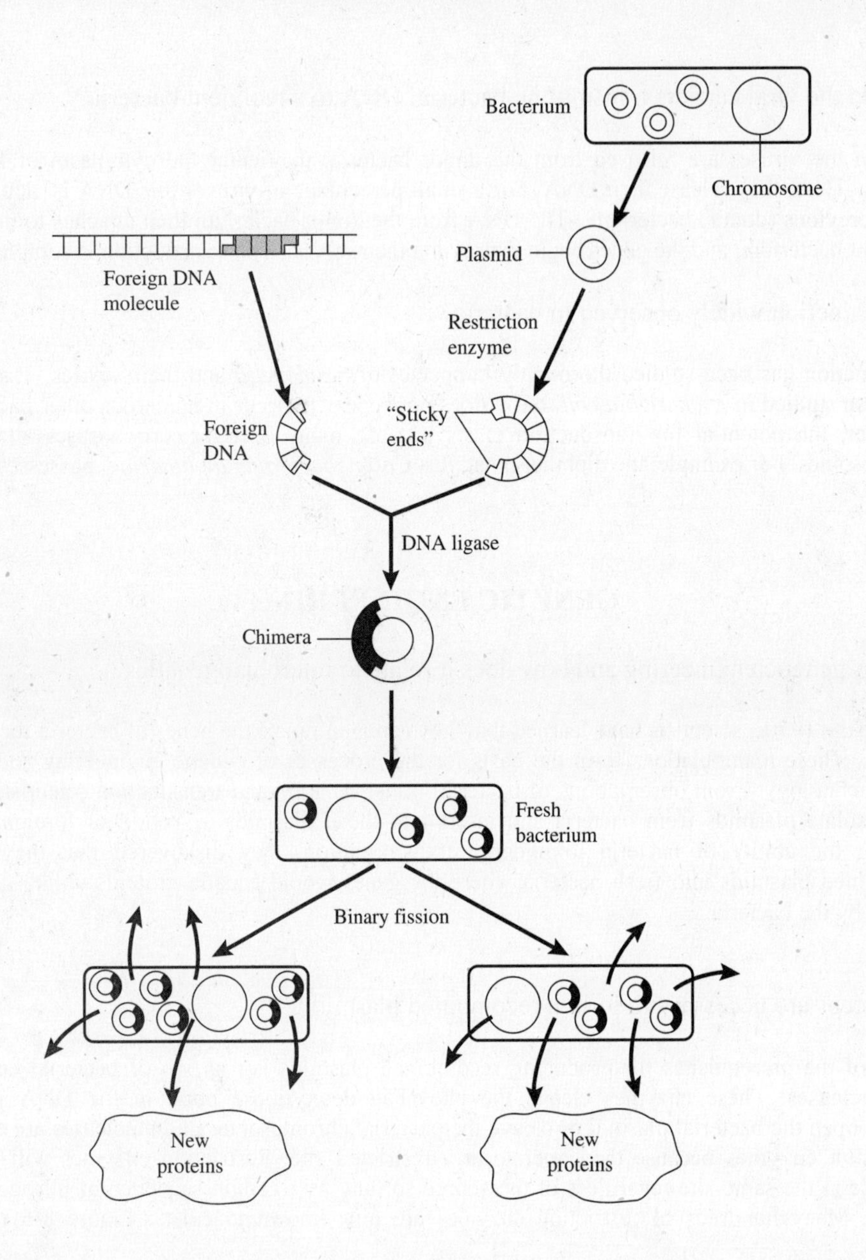

Fig. 8-6 The process of genetic engineering in bacteria. Plasmids are obtained from bacteria, and restriction enzymes are used to open the molecule at a restricted site. DNA is obtained from a foreign organism, and the restriction enzyme opens it at the same site. The enzyme DNA ligase now combines the open bacterial plasmid with the foreign DNA to form a recombined DNA molecule called a chimera. When the chimeras are introduced into fresh bacteria, they reproduce themselves and encode the normal proteins plus the proteins specified by the foreign DNA.

8.33 Can genes from human cells be inserted into plasmids in recombinant DNA technology?

Since the 1980s, it has been possible to insert human genes into bacterial plasmids. Among the first genes inserted were those for the pancreatic hormone **insulin**. Once the genes were inserted into plasmids and the plasmids placed into bacteria, the bacteria began producing human insulin. This insulin is now commercially available for use in diabetic individuals.

8.34 Which other human proteins can be produced by genetically engineered bacteria?

During the last two decades, many other human proteins have been derived from bacteria. For example, the antiviral protein **interferon** has been obtained from reengineered bacteria. Also, the genes for **human growth hormone** have been inserted into bacteria, and this hormone is now being manufactured for treating children with growth deficiencies. The clotting **factor VIII** and the clotting substance **tissue plasminogen activator** (TPA) have also been produced by altered bacteria. Vaccines have been produced for hepatitis B and other diseases by inserting the genes for viral proteins into yeast plasmids and inducing the yeast cells to produce the proteins.

8.35 Are there uses for genetic engineering other than for medicine?

Genetically engineered bacteria will have substantial impact on future science. For example, bacterial genes have been placed into tobacco plants to permit better tolerance for herbicides. Genes for viruses have been inserted into plant cells to increase their resistance to viral infection. And genes for bacterial insecticides can be obtained and inserted to plants to encourage their resistance to destructive insect pests.

8.36 Are there applications for genetic engineering technology outside of microbiology?

The experiments in gene technology and recombinant DNA have bred the modern science of **biotechnology**. Biotechnology is currently being used for such fields as DNA fingerprinting, gene therapy, and agriculture. For instance, genes have been introduced to tomato cells to help them resist excessive ripening, and genes have been inserted to animals to stimulate their growth. Biotechnology is used to identify in humans such genetic diseases as cystic fibrosis, Alzheimer's disease, and Huntington's disease. In addition, the effort to determine the sequence of all the nitrogenous bases in all the human genes has begun. Biotechnology represents one of the most innovative and imaginative uses of biological research in the history of science.

Review Questions

Completion. For each of the following, add the term that completes the statement best.

1. A mutation in which a nitrogenous base is replaced by a different base is known as a ____________.

2. Those mutations brought about in the laboratory when researchers apply mutagens to microorganisms are known as ____________.

3. A mutation may be reversed in the presence of a repair enzyme and ____________.

4. Mutation refers to a permanent change in the DNA of a microorganism, but the acquisition of new segments is called ____________.

5. When a recipient bacterium takes up a segment of DNA from its surrounding environment, the process is referred to as ____________.

6. Those bacteria able to take up DNA fragments from the local environment are said to be ____________.

7. The segments of DNA acquired by a bacterium from the environment may be extrachromosomal loops of DNA known as ____________.

8. Some of the original experiments in bacterial transformation were performed by Frederick Griffith on the bacterium ____________.

9. The designation of a donor cell in the process of conjugation is ____________.

10. A bacterium that receives new genes or segments of DNA is said to be ____________.

11. Those genes that are able to migrate from plasmid to plasmid or from plasmid to chromosome have come to be called ____________.

12. An essential element in a transduction reaction occurring between two bacteria is a particle known as a ____________.

13. The form of transduction in which bacterial fragments are packaged and carried into a recipient bacterium is called ____________.

14. The process of transduction has been thoroughly studied in species of ____________.

15. One of the essential elements of genetic engineering is a set of DNA-cleaving enzymes called endonucleases, or ____________.

16. In order to chemically stitch together the ends of a plasmid in genetic engineering experiments, researchers use the enzyme ____________.

17. When a plasmid has been engineered to contain foreign DNA, that plasmid is referred to as a ____________.

18. Among the first human genes inserted to bacterial plasmids were the genes for the human hormone ____________.

19. Genetically engineered bacteria have been cultivated with the ability to produce the antiviral protein ____________.

20. The novel experiments in gene technology and recombinant DNA have led to the development of the modern science of ____________.

Multiple Choice. Select the letter of the word or phrase that best completes each of the following.

1. When a mutation occurs in bacteria, the effects are seen on the (*a*) rate of ATP utilization (*b*) rate of nutrient ingestion (*c*) proteins synthesized (*d*) ATP taken into the cell.

2. A deletion mutation comes about (*a*) when an entire section of chromosome is lost (*b*) when a new sequence of nucleotides is added to the chromosome (*c*) when a segment of DNA reverses its order (*d*) when a plasmid joins the chromosomal DNA.

3. Ultraviolet light causes mutations in bacteria (*a*) by breaking the chromosome (*b*) by causing a frame shift (*c*) by binding together adjacent thymine bases (*d*) by reversing a segment of DNA.

4. A recombination process occurring in bacteria always refers to the (*a*) affect of a physical nature changing the nature of the genes (*b*) reversal of a microbial mutation (*c*) loss of genes from a cell (*d*) acquisition of DNA by an organism.

5. The process of transformation in bacteria does not require (*a*) a competent recipient cell (*b*) a disrupted donor cell (*c*) a virus (*d*) a series of enzymes.

6. All the following apply to plasmids except (*a*) they are self-replicating loops of DNA (*b*) they have 10 to 50 genes (*c*) they are essential for the survival of a bacterium (*d*) they are acquired in transformations.

7. The DNA molecule that transforms a bacterium (*a*) is single stranded (*b*) contains uracil (*c*) has no functional genes (*d*) contains three strands of DNA.

8. Which of the following is necessary for the process of conjugation in bacteria (*a*) one live recipient cell and one dead recipient cell (*b*) a large concentration of glucose molecules (*c*) viruses that multiply within bacteria (*d*) two live bacteria.

9. An Hfr bacterium is one that contains (*a*) many unusual plasmids (*b*) chromosomal material acquired from a recipient cell (*c*) a plasmid integrated into its chromosome (*d*) the ability to undergo transduction.

10. When the F factor enters a recipient cell during conjugation, the recipient cell (*a*) undergoes lysis (*b*) multiplies rapidly (*c*) becomes a donor cell (*d*) develops mitochondria.

11. Bacterial conjugation has been observed in numerous (*a*) spirochetes (*b*) Gram-positive bacteria (*c*) Gram-negative bacteria (*d*) cyanobacteria.

12. When a virus remains with the chromosome of a host bacterium for a long period of time, the viral DNA is called a (*a*) adenovirus (*b*) provirus (*c*) baculovirus (*d*) enterovirus.

13. Which of the following characteristics is associated with specialized transduction? (*a*) the virus attaches to the bacterial chromosome (*b*) a virus immediately replicates within a host bacteria (*c*) the virus fails to replicate within the host bacteria (*d*) the viral DNA is destroyed immediately on entering the bacterium.

14. A bacterium that has been transduced contains (*a*) an enhanced supply of ATP (*b*) numerous new plasmids (*c*) fewer DNA segments than it originally had (*d*) new DNA segments.

15. Bacterial recombinations are an essential feature of the process of (*a*) photosynthesis (*b*) chemiosmosis (*c*) genetic engineering (*d*) anaerobic fermentation.

16. The restriction enzymes used in biotechnology function by (*a*) binding nucleotide molecules together (*b*) cleaving the DNA molecule at certain points (*c*) binding amino acids together to form proteins (*d*) restricting the entry of plasmids into cells.

17. Chimeras are DNA molecules that contain (*a*) segments of RNA (*b*) segments of foreign DNA (*c*) interwoven glucose molecules (*d*) extra amino acids.

18. In order to perform a successful genetic engineering experiment, all the following are required except (*a*) fresh bacterial cells (*b*) the enzyme DNA ligase (*c*) bacterial plasmids (*d*) molecules of starch and glycogen.

19. Among the many products produced by genetically engineered bacteria are (*a*) sucrose and starch (*b*) interferon and insulin (*c*) ATP and ADP molecules (*d*) sodium and chloride ions.

20. One of the remarkable observations in genetic engineering is that (*a*) viruses can be made to replicate within bacteria (*b*) bacteria can take up segments of DNA from the environment (*c*) bacteria can conjugate with one another (*d*) foreign genes can be placed in bacteria and can be expressed.

True/False. Indicate with a "T" or "F" whether the statement is true or false in its entirety.

____**1.** Mutations are temporary changes occurring in the chromosome of bacteria.

____**2.** Inversion mutations are those in which a fragment of DNA is replaced by an identical piece of DNA, but in reverse order.

____**3.** Among the mutagenic agents for bacteria are urea, phenol, and glucose.

____**4.** Spontaneous mutations are those that occur in the laboratory as a result of scientific research.

____**5.** In a frameshift mutation, a mutagen causes a base or two to be lost from the DNA molecule and the reading frame to shift.

____**6.** Ultraviolet light causes mutations in bacteria by binding together adjacent guanine molecules to form dimers.

____**7.** Recombinant DNA is DNA in which new genes for gene combinations have been added to those already present in a microorganism.

____**8.** Recombination methods in fungi and protozoa have been well researched, and the processes of recombination have been well-documented.

____**9.** In order for transformation to occur in bacteria, two live bacteria must be available.

___**10.** A competent bacterium is one that contributes DNA to a recipient bacterium in transformation.

___**11.** Plasmids are loops of DNA existing outside the chromosome of the cell.

___**12.** In the process of conjugation, an entire chromosome normally passes from a donor cell to a recipient cell.

___**13.** The fertility (F) factor involved in conjugation processes is a plasmid.

___**14.** During conjugation processes, the plasmids move to the recipient cell and leaves the donor cell without any plasmids.

___**15.** It is possible that a recipient cell can become a donor cell in conjugation if it receives the F factor.

___**16.** Jumping genes are genes able to migrate between plasmids and between plasmids and chromosomes.

___**17.** In order for transduction to occur, it is always necessary that the viral DNA attach itself to the bacterial chromosome.

___**18.** It would be impossible for transduction to occur without the intervention of a virus.

___**19.** Genetic engineering is a purely research tool without any practical benefits for humans.

___**20.** Among the key elements for recombining plasmids is the enzyme DNA ligase, which opens DNA molecules at certain selected sites.

Chapter 9

Control of Microorganisms

OBJECTIVES

The control methods used against microorganisms may consist of physical methods such as heat, chemical methods such as iodine and alcohol, and antibiotics such as penicillin. These control methods are used for preventing microorganisms from reaching the body surface or killing them on its surface and within its tissues. This chapter explores the various methods available for microbial control. The chapter's objectives are to:

1. Identify the factors that are associated with a successful control agent for microorganisms.
2. Distinguish various forms of antimicrobial control such as sterilization, disinfection, and sanitization.
3. Discuss the various methods by which heat can be used to achieve sterilization of instruments, fluids, and other materials.
4. Study the antimicrobial effects that drying and radiation have on microorganisms.
5. Summarize the chemicals and gases available for achieving sterilization.
6. Compare the antimicrobial effects of alcohol, heavy metals, dyes, acids, halogens, hydrogen peroxide, and phenol products.
7. Identify the origin of antibiotics and indicate how they work to inhibit the growth of microorganisms.
8. Discuss the mechanisms by which antibiotic resistance develops in microorganisms.
9. Compare the modes of action and origin of such antibiotics as penicillin, cephalosporins, aminoglycosides, and tetracyclines.
10. Specify the side effects attributed to various antibiotics and indicate why their use is restricted.
11. Recognize which antibiotics are effective against fungi, viruses, and protozoa.

THEORY AND PROBLEMS

9.1 Why is microbial control necessary and what are some of the factors that go into the choice of a particular antimicrobial agent?

The control of microorganisms is an important way of preventing pathogens from reaching the body. Sterilizing laboratory equipment, hospital supplies, and industrial apparatus helps contains contamination. The choice of a particular antimicrobial agent depends on such things as the kind of material to be treated (living or nonliving); the kind of microorganism to be controlled (bacteria or virus or other); the environmental conditions existing at the time of the agent's use (for example, the temperature of the environment and concentration of microorganisms); the acidity or alkalinity of the area; and the presence of organic matter.

9.2 In how many different ways can microorganisms be inhibited or killed by antimicrobial agents?

There are various modes of action for antimicrobial control agents. Certain agents destroy cells by coagulating the cytoplasmic contents of the microorganism, causing them to separate into a liquid portion and an insoluble mass. Other agents cause the breakdown of large molecules into smaller ones by the loss of electrons, a process called **oxidation**. For some antimicrobial agents, damage occurs in the microbial cell wall or cell membrane, and the physical damage leads to death. Finally, certain antimicrobial agents interfere with essential metabolic pathways in microorganisms.

9.3 How does organic matter influence the effectiveness of a control agent?

The absence of **organic matter** is an essential feature in the successful use of an antimicrobial agent. Chemical agents can combine with organic matter such as feces, urine, or blood, and become unavailable for microbial control. In addition, organic matter coats microorganisms and protects them from the antimicrobial agent.

9.4 Is there a difference between sterilization and other forms of antimicrobial control?

Sterilization is the process in which all living organisms, including microbial endospores, are destroyed. Sterilization is an absolute term that implies the complete and total removal of all living things.

9.5 How does sterilization compare to disinfection?

Where sterilization implies the destruction of all living things, disinfection only implies the removal of pathogenic organisms. **Disinfection** can be accomplished by a **disinfectant**, which is an antimicrobial agent used on lifeless object such as an instrument, or an **antiseptic**, which is an antimicrobial agent used on the skin surface or contacts the body in some way.

9.6 How does an antibiotic compare with an antiseptic or disinfectant?

Where antiseptics and disinfectants are used outside the body, **antibiotics** are used within the body environment. Antibiotics must be nontoxic to the host, work at very low concentrations, and not cause allergic responses. Because they work inside the body, antibiotics can cause side effects, and their antimicrobial effect is generally less than antiseptics or disinfectants.

9.7 How does a sanitizing agent relate to an antiseptic, disinfectant, or sterilizing agent?

A **sanitizing agent** is one that reduces the microbial population to a safe level as determined by public health agencies. It does not remove all pathogens as an antiseptic or disinfectant does, nor does it remove all microorganisms as a sterilizing agent does. **Sanitization** may be accomplished by cleaning with detergents, heat, or any agent that lowers the microbial content of an environment.

PHYSICAL AGENTS

9.8 What is the easiest way to inhibit or destroy microorganisms?

Probably the easiest way to control a population of microorganisms is by using **heat**. Dry heat such as from a flame is rapid and efficient. However, the material may not be reused if it is placed into an incinerator or a fire.

9.9 Can boiling water be used as an antimicrobial agent?

Exposure to **boiling water** is an effective antimicrobial agent, but it may not always achieve sterilization. This is because some bacterial spores can resist over two hours of exposure to boiling water. However, nonsporeforming bacteria, which are the most common, die within a few seconds in boiling water. Fungi, protozoa, viruses, and simple algae are equally susceptible.

9.10 Is there any instrument that can be used to kill bacterial spores?

When water boils, it converts to steam at 100°C. If this steam is continually heated to 121°C, its pressure will rise to 15 pounds per square inch (psi). Under these conditions of temperature and pressure, bacterial endospores are killed after an exposure of about 15 minutes. The instrument that uses these conditions is an extremely valuable piece of equipment known as the **autoclave** shown in Figure 9-1. The autoclave may be a desktop model, or it may be of considerable size. It is used for instruments, blankets, bacteriological media, salt solutions, and equipment.

9.11 Can an oven be used for sterilization purposes?

In an ordinary **hot air oven,** the dry heat radiates within and around microorganisms and kills them. At a temperature of 160°C an exposure of about 90 minutes will result in sterilization. Dry heat is preferred to moist heat (such as in the autoclave) because dry heat can be used to sterilize powders, oily materials, dry glassware and instruments, and other materials that do not mix easily with water.

9.12 What is the effect on microorganisms when moist heat is used as compared to dry heat?

Moist heat generally coagulates and causes denaturation in microorganisms. In **denaturation**, proteins separate as an insoluble mass as they revert from their three-dimensional structure to a two-dimensional structure. In dry heat, by contrast, the primary effect on microorganisms is due to **oxidation** of large molecules, a process that breaks them down into smaller molecules. Since oxidation is a less efficient process it generally requires a longer period of time.

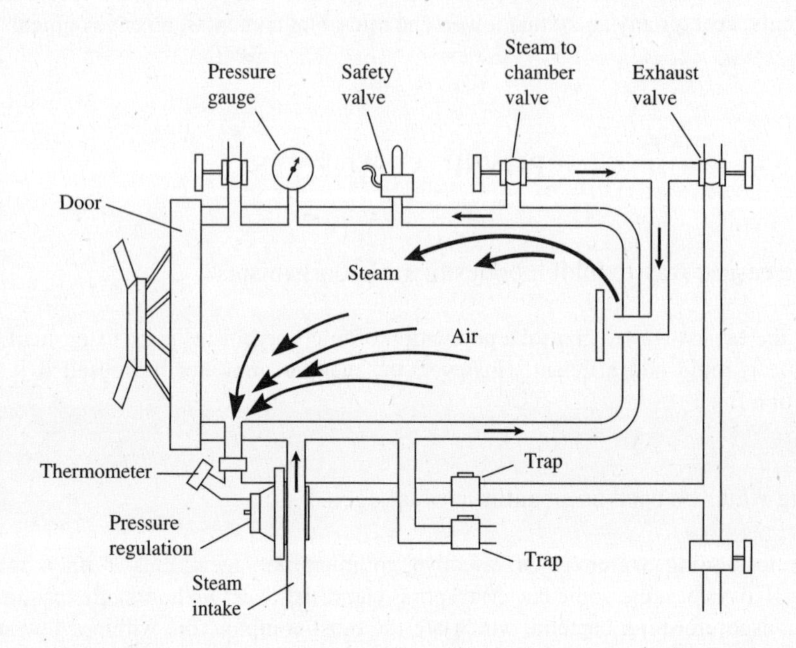

Fig. 9-1 A schematic view of a steam-jacketed autoclave. The entering steam displaces the air to a port in the bottom of the chamber and saturates the environment with steam. The pressure rises to 15 psi and the temperature rises to 121°C. Sterilization is achieved in approximately 15 minutes.

9.13 Are there any sterilization methods available for delicate instruments that cannot be exposed to long periods of heating at extremely high temperatures?

For certain delicate liquids, sterilization can be performed by exposing the liquid to free-flowing steam at 100°C for a 30 minute period of time on each of three successive days. The process is called **tyndallization**. The alternate heating periods shock the spores and they germinate into vegetative cells, and they are killed during the next exposure to steam.

9.14 Is the process known as pasteurization effective for achieving sterilization in milk and other dairy products?

Pasteurization is not a sterilization process. Pasteurization lowers the level of microorganisms in milk and dairy products and kills any pathogenic microorganisms present. Pasteurization may be conducted in a hot water bath at 62.8°C for 30 minutes (the holding method); or the milk may be exposed to heat at 71.8°C for 15 to 17 seconds (the flash method); or it may be exposed to heat at 82°C for three seconds (ultrapasteurization). For milk, beer, and fruit juice pasteurization is an effective way of preserving the products, but it does not sterilize them.

9.15 Can sterilization be achieved from withholding heat from microorganisms?

Lowering the temperature of the environment such as in a **refrigerator** is an effective way of inhibiting microbial growth, but not of killing microorganisms. In the **freezer**, the temperature is cooled to below zero, and the metabolism of microorganisms is also slowed. There may be some killing due to ice crystals forming in the freezer, but freezing is not considered a sterilization method. Deep freezing at -20°C does not completely eliminate a microbial population either.

9.16 Can drying be used as an effective control process?

All living things including microorganisms need water in order to carry out the chemical reactions of life. By drying foods by sunlight or machinery, spoilage can be prevented by lowering the microbial population. Such foods as flour, macaroni products, and dry cereals remain free of spoilage because they are dry. Foods can also be dried out by exposing them to high concentrations of salt. The salt draws water out of the food by **osmosis**, and microorganisms do not multiply in the dry environment. Highly sugared foods remain free of spoilage for the same reason. However, it should be noted that the food has not been sterilized, but preserved.

9.17 What types of radiation can be used as antimicrobial control agents?

There are many different kinds of radiations having antimicrobial effects. **Ionizing radiations** include x-rays and gamma rays. These radiations form free radicals in cytoplasm and the free radicals destroy microbial proteins and DNA. Spores generally resist the effect of ionizing radiations, but most other organisms are killed. Various foods, drugs, and plastic surfaces are now treated with ionizing radiations. Another important form of radiation is **ultraviolet radiations**. These radiations effect nucleic acids by binding together adjacent thymine bases. Microbial death follows quickly, because the DNA cannot function or replicate itself. Ultraviolet light is used to control microbial populations on operating room tables, hair brushes, and in the air.

9.18 Can microwaves be used as a type of sterilizing agent for foods and other materials?

Microwaves are a form of long-wavelength radiation. Sometimes known as infrared rays, microwaves generate heat by stimulating water molecules to vibrate within foods. These vibrations cause the heat that cooks food, and the heat reduces the microbial population. Microwaves have no direct effect on microorganisms, however, and their use as antimicrobial agents is still in the developing stage.

9.19 Are sound waves an effective agent for producing antimicrobial control?

High-frequency sound waves are known as **ultrasonic sound vibrations**. These vibrations coagulate cellular proteins and disintegrate cellular components. Ultrasonic vibrations are used commonly as cleaning agents for laboratory materials and as cell disrupters, but they have not been adapted for controlling populations of microorganisms.

9.20 Can filters be used to achieve sterilization?

To use a filter for sterilization, it would have to have pores the size of the smallest possible viruses. This pore size could be achieved but the flow of liquid through the filter would be extremely slow. Filters with larger pore sizes are currently used to remove bacteria and other microorganisms from consumable products such as juices and beer, and from blood sera and certain microbiological media.

CHEMICAL AGENTS

9.21 Are there any chemicals or gases that are able to achieve sterilization?

Among the chemical agents there are at least three gases that can achieve sterilization when they are used in closed chambers. The three gases are **ethylene oxide**, **formaldehyde**, and **beta propiolactone**. ETO is widely used to sterilize plastics (such as Petri dishes), rubber, and paper. BPL is used to sterilize liquids, and formaldehyde can be used for various materials. However, all three are irritating to the skin, toxic, and considered carcinogenic. Therefore they are used only within sealed environments.

9.22 Can alcohol be used as an antimicrobial control agent?

Alcohol is used as a chemical control agent in the form of **ethyl alcohol** (70 percent) and **isopropyl alcohol** (70 percent rubbing). Alcohol denatures proteins and can be used against spores, but complete immersion must take place and a minimum of 20 minutes exposure must be observed. Used as an antiseptic, alcohol has minimal activity because it dries very quickly. The chemical structure of alcohol and other disinfecting agents are displayed in Figure 9-2.

9.23 Which heavy metals are valuable as chemical control agents for microorganisms?

Two heavy metals, mercury and silver, are used as antimicrobial agents. **Mercury** is used as mercuric chloride (bichloride of mercury) and as a component of Mercurochrome, and merthiolate. **Silver** is commonly used as silver nitrate, silver picrate, and silver lactate. It is available in stick form to cauterize wounds. Silver nitrate is used in the eyes of newborns to prevent gonorrhea.

9.24 How can a change of pH be used to control microorganisms?

A pH change is deleterious to microorganisms. Calcium hydroxide (CaOH), known as slaked lime, is used to raise the pH into the alkaline range and is employed as a disinfectant for destroying pathogens in fecal material. Sodium hydroxide and potassium hydroxide are other bases that kill microorganisms. At the lower end of the pH scale, acetic acid, propionic acid, and benzoic acids are inhibitory to microorganisms. They are added to various foods such as breads and fruit juices. Benzoic, undecylenic, and salicylic acids are used in antifungal ointments to treat athlete's foot.

9.25 Are there any dyes that exhibit antimicrobial effects and are used for controlling microbial populations?

Gentian violet is used as an antiseptic. This dye is essentially similar to crystal violet used in Gram staining. The dye is sometimes employed as an antifungal agent for mouth infections (thrush) caused by *Candida albicans*.

9.26 Which aldehydes are available for use against microorganisms?

The two most widely used antimicrobial aldehydes are **formaldehyde** and **glutaraldehyde**. Formaldehyde is used in a 37 percent concentration known as formalin, but it is very irritating to the skin and eyes and in some individuals it causes an allergic response. Glutaraldehyde is used as a 2 percent solution to treat instruments, and under certain conditions it will achieve sterilization. Glutaraldehyde kills vegetative bacteria, bacterial spores, and a wide variety of other microbial forms, and is available in its trade name Cidex.

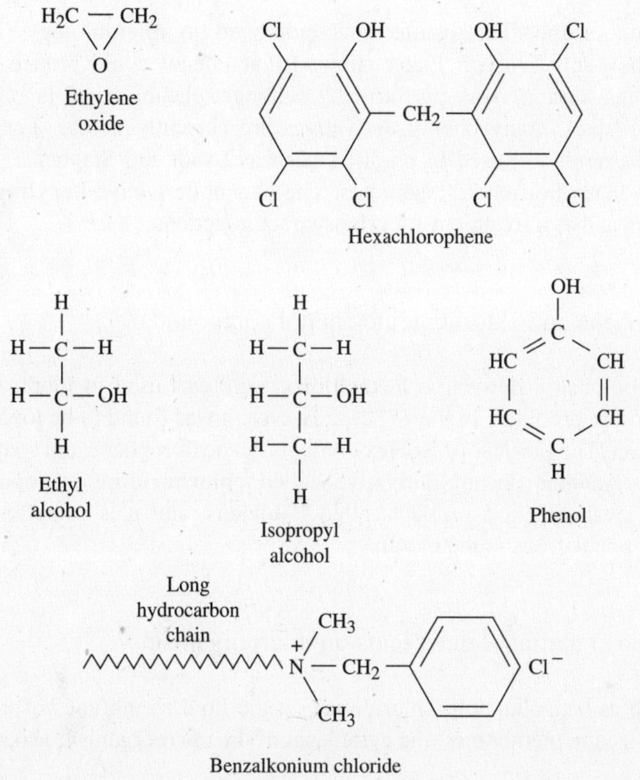

Ethylene oxide

Hexachlorophene

Ethyl alcohol

Isopropyl alcohol

Phenol

Long hydrocarbon chain

Benzalkonium chloride

Fig. 9-2 The chemical structures of various substances used as disinfectants and antiseptics.

9.27 Can halogens be used as antimicrobial agents?

Antimicrobial agents include two important halogens: **chlorine** and **iodine**. Both of these halogens are oxidizing agents that react with the amino acids in proteins and change the nature of the protein. Chlorine is used in its gaseous form in water purification and as sodium hypochlorite, the major component of household **bleach**. Iodine is available as a 2 percent alcohol solution known as a **tincture of iodine** and in an organic form called an **iodophore**. Such products as Betadine, Wescodyne, and Prepodyne are iodophores. Both chlorine and iodine react with organic matter, so materials must be thoroughly cleaned before disinfection is attempted.

9.28 How does hydrogen peroxide act as an antiseptic when it is applied to a wound?

Hydrogen peroxide (H_2O_2) is available as a 3 percent solution from local pharmacies. When it is applied to a wound, the enzyme catalase in the tissue decomposes the hydrogen peroxide into water and free oxygen. The oxygen causes the wound tissues to bubble, and the bubbling removes microorganisms mechanically. Also, the sudden release of oxygen brings about chemical changes in certain microorganisms, and these changes lead to microbial death.

9.29 Can phenol products be used as antiseptics or disinfectants?

Phenol was one of the first disinfectants employed in microbiology. In a landmark series of experiments in the 1860s, Joseph Lister showed that phenol could reduce the possibility of wound contamination when surgery was performed. Although phenol itself is rarely used anymore (it is irritating and corrosive), many phenol derivatives are currently in use. For example, several phenol derivatives called **cresols** are used in products such as Lysol and Staphene. **Pine oil** is also a phenol derivative used in many household detergents. The phenol derivative **hexylresorcinol** is widely used in detergent solutions and as a treatment for urinary tract infections.

9.30 Are hexachlorophene and chlorhexidine phenol compounds?

In the 1960s, the phenol derivative **hexachlorophene** was used in toothpaste, underarm deodorant, soaps, and many other products. In the 1970s, however, it was found to be toxic, and it is available today only by prescription. The product **phisoHex** contains hexachlorophene and soap. It is used in hospitals as a surgical scrub. Another phenol derivative called **chlorhexidine** is popular as an antiseptic and disinfectant. It is available in a product called **Hibiclens**, and it is sometimes used in toothpastes to reduce microbial populations in the mouth.

9.31 What is the mode of action of detergents on microorganisms?

Detergents such as benzalkonium chloride act on the lipid membranes of microorganisms. After the detergents solubilize the membranes, the cytoplasm of the microorganism leaks out, and death follows.

ANTIBIOTICS

9.32 How do the antibiotics compare with the antiseptics and disinfectants?

Antibiotics are products of microorganisms that react with and inhibit the growth of other microorganisms. They are not laboratory chemicals as are the antiseptics and disinfectants. In some cases, antibiotics have been synthesized by chemists, in which case they are known as **chemotherapeutic agents.** However, the term antibiotic is better known by the lay public.

9.33 Who performed the early experiments in the discovery and development of antibiotics?

Paul Ehrlich, shown in Figure 9-3, was among the first to show that chemical substances could destroy microorganisms in the body. His so-called "magic bullet," which was developed in the early 1900s, was an arsenic derivative used to treat syphilis. **Alexander Fleming** reported on the antimicrobial effects of penicillin in 1928, but his discovery work was not developed for practical use until the late 1930s. The age of antibiotics blossomed in the period after World War II with the discovery of streptomycin, chloramphenicol, tetracyclines, and other mainstays of modern medicine.

9.34 How do antibiotics work to inhibit the growth of microorganisms and control their populations?

Antibiotics inhibit microbial populations by any of five major methods. They disrupt cell wall synthesis; they interfere with cell membrane function; they inhibit protein synthesis or interfere with its completion; they disrupt the functioning of nucleic acids; and they interrupt selected metabolic pathways.

9.35 What are the properties of an ideal antibiotic?

An antibiotic should be selectively toxic to pathogenic microorganisms. It should not incite an allergic response in the body by stimulating the formation of antibodies. It should not upset the normal microbial population of various body sites, and its use should not foster the development of drug resistance. If these conditions are met, the antibiotic is well-suited to the condition at hand.

9.36 What are the major kinds of antibiotics available?

There are two major kinds of antibiotics available: narrow-spectrum antibiotics and broad-spectrum antibiotics. **Narrow-spectrum antibiotics** control a narrow range of microorganisms, such as Gram-positive or Gram-negative bacteria, but not both. A narrow-spectrum antibiotic may be useful against fungi but not protozoa. **Broad spectrum antibiotics** kill a wide range of microorganisms such as both Gram-positive and Gram-negative bacteria. An example of a narrow spectrum antibiotic is penicillin; an example of a broad-spectrum antibiotic is tetracycline.

Fig. 9-3 A photograph of Paul Ehrlich, the German scientist who showed that chemical substances could be used to kill microorganisms in the body.

9.37 How can drug resistance develop in microorganisms?

The use of antibiotics over the last 50 years has led to the development of drug-resistant strains of bacteria. These bacterial strains always existed in the microbial population but they never needed to use their resistance mechanisms because they were never confronted with the antibiotic. With widespread antibiotic use, the susceptible bacteria died off rapidly and the surviving bacteria were those with resistance. They quickly multiplied to form populations of drug-resistant microorganisms.

9.38 What are the various ways that resistance can exist in microorganisms?

Various types of microorganisms can exhibit resistance in various ways. For example, they can release enzymes to inactivate the antibiotic before the antibiotic kills the microorganism. Furthermore, they can stop producing the drug-sensitive structure or modify the structure so that it is no longer sensitive to the drug. Microorganisms can also change the structure of the plasma membrane so that the antibiotic cannot pass to the cytoplasm.

9.39 Which are the major antibiotics in use in contemporary medicine?

There are many antibiotics currently in use, and the antibiotics fall into various "families." Various families of antibiotics are used for various types of microbial infections. For example, the penicillin family of antibiotics is used for Gram-positive bacteria.

9.40 How do the penicillin family of antibiotics inhibit or control a microbial population?

Penicillin antibiotics act on the cell walls of Gram-positive bacteria. They prevent these bacteria from forming peptidoglycan, the major component of the cell wall. Without peptidoglycan, the cell wall cannot form, and internal pressures cause the bacterium to swell and burst.

9.41 Are there any serious side effects to the use of penicillins?

Some people are allergic to penicillin antibiotics, and they may suffer localized allergic reactions or whole body allergic reactions known as anaphylaxis. This can be a deadly experience, so penicillin sensitivity must be recognized before the antibiotic is used. Long-term use of penicillin also encourages the emergence of penicillin-resistant bacteria.

9.42 Which antibiotics belong to the penicillin family?

The penicillins include penicillin G, penicillin X, penicillin O, and many synthetic and semisynthetic derivatives such as ampicillin, amoxicillin, nafcillin, and ticarcillin illustrated in Figure 9-4. Although originally derived from the mold *Penicillium*, most penicillins are synthetically produced. All are effective against Gram-positive bacteria and a few are used against Gram-negative bacteria.

9.43 Are there any other antibiotics that interfere with cell wall synthesis?

Members of the **cephalosporin** family of antibiotics are effective in preventing the synthesis of bacterial cell walls. Most traditional cephalosporins are useful against Gram-positive bacteria, but the newer cephalosporin antibiotics are also effective against Gram-negative bacteria. Cephalosporin antibiotics include cefazolin, cefoxitin, cefotaxime, cefuroxime, and moxalactam.

9.44 What is the difference between first- , second-, and third-generation cephalosporins?

Cephalosporin antibiotics have been produced in various forms over the decades, and the generations refer to the older or newer cephalosporins, the third-generation cephalosporins being the most recent. The newer cephalosporins are especially useful against penicillin-resistant bacteria and are often used as substitutes for penicillin. However, they may cause allergic reactions in the body, so they should be used with care. The original cephalosporin antibiotics were produced by the fungus *Cephalosporium.*

9.45 What are the aminoglycoside antibiotics and what are some examples of them?

The **aminoglycoside antibiotics** are inhibitors of protein synthesis in bacteria. Members of this family of antibiotics all have amino groups and carbohydrate molecules associated with the main molecules. The members include gentamicin, kanamycin, tobramycin, and streptomycin.

9.46 Which organism produces the aminoglycoside antibiotics and against which bacteria are the antibiotics useful?

The aminoglycoside antibiotics were originally isolated from members of the genus *Streptomyces*. This is a moldlike bacterium of the soil. The aminoglycosides are now produced synthetically or semisynthetically. They are primarily effective against Gram-negative bacteria and against certain Gram-positive bacteria. One aminoglycoside, streptomycin, is effective against the organism of tuberculosis.

9.47 Are there any side effects to use of the aminoglycoside antibiotics?

Many aminoglycoside antibiotics have an effect on the ear. They interrupt the physiology of hearing, and certain ones, especially streptomycin, can lead to deafness. Thus, they should be used with care.

9.48 Are the tetracycline antibiotics valuable in chemotherapeutic medicine?

The **tetracycline antibiotics** are extremely valuable because they inhibit the growth of Gram-negative bacteria, rickettsiae, chlamydiae, and certain Gram-positive bacteria. Therefore, they are broad-spectrum antibiotics.

9.49 Are there any drawbacks to the use of tetracycline antibiotics?

Compared to other antibiotics, the tetracycline antibiotics have relatively mild side effects. They may cause some nausea and vomiting, but the most obvious side effect is the destruction of helpful bacteria in the body. Also, the tetracycline antibiotics appear to interfere with calcium deposits in the body, so they should not be used in very young children, whose bones and teeth are forming.

9.50 What is the mode of action of tetracycline antibiotics and how are they formed?

The tetracycline antibiotics are inhibitors of protein synthesis in bacteria. They were originally isolated from members of the genus *Streptomyces*. They include such antibiotics as minocycline, doxycycline, and tetracycline.

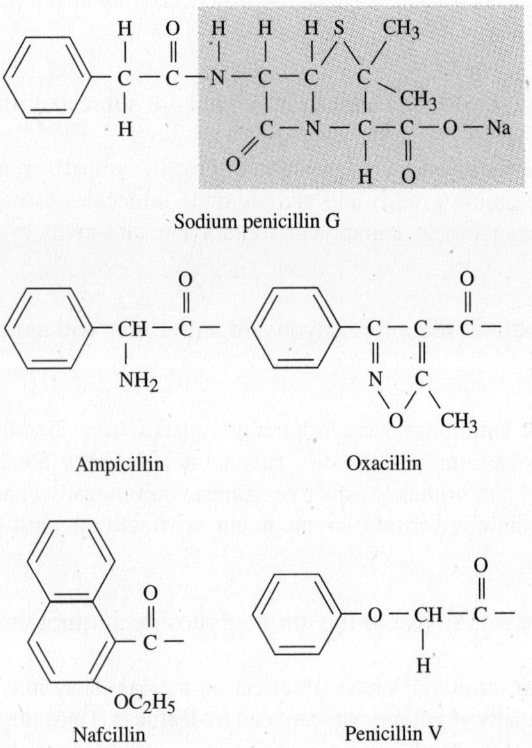

Sodium penicillin G

Ampicillin

Oxacillin

Nafcillin

Penicillin V

Fig 9-4 The chemical structure of the basic penicillin molecule and the side groups that form new penicillin antibiotics when added to the basic penicillin ring structure.

9.51 What are the major uses for the antibiotic erythromycin?

The antibiotic **erythromycin** is commonly used as a substitute for penicillin when penicillin sensitivity or penicillin allergy is displayed by the patient. Erythromycin is useful against Gram-positive bacteria and has been used against the organisms of Legionnaire's disease and mycoplasmal pneumonia. It is an inhibitor of protein synthesis.

9.52 Which are the major antituberculosis drugs now available for patient use?

Tuberculosis is a particularly difficult disease to deal with because the etiologic agent is an extremely resistant bacterium named *Mycobacterium tuberculosis*. Five drugs are currently useful for treating tuberculosis: **rifampin, ethambutol, streptomycin, para-aminosalicylic acid**, and **isoniazid**. The last antibiotic, isoniazid, is the most often prescribed, but resistance has emerged in the organism, and the other antibiotics are also used.

9.53 Are any antibiotics available for over-the-counter use on the skin?

Bacitracin is commercially available for the treatment of skin infections due to Gram-positive bacteria. This antibiotic inhibits cell wall synthesis in bacteria. It can be used internally by prescription but it may cause kidney damage.

9.54 What are the major uses for the antibiotic vancomycin?

Vancomycin is currently used for bacteria displaying resistance to penicillin, cephalosporin, and other antibiotics. Vancomycin is useful against Gram-positive bacteria, such as staphylococci, but it has serious side effects and is used only in life-threatening situations. It interferes with cell wall formation.

9.55 Are there any side effects to use of the antibiotic chloramphenicol?

Chloramphenicol, also known commercially as Chloromycetin, is effective against a broad range of bacteria including Gram-positive and Gram-negative bacteria, as well as rickettsiae and chlamydiae. However, it causes serious side effects such as aplastic anemia (blood cells without hemoglobin), and it may induce the gray syndrome in babies. Therefore, it is used on a restricted basis for only the most serious bacterial infections such as typhoid fever and meningococcal meningitis.

9.56 How do the sulfa drugs act to inhibit bacterial growth?

Sulfa drugs such as **sulfanilamide** and **sulfamethoxazole** are effective against bacteria that produce the substance folic acid. These bacteria are generally Gram-positive bacteria. In the production of folic acid, the bacterium obtains from the environment a substance called para-aminobenzoic acid (PABA). The sulfa drug resembles the PABA in structure and is taken up erroneously by the enzyme in folic acid production, as demonstrated in Figure 9-5. The enzyme is structurally altered by the sulfa drug and cannot function in the production of folic acid. Without folic acid, the bacterium soon dies. This is an example of how an antibiotic interferes with an important metabolic pathway in a bacteria.

9.57 Are there any antibiotics that are effective against fungi?

There are several antifungal antibiotics currently used in medicine. One example is **griseofulvin,** which is used against the fungi of ringworm and athlete's foot. Other examples are **nystatin, clotrimazole, ketoconazole,** and **miconazole,** all of which are used against vaginal infections due to *Candida albicans*. For systemic fungal infections such as cryptococcosis, the antibiotic **amphotericin B** is useful.

9.58 Are there any useful antibiotics for viruses?

Developing antibiotics for viruses is a difficult task because viruses have few functions or structures with which antibiotics can interfere. For example, they have no cell wall, they perform no protein synthesis, and they have no plasma membranes. However, there are certain antibiotics that interfere with viral replication and can be used for therapeutic purposes.

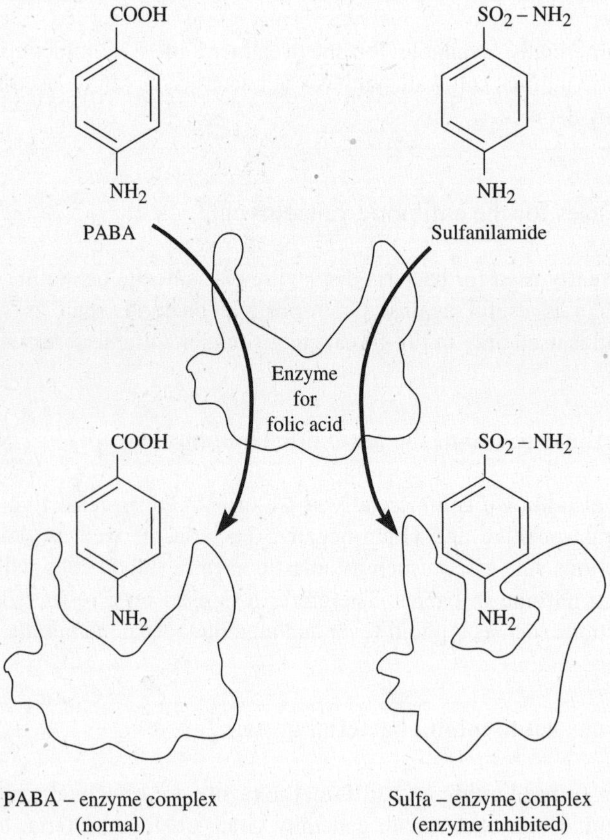

Fig 9-5 The mode of action of sulfa drugs. Both PABA and sulfanilamide have very similar chemical structures, so that the enzyme takes up the sulfanilamide erroneously, since it is in large supply. The enzyme is altered by the sulfa drug and becomes inoperative. Folic acid cannot be produced any longer.

9.59 What are some examples of antiviral antibiotics currently in use?

Among the antiviral antibiotics currently in use are **azidothymidine (AZT)**, which is used to interrupt the replication of human immunodeficiency virus (HIV); **acyclovir**, which is used against herpes viruses and chickenpox viruses; and **ganciclovir,** which is used against the cytomegalovirus. In addition, **amantadine** is prescribed against influenza viruses, and **interferon** has been used against rabies viruses and certain cancer viruses.

9.60 Are there any useful antibiotics for protozoal infections?

Many of the antibiotics used against bacteria are also useful against protozoa. For example, tetracycline is useful against various protozoa. Another drug, metronidazole (Flagyl), is useful against *Trichomonas vaginalis*. Quinine is used against malaria, and pentamidine isethionate is valuable against *Pneumocystis carinii*.

Review Questions

Multiple Choice. Select the letter of the item that correctly completes each of the following statements.

1. For purposes of microbial control, the term sterilization implies (*a*) the removal of pathogenic forms of microorganisms (*b*) the lowering of the microbial count (*c*) the destruction of all forms of life (*d*) the destruction of microorganisms only on the body surface.

2. The autoclave uses all the following conditions to kill microorganisms except (*a*) a steam temperature of 121° C (*b*) a time of 15 minutes (*c*) a pressure of 15 pounds per square inch (*d*) a volume of ten atmospheres.

3. The difference between a disinfectant and an antiseptic is that (*a*) a disinfectant is used on lifeless objects only, while an antiseptic is used on the skin surface (*b*) a disinfectant kills bacterial spores, while an antiseptic does not (*c*) a disinfectant does not achieve sterilization, and an antiseptic does (*d*) disinfectants are chemical substances, while antiseptics are not chemical substances.

4. Anything that lowers the microbial count of an environment is a (*a*) degerming agent (*b*) sanitizing agent (*c*) intermediary agent (*d*) oxidizing agent.

5. Dry heat is often used to sterilize (*a*) saline solutions (*b*) bacterial media (*c*) oily materials (*d*) hospital blankets.

6. The primary effect of ultraviolet radiations is on microbial (*a*) carbohydrates (*b*) enzymes (*c*) nucleic acids (*d*) cell walls.

7. In boiling water, nonsporeforming bacteria die within (*a*) one hour (*b*) a few seconds (*c*) three hours (*d*) four hours.

8. The process of tyndallization requires all the following conditions except (*a*) steam at 100°C (*b*) pressure of 10 pounds per square inch (*c*) a 30 minute period of time (*d*) exposure on each of three successive days.

9. Ethylene oxide is widely used to (*a*) sterilize plastics (*b*) disinfect table tops (*c*) kill bacteria on the skin surface (*d*) sterilize chemical solutions.

10. In order for alcohol to be used effectively as a control agent, it must be used as a chemical agent (*a*) a 10 percent solution must be used (*b*) a 5 minute exposure must take place (*c*) butyl alcohol must take place (*d*) complete immersion must take place.

11. In the clinical situation, silver nitrate is sometimes used (*a*) to prevent gonorrhea in the eyes of newborns (*b*) to prevent passage of streptococci into the blood after a strep throat (*c*) to sterilize the intestine before surgery (*d*) to flush out the respiratory tract of tubercle bacilli.

12. Mouth infections caused by *Candida albicans* are sometimes treated with the antiseptic (*a*) calcium hydroxide (*b*) formaldehyde (*c*) glutaraldehyde (*d*) Gentian violet.

13. Betadine, Wescodyne, and Prepodyne are all antimicrobial products containing (*a*) formaldehyde (*b*) hydrogen peroxide (*c*) calcium hydroxide (*d*) iodine.

14. Chlorhexidine and hexachlorophene share the characteristic of being (*a*) most useful against viruses (*b*) phenol derivatives (*c*) sterilizing agents (*d*) antibiotics.

15. All the following may be methods for the inhibition of microorganisms by antibiotics except (*a*) antibiotics disrupt cell wall synthesis (*b*) antibiotics interfere with cell membrane function (*c*) antibiotics prevent the release of energy from ATP (*d*) antibiotics inhibit the synthesis of protein.

16. The penicillin family of antibiotics is used primarily against (*a*) viruses (*b*) fungi (*c*) Gram-negative bacteria (*d*) Gram-positive bacteria.

17. The antibiotics kanamycin, streptomycin, and gentamicin all belong to the group known as (*a*) cephalosporins (*b*) moxalactams (*c*) aminoglycosides (*d*) tetracyclines.

18. The most serious side effect associated with tetracycline antibiotics is (*a*) severe allergy reactions (*b*) hearing difficulties (*c*) destruction of helpful bacteria in the body (*d*) kidney and liver damage.

19. Among the currently used drugs for treating tuberculosis are (*a*) isoniazid (*b*) erythromycin (*c*) vancomycin (*d*) chloramphenicol.

20. The drugs nystatin, chlotrimazole, and miconazole are all used against (*a*) viral infections (*b*) Gram-negative bacteria only (*c*) fungi (*d*) all types of bacteria.

Completion. Complete the following phrases with the word or words that best complete the sentence.

1. Before a chemical agent can be used effectively, the material to be treated must be free of ____________.

2. A disinfectant kills pathogenic microorganisms, but an agent that only reduces the microbial population to a safe level is known as a ____________.

3. In an ordinary oven, the amount of time necessary to effect sterilization at 160° C is about ____________.

4. Boiling water is an effective antimicrobial agent, but resistance may be displayed for over two hours by bacterial ____________.

5. In dry heat, the primary effect of the heat is to bring about the conversion of large molecules into smaller molecules by ____________.

6. Steam under 15 pounds per square inch pressure will kill microorganisms in 15 minutes in the instrument called the ____________.

7. Alternate heating periods shock spores and convert them into vegetative cells in the process of ____________.

8. Pasteurization is not a sterilizing process, but is used to lower the level of microorganisms in ___________.

9. When large amounts of sugar are added to food, the water is drawn out and a dry condition is established by the process of ___________.

10. Among the three gases able to achieve sterilization are ethylene oxide, beta propiolactone, and ___________.

11. Both isopropyl and ethyl alcohol can be used as antimicrobial agents at concentrations of ___________.

12. The metal available in stick forms to cauterize wounds is ___________.

13. Benzoic acid, salicylic acid, and undecylenic acids are all used as ointments to prohibit the growth of ___________.

14. Oxygen is suddenly released when a wound is treated with ___________.

15. An antibiotic used to kill a wide range of microorganisms is known as a ___________.

16. The cephalosporin family of antibiotics works against bacteria by preventing the synthesis of ___________.

17. Among the important producers of antibiotic is the moldlike bacterium of the soil belonging to the genus ___________.

18. Pentamidine isethionate is a valuable antibiotic for treating disease caused by ___________.

19. When treating viral diseases, the physician may use such drugs as AZT, ganciclovir, amantadine, and ___________.

20. Sulfa drugs such as sulfamethoxazole are effective against those bacteria that produce the substance ___________.

True/False. Enter the letter "T" if the statement is true in its entirety, or the letter "F" if the statement or any part of the statement is false.

___ **1.** The choice of a particular antimicrobial agent depends on such things as the kind of material to be treated, the kind of organism to be controlled, and the environmental conditions existing at the time of use.

___ **2.** Organic matter enhances the successful use of an antimicrobial agent.

___ **3.** Disinfection implies the removal of all living organisms, including microbial endospores.

___ **4.** The antimicrobial effect of antibiotics is generally less than that of disinfectants and antiseptics.

___ **5.** Exposure to boiling water brings about sterilization in a period of 30 minutes.

___ **6.** Dry heat is preferred to moist heat for the control of microorganisms when the material treated includes such things as powders, dry glassware and instruments, and oily materials.

___ **7.** Dry heat kills microorganisms by causing their proteins to separate as an insoluble mass as they revert from their three-dimensional structure to a two-dimensional structure.

___ **8.** Pasteurization is an effective way for preserving products such as milk, beer, and fruit juice, but it does not effect sterilization.

___ **9.** Microwaves have a very severe effect on microorganisms because they react with the nucleic acids of the organisms and change their structure.

___ **10.** Alcohol has minimal activity as an antiseptic because it dries very quickly.

___ **11.** Among the acids used to preserve foods and control the growth of microorganisms are acetic acid, propionic acid, and benzoic acid.

___ **12.** Chlorine is an oxidizing agent that reacts with amino acids in proteins and changes the nature of the protein, thereby killing microorganisms.

___ **13.** When hydrogen peroxide is applied to a wound, the enzyme amylase decomposes the hydrogen peroxide to form water and free oxygen.

___ **14.** Among the first disinfectants employed in microbiology was hexachlorophene, a chemical first used by Joseph Lister in the 1860s.

___ **15.** Those antibiotics that have been synthesized by chemists are known as chemotherapeutic agents.

___ **16.** Antibiotic-resistant bacteria do not exist in a microbial population until that population is confronted by an antibiotic, at which time the resistant strains form.

___ **17.** All antibiotics described as aminoglycosides act on bacteria by destroying their cell walls.

___ **18.** Doxycycline and minocycline are among the effective tetracycline antibiotics now in medical use.

___ **19.** Only one antibiotic has been found useful for treating tuberculosis, a serious disease of the human lungs.

___ **20.** Both bacitracin and vancomycin are useful against Gram-positive bacteria, while griseofulvin and nystatin are valuable when used against fungi.

Chapter 10

The Major Groups of Bacteria

OBJECTIVES

Among the bacteria, there is a wide variety of forms having an equally wide variety of pathological and physiological features. Many of those bacteria are surveyed in this chapter, and its objectives are to:

1. Identify the key features of the spiral bacteria and name several diseases caused by spirochetes and spirilla.

2. Explain the significance of species of *Pseudomonas, Legionella,* and *Brucella* species in medical bacteriology.

3. Define the critical role played by *Rhizobium* and *Azotobacter* species in agriculture.

4. Name several species of Gram-negative cocci and denote their significance in medical microbiology.

5. Compare the characteristics of bacterial species belonging to the broad family Enterobacteriaceae.

6. Discuss the diseases caused by species in the genera *Staphylococcus* and *Streptococcus.*

7. State the important characteristics of two species of bacterial sporeformers and one genus of acid-fast bacteria.

8. Discuss the importance of chemoautotrophic bacteria and photoautotrophic bacteria in the cycle of elements on Earth.

9. Recognize the important characteristics of rickettsiae, mycoplasmas, and chlamydiae that set these groups apart from other bacteria.

THEORY AND PROBLEMS

10.1 Which major text lists the species of bacteria and their characteristics?

The classification of bacteria is complex and extensive and thousands of species are involved. The official manual of bacteria is called ***Bergey's Manual of Systematic Bacteriology.*** Last published in 1984, the manual is divided into four volumes. It divides bacteria into certain logical groups based on such characteristics as form, shape, Gram reaction, and physical requirements for growth. We shall follow that general pattern in this chapter.

SPIROCHETES

10.2 What is the general description of spirochetes?

Spirochetes are coiled bacteria that resemble a telephone cord or spring. In some cases the spirochete is tightly coiled, while in other cases a spirochete resembles a stretched-out spring. Spirochetes may be located in the soil, in sewage and decaying matter, and in the human and animal body.

10.3 Bacterial rods move by rotating their flagella. Does this activity apply to spirochetes? If not, how do spirochetes move?

Spirochetes move freely by using a set of ultramicroscopic fibers called **axial filaments** located between the outer sheath and body of the spirochete. As axial filaments rotate, the cell rotates in the opposite direction and move through liquids. Many spirochetes move over 100 times their body length per second.

10.4 What are some of the general characteristics of spirochetes and how do those characteristics compare to other bacteria?

Spirochetes live under various environmental conditions. Many species are aerobic; other species are facultatively anaerobic; still others live anaerobically. Spirochetes are neither Gram-positive nor Gram-negative because they do not accept the dyes used in the Gram-stain procedure. They do not stain easily by any methods and are generally difficult to see in the laboratory. Darkfield microscopy is commonly used to observe them in live specimens.

10.5 Do spirochetes cause any human diseases of significance?

There are four that are caused by spirochetes. Members of the genus *Treponema* shown in Figure 10-1 cause **syphilis** and **yaws**; members of the genus *Borrelia* cause **relapsing fever** and **Lyme disease** (both of these diseases are transmitted by arthropods); members of the genus *Leptospira* cause **leptospirosis**, a blood disease found in animals.

10.6 What are the important differences between spirochetes and spirilla?

Some helical bacteria resemble spirochetes but do not have axial filaments. Rather, they possess flagella for motility. There may be a single flagellum at the pole of the bacterium or tufts of flagella at one or both poles. The helical bacteria are called **spirilla** (singular, spirillum).

10.7 What are some characteristics that distinguish the spirilla?

The spirilla are commonly found in water environments. Some spirilla can be unusually long, reaching up to 60 micrometers in length. Another spirillum, a member of the genus *Azospirillum*, lives in close association with many plants where it fixes nitrogen for the plant.

10.8 Are there any pathogenic spirilla?

Pathogenic spirilla include the species *Campylobacter jejuni*, the agent of **campylobacteriosis**, an intestinal disease accompanied by diarrhea and transmitted by contaminated milk, water, and food. Another species, *Helicobacter pylori*, has been implicated as a cause of stomach **ulcers** in humans. And *Spirillum minor* is the agent of **rat bite fever,** a blood disease transmitted during a rodent bite.

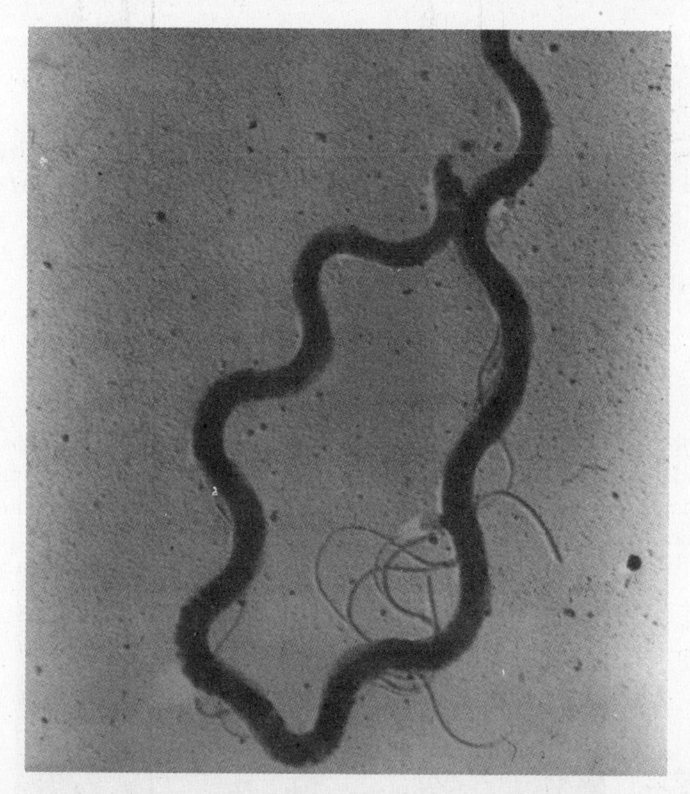

Fig. 10-1 A scanning electron micrograph of *Treponema pallidum*, the etiologic agent of syphilis. The axial filaments of the organism may be seen as hairlike bodies adjacent to the cell.

10.9 How do vibrios compare with other spirochetes?

Certain helical bacteria referred to as **vibrios** are curved bacteria in which the helix fails to make a complete turn. The bacteria resemble commas. One species, *Vibrio cholerae*, is well-known as the agent of **cholera** a serious intestinal disease with extensive diarrhea. Another genus of vibrios called *Bdellovibrio* are particularly interesting for their ability to multiply within other bacteria.

GRAM-NEGATIVE BACTERIA

10.10 What are some general characteristics of *Pseudomonas* species?

Among the Gram-negative rods are many species of aerobic rods, including members of the genus *Pseudomonas*. These bacteria are rod-shaped with flagella occurring at the poles. Many *Pseudomonas* species excrete extracellular pigments, which diffuse into the local environment. For example, *Pseudomonas aeruginosa* produces a blue-green pigment seen in burnt tissue and detected by ultraviolet radiation. Other species of *Pseudomonas* secrete their pigments in foods (causing blue-green rot in eggs). Among the unusual biochemical characteristics of *Pseudomonas* species is the ability to use nitrate instead of oxygen as a final electron acceptor. This process of anaerobic respiration yields considerable energy and can be a problem in agricultural circumstances because *Pseudomonas* species rob the soil of nitrates in this metabolism.

10.11 Why are members of the genus *Legionella* important in bacteriology?

At least six species of the Gram-negative rod *Legionella* have been identified since the species was first recognized in 1976 during the outbreak of **Legionnaires' disease** due to *Legionella pneumophila.* Legionellae as pictured in Figure 10-2 are cultivated in the laboratory with great difficulty and are known to be relatively common in the environment where water collects, such as in cooling towers, air conditioning systems, stagnant pools, and humidifiers.

10.12 What is the significance of members of the genus *Brucella*?

Members of the genus *Brucella* are well-known as the organisms of **brucellosis**. *Brucella* species are relatively common in animals, especially cows, pigs, and sheep. They are transmitted to humans in contaminated meat and animal products and they cause **undulant fever,** a blood disease accompanied by high and low periods of fever.

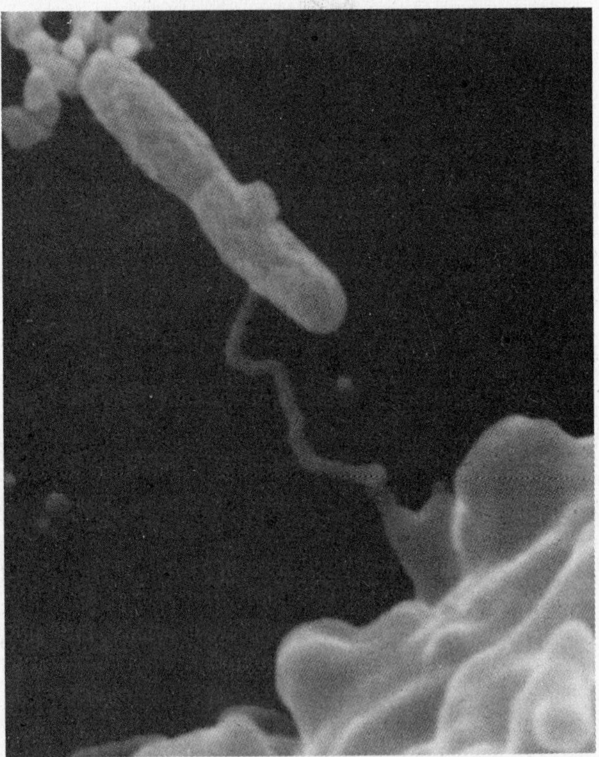

Fig. 10-2 A scanning electron micrograph of *Legionella pneumophila*, the etiologic agent of Legionnaires' disease. In this electron micrograph, the bacillus is attached to an appendage of a phagocyte as the first step to being engulfed. Phagocytosis is an essential body defense.

10.13 What critical role do *Rhizobium* species play in agriculture?

One of the most important Gram-negative rods in agriculture is the genus *Rhizobium*. *Rhizobium* species are **symbiotic**, that is, they live on the roots of legume plants (peas, beans, alfalfa, and clover) where they perform nitrogen fixation. In this process they trap nitrogen from the air and fix it into nitrogen compounds the plants can use in their metabolism. The plants form amino acids from the nitrogen compounds and become rich in protein. This protein is a source of food for all animals and humans.

10.14 How do *Azotobacter* species compare to *Rhizobium* species?

Another important nitrogen-fixing bacterium is species of *Azotobacter*. Unlike *Rhizobium* species, *Azotobacter* species live free in the soil. Here they also trap nitrogen and make it available to plants in usable nitrogen compounds. *Azotobacter* is a key element in the nitrogen cycle of the soil. Another Gram-negative rod, *Azomonas*, functions in a similar capacity.

10.15 Are any Gram-negative rods pathogenic to plants?

One plant pathogen is the Gram-negative rod *Agrobacterium tumefaciens*. This bacterium infects plants and causes a tumor-like growth called **crown gall**. DNA technologists have used the ability of *Agrobacterium* to insert its genes in the plant and have isolated its plasmids to deliver foreign genes to a plant.

10.16 Which Gram-negative rod causes pertussis?

Another Gram-negative rod of significance is *Bordetella pertussis*, the etiologic agent of **pertussis (whooping cough)**. Pertussis is a serious disease of young children characterized by blockage of the respiratory tract and possible suffocation. Two vaccines are currently available (DPT and DTaP), and the disease is considered under control in the United States.

10.17 Why is *Francisella tularensis* important in bacteriology?

The final Gram-negative rod in this section is *Francisella tularensis*, the cause of **tularemia**. Tularemia is a disease of the human blood, transmitted by arthropods in the fur of small game animals such as chipmunks, squirrels, and rabbits. The disease is not a fatal disease and can be treated with antibiotics.

10.18 Are any Gram-negative cocci significant in medical bacteriology?

Among the important Gram-negative bacteria are several genera of cocci. An example is the genus *Neisseria*. One species of neisseriae, *Neisseria gonorrhoeae,* is the cause of **gonorrhea**, the most reported disease in the United States today with approximately 650,000 cases per year. This sexually transmitted disease is not fatal, but it can lead to blockages of the reproductive tubes resulting in sterility. Another important *Neisseria* species is *Neisseria meningitidis*. This is the etiologic agent of **meningococcal meningitis**, a very serious and lethal disease of the spinal coverings. Transmitted by respiratory droplets, meningococcal meningitis must be treated early and aggressively to avoid death.

10.19 What significance do *Moraxella* species have in microbiology?

Moraxella species are Gram-negative somewhat coccobacillary organisms. One species, *Moraxella lacunata*, is the etiologic agent of **conjunctivitis (pinkeye)**, an inflammation of the conjunctiva.

10.20 Which anaerobic Gram-negative cocci cause human disease?

Veillonella species are Gram-negative cocci that live anaerobically within the **plaque** found among the teeth and gingival crevices in the oral cavity. The bacteria are not known to be pathogens, but when the tissues are traumatized, they can invade and cause gingival disease.

10.21 Are there any medically important facultatively anaerobic Gram-negative rods?

There are numerous species of Gram-negative rods that are facultatively anaerobic, that is, they grow in the absence of oxygen but many also grow in the presence of oxygen. Many genera cause disease of the gastrointestinal tract and other organs of the human body. The group is a large one and includes many well-known pathogens. Many of these pathogens belong to the broad family of bacteria called **Enterobacteriaceae**. These intestinal bacteria are commonly known as **enteric bacteria.**

10.22 Which facultatively anaerobic bacterium inhabits the human intestine?

The familiar bacterium *Escherichia coli* (*E. coli*) is a type of enteric bacterium. Most strains of *E. coli* live as harmless commensals in the human intestine, but there are certain strains that are considered pathogenic because they invade the tissues and produce toxins. These strains are said to be enteroinvasive and enterotoxic strains, respectively. One strain, *E. coli* 0157:H7, has been implicated in food-related outbreaks of intestinal disease in recent years.

10.23 What are some characteristics of members of the genus *Salmonella*?

Members of the genus *Salmonella* are among the most well-known members of the Gram-negative rods of the intestine. Most members in this genus are pathogenic, and such diseases as **typhoid fever** and **salmonellosis** (sometimes called salmonella food poisoning) are caused by them. The genus members are technically not known as species, but rather as serological types. Most have flagella, and many have capsules.

10.24 Which etiologic agent of bacterial dysentery is classified as an enteric bacterium?

Several species of *Shigella* are responsible for the intestinal disease known as **bacterial dysentery**, also known as **shigellosis**. Traveler's diarrhea may be due to this organism, and life-threatening illness can develop. Food and water are implicated in transmission.

10.25 Which pigmented bacterium is considered an enteric bacterial rod?

The small Gram-negative rod *Serratia marcescens* is notable for its production of **prodigiosin**, a bright red pigment that accumulates at room temperature. This organism, because of its pigment, has been used for many years as a test bacterium in experiments studying air currents. In recent decades, the organism has been identified as a cause of urinary and respiratory tract infections, so its use is restricted.

10.26 Which other pathogens are important Gram-negative facultatively anaerobic rods of the intestine?

Several other genera of bacteria are considered enteric bacteria. For example, *Klebsiella pneumoniae*, an encapsulated rod, is an important cause of respiratory tract, intestinal tract, and blood infections. Another rod, *Proteus vulgaris*, is an actively motile organism used in microbiology laboratories as a test bacterium to show motility. The organism *Yersinia pestis* is an enteric bacterium that causes **bubonic plague**. Members of the genus *Erwinia* are plant pathogens, and some cause soft-rot diseases. A final organism, *Enterobacter cloacae*, has been implicated in urinary and intestinal tract infections.

10.27 Which genus of enteric bacteria are notable for their slightly curved shape?

Members of the genus *Vibrio* are Gram-negative enteric bacterial rods with a slightly curved shape. Important pathogens in this group include *Vibrio cholerae,* the etiologic agent of **cholera,** and *Vibrio parahaemolyticus,* a cause of an intestinal tract infection.

10.28 Are there facultatively anaerobic rods that do not belong to the enteric bacteria group?

There are at least three genera of bacteria that are Gram-negative and facultatively anaerobic and live outside the gastrointestinal tract . Members of the genus *Pasteurella* are bloodborne and cause diseases in cattle, chickens, and dogs and cats. Members of the genus *Haemophilus* include the organisms of **bacterial meningitis** and **soft chancre**: The first disease occurs in the blood and central nervous system, while the second is a type of sexually transmitted disease. Finally, a member of the genus *Gardnerella* causes a sexually transmitted disease of the vagina called **vaginitis**.

10.29 Which Gram-negative rods live in a strictly anaerobic environment and have medical significance to humans?

Among the Gram-negative bacteria anaerobic species that have significance are members of the genera *Bacteroides* and *Fusobacterium. Bacteroides* species live in the oral cavity and gastrointestinal tract, where they can cause infections associated with wounds and surgery. Species of *Fusobacterium* are long and spearlike. They are often associated with infections of the gum and dental tissues developing under anaerobic conditions.

10.30 Which Gram-negative anaerobic rods are important in the soil environment ?

A number of Gram-negative rods live in anaerobic mud and sediment and participate in the **sulfur cycle**. Members of the genus *Desulfovibrio* are well-known in this respect. These organisms use sulfur rather than oxygen as a hydrogen acceptor and recycle the sulfur in soils by forming hydrogen sulfide (H_2S).

GRAM-POSITIVE BACTERIA

10.31 Which two genera of Gram-positive cocci have medical significance in humans?

Two genera of Gram-positive cocci important to humans are *Staphylococcus* and *Streptococcus.* Members of the genus *Staphylococcus* occur in grapelike clusters and include the organisms *Staphylococcus aureus*. This organism may be the cause of **abscesses, boils, and carbuncles**, as well as **toxic shock syndrome**. Members of the genus *Streptococcus* occur in chains and are the agents of **strep throat, scarlet fever**, **endocarditis**, and a type of pneumonia. Species of streptococci also are involved in **tooth decay,** and certain nonpathogenic species are used in the dairy industry to produce yogurt and other dairy products. Species of *Streptococcus* and *Staphylococcus* can be aerobic, facultatively anaerobic, or anaerobic.

10.32 Which genera of bacteria form endospores?

Endospore formation in bacteria occurs in two important genera: *Bacillus* and *Clostridium. Bacillus* species as presented in Figure 10-3 are aerobic or facultatively anaerobic Gram-positive rods that include the etiologic agent of **anthrax** (*Bacillus anthracis*). Another species is used as an important biological insecticide (*Bacillus thuringiensis*). The clostridia are obligate anaerobic Gram-positive rods whose members include the etiologic agent of **tetanus** (*Clostridium tetani*), **botulism** (*C. botulinum*), and **gas gangrene** (*C. perfringens*). These three diseases are among the most significant causes of death in humans.

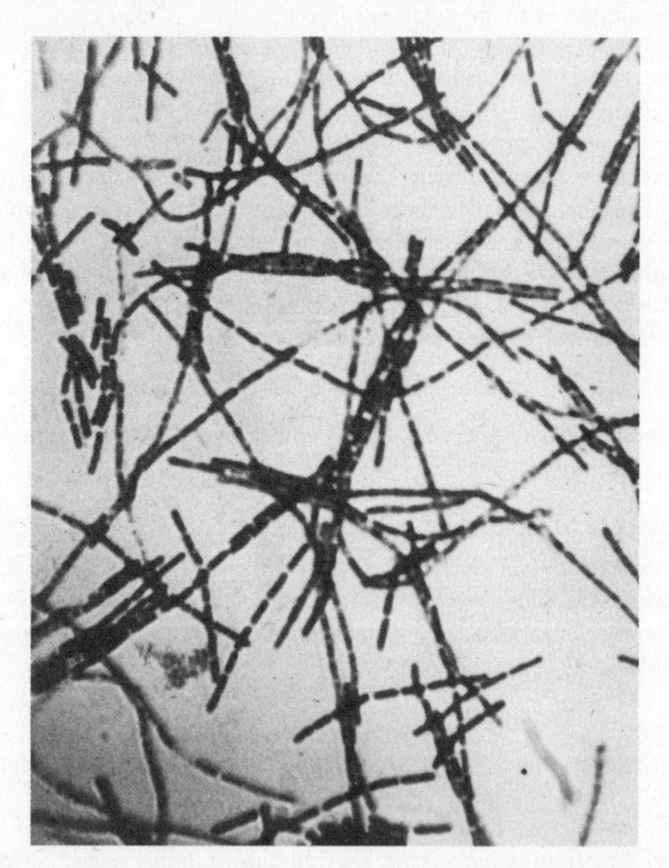

Fig. 10-3 A light photomicrograph of *Bacillus anthracis*, the etiologic agent of anthrax. The cells are seen in long chains, and oval spores are visible within some of the cells.

10.33 Are any Gram-positive rods unable to form spores?

There are several genera of Gram-positive rods that cannot form endospores. An important example is the genus *Lactobacillus*. Lactobacilli commonly grow under anaerobic conditions and produce lactic acid. In dairy processes, this lactic acid converts condensed milk to **yogurt**. Lactobacilli also grow in the human vaginal and intestinal tracts and produce acids that restrict the growth of other microbial populations. Lactobacilli are utilized in the commercial production of **sauerkraut** and **pickles**.

10.34 Are there any Gram-positive nonsporeforming pathogens?

There are several Gram-positive bacteria that cannot form endospores and are pathogenic. An example is *Listeria monocytogenes*, the agent of **listeriosis**. Dairy products are sometimes contaminated with this organism. Another example is *Corynebacterium diphtheriae*, the agent of **diphtheria**, which is a disease of the epithelial tissues of the respiratory tract. A final example is *Actinomyces israelii*, the cause of **actinomycosis**. This organism causes infections of the mouth and throat in both animals and humans.

ACID-FAST AND OTHER BACTERIA

10.35 What are the distinctive characteristics of members of the genus Mycobacterium?

Members of the genus *Mycobacterium* are neither Gram-positive nor Gram-negative. The bacteria are acid-fast, which means that after they have been stained with rigorous methods, they will not lose the stain even when washed with a dilute solution of acid-alçohol. Classified in the genus are two important pathogens: *Mycobacterium tuberculosis,* the agent of **tuberculosis,** and *Mycobacterium leprae,* the cause of **leprosy**. Members of this genus have mycolic acid and other fatty substances in their cell walls that prevent staining by ordinary staining methods. They are extremely difficult to cultivate in the laboratory and are generally difficult for researchers to work with. Figure 10-4 shows how they appear under the light microscope.

10.36 Are there any other acid-fast bacteria besides members of the genus Mycobacterium?

Members of the genus *Nocardia* are also acid-fast. These are anaerobic filaments of bacteria, which include the species *Nocardia asteroides.* Pulmonary infections due to this organism sometimes occur in AIDS patients.

10.37 Are any bacterial species capable of forming appendages?

Several genera of bacteria form stalks and buds as part of their life cycles. These stalks and buds protrude from the main organisms and are known as **prosthecae**. The stalk may anchor the organism to a surface and increase its ability to take up nutrients. Members of the genus *Caulobacter* are stalked. An unusual reproductive pattern is also seen in this genus.

10.38 Within the bacterial kingdom which genera has members able to glide or form buds?

Among the bacteria, there are at least two genera whose members glide over surfaces within a layer of slime. Species of *Cytophaga* and *Beggiatoa* belong with the gliding group. Another group of gliding bacteria called the **myxobacteria** are Gram-negative bacteria that move toward one another and converge to form an aggregate of cells. A fruiting body in the form of a stalk grows from the aggregate and forms spores known as myxospores. Each myxospore can germinate to form a new gliding cell.

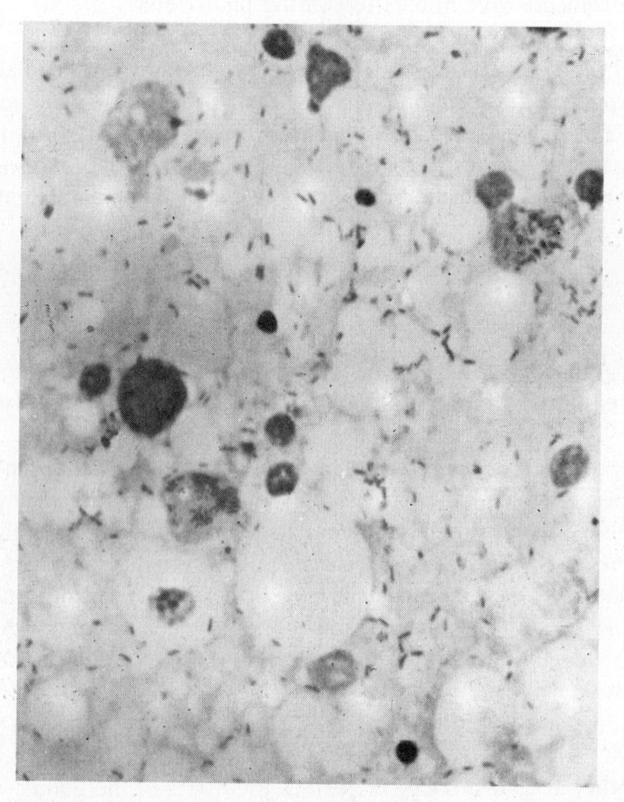

Fig. 10-4 A light photomicrograph of *Mycobacterium tuberculosis*, the etiologic agent of tuberculosis. The bacilli are seen in lung tissue after being stained by the acid-fast technique.

10.39 Are there any genera of bacteria whose members form buds or sheaths?

Among the bacteria, members of the genus *Hyphomicrobium* reproduce by forming tiny extensions called **buds**. The bud separates from the parent cell to produce a new cell during the reproduction process. Members of the genus *Sphaerotilus* are able to form a hollow sheath in which to live. The sheath is synthesized in fresh water and sewage, and it forms an important aspect of the organism's life cycle.

10.40 What is the importance of chemoautotrophic bacteria in the recycling of elements?

Chemoautotrophic bacteria synthesize their own foods and use energy derived from chemical reactions in the synthesis. The nitrifying bacteria are particularly important because they participate in nitrogen conversions in the soil and make the **nitrogen cycle** possible. Members of the genera *Nitrobacter* and *Nitrosomonas* are important in this respect. Another significant chemoautotroph is *Thiobacillus*. This bacterium uses sulfur compounds and participates in the **sulfur cycle** in the soil.

10.41 What are examples of the photoautotrophic bacteria and why are they important in elemental cycles?

Phototrophic bacteria are bacteria that synthesize their own foods and utilize sunlight energy for the synthesis reactions. They include the **purple bacteria**, the **green bacteria**, and the cyanobacteria. The purple and green bacteria are anaerobic species living in lake and pond sediments. They carry out photosynthesis in the absence of oxygen and they use sulfur compounds instead of water in the synthesis. Granules of sulfur are commonly present in their cytoplasm. The purple nonsulfur and green nonsulfur bacteria use acids and carbohydrates instead of sulfur compounds during the synthesis.

10.42 How do the cyanobacteria compare to the purple and green bacteria in their chemistry and physiology?

The **cyanobacteria** are aerobic species that live in lakes, ponds, and fresh and salt waters. They produce their carbohydrates using the energy of the sun through the process of photosynthesis. Specialized cyanobacterial cells called **heterocysts** contain enzymes that trap nitrogen gas and convert it into compounds for use by the cyanobacteria. The organisms divide by binary fission and are unicellular forms, although they may form colonies and filaments. Cyanobacteria (formerly called blue-green algae) are extremely important in the history of the Earth, because they produced much of the oxygen currently present in the atmosphere.

10.43 Is there any special importance to bacteria known as actinomycetes?

Actinomycetes are bacteria existing in long, tangled filaments. They are commonly found in the soil where their filaments improve the nutritional efficiency of the bacteria. Members of the genus *Frankia* fix nitrogen in the soil and contribute to the recycling of nitrogen, and members of the genus *Streptomyces* are producers of many important antibiotics.

10.44 What characteristics of rickettsiae set this group of bacteria apart from the other bacteria?

Rickettsiae are very small bacteria that measure about 0.45 μm in diameter. They occur in many shapes and are said to be **pleomorphic**. The organisms are nonmotile, they divide by binary fission, and they are usually transmitted by arthropods such as lice and ticks. Rickettsiae are important human pathogens and are the etiologic agents of **Rocky Mountain spotted fever**, **typhus**, and other diseases. One species, *Coxiella burnetii,* is an important contaminant of dairy products and causes **Q fever.**

10.45 What is the significance of the chlamydiae in medical microbiology?

Chlamydiae are a group of tiny bacteria measuring about 0.25 μm in diameter. They cause several important diseases such as **chlamydia** (a sexually transmitted disease), **psittacosis** (a respiratory disease), and **chlamydial pneumonia**. The organisms are pleomorphic and are difficult to cultivate in the laboratory. They have a developmental cycle that includes an infectious form called the **elementary**

body and a larger, less infectious form called the **reticulate body**. Chlamydiae multiply only within living cells.

10.46 What are the notable features of mycoplasmas?

Mycoplasmas are a group of very tiny bacteria measuring about 0.15 μm in diameter. They are pleomorphic and they lack a cell wall. Since they lack a cell wall, they assume myriad shapes as Figure 10-5 demonstrates, and they are not susceptible to the activity of penicillin. Mycoplasmas cause a form of walking pneumonia as well as a sexually transmitted disease of the reproductive tract called mycoplasmal urethritis.

10.47 What are the distinctive features of the archaeobacteria and why are these bacteria of particular interest to archeologists?

Archaeobacteria are a group of bacteria whose origins appear trace to back over three billion years. They are extremely tolerant to heat, acid, and toxic gases, and they are believed to have lived during the formative years of the Earth. Because their cell walls lack peptidoglycan, they are separated from most other bacterial groups and are sometimes placed in their own kingdom, Archaea. Archaeobacteria are found in environments that contain much methane, sulfur, acid, or heat. Many of these conditions existed during the early eons of the earth's existence.

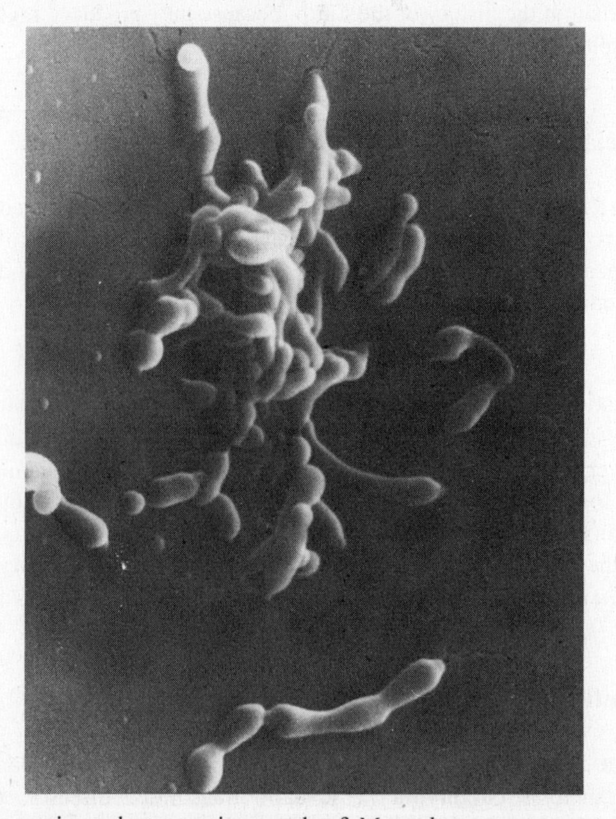

Fig. 10-5 A scanning electron micrograph of *Mycoplasma pneumoniae,* the cause of walking pneumonia. These bacteria have no cell walls, and they assume an infinite variety of shapes. Organisms as these are said to be pleomorphic.

Review Questions

Multiple Choice. Select the letter of the item that correctly completes each of the following statements.

1. All the following human diseases are caused by spirochetes except (*a*) syphilis (*b*) Lyme disease (*c*) food poisoning (*d*) leptospirosis.

2. The species *Campylobacter jejuni* causes (*a*) a blood disease with skin rash (*b*) an intestinal disease accompanied by diarrhea (*c*) a skin disease with local degeneration of tissue (*d*) a nervous system disease accompanied by paralysis.

3. All the following apply to members of the genus *Pseudomonas* except (*a*) they are Gram-positive cocci (*b*) they excrete extracellular pigments that diffuse into the local environment (*c*) they cause blue-green rot in eggs (*d*) they use nitrate instead of oxygen as a final electron acceptor.

4. Members of the genera *Azotobacter* and *Rhizobium* are important in nature because they (*a*) break down sulfur compounds (*b*) convert garbage into compost (*c*) cause serious diseases in plants (*d*) trap nitrogen from the air and fix it into usable compounds.

5. Members of the genus *Streptococcus* are responsible for cases of (*a*) neurological disease and plague (*b*) Scarlet fever and endocarditis (*c*) meningitis and diphtheria (*d*) food poisoning and boils.

6. Species of *Cytophaga* and Myxobacteria are notable for their (*a*) ability to produce pathogenic toxins (*b*) tendency to produce light-absorbing pigments (*c*) ability to glide (*d*) ability to form spores.

7. Chemoautotrophic bacteria are bacteria that derive their energy from chemical reactions and (*a*) use preformed organic matter in their metabolism (*b*) synthesize their own foods (*c*) are difficult to stain (*d*) cause human diseases such as tularemia.

8. All the following human diseases are due to members of the genus *Clostridium* except (*a*) botulism (*b*) tetanus (*c*) gangrene (*d*) tuberculosis.

9. All the following characteristics apply to the lactobacilli except (*a*) they grow in the human vaginal and intestinal tracts (*b*) they produce lactic acid in dairy products (*c*) they are aerobic Gram-negative cocci (*d*) they are used for the production of sauerkraut.

10. Members of the genus *Mycobacterium* share the characteristic of being (*a*) acid-fast (*b*) easy to cultivate in the laboratory (*c*) cocci in pairs (*d*) spirochetes.

11. The chlamydiae are distinguished by all the following characteristics except (*a*) they are tiny bacteria (*b*) their developmental cycle includes an elementary body and a reticulate body (*c*) they cause Rocky Mountain spotted fever and typhus (*d*) they multiply only in living cells.

12. Those bacteria believed to have lived during the earliest years of the Earth belong to the group (*a*) eubacteria (*b*) rickettsiae (*c*) archaeobacteria (*d*) clostridia.

13. Members of the genus *Desulfovibrio* are well-known for their participation in the (*a*) nitrogen cycle (*b*) phosphorous cycle (*c*) sulfur cycle (*d*) water cycle.

14. Enteric bacteria are those bacteria that live (*a*) in the soil (*b*) in the human nervous system (*c*) in the human respiratory tract (*d*) in the human intestinal tract.

15. The common characteristic among the organisms *Bordetella pertussis* and *Francisella tularensis* is that both bacteria (*a*) are Gram-positive (*b*) cause human disease (*c*) participate in the nitrogen cycle in the soil (*d*) are types of rickettsiae.

16. All the following characteristics apply to spirochetes except (*a*) they form spores (*b*) they do not stain easily in the laboratory (*c*) certain species cause human disease (*d*) they move freely by moving axial filaments.

17. Legionnaire's disease is caused by a bacterium that is common (*a*) where garbage collects (*b*) in the fecal matter (*c*) where water collects (*d*) in the upper reaches of the atmosphere.

18. *Agrobacterium tumefaciens* is noted for its ability to cause (*a*) decay of the tooth enamel (*b*) a lake to dry rapidly (*c*) plant disease (*d*) sulfur compounds to accumulate in the soil.

19. All members of the genus *Bacillus* are known for (*a*) their ability to break down sulfur-containing compounds (*b*) their ability to form spores (*c*) their ability to live in the absence of oxygen (*d*) the capsules they possess.

20. Bacteria of the genus *Sphaerotilus* are able to (*a*) convert water into acid in the environment (*b*) move by means gliding within slime (*c*) synthesize their own foods (*d*) form a hollow sheath to live in.

True/False. Enter the letter "T" if the statement is true. If the statement is false, replace the underlined word to make it true.

___ **1.** A member of the genus *Mycobacterium* is notable for its ability to cause <u>diphtheria</u>.

___ **2.** The rickettsiae are very small bacteria that occur in many shapes and are therefore said to be <u>autotrophic</u>.

___ **3.** Cyanobacteria are aerobic species of bacteria that produce their carbohydrates using the sun's energy through the process of <u>glycolysis</u>.

___ **4.** The bacteria causes of meningococcal meningitis and gonorrhea belong to the genus <u>*Staphylococcus*</u>.

___ **5.** The purple and green bacteria are anaerobic species that carry out photosynthesis in the absence of <u>oxygen.</u>

___ **6.** Among the important chemoautotrophs of the soil is Thiobacillus, which participates in the <u>sulfur</u> cycle.

___ **7.** The etiologic agent of bubonic plague is a Gram-negative bacterium known as <u>*Vibrio cholerae.*</u>

___ **8.** Members of the genus *Nocardia* are neither Gram-positive nor Gram-negative, but instead they are <u>acid-fast</u>.

___ **9.** Blockage of the respiratory tract and possible suffocation can occur in young children who are suffering from <u>salmonellosis</u>.

___ **10.** The plaque found among the teeth and gingival crevices in the human oral cavity often harbor species of <u>*Gardnerella*</u>.

__ **11.** Those bacteria commonly found in the human intestinal tract are known as pneumotropic bacteria.

__ **12.** The official manual listing known species of bacteria is *Bergey's Manual of Systematic Bacteriology*.

__ **13.** Spirochetes move about in the environment by using a set of ultramicroscopic fibers known as cilia.

__ **14.** Species of *Brucella* are relatively common causes of disease in plants.

__ **15.** Members of the genus *Azotobacter* and *Rhizobium* are well-known for their ability to fix nitrogen.

__ **16.** One of the causes of conjunctivitis (pink eye) in humans is the organism *Corynebacterium diphtheria.*

__ **17.** The Gram-negative rod *Serratia marcescens* is notable for its production of prodigiosin, a bright red pigment.

__ **18.** Archaeobacteria are separated from other bacterial groups in part because their cell walls lack glucose.

__ **19.** Long, tangled filaments of cells characterize the group of bacteria known as cyanobacteria.

__ **20.** Several species of the bacterial genus *Caulobacteria* are able to form a stalk to anchor the organism to a surface.

Matching. Match the genus of bacteria in Column A with the characteristic in Column B by placing the correct letter next to the genus name.

Column A	Column B
_____ 1. *Vibrio*	(*a*) Reproduces with buds
_____ 2. *Rhizobium*	(*b*) Actinomycete; produces antibiotics
_____ 3. *Serratia*	(*c*) Causes soft chancre and meningitis
_____ 4. *Veillonella*	(*d*) Cause crown gall in plants
_____ 5. *Treponema*	(*e*) Human stomach ulcers
_____ 6. *Bacillus*	(*f*) Measures 0.25 μm in diameter
_____ 7. *Haemophilus*	(*g*) Harmless commensal of intestine
_____ 8. *Spirillum*	(*h*) Produces the red pigment prodigiosin
_____ 9. *Corynebacterium*	(*i*) Converts milk to yogurt
____ 10. *Helicobacter*	(*j*) Lives anaerobically in dental plaque
____ 11. *Borrelia*	(*k*) Causes rat bite fever, spirochete
____ 12. *Hyphomicrobium*	(*l*) Gram-negative; slightly curved
____ 13. *Coxiella*	(*m*) Aerobic sporeformer; causes anthrax
____ 14. *Azotobacter*	(*n*) Lives in legume plant roots
____ 15. *Escherichia*	(*o*) A rickettsia; causes Q fever
____ 16. *Streptomyces*	(*p*) Causes diphtheria
____ 17. *Agrobacterium*	(*q*) Test organism; actively motile
____ 18. *Proteus*	(*r*) Causes syphilis; spirochete
____ 19. *Chlamydia*	(*s*) Lives free in soil; traps nitrogen
____ 20. *Lactobacillus*	(*t*) Causes Lyme disease

Chapter 11

The Fungi

OBJECTIVES

Fungi are a group of eukaryotic microorganisms that have industrial as well as medical importance. The objectives of this chapter are to:

1. Discuss the broad characteristics that apply to all fungi.

2. Examine the microscopic and macroscopic structures found in molds and yeasts.

3. Describe the nutritional patterns and environmental characteristics necessary for the proliferation of fungi.

4. Summarize the reproductive methods of fungi and recognize the importance of spores in the life of the fungus.

5. Compare the important characteristics of the six classes of fungi.

6. Briefly discuss several diseases of fungal origin.

THEORY AND PROBLEMS

11.1 How do the fungi relate to bacteria and other microorganisms?

Fungi are considered eukaryotes and are grouped in a different category than bacteria, which are prokaryotes. As noted in Chapter 4, the cells of eukaryotes have nuclei with multiple chromosomes and a nuclear membrane, as well as a series of organelles, and other distinguishing features. In this regard, fungi are in the same category of organisms as simple algae (Chapter 13) and protozoa (Chapter 12).

11.2 Which types of organisms are considered fungi?

Various organisms are considered fungi and are classified in the Kingdom Fungi in the Whittaker classification system. The organisms include yeasts, molds, truffles, morels, mushrooms, and numerous other organisms. In addition, many biologists classify the slime molds with fungi (although many other biologists consider them more like protozoa).

CHARACTERISTICS OF FUNGI

11.3 Which characteristics apply to fungi?

In general terms, fungi are distinguished from other eukaryotes because they possess rigid cell walls. Also, they have a type of metabolism in which energy is obtained from chemical reactions (rather than from photosynthesis); moreover, they have a heterotrophic mode of nutrition based on absorption of simple molecules from the external environment; and they have no chlorophyll. Some fungi are unicellular, while others are multicellular. They are generally nonmotile. An exception to this pattern is the slime molds, which have a motile stage reminiscent of protozoa. They will be considered briefly in this chapter.

11.4 How do the cell walls of fungi compare to those of other microorganisms?

The cell walls of fungi are chemically unique. They contain **chitin**, which is a polymer of acetylglucosamine, as well as a small amount of cellulose and polymers of glucose called glucans.

11.5 Are microscopes necessary to observe the fungi?

Some fungi such as yeasts are microscopic, while other fungi such as molds and mushrooms are macroscopic, and a microscope is not required to observe their morphological details. Yeasts are unicellular eukaryotes, while molds and other fungi are multicellular. In multicellular fungi, the cells occur in branching filaments called **hyphae**. A mass of filaments is called a **mycelium**. Cross walls separate the cells in the hypha of many species of fungi.

11.6 What are the different kinds of hyphae in a mold?

Two basic kinds of hyphae exist: the **vegetative hyphae**, which anchor the mold to a substratum and are used for absorbing nutrients; and the **aerial hyphae**, which forms the reproductive structures of the fungus. These structures are shown in Figure 11-1, together with other general details of a typical fungus. Spores formed at the tip of the aerial hypha usually give color to the fungal mycelium.

11.7 Is the multicellular nature of the fungus similar to that of plants?

The cross walls between fungal cells are called **septa** (singular, **septum**). The septa are often perforated and the cytoplasm mingles among adjacent cells. The nuclei may also move among the cells, so the multicellular nature of a fungus is not analogous to that of the plant, where the cells are fully separate and independent.

11.8 **Do all fungi have divisions between the cells in their hyphae?**

Not all species of fungi have septa. Some fungi lack these divisions, and are called **nonseptate**, or **coenocytic** hyphae. The common bread mold *Rhizopus stolonifer* is an example of a nonseptate (coenocytic) fungus.

11.9 What is a general description of yeast cells?

Yeast cells are unicellular fungi that do not form hyphae or mycelia. Yeast cells are oval cells measuring 5 to 10 μm in diameter. Their structures are visualized only with the microscope. Yeast reproduce sexually as well as asexually.

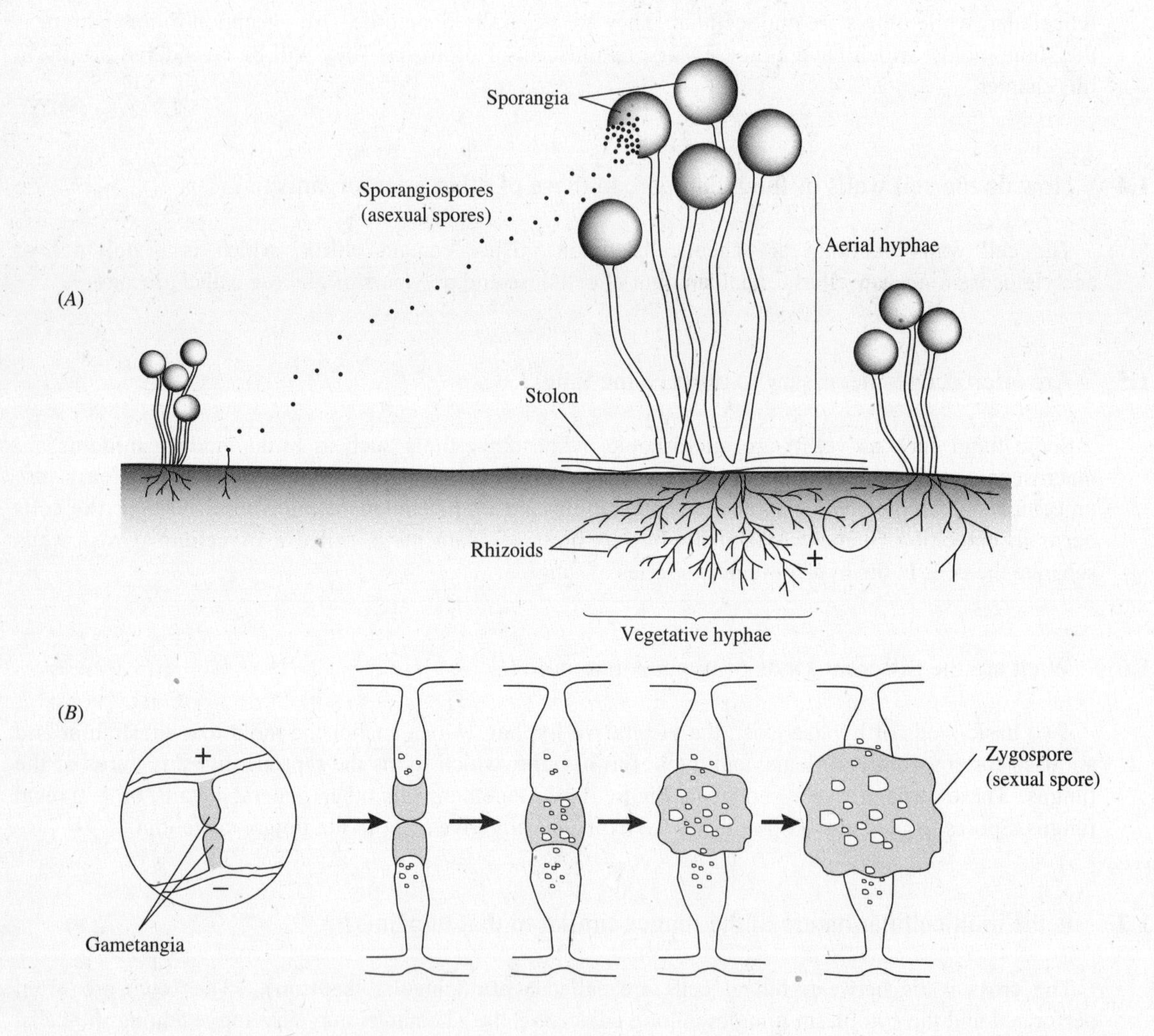

Fig. 11-1 Some of the general characteristics applying to fungi. (*A*) The vegetative and aerial hyphae of a zygomycete are shown. The asexual spores are each capable of reproducing the fungus. (*B*) The sexual reproductive stage of a fungus in which vegetative hyphae unite and a sexual spore forms at the point of the fusion. This form of reproduction occurs in the class Zygomycetes.

11.10 Which fungi have both yeast and mold forms?

Certain fungi can form a mycelium under certain environmental conditions, then revert to a yeast form under other environmental conditions. These fungi are said to be **biphasic** or **dimorphic**. An example of a biphasic fungus is ***Candida albicans,*** the cause of "yeast disease" in females. In the body, this organism grows in oval, microscopic yeasts, but in the laboratory it forms hyphae.

PHYSIOLOGY AND REPRODUCTION OF FUNGI

11.11 Which nutritional pattern is present in the fungi?

Fungi are **heterotrophic** microorganisms, that is, they utilize preformed organic molecules for their nutrition. Fungal species secrete digestive enzymes into the surrounding environment and break down large organic molecules into small ones that can be absorbed. They obtain their energy from chemical reactions occurring in the cell. Most fungi live in the presence of oxygen.

11.12 Can fungi tolerate extreme environments?

Certain species of fungi tolerate high sugar concentrations, high acid environments, and extremely cold temperatures. It is not unusual, for example, to find a fungus growing in a cup of sour cream in the refrigerator.

11.13 Are there any anaerobic fungi?

Among the most familiar anaerobic fungi are the yeasts. These organisms are facultative anaerobes, that is, they metabolize nutrients in the presence of oxygen, but when oxygen is lacking they can revert to perform their metabolism anaerobically and utilize the process of **fermentation**. Fermentation is a chemical process in which energy is obtained during the anaerobic breakdown of glucose molecules through the chemistry of glycolysis.

11.14 Are there any important industrial products of the fermentation process occurring in yeast cells?

When yeast cells metabolize glucose during fermentation they break down the carbohydrate, release its energy, and form ethyl alcohol and carbon dioxide. In the baking process the carbon dioxide helps dough to rise, and in the liquor industry, the ethyl alcohol is the basis for beer, wine, and liquor. Thus, the yeasts are essential organisms in many industrial processes. An important species is ***Saccharomyces cerevisiae,*** the common baker's and brewer's yeast.

11.15 How do fungi reproduce?

Most fungi reproduce by sexual as well as asexual modes. In both cases, the important feature of the reproduction is a single cell known as a **spore**. In contrast to the bacterial spore used as a resistance mechanism, the fungal spore is essential to reproduction. A single spore is capable of regenerating the entire mycelium of a fungus.

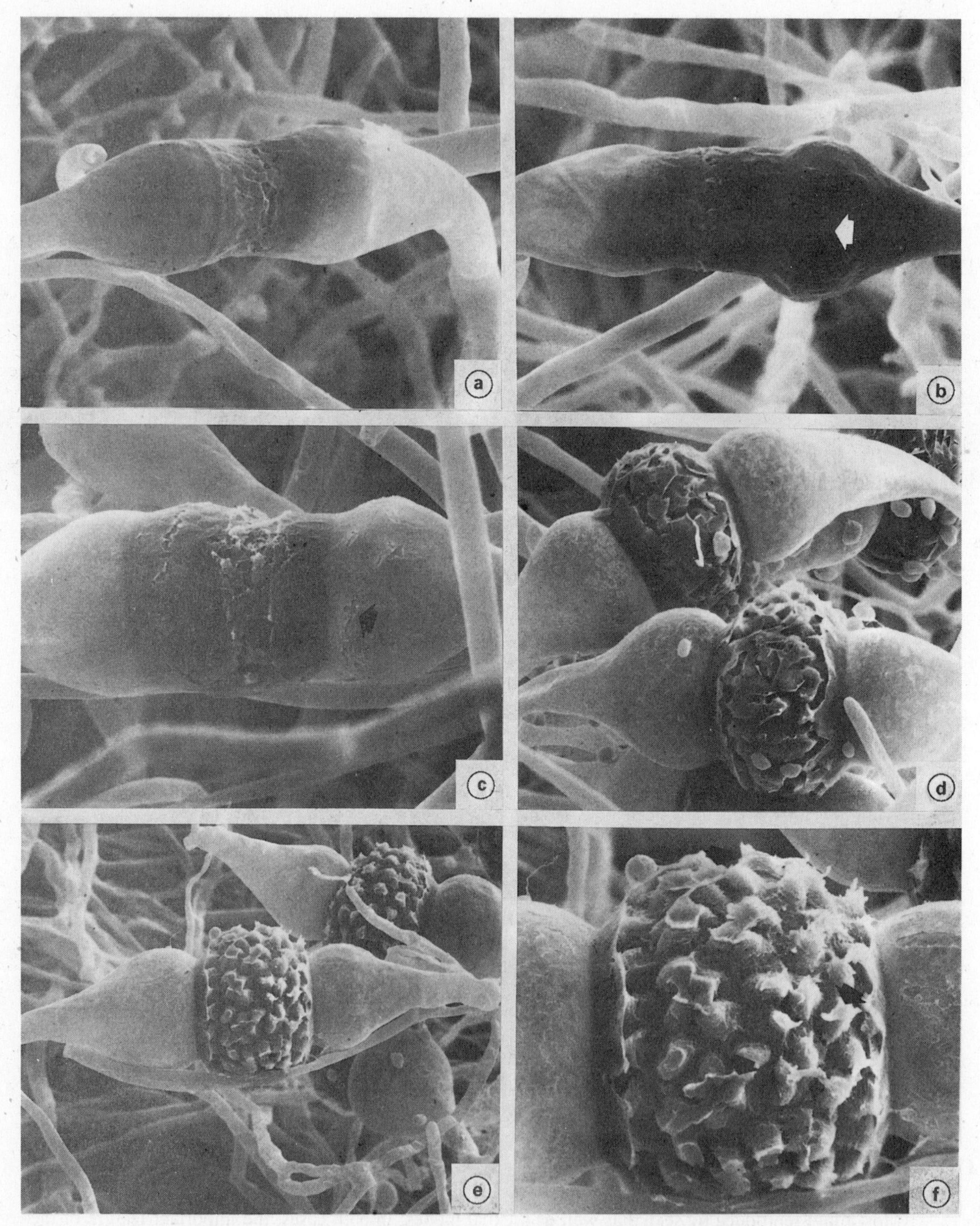

Fig. 11-2 A series of scanning electron micrographs showing the development of the zygote. The zygote is formed during sexual reproduction in the fungus. It is a diploid cell formed from the fusion of haploid gametes produced by parent fungal cells.

11.16 How does the asexual mode of reproduction occur in fungi?

The **asexual reproductive** mode in fungi occurs at the tips of aerial hyphae. Mitosis occurs here, and the daughter cells develop into spores that move off on wind currents and land at distant locations. The spore then germinates, and new cells emerge to form filaments and reproduce the fungus. Asexual reproduction occurs without a union of sexual cells, and asexually produced spores are generally very numerous. All the spores are genetically identical.

11.17 What are the various kinds of asexual spores that a fungus can produce?

A fungus can produce various kinds of asexual spores depending on its genes and their expression. For example, an **arthrospore** is a spore formed by fragmentation of the tip of the hyphae, while a **blastospore** is produced as an outgrowth along a septate hypha. **Conidiospores** are unprotected spores formed by mitosis at the tips of the hyphae, and **sporangiospores** are spores produced within a sac called a sporangium.

11.18 Is spore formation the only mode of asexual reproduction in the fungi?

A type of asexual reproduction not involving spores occurs in yeast cells. The yeast cell undergoes mitosis and forms a tiny cell at its border. This cell, called a **bud**, increases in size and eventually separates from the parent cell. The process is known as **budding**.

11.19 Describe the general process for sexual reproduction in species of fungi?

Sexual reproduction begins with the production of haploid gametes, each containing a single set of chromosomes. The gametes are produced by parent fungal cells of opposite sexual types. Gamete production occurs by the process of meiosis. The gametes then fuse to form a diploid cell called a **zygote**. Details of the fusion and the development of the zygote are shown in Figure 11-2. The zygote sometimes remains with the parent mycelium and undergoes multiple cell divisions by mitosis to develop into a genetically different mycelium. Sometimes the zygote will break free and establish a new mycelium elsewhere.

11.20 Can separate gametes form in a single hypha?

A single mycelium may contain both male and female cells, the sexually opposite cells that produce male and female gametes. The terms "plus" and "minus" are used interchangeably with "male" and "female." In some fungi, different mating types exist, so an individual mold may be of the male (plus) or female (minus) type.

CLASSIFICATION OF FUNGI

11.21 How is the reproductive process related to the classification system for fungi?

The fungi are so diverse and complex that one classification scheme is not universally accepted. A commonly used classification scheme is based on how spores are formed during sexual reproduction.

According to one widely accepted scheme, fungi are separated into the divisions **Eumycophyta** and **Myxomycophyta**. Members of the Myxomycophyta are slime molds. The Eumycophyta, or true fungi, are subdivided into six classes depending on the method for producing their sexual spores.

11.22 What are the six named classes of the Eumycophyta?

The division Eumycophyta contains the following classes: Oomycetes, Chytridiomycetes, Zygomycetes, Ascomycetes, Basidiomycetes, and Deuteromycetes. Sexual reproduction occurs in the first five classes, but it has not been observed in members of the class Deuteromycetes. Species in this class multiply by an asexual method only.

11.23 What are the characteristic features of members of the class Oomycetes?

Members of the class **Oomycetes** are waterborne fungi that produce motile, flagellated **zoospores** during asexual reproduction. Certain species also produce motile zoospores during the sexual mode of reproduction, but other nonmotile species also exist in the class. In the asexual process, the zoospore is formed within a sporangium. A sexual spore called the **oospore** is the major reproductive structure of members of the this group. The potato blight related to the Irish potato famine was caused by a member of the Oomycetes group called ***Phytophthora infestans.***

11.24 How do the features of the Chytridiomycetes differ from those of the Oomycetes?

Like the Oomycetes, the **Chytridiomycetes** form zoospores that are flagellated and motile. In the asexual life cycle, a chytridiomycete zoospore settles on a solid surface and forms a special branching system of hyphae called **rhizoids.** Rhizoids anchor the fungus to a land environment. Oomycetes, by comparison, are aquatic. Growth of the chytridiomycete is accompanied by formation of a spherical zoosporangium. **Zoospores** develop within the zoosporangium. Each zoospore then germinates to form a new mycelium. Sexual reproduction in the Chytridiomycetes involves male and female gametes that unite and form a sexual spore called the oospore, as in Oomycetes.

11.25 Which are the typical characteristics of fungi of the Zygomycetes class?

Fungi of the class **Zygomycetes** have nonseptate hyphae and produce **sporangiospores** within sacs called sporangia. In the sexual reproductive cycle, the cells of sexually opposite hyphae fuse and develop to sexual spores called **zygospores**. The zygospores then move off to a surface such as a slice of bread and germinate to reproduce the fungus. ***Rhizopus stolonifer***, the common bread mold, is a zygomycete having anchorlike rhizoids and strong surface structures called stolons.

11.26 Are there any industrial uses for members of the Zygomycetes class of fungi?

Various members of the Zygomycetes, including members of the genus ***Rhizopus***, are used by industrial organizations. The fungi are used to ferment soy beans into soy sauce, to modify steroids in the production of cortisone and contraceptives, and in the production of organic acids such as citric and acetic acids. In these industrial processes, scientists take advantage of unique enzyme systems in the fungus or metabolic deficiencies that lead to accumulation of a particular end product.

11.27 Which features distinguish the Ascomycetes?

The **Ascomycetes** produce a sexual spore called the **ascospore**. Ascospores form within microscopic sacs called asci (singular, **ascus**), which are found on macroscopic structures known as **ascocarps**. Typically, there are eight ascospores per ascus. The ascus develops from the fusion of sexually opposite hyphae. Within the Ascomycetes group are *Saccharomyces cerevisiae* (the brewing and baking yeast) and the organism that causes ergot disease, ***Claviceps purpurea.*** Many dimorphic fungi are classified as Ascomycetes.

11.28 Which characteristics separate the Basidiomycetes from the other fungi?

The **Basidiomycetes** are a diverse class of mushrooms, puffballs, morels, and truffles known collectively as **basidiocarps**. During their sexual reproductive cycles, the fungi in this group form spore-bearing structures called **basidia** (singular, **basidium**). On these structures, a number of sexual spores called **basidiospores** form. Basidiospores are spread about in the wind and germinate to reproduce the fungus.

11.29 Does the common mushroom fit the description of a fungus?

The common **mushroom** is a tightly packed mycelium formed from innumerable hyphal strands. On the underside of the mushroom cap, a series of basidiospores form on basidia in the reproductive cycle. As the spores spread about, each germinates and reproduces the entire mushroom.

11.30 Are all mushrooms safe to eat?

Mushrooms can be safe to eat or they can be extremely poisonous. For example, members of the genus ***Amanita*** are very toxic to humans and cause deadly mushroom poisoning, while members of the genus *Agaricus* are safe to eat. ***Agaricus*** species are commonly sold as supermarket mushrooms.

11.31 Are there any human pathogens in the Basidiomycetes group?

Many agricultural pests including the **rust** and **smut** fungi are classified as Basidiomycetes. Rust and smut diseases occur in corn, wheat, and other agricultural crops. Other than the toxic mushrooms, there are very few human pathogens in the Basidiomycetes group.

11.32 How does the class Deuteromycetes differ from the other fungi?

Fungi currently classified in the class **Deuteromycetes** do not have a known sexual stage of reproduction. The sexual cycle may exist, but scientists have not observed it. Usually when the cycle has been described, the fungus is placed in one of the other classes. Members of the Deuteromycetes reproduce solely by an asexual method. This may involve fragmentation of the hyphae or the production of asexual spores. Because the fungi have no known sexual cycle, they are referred to as "imperfect fungi" or Fungi Imperfecti.

11.33 In which classes of fungi are many of the human pathogens classified?

Pathogens are found in all classes of the fungi. Numerous human pathogens are currently classified as Deuteromycetes, including *Candida albicans*, the cause of thrush, "yeast" infection, and other maladies. Also classified here is ***Cryptococcus neoformans***, the cause of spinal meningitis, and the fungi that cause histoplasmosis, blastomycosis, and coccidioidomycosis.

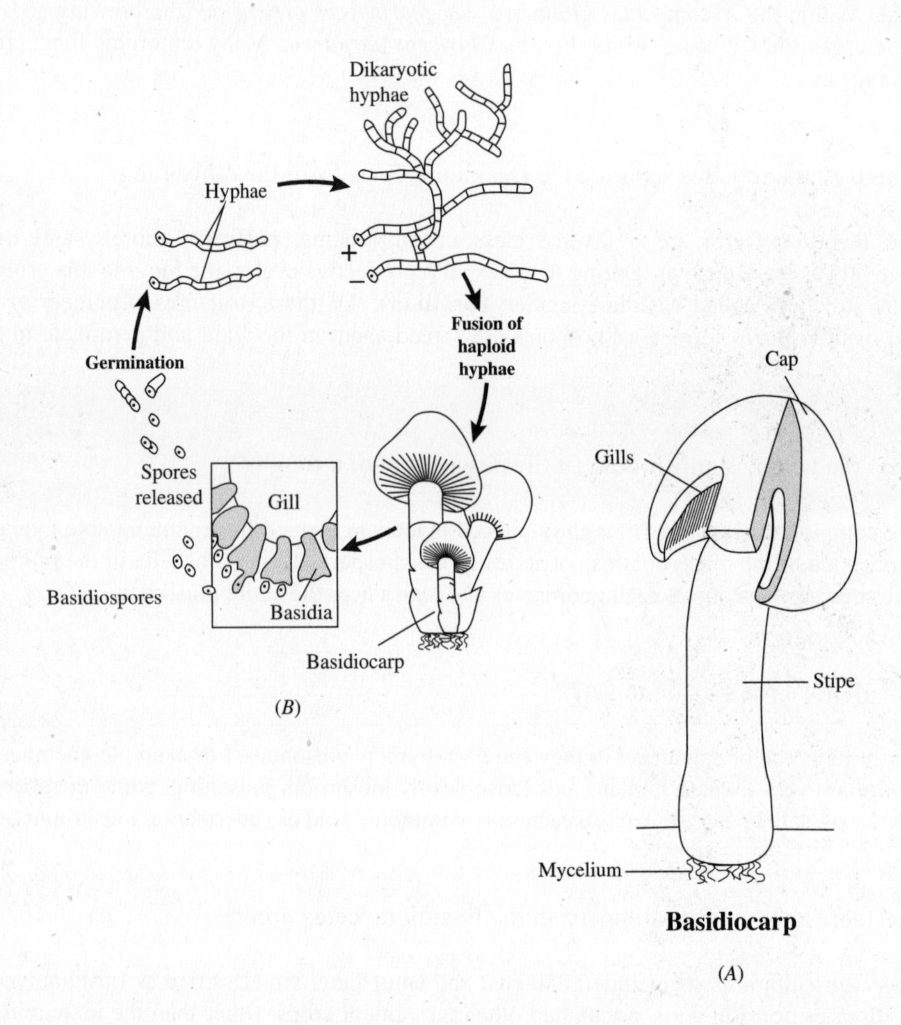

Fig. 11-3 Details of the common mushroom, a member of the class Basidiomycetes. (*A*) The macroscopic features of a mushroom showing the mass of tightly packed mycelium called the stipe. The basidia are located along the gills on the underside of the mushroom cap. (*B*) The life cycle of the mushroom beginning with fusion of sexually opposite hyphae and the development of a mycelium and a basidiocarp. Basidiospores are sexually produced spores typical of the Basidiomycetes.

11.34 Do any fungal species live in association with other organisms?

Many species of fungi exist together in nature with other organisms in symbiotic relationships. This symbiosis, called **mutualism,** is mutually beneficial to the fungus and the other organism. One example of mutualism is the **lichen**. In this relationship, fungi coexist with simple algae. The algal cells perform photosynthesis and provide carbohydrates for themselves and the fungus; the fungus shields the algal cells from high intensity light and protects it from drying out, while providing valuable nutrients. Lichens are crusty, colored growths found on many rocks and tree trunks.

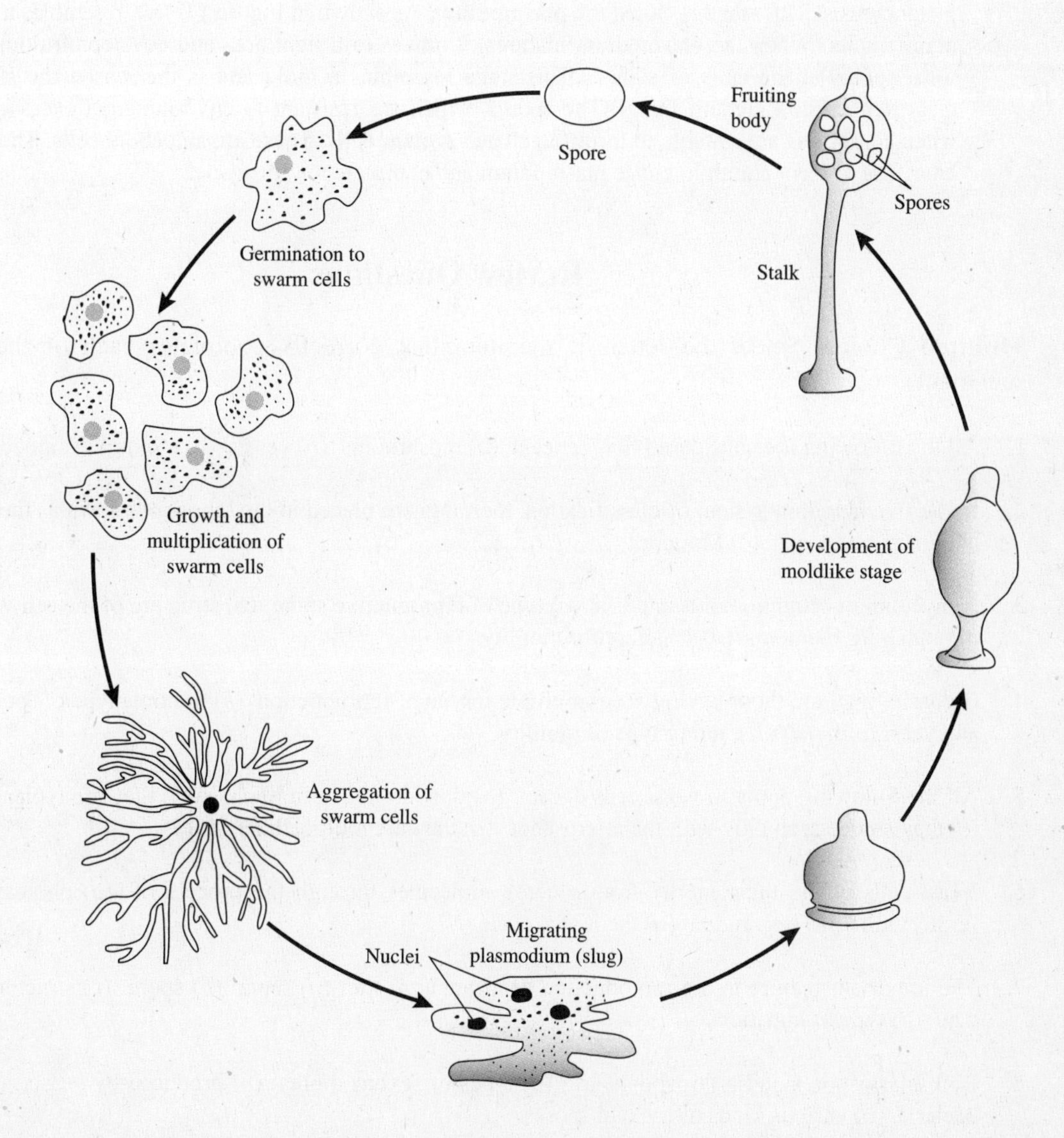

Fig. 11-4 Life cycle of the slime mold.

11.35 What is another example of fungi in a mutualistic relationship?

Another example of mutualism is the **mycorrhiza**, an association between a fungus and the root of a plant. Mycorrhizae happen on the roots of many forest trees. The fungus obtains many nutrients from the plants, and, in turn, helps the plant root absorb soil nutrients.

11.36 What are the general characteristics of members of the division Myxomycophyta?

Members of the division Myxomycophyta are **slime molds**. Under certain conditions, the slime mold exists as a mass of cytoplasm similar to an amoeba. It moves over rotting logs or leaves and feeds by phagocytosis. This stage is called the **plasmodium**. As shown in Figure 11-4, it resembles a slug and has many nuclei. When the plasmodium matures, it moves to a light area and develops fruiting bodies that form spores at the ends of stalks. This stage resembles a mold and is the reason the slime mold is classified in the kingdom Fungi. The spores, which are resistant to environmental excesses, germinate when conditions are suitable to form flagellated **swarm cells**. These are amoeboid cells. The swarm cells later fuse to form a multinucleate plasmodium and complete the life cycle.

Review Questions

Multiple Choice. Select the letter of the item that correctly completes each of the following statements.

1. All the following are considered fungi except (*a*) mushrooms (*b*) yeasts (*c*) molds (*d*) amoebas.

2. In the five kingdom system of classification, the fungi are placed in the kingdom known as the (*a*) Protista (*b*) Fungi (*c*) Algae (*d*) Monera.

3. A mycelium occurs in molds and is a (*a*) type of reproductive spore (*b*) structure of the cell wall (*c*) mass of branching filaments (*d*) structure for motility.

4. Biphasic fungi are those having a (*a*) multiple format of reproduction (*b*) multiple type of spores (*c*) mold and yeast form (*d*) alternating type of motility.

5. All the following apply to yeast cells except (*a*) they do not form hyphae (*b*) they are typically oval cells (*c*) they can be seen only with the microscope (*d*) they are multicellular fungi.

6. Yeast cells obtain their energy from glucose molecules through the process of (*a*) photosynthesis (*b*) fermentation (*c*) respiration (*d*) the Krebs cycle.

7. The important feature in the reproduction of a fungus is the (*a*) septa (*b*) spore (*c*) structure of the cell wall (*d*) type of nutrition.

8. Both blastospores and arthrospores are (*a*) types of sexual spores (*b*) produced by yeasts (*c*) similar to bacteria (*d*) various kinds of asexual spores.

9. The process of mitosis occurring at the tips of aerial hyphae accounts for the (*a*) motility of the fungus (*b*) synthesis of energy compounds by the fungus (*c*) production of asexual spores (*d*) production of ATP in the fungus.

10. Members of the fungal groups Oomycetes and Chytridiomycetes are both characterized by (*a*) production of zoospores (*b*) no known sexual reproductive cycle (*c*) photosynthesis (*d*) prokaryotic structure.

11. Six classes of the fungi are separated according to the (*a*) means for locomotion (*b*) method for sexual reproduction (*c*) ability to form cross walls (*d*) presence of varying numbers of chromosomes.

12. Which of the following characteristics applies to fungi of the class Zygomycetes (*a*) the fungi are prokaryotic (*b*) the fungi have chlorophyll pigments (*c*) *Rhizopus* is a member of the class (*d*) the sexual spore is called an ascospore.

13. The technical name of the common brewing and baking yeast is (*a*) *Candida albicans* (*b*) *Escherichia coli* (*c*) *Amanita toxicans* (*d*) *Saccharomyces cerevisiae.*

14. The common mushrooms, puffballs, and truffles belong to the class of fungi called (*a*) Ascomycetes (*b*) Basidiomycetes (*c*) Oomycetes (*d*) Deuteromycetes

15. The cause of thrush, yeast infection, and other maladies in humans is the fungus (*a*) *Cryptococcus neoformans* (*b*) *Agaricus nigricans* (*c*) *Candida albicans* (*d*) *Rhizopus stolonifer.*

16. Fungi of the class Deuteromycetes are notable because they (*a*) undergo photosynthesis (*b*) lack septa (*c*) produce basidiospores (*d*) lack a known sexual cycle of reproduction.

17. All the following characteristics can be found in fungi except (*a*) they lack chlorophyll (*b*) they have a heterotrophic mode of nutrition (*c*) they are generally motile (*d*) they are unicellular or multicellular.

18. The unique substance chitin is found in the fungal (*a*) cytoplasm (*b*) spore (*c*) cell wall (*d*) mitochondrion.

19. Vegetative hyphae are those hyphae used for (*a*) producing reproductive structures (*b*) absorbing nutrients (*c*) forming mitochondria (*d*) forming septa.

20. Most fungi live in the presence of (*a*) oxygen gas (*b*) high alkaline environments (*c*) oxygen-free environments (*d*) environments rich in carbon dioxide.

21. The carbon dioxide released by yeast cells during their metabolism is used (*a*) for producing silk (*b*) to flavor liqueurs (*c*) to produce spores (*d*) to make bread rise.

22. Asexually produced spores are (*a*) nutritionally unable to survive (*b*) similar to basidiospores (*c*) genetically identical (*d*) not produced by Deuteromycetes.

23. The asexual reproductive process of budding occurs in (*a*) all fungi (*b*) yeasts (*c*) fungi undergoing sexual reproduction (*d*) none of the fungi.

24. Sporangia are fungal structures (*a*) in which spores are formed (*b*) used for obtaining nutrients from the environment (*c*) where energy is released from carbohydrates (*d*) where ribosomes are synthesized.

25. The common edible mushroom is a (*a*) mass of fungal spores (*b*) type of hypha (*c*) tightly packed mycelium (*d*) structure used for producing asexual spores.

Completion. Add the word or words that correctly complete each of the following statements.

1. Those fungi which are biphasic occur as yeasts or as ____________.

2. Fungi tolerate environments that are high in sugar and high in ____________.

3. Yeasts are notable for their ability to live in the presence or absence of ____________.

4. Where the bacterial spore is used for resistance, the fungal spore is used for ____________.

5. Asexual spores formed within a sporangium are known as ____________.

6. Fungi display a mode of nutrition that is ____________.

7. The cell wall of fungi contains cellulose, chitin, and polymers of glucose known as ____________.

8. Those hyphae that form the reproductive structures of the fungus are called ____________.

9. The typical shape observed in a common yeast cell is the ____________.

10. The metabolic process performed by yeasts and used in the beer and wine industries is ____________.

11. In the sexual reproductive processes of a fungus, the terms "male" and "female" are used interchangeably with the terms ____________.

12. Fungi are generally separated into two major divisions, with the true fungi placed into the division ____________.

13. Motile zoospores are produced by fungi of the classes Chytridiomycetes and ____________.

14. The fungus used to ferment soybeans to soy sauce, to modify steroids, and in the production of organic acids is ____________.

15. The mushroom *Agaricus* is safe to eat, but the poisonous mushroom belongs to the genus ____________.

Matching. Select from Column B the class of fungi that fits the description in Column A.

Column A	Column B
_____ 1. Form ascospores	(*a*) Oomycetes
_____ 2. Common mushroom	(*b*) Chytridiomycetes
_____ 3. "Imperfect fungi"	(*c*) Zygomycetes
_____ 4. Zygospore forms	(*d*) Ascomycetes
_____ 5. Waterborne fungi	(*e*) Basidiomycetes
_____ 6. Includes the fermentation yeast	(*f*) Deuteromycetes
_____ 7. No known sexual cycle	
_____ 8. Spore-bearing sexual structures	
_____ 9. *Candida albicans* a member	
____ 10. Species causes ergot disease	
____ 11. Nonseptate hyphae with sporangiospores	
____ 12. Rhizoids for anchorage	
____ 13. Includes *Rhizopus*	
____ 14. Reproduce solely by asexual method	
____ 15. Include rust and smut fungi	

Chapter 12

The Protozoa

OBJECTIVES

The protozoa are a group of eukaryotic microorganisms in which many species have the ability to move independently. Amoebas, flagellates, and ciliates are types of protozoa considered in this chapter. The objectives of the chapter are to:

1. Understand the environments in which protozoa live and the type of nutrition they display.
2. Identify some of the unique features of the protozoal cell.
3. Define the reproductive patterns and forms of reproduction occurring in protozoa.
4. Summarize the characteristics of four major classes of protozoa.
5. Briefly note some of the diseases caused by various species of protozoa.

THEORY AND PROBLEMS

12.1 How do the protozoa compare to the other groups of microorganisms?

Protozoa are eukaryotic microorganisms having "animal-like" characteristics. Like other eukaryotes, their cells have a nucleus, nuclear membrane, organelles, multiple chromosomes, and mitotic structure during reproduction.

12.2 In what group are the protozoa classified?

The protozoa are currently classified in the Kingdom Protista, together with simple algae. They are considered a basic stock of evolution for the fungi, plants, and animals.

CHARACTERISTICS OF PROTOZOA

12.3 What is the cellular composition of the protozoa and what is the general pattern of nutrition exhibited by these organisms?

Protozoa are unicellular microorganisms as compared to the fungi, plants, and animals, which are multicellular. The protozoa display a heterotrophic mode of nutrition in which they obtain their food by ingesting small particles of organic matter. The particles are then digested within the cell, and the products of digestion are used for synthesis reactions.

12.4 In which environments do the protozoa live?

The protozoa are generally inhabitants of aquatic environments. Although some species are found in the soil, the major populations of protozoa live in salt and fresh water, such as in oceans, ponds, and lakes. Some species of protozoa are parasites, such as the organism of malaria and sleeping sickness. These organisms live in human and animal tissues, where they derive their nutrients from living cells.

12.5 Describe the general morphology of protozoal cells.

Protozoa have no cell walls, and are therefore different than bacteria, fungi, and plants. Their cells are surrounded by a cell membrane that encloses the cytoplasm of the organism, and in this respect they resemble animal cells. Some species of protozoa are encased in shells, and some can form highly resistant **cysts** to withstand harsh environments. Internally, the protozoal cytoplasm is divided into an inner zone of **endoplasm** and an outer zone of **exoplasm**. Various organelles, including mitochondria, Golgi bodies, lysosomes, and vacuoles, are also found within the cytoplasm.

12.6 What is the function of the contractile vacuole and the cytoskeleton in protozoa?

Many protozoa form **contractile vacuoles**, which are membrane-bound vesicles that accumulate fluids and expel them to maintain optimal fluid levels within the cytoplasm. An internal cytoskeleton is a maze of microfilaments and microtubules that help maintain the integrity of the protozoal cell.

12.7 Which unique features are found in the nuclei of protozoal cell?

Many species of protozoa are unique because they possess two different nuclei: a large **macronucleus** and one or more small **micronuclei.** Different functions are associated with the macronucleus and the micronucleus during the reproductive process, to be discussed presently.

PHYSIOLOGY AND REPRODUCTION OF PROTOZOA

12.8 Are protozoa able to move?

Many species of protozoa have organs for motility. Indeed, the form of motion is an important criterion in the classification of protozoa. To move, certain species possess cytoplasmic extensions called **pseudopodia**, as displayed in Figure 12-1. Other species have **flagella**. Still other species move by means of the coordinated activity of **cilia**.

12.9 How does nutrition occur in protozoa?

Protozoal nutrition begins with the uptake of materials from the external environment by the process of **endocytosis**. If large molecules in solution are taken up, the process is called **pinocytosis**. If particles of food are obtained from the environment, the process is known as **phagocytosis**. Once organic materials are inside the cytoplasm, enzymes from the lysosome digest the complex carbohydrates, proteins, and lipids into smaller forms for use in the synthesis reactions taking place in the protozoa.

12.10 Do protozoa have any special food-gathering devices to aid their nutrition?

In several species of protozoa, cilia and flagella create water currents that bring food particles close to the cell where phagocytosis occurs. Other species of protozoa have tentacle-like appendages that stick to particles of food, a process that assists phagocytosis. Several species of protozoa have a structure called the **cytopharynx**, which is a channel through which food particles pass into the cytoplasm. These devices are not seen in other microorganisms such as bacteria or fungi.

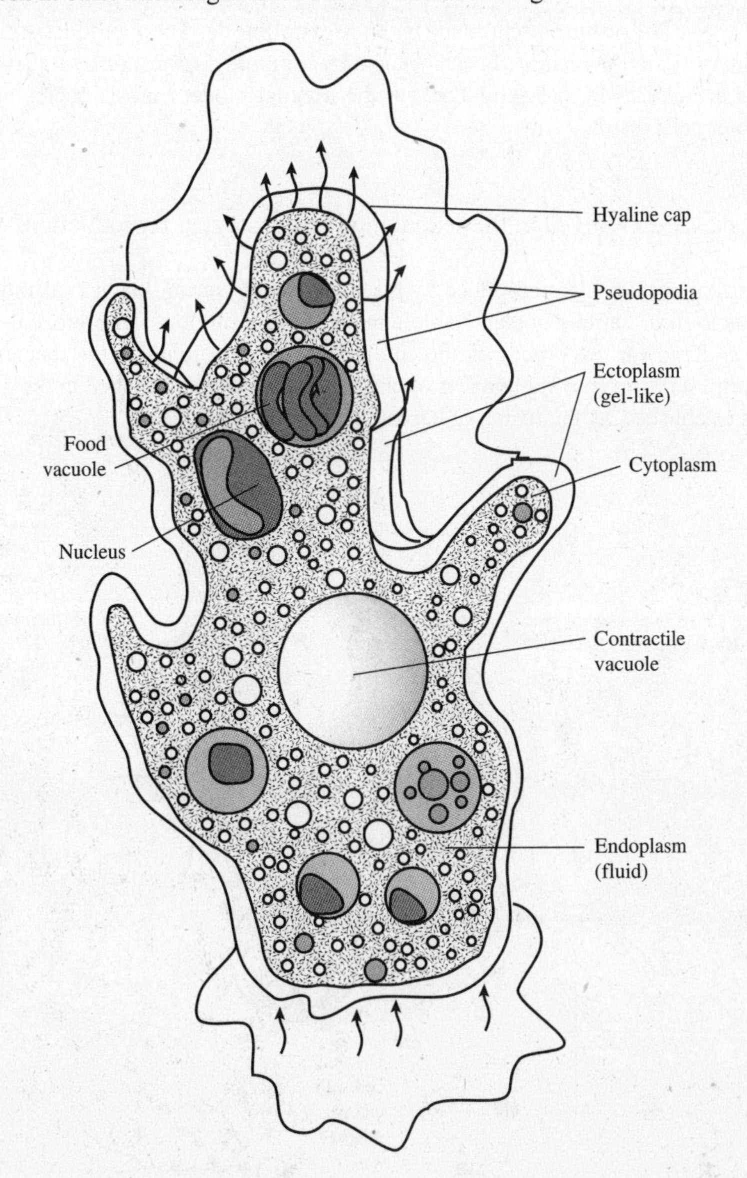

Fig. 12-1 Cytoplasmic details of an amoeba, a member of the class Sarcodina. The pseudopodia can be seen and the distinction between ectoplasm and endoplasm is clear. Numerous other unique structures of the cytoplasm are visible.

12.11 What reproductive patterns are displayed by the protozoa?

Many protozoa have complex life cycles involving various nutritional or morphological forms. For example, the protozoa may exist in a feeding form known as the **trophozoite**. Then they may revert to a dormant, resistant form called the **cyst.** These alternative forms are somewhat similar to the hypha and spores of a fungus. The cysts may have a protective function or a reproductive function. In the reproductive function, they assist the propagation of the protozoa. Reproductive cysts, for example, are known to undergo sexual fusions.

12.12 In which ways does asexual reproduction occur in the protozoa?

Virtually all species of protozoa reproduce by an **asexual mode**. The reproduction may takes place by longitudinal binary fission, by simple binary fission, by multiple fissions, or by transverse binary fission. These methods are displayed in Figure 12-2. In the asexual mode, mitosis occurs and two genetically identical daughter cells result.

12.13 What processes are involved in the sexual mode of protozoal reproduction?

Certain protozoal species also reproduce by a sexual mode. During the **sexual mode**, protozoal cells undergo meiosis to form haploid nuclei. Haploid nuclei have a single set of protozoal chromosomes. The haploid nuclei undergo various processes and configurations, depending on the species of protozoan, and eventually the diploid form (having two sets of chromosomes) is reestablished in the new individual. The diploid form is established by the fusion of haploid nuclei.

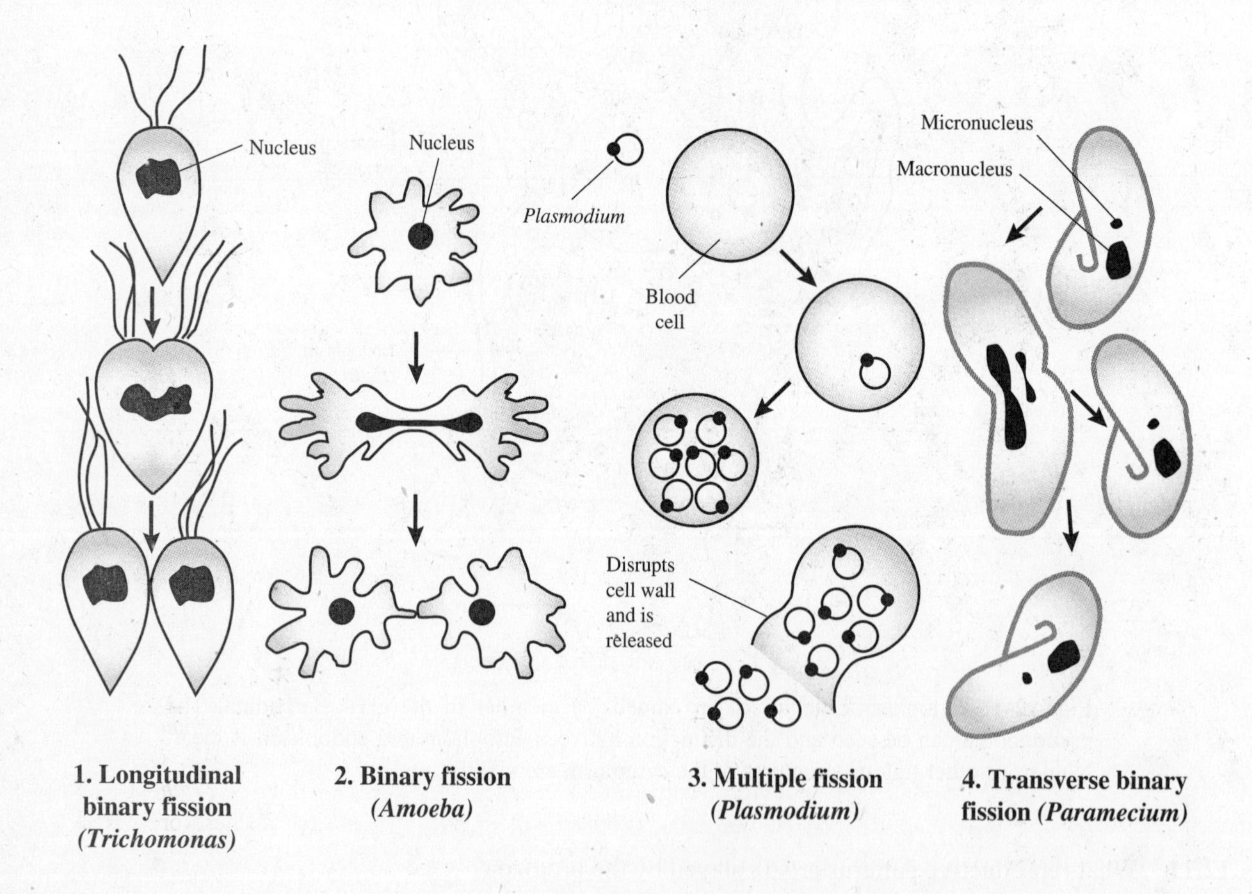

Fig. 12-2 Four different types of asexual reproduction occurring in various species of protozoa.

CLASSIFICATION OF PROTOZOA

12.14 What method is used for classifying protozoa?

There is no one classification scheme for protozoa that is accepted by all biologists. However, a convenient method for classifying the protozoa is based on the type of motility they exhibit. In this scheme, the amoeboid protozoa are placed in the class (or subphylum) Sarcodina. The flagellated protozoa are placed in the class Mastigophora. The ciliated protozoa are placed in the class Ciliophora. The generally nonmotile protozoa are placed in the class Sporozoa, also known as Apicomplexa.

12.15 What are some characteristics of members of the class Sarcodina?

Protozoa of the class **Sarcodina** are amoebas that move by means of pseudopodia. **Pseudopodia** ("false feet") are extension of the cytoplasm of the organism. The pseudopod is thrust out into the environment, where it gels. The cytoplasm then flows into the pseudopod and the protozoan moves into it. Certain amoeboid protozoa are surrounded by hard shells or skeletal coverings. For example, the **foraminifera** are amoebas surrounded by multichambered shells composed of calcium salts. The **radiolaria** are amoebas surrounded by radiating shells composed of silicon. Both foraminifera and radiolaria are found in marine environments such as the oceans.

12.16 Are there any pathogenic protozoa in the Sarcodina class?

Most of the amoeboid protozoa are free-living species not involved with disease. However, one species, ***Entamoeba histolytica,*** causes a disease called amoebic dysentery. This is a serious intestinal disease accompanied by deep bleeding ulcers and extensive fluid loss. It is transmitted by contaminated food and water, and it occurs primarily in tropical regions.

12.17 What are some of the distinguishing characteristics of the class Mastigophora?

Protozoa of the class **Mastigophora** move by means of whiplike flagella. They reproduce by binary fission in the longitudinal plane and obtain their food by ingesting large particles and absorbing small particles. The well-known organism ***Euglena*** is in the Mastigophora class. The organism possesses chlorophyll within cytoplasmic bodies called **chloroplasts**. The chlorophyll allows the organism to trap sunlight energy and convert it to energy in carbohydrates during the process of photosynthesis. *Euglena* also has the enzymes systems allowing it to digest carbohydrates and release the energy by cellular respiration.

12.18 Do any pathogenic species belong to the class Mastigophora?

There are several human pathogens found in the class Mastigophora. One example is ***Trypanosoma brucei,*** the cause of African sleeping sickness. Another pathogen is ***Giardia lamblia***, the cause of giardiasis, an intestinal disease accompanied by severe diarrhea. Also in the group is ***Trichomonas vaginalis,*** the cause of trichomoniasis, a sexually transmitted disease in humans. All these organisms move by means of flagella.

12.19 Do any members of the Mastigophora class form cysts?

Certain species of the Mastigophora form cysts, the highly resistant structures that encourage survival in the external environment. For instance, the pathogen *Giardia lamblia* forms cysts and remains alive outside the body in contaminated water until it reenters the body. *Trichomonas vaginalis* forms no cysts and must pass from individual to individual to remain alive.

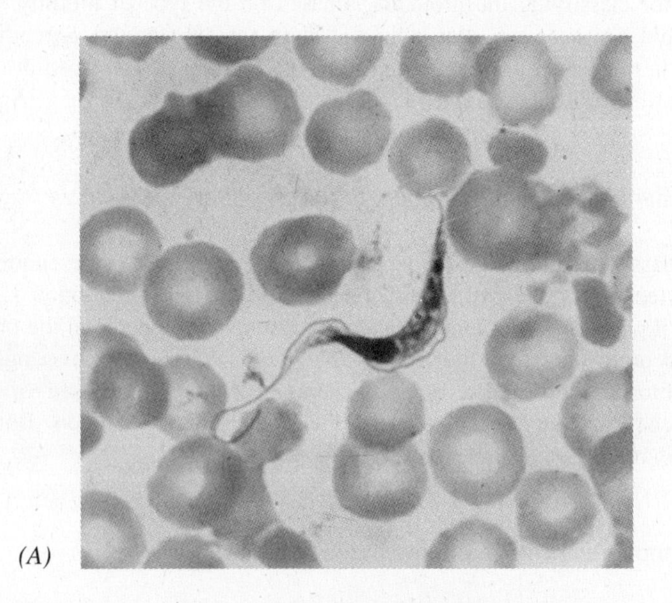

(*A*)

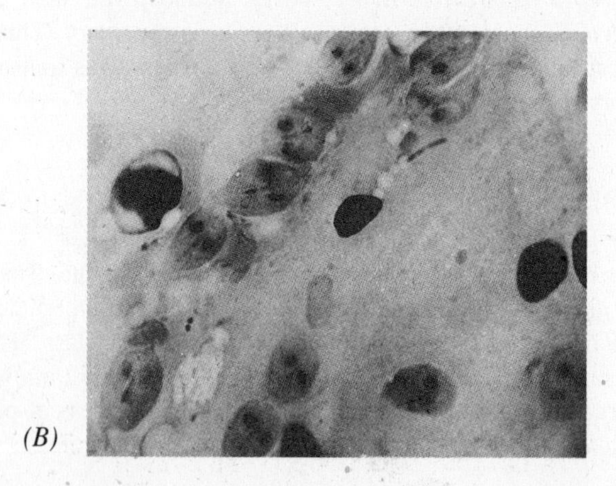

(*B*)

Fig 12-3 Two protozoa involved in human disease. (*A*) *Trypanosoma brucei*, the cause of African sleeping sickness. (*B*) *Giardia lamblia*, the cause of giardiasis, a serious intestinal disease.

12.20 Which organism is the most well-known member of the Ciliophora class?

Probably the most recognized member of the **Ciliophora** class is the organism ***Paramecium***. This slipper-shaped organism is surrounded by cilia that beat synchronously and allow the organism to move in several directions through fluids. This organism and others in the class have both a micronucleus and a macronucleus.

12.21 Does sexual reproduction occur in *Paramecium* species?

Species of *Paramecium* display a type of sexual reproduction in which two organisms come together and undergo conjugation. The micronuclei replicate in the two individuals, then move from one organism into the other. This form of sexual reproduction changes the genetic characteristic of the two individuals since each has acquired the other's genes.

12.22 Which other organisms in the Ciliophora class have been extensively studied by microbiologists?

The class Ciliophora contains a number of protozoa extensively investigated as part of research projects. For example, species of ***Tetrahymena*** have been studied for their unique form of sexual reproduction involving conjugation of compatible mating types and nuclear exchanges. This reproduction involves the fusion of haploid nuclei to form a diploid nucleus and succeeding mitotic divisions to form haploid nuclei once again. Members of the genera ***Stentor*** and ***Vorticella*** are also well studied because of their large size and complex cellular features reminiscent of multicellular organisms.

12.23 Are there any pathogens in the class Ciliophora?

A small number of pathogenic species are classified within the class Ciliophora. A notable pathogen is ***Balantidium coli.*** This organism inhabits the human intestinal tract where it may cause mild to severe diarrhea. Contaminated water and food, especially pork, transmit the organism among humans.

12.24 What are the characteristic features of protozoa in the class Sporozoa?

Members of the class **Sporozoa (Apicomplexa)** have no organs for motion in the adult form, and therefore they differ from other protozoal species. The members also have complex life cycles involving both asexual and sexual stages, often occurring in different hosts. Various stages in the life cycle may include forms known as merozoites, sporozoites, and oocysts.

12.25 Is the organism of malaria classified in the Sporozoa group?

Malaria is caused by several species of ***Plasmodium*** classified as Sporozoa. The protozoa are transmitted among humans by the female *Anopheles* mosquito. They undergo their life cycles in human red blood cells and destroy the latter, causing chills and extremely high fever in alternating sequence.

12.26 Are there any other important pathogens classified as Sporozoa?

Several other important pathogens of humans are classified as Sporozoa. One example is ***Toxoplasma gondii.*** This organism causes toxoplasmosis, a disease of white blood cells transmitted to humans by domestic housecats. In AIDS patients, toxoplasmosis is an important opportunistic disease. Another important pathogen is ***Cryptosporidium coccidi.*** In AIDS patients, this organism induces extensive diarrhea that can lead to dehydration and emaciation.

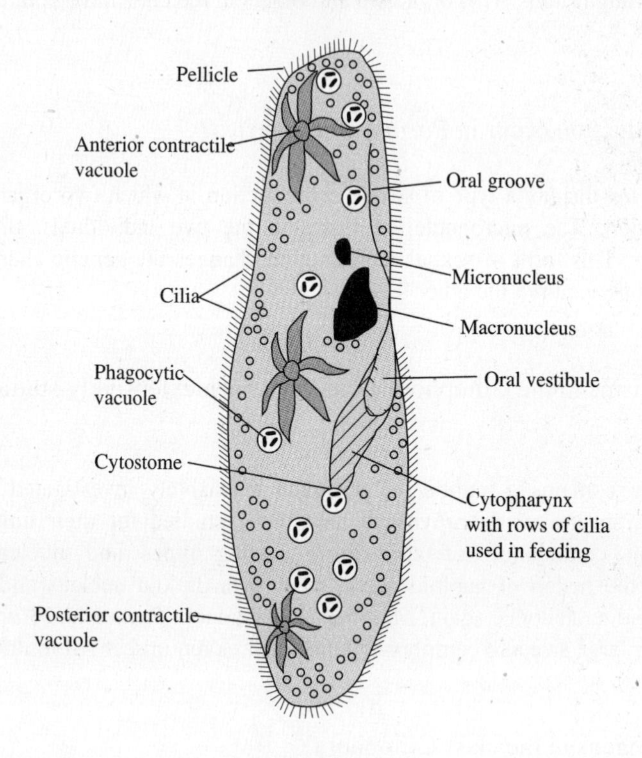

Fig. 12-4 *Paramecium*, a member of the class Ciliophora. The cellular details of this highly specialized ciliate are illustrated.

12.27 Is the organism *Pneumocystis carinii* classified in the Sporozoa group?

There is some controversy as to whether ***Pneumocystis carinii*** is a fungus or a protozoan. The ribosomal composition of the organism suggests that it is a fungus, but the organism has traditionally been considered a protozoan since its other characteristics of morphology, biochemistry, and nutrition are more like the protozoa. It lives in the lungs of persons with AIDS, where it induces a severe form of pneumonia that can be fatal. Indeed, fully half the deaths associated with AIDS are due to disease caused by *Pneumocystis carinii.*

12.28 What are some practical importances ascribed to the protozoa in nature?

In nature, protozoa serve as important components of the food chains. They consume bacteria and, in turn, they are consumed by microscopic invertebrates and larger species. In this way, they bring the organic materials of bacteria into the food chain and provide important nutrients for other organisms. In ruminant animals and termites, certain protozoal species produce cellulase, which breaks down cellulose and releases its glucose. This activity brings glucose into the food chains and makes it available to larger organisms for their energy metabolisms.

Review Questions

True/False. For each of the following statements, mark the letter "T" next to the statement if the statement is true. If the statement is false, change the underlined word to make the statement true.

___ **1.** The contractile vacuole of protozoa is a membrane-bound vesicle that accumulates food particles.

___ **2.** As a resistance mechanism, some species of protozoa form highly resistant spores.

___ **3.** All species of protozoa reproduce by the asexual mode.

___ **4.** In the life cycle of a protozoan, the feeding form is referred to as the cyst.

___ **5.** Protozoa belonging to the class Sarcodina move by means of flagella.

___ **6.** The well-known organism *Euglena* possesses hemoglobin molecules within its chloroplasts.

___ **7.** A severe disease of the blood may be caused by the protozoan *Giardia lamblia.*

___ **8.** Members of the class Ciliophora are characterized by the presence of cilia surrounding the cell.

___ **9.** The organism *Pneumocystis carinii* causes severe disease of the lungs in persons who have hepatitis.

__ **10.** Protozoa are prokaryotic microorganisms having animal-like characteristics.

__ **11.** The classes of protozoa are distinguished from one another on the type of nutrition exhibited by the organisms.

__ **12.** Multichambered shells composed of calcium salts are found in protozoa known as foraminifera.

__ **13.** An alternative expression for the class Sporozoa is Sarcodina.

__ **14.** Members of the class Mastigophora reproduce in the longitudinal plane by the process of binary fission.

__ **15.** The flagellated protozoan *Trichomonas vaginalis* causes a waterborne disease in humans.

__ **16.** The slipper-shaped organism surrounded by cilia and classified in the Ciliophora is *Euglena*.

__ **17.** Merozoites, sporozoites, and oocysts are various stages in the life cycle of protozoa belonging to the class Mastigophora.

__ **18.** Toxoplasmosis is a protozoal disease often transmitted to humans by domestic housecats.

__ **19.** Some controversy exists as to whether *Pneumocystis carinii* is a protozoan or a bacterium.

__ **20.** Cases of malaria are transmitted among humans by dogs.

Matching. Match the characteristic in Column A with the correct protozoan in Column B.

Column A	Column B
_____ 1. Slipper-shaped ciliate	(*a*) *Pneumocystis carinii*
_____ 2. Radiating shell	(*b*) *Euglena*
_____ 3. Flagellate; causes diarrhea	(*c*) *Paramecium*
_____ 4. Causes African sleeping sickness	(*d*) *Giardia lamblia*
_____ 5. Green flagellate	(*e*) radiolarian
_____ 6. Sporozoa; intestinal disease	(*f*) *Balantidium coli*
_____ 7. Causes pneumonia	(*g*) *Cryptosporidium coccidi*
_____ 8. Displays sexual reproduction	(*h*) *Entamoeba histolytica*
_____ 9. Amoeba with silicon shell	(*i*) *Trypanosoma brucei*
_____ 10. Transmitted in pork	(*j*) foraminiferan
_____ 11. Multichambered shell	
_____ 12. Possibly a fungus	
_____ 13. Moves by flagella; photosynthetic	
_____ 14. Mastigophora; intestinal disease	
_____ 15. Intestinal ciliate	
_____ 16. Amoeba with calcium shell	
_____ 17. Lung disease in AIDS patients	
_____ 18. Micronucleus and macronucleus	
_____ 19. Has chlorophyll	
_____ 20. Forms cyst; flagellate	

Multiple Choice. Select the letter of the item that correctly completes each of the following statements.

1. All the following characteristics apply to the protozoa except (*a*) protozoa are unicellular (*b*) protozoa display a heterotrophic mode of nutrition (*c*) many species of protozoa have the ability to move (*d*) protozoa have chlorophyll dissolved in their cytoplasm.

2. The characteristics displayed by protozoa are most closely similar to the characteristics of (*a*) plants (*b*) animals (*c*) bacteria (*d*) viruses.

3. All the following mechanisms are used as means of motion in various species of protozoa except (*a*) pili (*b*) cilia (*c*) pseudopodia (*d*) flagella.

4. Both pinocytosis and phagocytosis are processes used by protozoa to (*a*) bring materials into the cytoplasm (*b*) perform chemiosmosis (*c*) synthesize carbohydrates using energy from the sun (*d*) synthesize proteins from amino acids.

5. The feeding form of a protozoan is known as the (*a*) cyst (*b*) heterocyst (*c*) tronchodon (*d*) trophozoite

6. The radiolaria and foraminifera are types of protozoa (*a*) that move by cilia (*b*) that are surrounded by hard shells (*c*) that have no mechanism for motility (*d*) that synthesize carbohydrates from carbon dioxide.

7. Members of the class Sarcodina include the protozoa that (*a*) display no organs for locomotion (*b*) have cell walls (*c*) form spores (*d*) move by pseudopodia.

8. The slipper-shaped *Paramecium* has both (*a*) spores and cysts (*b*) cilia and flagella (*c*) a macronucleus and a micronucleus (*d*) the ability to cause disease in humans.

9. Fully half the deaths associated with AIDS are due to infection by the protozoan (*a*) *Euglena gracilis* (*b*) *Pneumocystis carinii* (*c*) *Giardia lamblia* (*d*) *Anopheles nigricans.*

10. The contractile vacuole of protozoa is used for (*a*) storing phosphate granules (*b*) maintaining optimal fluid levels (*c*) producing disease-causing toxins (*d*) motility.

11. Members of the class Sporozoa differ from other protozoa because the members of the class (*a*) engage in photosynthesis (*b*) have cell walls (*c*) cause disease (*d*) are unable to move.

12. Domestic housecats are important in the transmission of the protozoan that causes (*a*) dysentery (*b*) malaria (*c*) toxoplasmosis (*d*) spinal meningitis.

13. The formation of a cyst permits a protozoan to (*a*) move (*b*) synthesize protein (*c*) survive in the external environment (*d*) remain alive in the tissues of mosquitoes.

14. The cause of African sleeping sickness belongs to the same class of protozoa as (*a*) *Plasmodium vivax* (*b*) *Giardia lamblia* (*c*) *Paramecium* (*d*) radiolaria.

15. A notable pathogen in the class Ciliophora is responsible for human disease of the (*a*) intestinal tract (*b*) respiratory tract (*c*) urogenital tract (*d*) skin.

16. Which of the following may not be found in protozoal cells (*a*) mitochondria (*b*) Golgi bodies (*c*) lysosomes (*d*) cell walls.

17. The fungi are currently classified in the kingdom (*a*) Monera (*b*) Cytophagia (*c*) Animalia (*d*) Protista.

18. All protozoa are known to possess a (*a*) cyst form (*b*) cytopharynx (*c*) set of cilia (*d*) cell membrane.

19. The organism *Stentor, Vorticella,* and *Tetrahymena* are all complex members of the class (*a*) Mastigophora (*b*) Monera (*c*) Ciliophora (*d*) Apicomplexa.

20. The human disease malaria occurs primarily in the (*a*) brain (*b*) red blood cells (*c*) tubules of the urinary system (*d*) sacs of the lungs.

Chapter 13

The Unicellular Algae

OBJECTIVES

Algae are a group of plantlike organisms that lack the quality of true plants. Within this group are a series of unicellular (single-celled) organisms that are considered in the microbial world. They are the subject matter of these pages, whose objectives are to:

1. Consider the roles played by algae in the natural environment.
2. Outline and discuss the various groups in which unicellular algae are classified.
3. Identify some of the important types of unicellular algae and describe their significance in nature.
4. Briefly mention the unicellular algae that may pose a danger to health.

THEORY AND PROBLEMS

13.1 What is a general explanation for the term algae?

The term "algae" is difficult to define in microbiology. It is used for a group of photosynthetic plantlike organisms that are not otherwise classified as true plants. Some algae are single-celled organisms, while others are multicellular organisms. The unicellular algae are considered microorganisms.

CHARACTERISTICS OF UNICELLULAR ALGAE

13.2 What important roles are played by algae in the natural environment?

Algae are extremely ubiquitous organisms that live in all aquatic habitats and numerous terrestrial habitats. The unicellular algae often occur as **phytoplankton**, the microscopic organisms that live near the surface of the oceans of the world. Phytoplankton are of immense ecological importance because they participate in photosynthesis and form carbohydrates, which represent the energy base of most marine and freshwater food chains.

13.3 Do unicellular algae produce oxygen?

Unicellular algae produce the great share of the molecular oxygen available on Earth. All animals that breathe, including marine and land forms, depend on this oxygen derived from photosynthesis and produced by phytoplankton. This factor together with their position in the food chains makes unicellular algae among the most important microorganisms in the world.

13.4 What is the cellular makeup of the unicellular algae?

The unicellular algae are eukaryotes. Each cell has a nucleus and a nuclear membrane. It also has organelles such as mitochondria, lysosomes, Golgi bodies, the endoplasmic reticulum, and ribosomes. Reproduction occurs by the process of mitosis, and the algal chlorophyll is concentrated within chloroplasts.

13.5 Are there any prokaryotic algae?

A generation ago, the prokaryotic cyanobacteria were known as "blue-green algae." The term blue-green algae is no longer used, and the cyanobacteria are now considered to be bacteria. Therefore, there are no prokaryotic algae.

13.6 In which groups are the algae classified?

Most biologists agree that there are six groups of algae known as Divisions, the classification group used for plants. Two of the divisions, Rhodophyta and Phaeophyta, contain multicellular algae exclusively. These algae are the red seaweeds and brown seaweeds, respectively. The third division, Chlorophyta, includes several unicellular species considered in microbiology. The remaining three divisions, Chrysophyta, Euglenophyta, and Pyrrophyta, contain unicellular algae exclusively.

CLASSIFICATION OF UNICELLULAR ALGAE

13.7 What are the important members of the division Chlorophyta?

The division **Chlorophyta** contains thousands of species of green algae, both freshwater and marine specimens. Among the single-celled green algae are the organisms *Chlorella* and *Chlamydomonas*. *Chlamydomonas* is easily grown, and its physiology and genetics have been studied in depth. The organism forms carbohydrates by photosynthesis, and it has opposite mating types that undergo union in a sexual form of reproduction. A zygote results from the fusion. The zygote undergoes meiosis to yield sexually opposite cells to complete the process, which is illustrated in Figure 13-1.

13.8 Are there any colonial forms of green algae?

Colonial forms of green algae are not considered multicellular because each cell operates independently. The aggregate of cells is known as a **colony**. The colonial forms may have been the predecessors of multicellular forms. An example of a colonial member of the division Chlorophyta is the organism *Volvox*. This is a spherical colony of identical cells, with little differentiation among the cells, although some cells appear to be specialized for reproduction.

13.9 Which algae are classified in the division Chrysophyta?

Algae in the division **Chrysophyta** are yellow-green algae and golden-brown algae. Both groups are important members of the phytoplankton. Their names are derived from the colors of their pigments. All members of the division have high concentrations of keratinoid pigments, one of which is **fucoxanthin**. Many species of chrysophytes are enclosed in coverings of silicon dioxide. Flagella are found in the cells of many species.

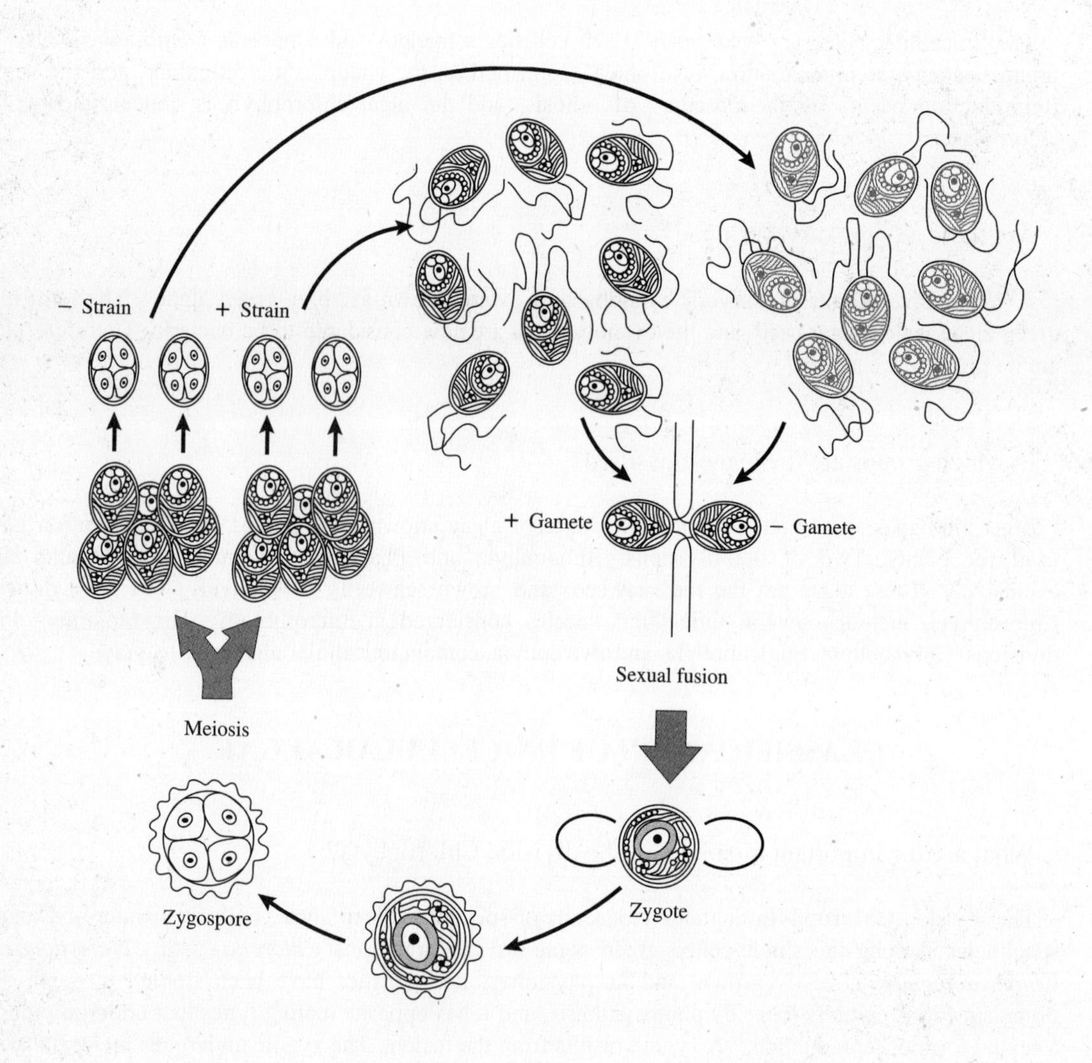

Fig. 13-1 The sexual reproduction cycle of the unicellular alga *Chlamydomonas*. Plus and minus strains resulting from meiosis multiply, then fuse to form the zygote. After several nuclear divisions, meiosis begins once again.

13.10 What are some examples of members of the Chrysophyta division?

Perhaps the most familiar chrysophytes are the **diatoms**. Diatoms are photosynthetic marine and freshwater algae distinguished by their intricate shells of silicon dioxide. The diatoms secrete the shells for protection. Perforated holes in the shells permit contact between the photosynthetic organisms and their environment. Because the shells are clear, light can penetrate through for photosynthesis.

13.11 Are there any economic values to the diatoms?

The shells of diatoms are extremely beautiful to the eye, but they also have economic importance as well. The sediments of many oceans floors consist largely of diatom shells, which form **diatomaceous earth**. Diatomaceous earth is harvested as a fine white powder for use in swimming pool filters, insulating materials, and toothpaste. Diatomaceous earth is also used as an abrasive in polishing compounds.

13.12 What important characteristics are associated with members of the division Pyrrophyta?

The term **Pyrrophyta** means "fire algae." Members of this division are dinoflagellates, which are single-celled photosynthetic algae having two flagella. The cell walls of dinoflagellates are stiff cellulose plates, which appear to enclose the alga like a suit of armor. A representative dinoflagellate is shown in Figure 13-2.

13.13 What is the environmental importance of dinoflagellates?

Like the diatoms, the **dinoflagellates** serve an extremely important role at the base of the marine food chains. They synthesize carbohydrates and provide food for microscopic organisms and invertebrates. These organisms utilize the dinoflagellates' products of photosynthesis as energy sources. Dinoflagellates reproduce primarily by binary fission, although sexual fusions do occur at times in the life cycle of certain species.

13.14 Are there any types of dinoflagellates that are dangerous to health?

Some species of dinoflagellates are responsible for the **red tide**. This condition occurs when the proper conditions of temperature, nutrients, and other factors exist. Species of reddish brown dinoflagellates multiply so extensively that they tint oceans and inland lakes a distinctive red color. Certain dinoflagellates in the red tide produce powerful nerve toxins that kill the fish, and if people consume shellfish such as clams, oysters, or muscles that have been exposed to the toxins, they will die of paralytic shellfish poisoning.

13.15 Are there any unique characteristics associated with the algae of the division Euglenophyta?

The **Euglenophyta** include species of the genus ***Euglena*** and other organisms known as euglenoids. Euglenoids have characteristics found in both animal and plant cells. They have flagella and are able to move independently, including a response to light by a photoreceptor at the base of the flagellum. They also have chlorophyll pigments similar to those of multicellular plants, and they are able to undergo photosynthesis using these chlorophylls. For this reason, the euglenoids are believed to be the basic stock of evolution for both the animals and the plants.

13.16 Are there any important research capabilities associated with *Euglena*?

Because of their animal and plant properties, euglenoids are extremely important to researchers. *Euglena* species have no cell walls, which adds to their animal-like qualities (Chapter 12). They store their carbohydrates in starch granules called **paramylon**, a form different than the starch granules in other plants. No sexual reproduction has been observed in *Euglena*, but asexual reproduction occurs by mitosis. Reproduction can occur rapidly, and euglenoids may impart a green color to the pond in which they live.

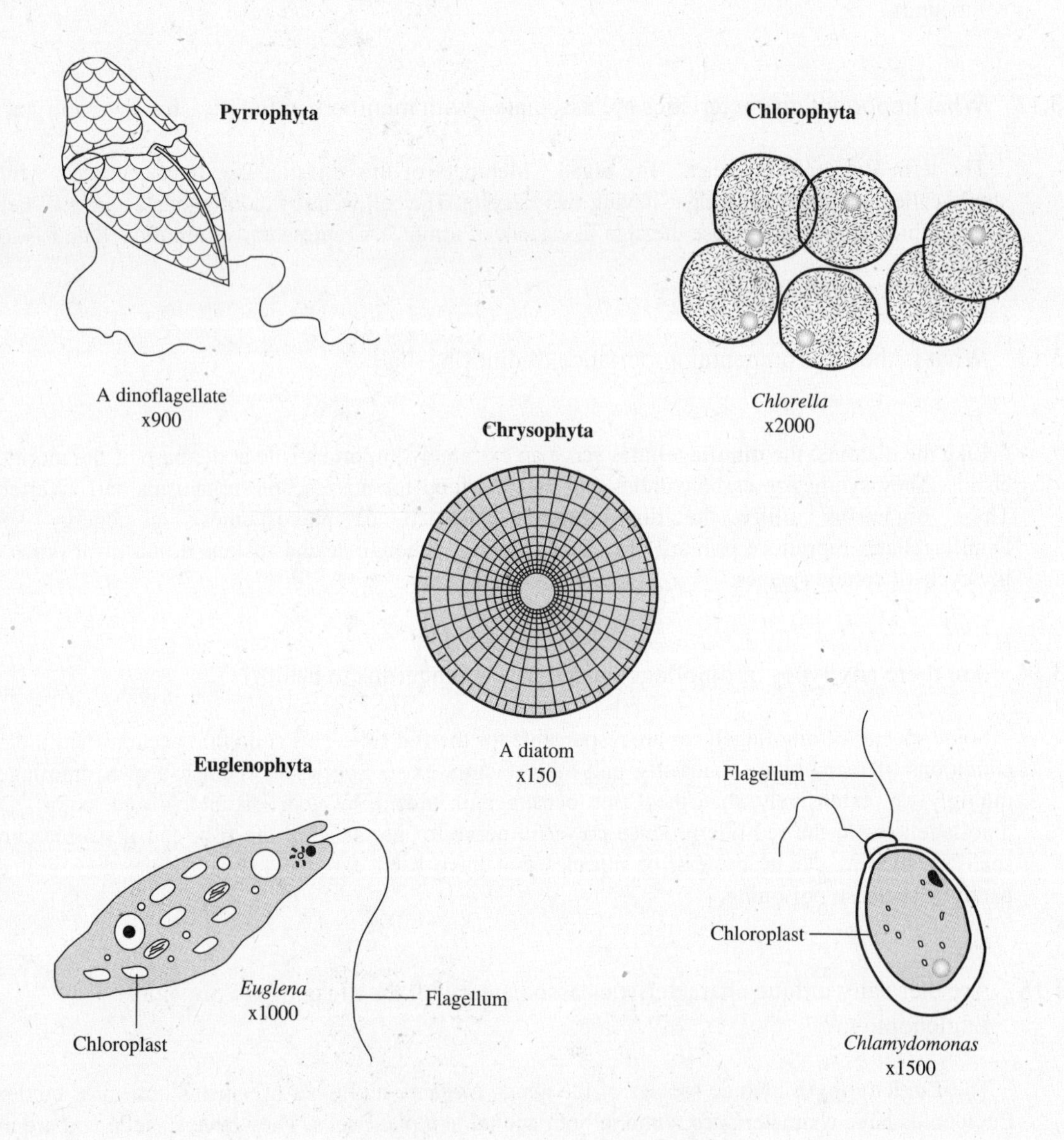

Fig. 13-2 Representative members of the four divisions of unicellular algae.

Review Questions

Multiple Choice. Select the letter of the item that correctly completes each of the following statements.

1. The unicellular algae include the (*a*) chlamydiae (*b*) eukaryotes (*c*) cyanobacteria (*d*) diatoms.

2. Unicellular algae are important in the natural environment because they (*a*) cause human disease (*b*) perform photosynthesis and produce oxygen (*c*) perform respiration and produce carbon dioxide (*d*) are natural parasites of fungi.

3. An important single-celled green alga widely used in research is (*a*) *Chlamydomonas* (*b*) *Chlamydia* (*c*) *Chlorophyta* (*d*) *Corynebacterium.*

4. The older term blue-green algae has been replaced by the term (*a*) cyanobacteria (*b*) eubacteria (*c*) eukaryote (*d*) schizomycophyta.

5. The algae are classified into six groups known as (*a*) phyla (*b*) classes (*c*) orders (*d*) divisions.

6. High concentrations of keratinoid pigments are found in the (*a*) green algae (*b*) red algae (*c*) golden-brown algae (*d*) blue algae.

7. Intricate shells of silicon dioxide are found in algae known as (*a*) *Volvox* (*b*) diatoms (*c*) *Euglena* (*d*) *Paramecium.*

8. Certain species of dinoflagellates are responsible for the (*a*) diseases of swimming fish (*b*) various forms of hepatitis (*c*) blue pigments in Arctic waters (*d*) red tides.

9. The shells of diatoms are widely used as (*a*) swimming pool filters (*b*) textile desizers (*c*) sources of ethanol (*d*) fillers for candies and confections.

10. The dinoflagellates belong to the division (*a*) Euglenophyta (*b*) Pyrrophyta (*c*) Chlorophyta (*d*) Phaeophyta.

11. Euglenoids are unique among the algae because of their ability (*a*) to synthesize proteins (*b*) to practice photosynthesis (*c*) to move (*d*) to produce carbon dioxide.

12. The paramylon formed in *Euglena* cells is a form of (*a*) amino acid (*b*) triglyceride (*c*) polysaccharide (*d*) tripeptide.

13. Paralytic shellfish poisoning can be caused by the toxins produced by (*a*) *Euglena* species (*b*) yellow-green algae (*c*) *Chlamydomonas* (*d*) dinoflagellates.

14. The chlorophyll pigments found in algae are usually concentrated within cytoplasmic bodies called (*a*) ribosomes (*b*) mitochondria (*c*) chloroplasts (*d*) lysosomes.

15. An example of a green algae occurring in a colonial form is the organism (*a*) *Volvox* (*b*) *Euglena* (*c*) *Chlamydomonas* (*d*) *Paramecium.*

True/False. Enter the letter "T" if the statement is true in its entirety, or the letter "F" if any part of the statement is false.

___ **1.** The dinoflagellates have an important role in nature because they lie at the base of food chains occurring in lakes and rivers.

___ **2.** All forms of algae are unicellular.

___ **3.** All algae are eukaryotes.

___ **4.** The most familiar chrysophytes are unicellular algae known as diatoms.

___ **5.** Certain algae such as the organism *Volvox* exist in colonial forms.

___ **6.** The cell walls of dinoflagellates consist of stiff plates of protein.

___ **7.** Members of the division Euglenophyta display characteristics of both plant and animal cells.

___ **8.** When they multiply rapidly, euglenoids such as the organism *Euglena* impart a red color to the water.

___ **9.** The great share of molecular carbon dioxide on Earth has been produced by the algae.

___ **10.** It is expected that one will find organelles such as mitochondria and lysosomes in the unicellular algae.

___ **11.** The cyanobacteria are a group of bacteria once known as dinoflagellates.

___ **12.** Members of the division Chlorophyta include thousands of species of brown and red algae.

___ **13.** Most reproductive methods among the algae are asexual methods.

___ **14.** Fucoxanthin is an example of a keratinoid pigment found in the euglenoids.

___ **15.** The dinoflagellates are able to move about because they possess two flagella per cell.

Completion. Add the word or words that correctly complete each of the following statements.

1. The algae produce the great share of molecular ___________ available on Earth.

2. Those algae once known as blue-green algae are now considered to be types of ___________.

3. The brown and red seaweeds make up divisions of ___________ algae.

4. Most unicellular algae reproduce by some form of ___________ reproduction.

5. Dinoflagellates are able to move in their environment because they possess ___________ flagella per cell.

6. *Euglena* cells are able to respond to ___________ by using a photoreceptor at the base of the flagellum.

7. The shells of ___________ accumulate at the ocean floor and form a fine white powder for use in filters.

8. Dinoflagellates are surrounded by stiff plates of ____________ which enclose the cell.

9. When a boom of dinoflagellates occurs in the ocean, the ____________ tide is often a result.

10. The organism *Chlamydomonas* has opposite mating types that unite to form a zygote in the ____________ form of reproduction.

Chapter 14

The Viruses

OBJECTIVES

Viruses are extraordinarily simple particles of matter with at least one property of living things, the ability to replicate. This chapter will survey some of their properties. The objectives of the chapter are to:

1. Compare viruses with other microorganisms in size and other properties.

2. Delineate the components of the virus and indicate the importance of each component.

3. Describe some of the types of symmetry found in viruses.

4. Summarize the general pattern of viral replication and conceptualize some aspects of the pattern.

5. Compare the replication patterns for RNA and DNA viruses.

6. Define the concept of lysogeny.

7. State the effect of viral replication on the host cell and define several diseases that may result.

8. Discuss several methods for detecting viruses in the laboratory and cultivating viruses.

9. Understand how drugs work against viruses and how protection is rendered by body defenses against viruses.

10. Explain two types of viral vaccines and the advantages of each type.

THEORY AND PROBLEMS

14.1 In what ways do viruses relate to other types of microorganisms?

Compared to eukaryotic and prokaryotic cells, viruses are extraordinarily simple in their physical and chemical properties. Viruses are noncellular. They generally consist of a fragment of nucleic acid encased in protein, and in some cases, a membranelike envelope.

14.2 Are viruses considered living organisms in view of their properties?

Since the discovery of viruses at the beginning of the 1900s, scientists have debated whether viruses are living things or lifeless particles. Viruses do not have a cellular composition, nor do they grow or metabolize organic materials. They produce no waste products, utilize no energy, and do not adapt to their environment. They reproduce, but only within a living cell. It would appear that viruses are not living things, but particles with at least one property of living things.

14.3 How do the viruses compare with other microorganisms in size?

Viruses are probably the smallest entities that cause disease in humans. The smaller viruses, such as the polio viruses, have a diameter of about 20 nanometers (nm). This size can also be expressed as 0.020 micrometers (μm). The larger viruses, such as the poxviruses, have a diameter of about 300 nm (0.30 μm). A bacterium such as a *Staphylococcus aureus* cell has a diameter of about 500 nm (0.5 μm). The smallest bacteria, the mycoplasmas, have a diameter of about 200 nm (0.20 μm).

14.4 Can any viruses be seen with the standard light microscope?

The extremely small size of viruses usually prevents their observation with standard light microscopes. Indeed, viruses could not be visualized clearly until the electron microscope was invented in the 1940s. A magnification of over 200,000 times with this instrument still does not show the virus clearly, as Figure 14-1 indicates.

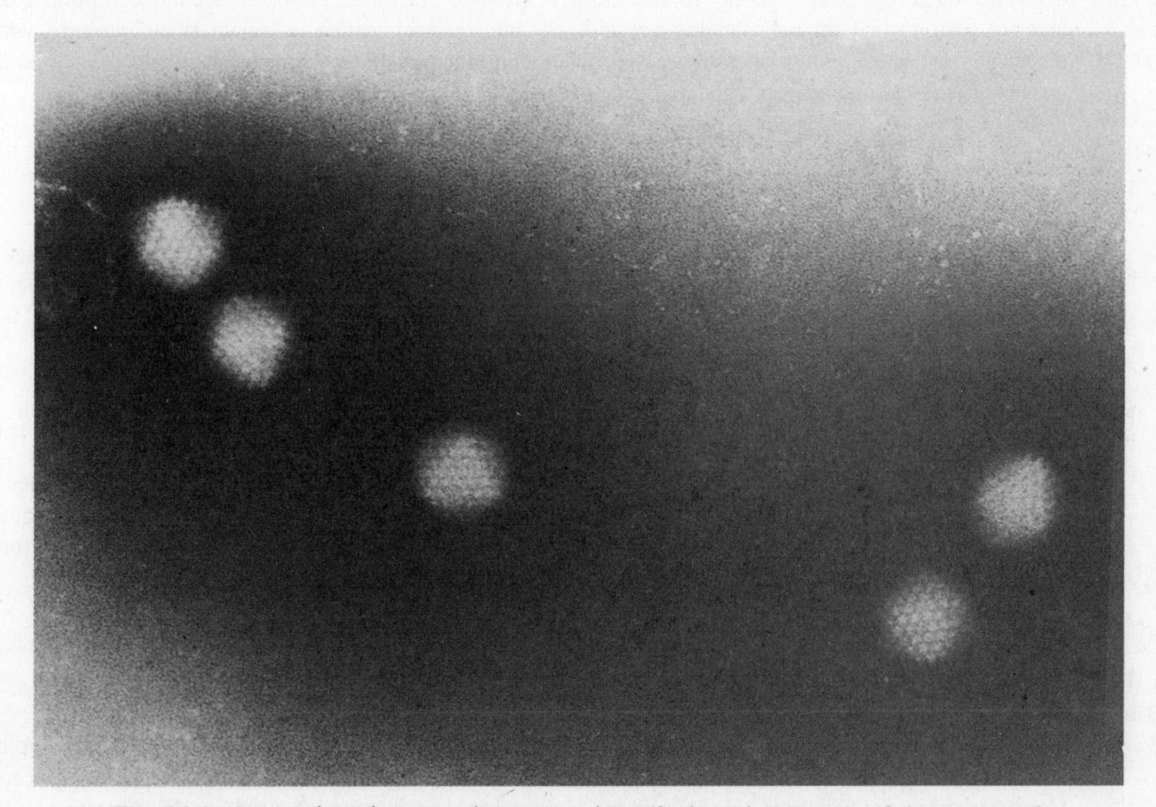

Fig. 14-1 A scanning electron microscope view of adenoviruses, one of the causes of common colds. The icosahedral shape of the viruses is visible. The magnification is 233,800 times.

14.5 How does the electron microscope permit scientists to see the viruses?

To see microorganisms, standard light microscopes use visible light having a relatively long wavelength. Because of the long wavelength, the light beam passes over the viruses. The **electron microscope**, in comparison, utilizes a beam of electrons having an extremely short wavelength. The electron beam bounces off objects even as small as viruses and forms images on projection screens.

14.6 Are viruses able to pass through filters that trap bacteria?

At the beginning of the 1900s, viruses were known as "filterable viruses." This term implies that they were able to pass through filters that could trap the smallest bacteria. Eventually the term filterable was dropped. Because of their small size, viruses pass through bacterial filters, and a fluid cannot be considered sterilized unless the smallest virus has been retained on the filter.

VIRAL STRUCTURE

14.7 What are the components of a virus?

All viruses are composed of at least two components: the first is a strand of **nucleic acid**, which can be deoxyribonucleic acid (DNA) or ribonucleic acid (RNA), but never both; the second component is a layer of **protein** that encloses the nucleic acid. The nucleic acid portion is known as the **genome**, while the protein covering is known as the **capsid**. The combination of the genome and capsid is called the **nucleocapsid**. Certain viruses also have a membranelike **envelope** enclosing the nucleocapsid.

14.8 In what forms may the nucleic acid occur in a virus?

The nucleic acid of a virus may occur in various forms. It may be composed of DNA or RNA, either of which can be single-stranded or double-stranded. The nucleic acid may occur as a continuous molecule, or it may be segmented.

14.9 What is the structure of the capsid that encloses the viral genome?

The capsid of a virus is composed exclusively of protein. The proteins are organized into a series of repeating units called **capsomeres**. To form a capsid, capsomeres are chemically bound together much like the patches of a quilt. The number of capsomeres in the capsid of a virus is distinctive for a particular virus. For example, the virus that causes herpes simplex has 252 capsomeres.

14.10 What is the chemical composition of the viral envelope?

Certain viruses, such as the viruses of chickenpox, infectious mononucleosis, and herpes simplex, have an **envelope**. The envelope is derived from the membrane of the cell in which the virus replicates. However, it is not identical to the cell membrane because it contains viral-specified proteins. The nucleocapsid obtains the envelope on leaving the cell at the conclusion of the replication process.

14.11 Is the viral envelope a continuous structure?

In many viruses, the envelope is a continuous membranous structure. However, in some viruses, the envelope contains projections known as **spikes**. These spikes often contain substances used by the virus tounite with the host cell during the replication process.

14.12 Are the chemical components of viruses susceptible to chemical destruction?

The genome, capsid, and envelope are susceptible to various chemical agents which can destroy the virus. For example, **formaldehyde** alters the structure of the genome and thereby inactivates the virus; **phenol** reacts with the protein capsid of the virus and brings about inactivation; a lipid solvent such as a **detergent** reacts with the envelope of the virus.

14.13 Do viruses replicate in various kinds of cell?

There are viruses capable of replicating in all conceivable types of cells. Certain viruses, for example, attack and replicate in human cells. Diseases such as yellow fever, AIDS, hepatitis, and measles are caused by these viruses. Plant cells are susceptible to viruses that cause plant diseases such as tobacco mosaic disease and tomato mosaic disease. In both cases, the plant leaves assume a mottled, mosaic appearance as the cells die. There are even viruses that attack bacterial cells. These viruses are known as **bacteriophages,** or **phages**. They bring about the destruction of bacterial cells as the replication process proceeds.

14.14 What shapes do different viruses appear in?

Enveloped viruses appear under electron microscopes in a spherical form because the amorphous envelope encloses the virus. However, unenveloped viruses take on the symmetry of the protein capsid. This symmetry can be icosahedral or helical.

14.15 What is meant by icosahedral symmetry?

Icosahedral symmetry is a symmetry in which the capsid is arranged in a geometric pattern consisting of 20 equilateral triangles. (Figure 14-2 illustrates this pattern.) The icosahedron has 20 triangular sides, 12 edges, and 12 points. The viruses that cause infectious mononucleosis, mumps, chickenpox, and herpes simplex have icosahedral symmetry.

14.16 What is the appearance of a virus having helical symmetry?

Certain viruses such as the rabies virus and the tobacco mosaic virus have **helical symmetry**. The genome consists of a spiral coil of nucleic acid; the capsid follows the spiral arrangement. Under the electron microscope, it may be difficult to see the coil, and the virus may appear as a rod.

14.17 Do all viruses have either icosahedral or helical symmetry?

There are some viruses that have neither icosahedral nor helical symmetry. Poxviruses, such as those of cowpox and smallpox, appear in a boxlike arrangement with ultramicroscopic rods covering the surface. Other viruses such as influenza viruses are composed of eight helical nucleocapsids enclosed in an envelope.

14.18 What shape do the bacteriophages have and how does this shape compare to other viral shapes?

Bacteriophages have a complex shape. A bacteriophage may have a head region in which the nucleic acid is stored, a contractile tail region, a set of tail fibers, and a base plate having tail pins. These structures are all utilized for attachment to the host cell bacterium. Animal and plant viruses are considerably more simple.

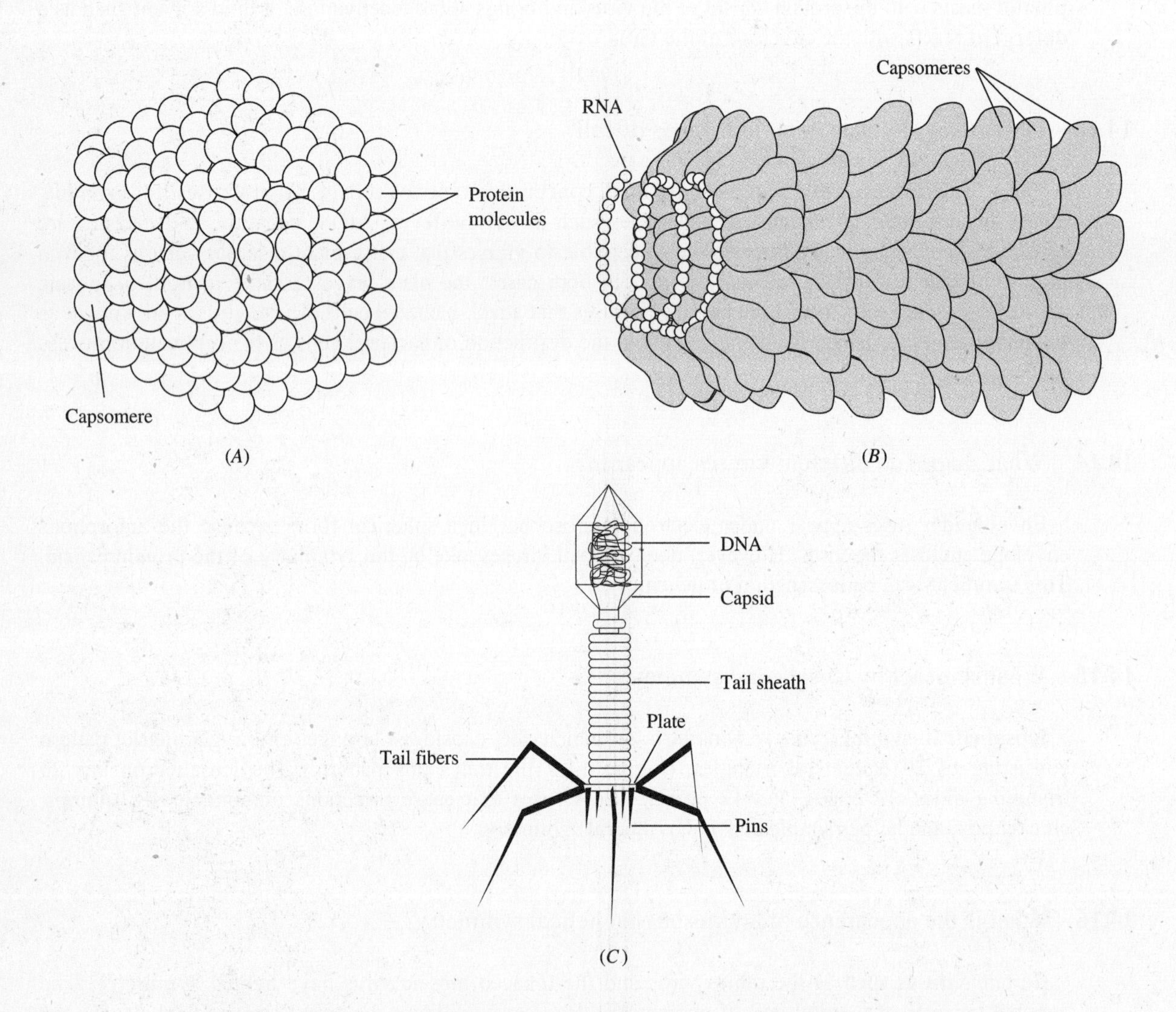

Fig. 14-2 Three possible shapes taken by viruses. (*A*) An icosahedral virus whose capsid is composed of 20 triangles organized in repeating units called capsomeres. (*B*) A helical virus showing the helical genome and capsomeres enclosing it. (*C*) The bacteriophage, a complex virus.

14.19 Is there a classification system presently in use for viruses?

At the present time, there is no universally accepted classification system for viruses. Viruses are categorized according to the type of nucleic acid they have (DNA or RNA), the single-strandedness or double-strandedness of the genome, the symmetry of the capsid, the presence or absence of an envelope, and the host range. Virus names have not been assigned, and viruses are still known by their common names such as poliovirus or poxvirus.

VIRAL REPLICATION

14.20 What is the general pattern displayed in viral replication?

Viruses cannot reproduce independently. They penetrate a host cell and utilize the metabolism of the host cell to produce copies of themselves. Viruses redirect the metabolic reactions of the cell and utilize resources of the cell to produce hundreds of new viruses. In doing so, they usually destroy the host cell and bring on viral disease.

14.21 What are the general steps of viral replication?

Most patterns of viral replication embody six steps: the attachment step in which a virus associates with its host cell; the penetration and uncoating steps in which the nucleocapsid penetrates into the cell cytoplasm and the viral capsid is lost; the synthesis step, where viral genomes and capsids are manufactured; the assembly step, in which the components are chemically joined to produce new viral particles; and the release step, where new viral particles exit the host cell. There are many variations of the steps depending upon where the replication takes place, but all steps generally occur. Figure 14-3 summarizes the process.

14.22 How does viral attachment to the host cell take place?

In the first event in viral replication, the virus attaches to the surface of a susceptible cell. This attachment may follow a random collision. It requires the specific interaction of **attachment sites** on the surface of the virus and **receptor sites** on the cell's surface. A proper fit must take place for attachment to occur. The envelope spikes of a virus often contain the substances to accomplish this fit. The attachment can be neutralized if the host organism produces antibodies to cover the attachment site before the virus adheres to the cell's receptor sites.

14.23 What are the various ways in which a virus can enter its host cell?

During the **penetration** step of viral replication, the virus may dissolve the cell membrane and propel its genome through the opening into the cell cytoplasm. This mode of penetration is observed when bacteriophages attach to their host bacterial cells. Nonenveloped viruses are often taken into their host cells by the process of phagocytosis. This process occurs in human and animal cells. In envelope viruses, a union may develop between the envelope and the membrane of the host cell. Where these two structures unite, a passageway opens for the nucleocapsid to enter the cell cytoplasm.

14.24 How does the uncoating step of viral replication happen in the host cell?

In order for a virus to replicate, its genome must be set free in the host cell cytoplasm. This **uncoating** step is achieved when cellular enzymes dissolve the protein of the viral capsid. The enzymes may be derived from lysosomes in the cellular cytoplasm. The protein digestion frees the viral nucleic acid and begins a stage of the replication known as the eclipse phase.

14.25 Which mechanisms underlie the replication of new viral particles in the host cell cytoplasm?

The nucleic acid of the viral genome directs the **synthesis** of new viral particles by encoding the synthesis of enzyme proteins that interfere with the expression of host genes. This interference leaves the viral nucleic acid as the sole director of the cell's metabolism. Still other genes specify proteins used in viral synthesis. And still other genes direct the synthesis of viral proteins that will become capsids. Essentially then, all the necessary genetic information for the synthesis of new viruses is brought into the cell by the viral genome.

14.26 What is the role of the host cell in viral synthesis?

Most of the organic compounds and structures required for viral synthesis are contained in the host cell. For example, the ribosomes necessary for protein synthesis are supplied by the host cell. Energy for synthetic processes is contained in ATP and other energy compounds within the cell. The building blocks for viral components are also supplied by the cell. These building blocks include amino acids for the proteins of viral capsids and nucleotides for the synthesis of viral genomes. Viruses can be seen within the cell in Figure 14-4.

14.27 What is the general pattern of replication in RNA-containing viruses?

When the genomes of **RNA viruses** are released in the cytoplasm, the RNA serves directly as a messenger RNA (mRNA) molecule and provides the genetic code for the synthesis of protein using the process of translation. The RNA molecule will proceed to the ribosome and encode the structural proteins and enzymes necessary for viral replication. In some cases, the RNA may be an antisense molecule. In this case, it serves as a template for the synthesis of a functional mRNA molecule. Protein synthesis can then proceed.

14.28 How does the pattern of replication for a DNA-containing virus differ than for an RNA containing virus?

When a **DNA virus** enters the host cell, the DNA genome provides a genetic code for the synthesis of mRNA molecules in the process of transcription. The mRNA molecules then proceed to the ribosomes and translate the code into an amino acid sequence in protein. If the DNA of the genome is double-stranded, only one strand will function in the protein synthesis. The other strand remains dormant.

14.29 Does the DNA always encode viral proteins immediately?

It sometimes happens that viral DNA does not function immediately. Instead of encoding proteins, the viral DNA proceeds to the cell nucleus and attach itself to one of the cell's chromosomes. This phenomenon is known as **lysogeny**. The virus is known as a **lysogenic virus,** or a **provirus**. At some time in the future, the provirus may be stimulated to encode viral proteins and encourage viral replication. For the time being, however, the provirus remains within the cell nucleus.

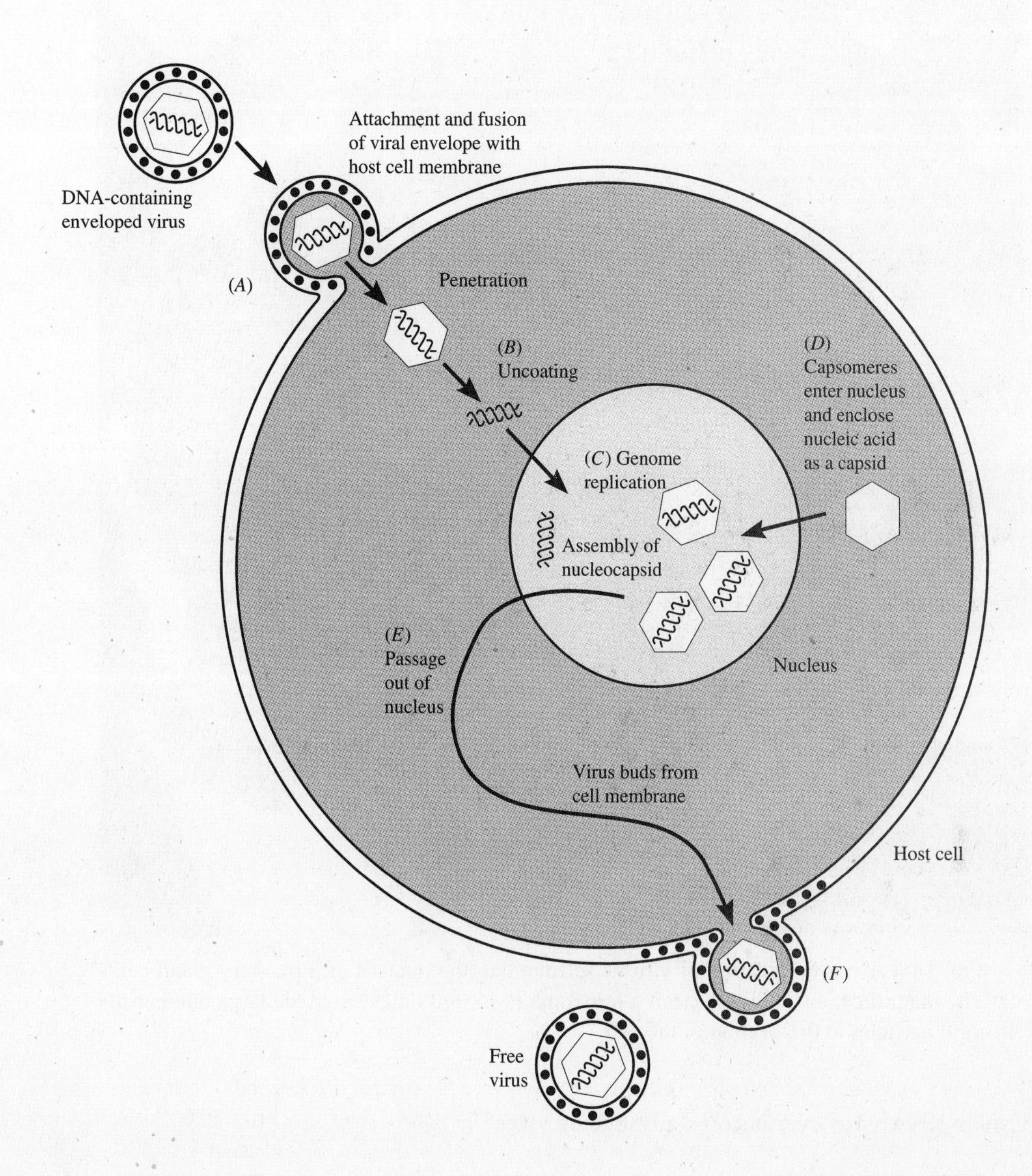

Fig. 14-3 The general steps of viral replication. (*A*) Attachment and fusion. (*B*) Uncoating of the viral genome. (*C*) Replication of the genome in the host cell nucleus. (*D*) Production of capsomeres in the host cell cytoplasm and passage of the capsomeres into the nucleus to form an enclosing capsid. (*E*) Passage of the nucleocapsid out of the nucleus into the cytoplasm of the host cell. (*F*) Budding of the nucleocapsid through the cell membrane to form an enveloped virus.

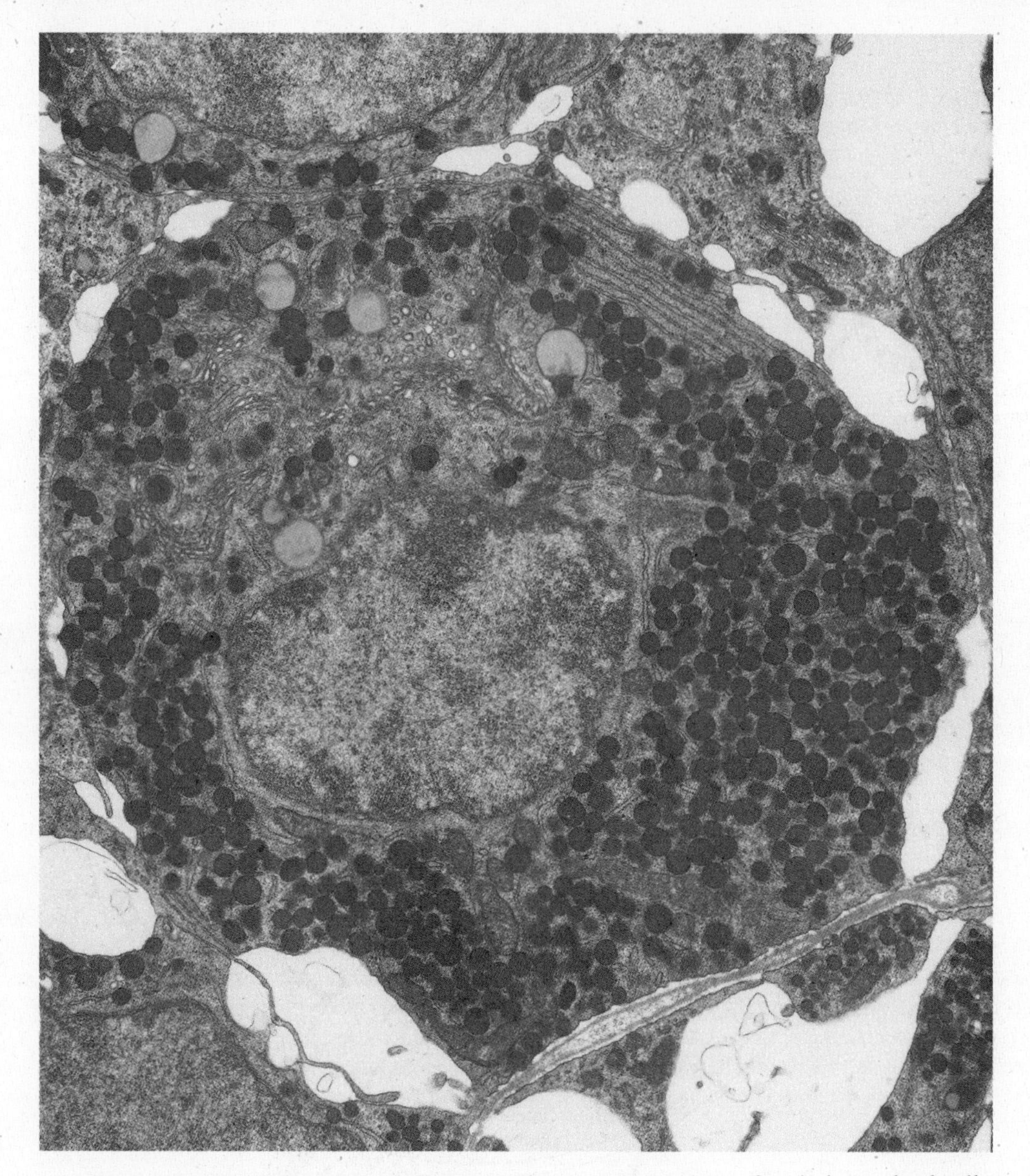

Fig. 14-4 An accumulation of viruses surrounding the nucleus of a pituitary gland cell. The magnification of this electron micrograph is 15,000 times. Note the large number of viral particles in this section of the cell.

14.30 Can an RNA virus ever become a lysogenic virus?

Certain RNA viruses can remain with the host cell and form a lysogenic relationship. In this case, the viral RNA is released in the cytoplasm and an enzyme called **reverse transcriptase** utilizes the RNA as a template to synthesize a complementary molecule of DNA. The DNA then proceeds to the cell nucleus and enters the nucleus as a lysogenic virus, or provirus. This pattern is followed by the human immunodeficiency virus (HIV) that causes **AIDS** in humans.

14.31 Where does the assembly phase of viral replication take place?

Different viruses direct the **assembly** of new viruses at different places in the cell. In many cases, the viral genomes and viral capsids are synthesized in the cytoplasm and assembly occurs here. For other viruses, genomes are synthesized in the cell nucleus and capsids in the cytoplasm. The capsids are then transported to the nucleus for assembly, then the nucleocapsids migrate back to the cytoplasm for release from the cell.

14.32 What are some of the characteristics of the release stage of viral replication?

There are various ways in which the **release** of new viruses can occur. In some viruses the enzyme lysozyme is synthesized, and this enzyme breaks down the cell wall of the host cell, such as the bacterium. The new viruses are then released during the breakdown of the cell. In other cases, metabolism of the cell is so interrupted that the cell undergoes spontaneous decay. This decay causes the cell to disintegrate and release its new viruses.

14.33 When do enveloped viruses acquire their envelopes?

Enveloped viruses acquire their envelopes during the release stage of replication. These viruses move to the membrane of the cell and force their way through the membrane in a process called **budding**. During budding, the nucleocapsid is surrounded by a portion of the membrane, which has been modified with viral proteins. The virus pinches off a portion of the membrane as the it escapes, after which the hole is sealed. If budding occurs too rapidly to seal the holes, the cytoplasm leaks out of the cells and they die.

14.34 What is the effect of viral replications on the host?

Normally, the effect of viral replication on the host cell is destruction. As the cells gradually die, hundreds and thousands of new viruses are produced to infect and destroy another generation of cells. The progressive deterioration of the tissues results in plant and animal disease. Indeed, bacteriophages are so-named for their ability to "eat" bacteria.

VIRAL PATHOLOGY

14.35 What diseases can viruses cause in humans?

Depending upon the tissues they infect, various viruses cause various human diseases. When viruses affect the liver, they cause diseases such as yellow fever or hepatitis (Figure 14-5). Infection of the respiratory tract may result in influenza, common colds, or respiratory syncytial disease. Brain infection can result in rabies, encephalitis, or polio. Infection of the gastrointestinal tract brings on gastroenteritis.

14.36 Can viruses be involved in cancers?

The integration of viral DNA into the cell nucleus may stimulate the cell to multiply without control. This uncontrolled replication results in a growth called a tumor and then a spreading of the **tumor** cells known as a **cancer**. Cancer cells have lost their sensitivity to signals that inhibit reproduction. The viruses that may induce this phenomenon are called oncogenic viruses. Certain viruses such as herpesviruses and Epstein-Barr virus have been known to induce tumors and cancers. Certain forms of leukemia are also known to be stimulated by viruses.

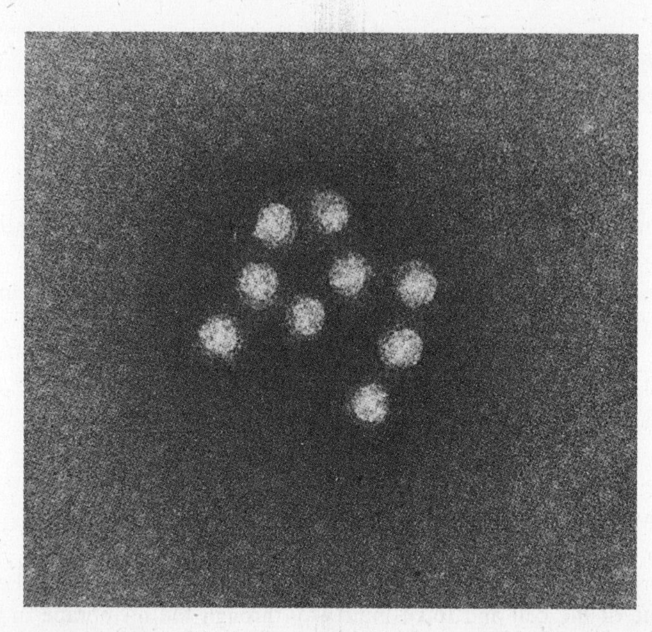

Fig 14-5 The viruses of hepatitis A observed with the transmission electron microscope and magnified 241,000 times.

14.37 What methods are used in the laboratory for cultivating viruses?

Strictly speaking, viruses are not "grown" in the laboratory because they are not alive. However, viruses can be cultivated and their numbers increased in cultures of living cells called **tissue cultures**. In addition, viruses can be cultivated in living animals such as mice and rabbits, and viruses can be cultivated in fertilized chicken eggs. Indeed, the different effects that different viruses have on the membranes of fertilized chicken eggs can be used as a detection method.

14.38 What signs and symptoms may be present that viruses have infected animals and humans?

Many symptoms of disease are accompanied by signs of viral presence. For example, brain cells infected with rabies viruses develop unique bodies called **Negri bodies**. In addition, white blood cells affected with mononucleosis viruses become granulated and foamy and form characteristic **Downy cells**. Moreover, various viruses cause unique symptoms such as characteristic skin rashes. The measles and smallpox are examples.

14.39 How can viruses be detected in the laboratory?

Because they are smaller than other microorganisms, the standard light microscope cannot be used to detect viruses. However, many indirect techniques do exist and tissue destructions are used as signals that viruses are present. For example, viruses may cause zones of cell destruction called **plaques** when viruses are cultivated in tissue culture or on lawns of bacteria, as Figure 14-6 displays. In addition, viruses may cause cells to fuse into larger bodies called **syncytia**, or they may cause cells to clump in a process called **agglutination**, or they may induce cells to develop intracellular aggregates called inclusions. These effects on cells are called **cytopathic effects**.

14.40 Are antibiotics or other drugs useful against viruses?

The well-known antibiotics such as penicillin, tetracycline, erythromycin, and other antibiotics are useless against viruses because viruses do not have the chemistry or metabolism with which these antibiotics interfere. For example, viruses have no cell walls and therefore they are resistant to penicillin (which interrupts the synthesis of the cell wall). However, there are certain drugs that can interfere with the replication of viruses within cells. One example is **acyclovir,** which interferes with the replication of herpes viruses in skin cells. Another example is **azidothymidine (AZT),** which interferes with HIV replication in lymphocytes. Still another drug is **amantadine**, which is used against influenza viruses.

14.41 How is interferon used against viruses?

Interferon is a protein produced by cells after they have been penetrated by viruses. The interferon cannot stop viral replication in that cell, but it is released to adjacent cells where it stimulates the production of antiviral factors. Interferon is now produced by genetic engineering techniques and it is possible that interferon will become a useful drug for antiviral therapy in future years.

14.42 How does resistance to viral disease occur in the body?

When viruses enter the body, they stimulate the immune system to produce long strands of protein called **antibodies**. The antibodies fill the body fluids and eventually concentrate in the region of viral replication. Here they unite with the capsids and/or envelopes of the viruses and prevent the attachment of viruses with host cells. In this way, the viruses cannot replicate any further and the infection comes to an end. The body's phagocytes then engulf and destroy clumps of viruses neutralized by the antibodies.

14.43 In what ways can protection against viral diseases be stimulated in the body?

Protection against viral diseases can be stimulated by encouraging the body to produce antibodies before the viral infection takes place. The body's immune system is stimulated by **vaccines**, which consist of viruses altered so they will not replicate in the body cells, but remain able to stimulate an immune response.

14.44 Which two kinds of viruses can be used in vaccines?

The two types of viruses currently used in vaccines are inactivated viruses and attenuated viruses. **Inactivated viruses** are viruses whose genomes have been destroyed by chemical or physical means. These viruses retain their capsids, and the capsids call forth a specific response by the immune system. **Attenuated viruses**, by contrast, are viruses that have been developed for their ability to replicate very slowly in the body. These viruses also elicit an antibody response by the immune system. The viruses are so weak that they do not cause disease, but they are active enough to stimulate an immune response.

14.45 What is the advantage of using inactivated viruses in a vaccine?

The advantage of inactivated viruses in a vaccine is that the viruses cannot multiply in the body, and therefore they cannot possibly cause disease. The drawback is that the inactivated viruses must be injected into the body because they will not survive the passage into the gastrointestinal tract if they were taken orally.

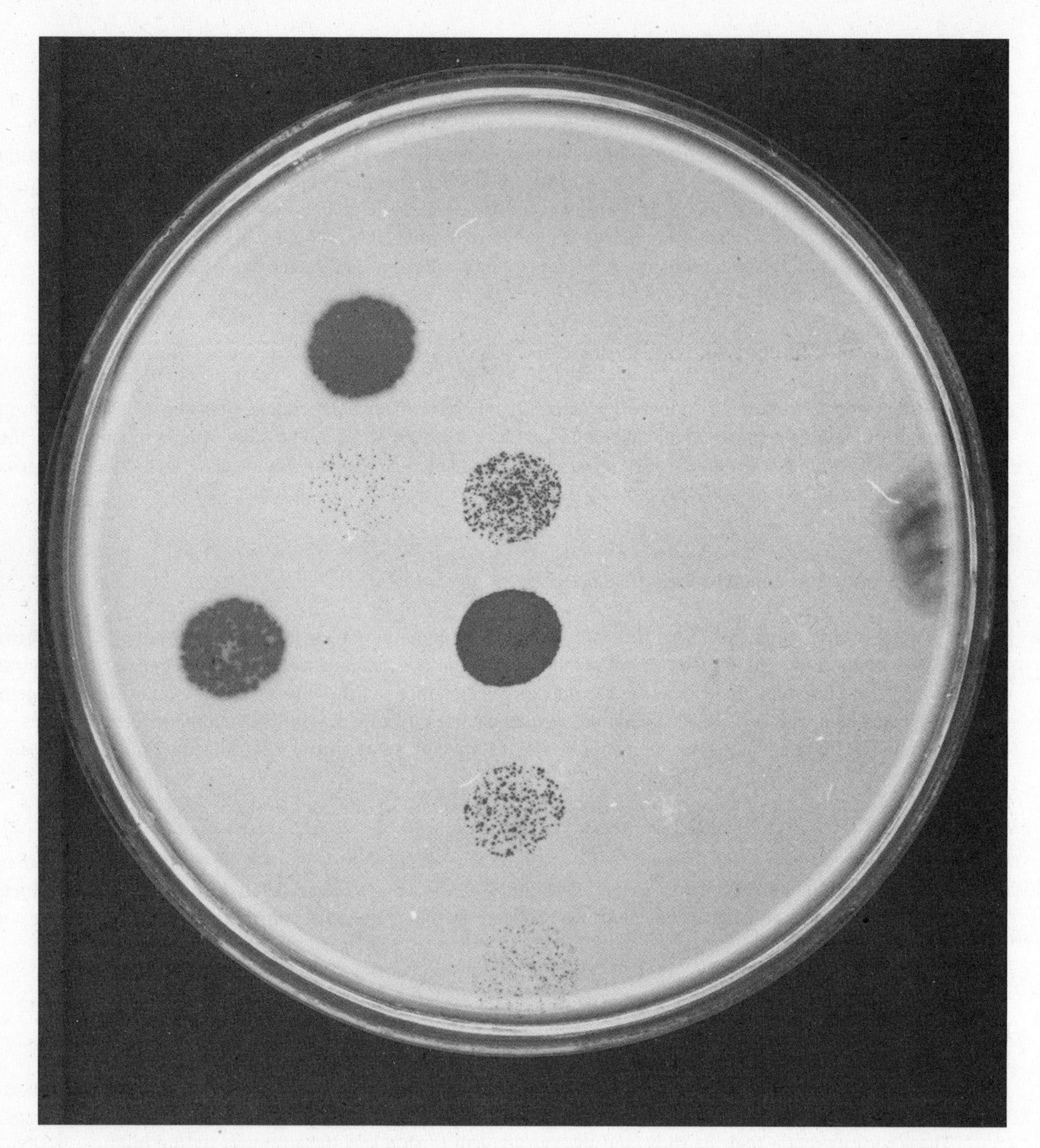

Fig. 14-6 Viral plaques. This plate of medium contains bacteria growing evenly throughout the surface. The dark areas represent places where a virus has multiplied within the bacteria and destroyed the bacteria to produce a moth-eaten area known as a plaque. Presence of five or more plaques can be seen.

14.46 What is the advantage of the attenuated viruses in a vaccine?

Attenuated viruses can be taken orally because they will survive passage into the gastrointestinal tract and will be absorbed into the blood stream. The method of administration is therefore easier than by injection. However, the drawback is that the viruses may on certain occasion induce mild symptoms of disease in the body and they therefore pose more of a risk than the inactivated viruses. They do call forth a more substantial immune response, however, and for this reason, they may be advantageous when an epidemic is taking place or is anticipated.

Review Questions

True/False. For each of the following statements, mark the letter "T" next to the statement if the statement is true. If the statement is false, change the underlined word to make the statement true.

___ **1.** The protein coat of a virus, known as the capsid, is composed exclusively of carbohydrate.

___ **2.** The two basic types of symmetry that most viruses have are helical symmetry and icosahedral symmetry.

___ **3.** An example of a virus having helical symmetry is the herpes simplex virus.

___ **4.** Bacteriophages are those viruses that replicate within bacteria.

___ **5.** When a virus associates with its host cell, an attachment takes place between the attachment site on the virus and the receiving site on the host cell.

___ **6.** Among the structures supplied by the host cell for the process of viral replication are the ribosomes.

___ **7.** The poxviruses are among the smaller viruses known to science.

___ **8.** The capsid of a virus is composed of smaller, repeating units known as telomeres.

___ **9.** The viruses of chickenpox, infectious mononucleosis, and herpes simplex are distinctive because they have a membrane-derived structure known as the genome.

__ **10.** To inactivate a virus, the chemical phenol can be used to alter the genome of the virus.

__ **11.** When the RNA of an RNA virus is released in the cytoplasm, the molecule serves as a messenger RNA molecule and provides the information for the synthesis of protein.

__ **12.** A lymphogenic virus is one whose DNA attaches itself to the DNA of the host cell chromosome.

__ **13.** An example of a virus that contains RNA in its genome is the AIDS virus.

__ **14.** A virus attacking the human brain may cause the disease hepatitis.

__ **15.** Zones of cell destruction known as plaques can be used as signals that viruses are present.

__ **16.** When the cells of the brain have been infected with viruses, they form Downy cells.

__ **17.** Viruses are resistant to the effects of penicillin because they do not possess genomes.

__ **18.** The substance acyclovir is a natural substance produced by human body cells after they have been infected by viruses.

__ **19.** The major mechanism for the body's resistance to viral disease is centered in strands of protein called antigens.

__ **20.** Inactivated viruses are viruses whose genomes have been destroyed by chemical or physical means for use in vaccines.

Multiple Choice. Select the letter of the item that correctly completes each of the following statements.

1. Which of the following statements applies to the viruses (*a*) they metabolize organic materials and grow actively (*b*) they are noncellular particles (*c*) they produce various waste products (*d*) they adapt quickly to their local environment.

2. The smaller viruses have a diameter of about (*a*) 20 mm (*b*) 300 μm (*c*) 300 nm (*d*) 20 nm.

3. In order to see the viruses (*a*) the oil immersion lens must be used (*b*) a darkfield microscope must be employed (*c*) an electron microscope is necessary (*d*) the low power objective is sufficient.

4. All of the following are possible components of viruses except (*a*) a capsid of protein (*b*) a genome of nucleic acid (*c*) a membranelike envelope (*d*) a cytoplasm with enzymes.

5. The genome of a virus may be composed of (*a*) DNA but not RNA (*b*) RNA but not DNA (*c*) both RNA and DNA together in the same virus (*d*) either DNA or RNA.

6. Substances used by the virus to unite with its host cell are often contained in the (*a*) genome (*b*) spikes (*c*) DNA portion of the virus (*d*) RNA portion of the virus.

7. The icosahedral shape assumed by many viruses resembles (*a*) a spiral coil (*b*) a boxlike arrangement (*c*) a geometric figure of 20 equilateral triangles (*d*) a rectangle with triangular corners.

8. In order for viruses to replicate in their host cells, all the following must occur except (*a*) the genome must be released in the host cell cytoplasm (*b*) ATP must be synthesized within the virus (*c*) the virus must unite with the correct host cell (*d*) the host cell must contain ribosomes for the synthesis of proteins.

9. All of the following are considered in categorizing a virus except (*a*) the type of nucleic acid it has (*b*) the single-strandedness or double-strandedness of the genome (*c*) the symmetry of the capsid (*d*) the Gram reaction.

10. Most of the organic compounds and structures necessary for the synthesis of new viruses are contained (*a*) in the viral capsid (*b*) in the host cell (*c*) in the viral genome (*d*) in the viral envelope.

11. A lysogenic virus is one which (*a*) remains in the host cell nucleus (*b*) multiplies immediately after it enters the host cell (*c*) contains its own ATP for replication (*d*) contains its own enzymes for replication.

12. An envelope is acquired by certain viruses when they (*a*) enter the host cell nucleus (*b*) combine the genome and capsid (*c*) bud through the host cell membrane (*d*) migrate to the Golgi body.

13. The type of disease caused by a virus is most dependent upon (*a*) the type of nucleic acid in the viral genome (*b*) the presence or absence of capsids in the virus (*c*) the tissue infected by the virus (*d*) whether there are spikes in the viral envelope.

14. Oncogenic viruses are those viruses believed involved in (*a*) cases of hepatitis (*b*) tumors and cancers (*c*) skin diseases in animals only (*d*) viral replication without the use of the genome.

15. Most protection against viral disease in the body takes place through the activities of (*a*) penicillin molecules (*b*) antibody molecules (*c*) antigen molecules (*d*) interferon molecules.

16. All the following can be used to cultivate viruses in the laboratory except (*a*) fertilized chicken eggs (*b*) living animals such as mice (*c*) slants of nutrient agar (*d*) tissue cultures.

17. Antiviral drugs are used to (*a*) inhibit the replication of viruses within cells (*b*) prevent the formation of enzymes within the viral genome (*c*) inhibit the formation of the viral envelope (*d*) stimulate the production of antibodies in the body.

18. Under the electron microscope, helical viruses appear as (*a*) tiny spheres (*b*) elongated triangles (*c*) rods (*d*) tadpoles.

19. An example of a virus that dwells within the nucleus of a host cell is the virus that causes (*a*) malaria (*b*) measles (*c*) AIDS (*d*) strep throat.

20. Cytopathic effects are observed during the (*a*) cultivation of viruses in tissues (*b*) dormant stage of the virus (*c*) assembly stage of new viral particles (*d*) inactivation of viruses by chemicals.

Completion. Add the word or words that correctly complete each of the following statements.

1. Viruses are able to reproduce only within ____________ .

2. The genome of a virus may be composed of RNA or ____________ .

3. The size of the largest viruses is approximately as the size of the smallest ____________ .

4. The capsid of a virus is composed of repeating units known as ____________ .

5. In certain types of viruses, the nucleocapsid leaves its host cell at the conclusion of the replication process and acquires its ____________ .

6. The name "phage" is an abbreviation for those viruses that attack and multiply in ____________ .

7. Because an envelope encloses the nucleocapsid, envelope viruses appear under the electron microscope in the form of a ____________ .

8. To inactivate a virus, the protein capsid may be treated with a chemical such as ____________ .

9. An alternate way of inactivating a virus is to treat its envelope with a lipid solvent such as a ____________.

10. A boxlike arrangement with ultramicroscopic rods at the surface describes the appearance of ____________ .

11. The attachment of a virus to its host cell may be neutralized by the presence of specific ____________ .

12. Enzymes for breaking down the capsid of a virus may be derived from the cell's ____________ .

13. The genetic information for the synthesis of new viruses is brought into the host cell by the viral ____________ .

14. When the RNA of an RNA virus enters the host cell cytoplasm, it will encode necessary proteins at the cell structure known as the ____________ .

15. The human immunodeficiency virus forms DNA using RNA as a template in the presence of a necessary enzyme called ____________ .

16. In the release of new viruses from the host cell, the cell wall can be broken down by the enzyme ____________ .

17. The effect of viral replication in host cell tissues usually is ____________ .

18. Although viruses can be cultivated in the laboratory, in the strict sense they cannot be ____________ .

19. Viruses such as certain herpesviruses and the Epstein-Barr virus have been implicated in the formation of ____________ .

20. One drug that is able to interfere with the replication of viruses such as the human immunodeficiency virus is the drug ____________ .

Chapter 15

The Host-Parasite Relationship

OBJECTIVES

The host-parasite relationship is a fundamental concept in the study of infectious disease. Learning the details of this concept is a main focus of this chapter, whose objectives are to:

1. Understand the essential difference between infection, contamination, and disease.

2. Recognize the existence and members of the normal flora found in various parts of the human body.

3. Summarize various prerequisites that must be met if infectious disease is to occur.

4. Identify a number of virulence factors that increase the potential for disease.

5. Explain various types of infections that can occur in the human body and several modes of transmission by which infectious organisms pass among individuals.

THEORY AND PROBLEMS

15.1 Which different relationships exist between hosts and parasites, and what is the importance of the host-parasite relationship in human disease?

There are many relationships that can exist between the human host and the parasites that live on or in it. For example, the relationship may be beneficial to both organisms, a condition known as **mutualism**; or the relationship may be one in which one organism damages the other, a relationship known as **parasitism**. Parasitism is the condition that is taking place when humans are invaded by infectious microorganisms. Therefore, studying the host-parasite relationship is fundamental to the study of infectious diseases.

15.2 Is there any difference between infection and disease?

Infection is a condition in which microorganisms penetrate the host defenses, colonize its tissues, and multiply furiously. The microbe may be eliminated by the body or become part of its normal flora. When the microorganisms display their pathogenicity (i.e., disease potential) and when the cumulative effects of the infection bring about damage in the tissue, the result is **disease** (Figure 15-1). Infection is a living together of the host and parasite, while disease is a deviation from the good health of the host. Despite this difference, the terms are often interchanged in medical literature.

15.3 Is contamination the same as infection?

There is a significant difference between contamination and infection. **Contamination** takes place when a potentially infectious microorganism exists in the body but has not yet invaded the tissue. The body is said to be contaminated under these circumstances. When the microorganism begins its invasion of the tissue and its rapid multiplication, then infection occurs.

15.4 Are there any microorganisms that inhabit the human body but do not cause disease?

There is a host of microorganisms that inhabit various regions of the body and cause no disease. These populations of microorganisms are alternatively called the **normal flora**, the **indigenous flora**, or the **microflora**. Numerous species of bacteria, protozoa, fungi, and viruses may comprise the normal flora.

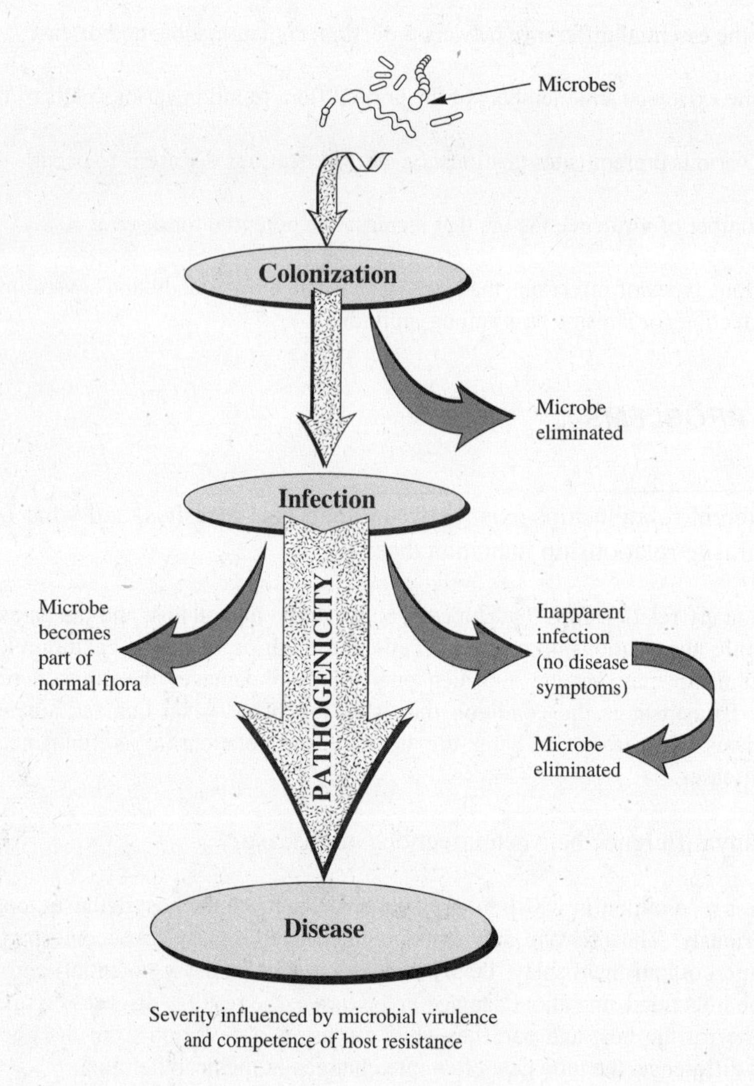

Fig. 15-1 The relationship between infection and disease following entry of microbes to the body and colonization in its tissues.

THE NORMAL FLORA

15.5 Explain the relationship between the normal flora and the human body?

In many cases the relationship is a mutualism (as noted previously), or it may be a **commensalism**. In commensalism one population benefits and there is no damage to the other population. Organisms that exist in a commensalism are referred to as **commensals**. An example of commensals is the population of microorganisms living in the human intestine that provide neither advantage nor harm to the human body.

15.6 What is the source of the normal flora in various parts of the human body?

The normal flora is acquired during the first human contact with the world after birth. Some microorganisms are acquired by human contact (such as the baby's contact with its mother), while others are acquired from foods such as baby formulas, or from contact with various objects such as toys. Microorganisms in the air also contribute to the normal flora of epithelial surfaces. The development continues and is refined over many years.

15.7 Which are the sites in the body that have a normal flora?

Virtually all surfaces exposed to the environment harbor a normal flora. These surfaces include the skin and mucous membranes, various parts of the gastrointestinal tract, the respiratory tract, the external portions of the eye and ear, and the external reproductive organs that come in contact with the air.

15.8 Are there any places in the human body where there is no normal flora?

The internal tissues and organs of the body and the cells within these tissues do not have a normal flora. Organs such as the heart, liver, brain, muscles, and reproductive organs have no normal flora. Neither is there a normal flora in the blood, cerebrospinal fluid, semen in the testes, or urine in the kidneys.

15.9 Which site in the human body has the most extensive normal flora?

Because the skin is the largest and most accessible organ, it has the most numerous and varied normal flora. The **transient population** of the skin consists of microorganisms that do not usually grow there. For example, contact with the soil will bring organisms to the skin. The **resident population**, by comparison, consists of microorganisms that are normally able to grow on and within the skin. Species of staphylococci are typical.

15.10 Which microorganisms inhabit the normal flora of the mouth?

The mouth contains one of the most diverse normal floras of the body. Aerobic species of streptococci live in the epithelial tissues of the mouth, while anaerobic species such as lactobacilli live in the sticky plaque that develops around the teeth. Numerous other species of bacteria, as well as many mold spores and yeasts, are also found in the mouth.

15.11 What types of bacteria are found in the normal flora of the large intestine?

The large intestine contains huge populations of microorganisms, so abundant that they make up almost 25% of the feces. Both anaerobic and aerobic bacteria are present in this flora, and many species break down otherwise indigestible foods and manufacture vitamins used by humans. Vitamins K, B_{12}, and riboflavin are synthesized by bacteria of the flora. *Escherichia coli* is a common member of the large intestine's normal flora.

15.12 Do members of the normal flora ever cause human disease?

Under normal conditions the body's defense mechanisms hold microbial populations in check. However, should the immune system be compromised, such as in AIDS patients, some "opportunistic" species of the flora may invade the tissue. For example, *Candida albicans* is a yeast found normally in the female vagina and the human intestine without any damaging effects. In the AIDS patient, however, the organism invade many internal tissues such as the esophagus, where it causes severe erosion . Also, excessive amounts of antibiotics taken for another disease kill the bacteria normally holding the *Candida albicans* in check, and the yeast causes "yeast disease." Another opportunistic organism is *Toxoplasma gondii*, the cause of brain seizures in AIDS patients. The organism is shown in Figure 15-2.

PATHOGENICITY

15.13 What is the difference between microorganisms of the normal flora and microorganisms that are pathogenic?

Pathogenic microorganisms possess qualities not possessed by microorganisms of the normal flora. These qualities permit pathogens to invade the tissue and bring about a change in good health. They also permit the organisms to overcome normal body defenses. The qualities add to the pathogenicity of the microorganisms.

15.14 Are there degrees of pathogenicity among microorganisms?

The term **virulence** refers to the degree of pathogenicity among microorganisms. Some organisms such as the typhoid bacilli are highly virulent, while others such as the measles virus are moderately virulent, and others, such as the common cold viruses are weakly virulent. Virulence refers to the potential for causing disease in the body.

15.15 Where do the opportunistic pathogens fit the scheme of pathogenicity?

Opportunistic microorganisms invade the tissue when given the "opportunity," as noted previously. Often this will happen when the individual is receiving therapy for cancer or is being given antirejection drugs following an organ transplant. In both cases the normal immune responses have been compromised and the opportunistic organisms invade the body tissues.

15.16 How important is the size of the inoculum to the establishment of disease?

The dose of invading microorganisms received by the host is referred to as the infectious dose. For certain microorganisms, such as typhoid fever bacilli, the **infectious dose** is very low, possibly as low as a few hundred bacteria. For other bacteria, such as those that cause cholera, the infectious dose is quite high. The reason that the infectious dose varies has to do with the resistance to the invading microorganisms taking place in various regions of the body. For human immunodeficiency virus (HIV), the infectious dose is high.

15.17 How can the infectious dose be increased in nature?

The mechanism of exposure to pathogenic microorganisms often reflects the infectious dose. For instance hepatitis B viruses may be well dispersed in water, and fish can be taken from that water and eaten safely. However, the viruses are concentrated in the gills of clams, oysters, and other shellfish, and if these foods are eaten raw, the infectious dose will be much higher. For this reason, raw seafood taken from contaminated water poses a danger to health.

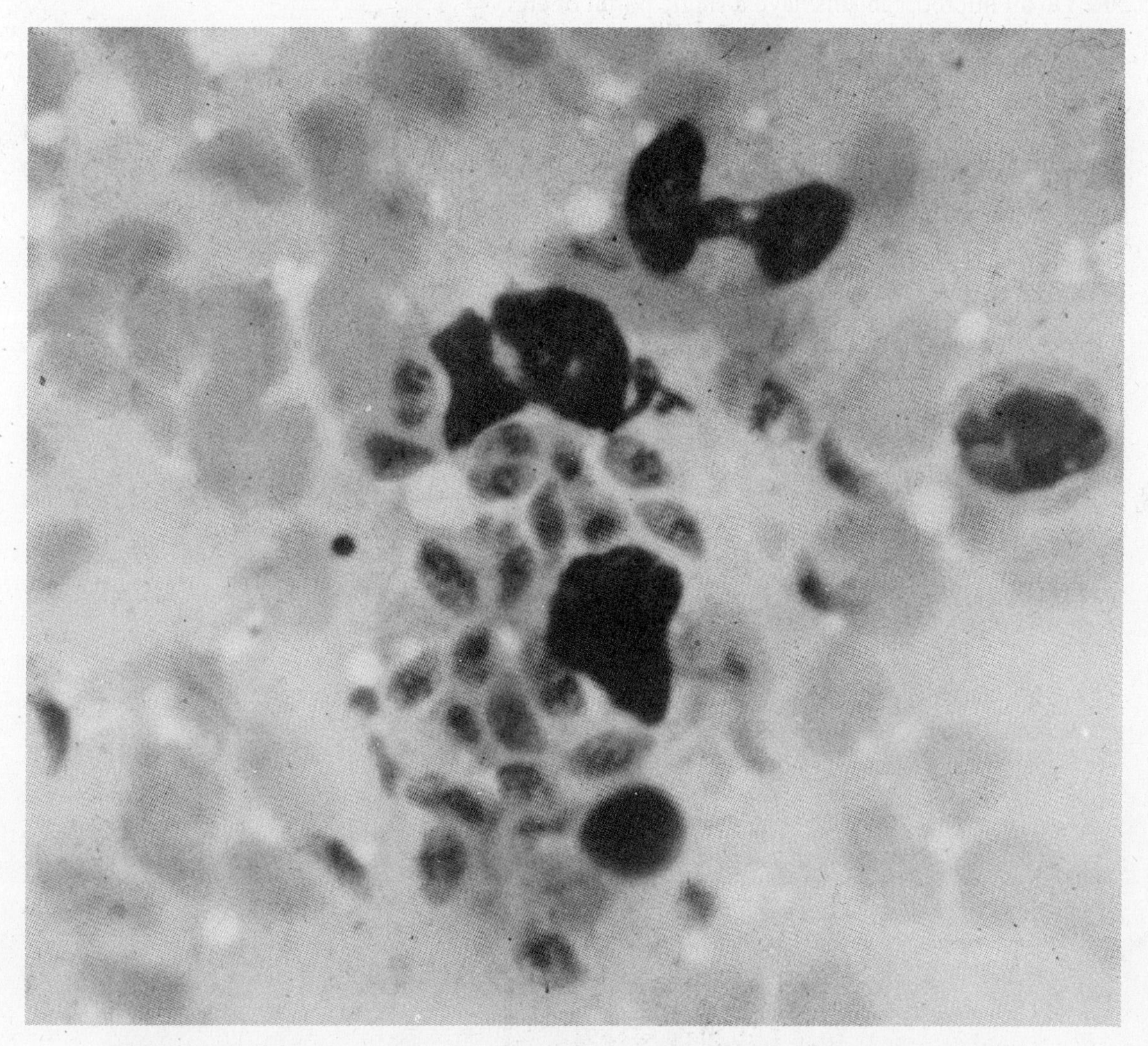

Fig. 15-2 A light micrograph of numerous crescent-shaped cells of *Toxoplasma gondii*. These organisms rarely cause serious disease in immune-competent individuals, but in immune-compromised people, they act as opportunists and invade numerous tissues such as the brain.

15.18 How important is the portal of entry to the establishment of infection in the body?

If a microorganism is pathogenic, it must gain entry to the body before disease can be established. The mechanism of entry is referred to as the **portal of entry**. Usually the microorganism enters from an **exogenous source**, that is, a source outside the body. On occasion, however, it enters from an **endogenous source**, one already within the body such as in the gastrointestinal tract.

15.19 Will infection occur if the portal of entry is not met?

Infection will probably not occur if the correct portal of entry has not been fulfilled. For example, common cold viruses must enter the respiratory tract for a common cold to develop. If these viruses were present on the skin, the portal of entry would not be fulfilled, and disease would probably not occur.

15.20 Do all microorganisms have a single portal of entry?

For many microorganisms, there is a single portal of entry. For instance, in order for malaria to occur, the malaria parasite must be injected into the bloodstream by a mosquito bite. For other diseases, multiple portals of entry may exist. Tuberculosis, can develop if the tubercle bacilli enter the respiratory or gastrointestinal tracts, and the tularemia bacillus enters the body through the eye, skin, respiratory tract, or directly into the blood. Several portals of entry in two individuals are shown in Figure 15-3.

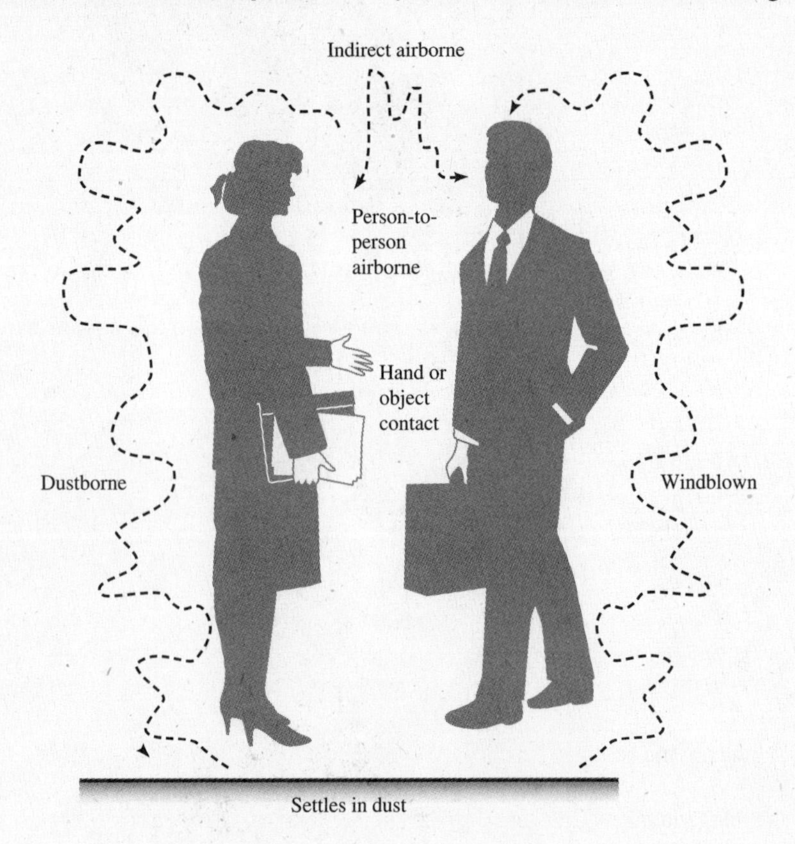

Fig. 15-3 Several methods for the transmission of microorganisms between two individuals by several portals of entry.

15.21 Which other virulence factors contribute to the establishment of an infection?

Once the pathogenic organisms have entered the correct portal of entry and are in sufficient numbers, they bind to the tissue and penetrate it to establish an infection. **Binding** is encouraged by the presence of villi on bacteria, and **penetration** is encouraged by enzymes produced by the pathogens. One example of penetration is found in *Entamoeba histolytica*, the cause the amoebiasis. This protozoal parasite produces enzymes that destroy the epithelial cells of the gastrointestinal lining and permit the parasite to invade and cause ulcers. Microbial capsules also encourage adhesion and binding, and viruses use specific binding molecules that attach to receptor sites on cells.

15.22 Can microorganisms produce enzymes to encourage their passageway through the tissues?

Certain microorganisms produce the enzyme **hyaluronidase**. This enzyme digests the polysaccharide hyaluronic acid, a type of "intercellular cement" which binds cells together in tissues. The enzyme encourages the pathogen to pass between the tissues. Another enzyme, **collagenase**, digests collagen fibers in connective tissues and thereby encourages passage.

15.23 What mechanisms do pathogens possess to interrupt cellular processes occurring in the body?

Many microorganisms produce **toxins**, a series of toxic chemical substances that interfere with the physiology and metabolism of cells and tissues. The presence of toxins in the tissues is referred to as **toxemia**, and a disease in which large amounts of toxin are produced is known as an **intoxication**.

15.24 Are there various kinds of toxins that can be produced in an infection?

Two general types of toxins can be produced by microorganisms. Toxins that are secreted by the microorganisms are referred to as **exotoxins**. Toxins retained by microorganisms and released on the death of the microorganisms are called **endotoxins**. Endotoxins are usually part of the cell wall of the microorganism.

15.25 Which effects can exotoxins cause in the body?

There are various types of exotoxins that can affect body cells. Certain microorganisms produce **hemolysins,** a series of exotoxins that destroy red blood cells and thereby reduce the body's oxygen-carrying capacity. Other microorganisms produce **leukocidins**, which are toxins that destroy white blood cells. White blood cells normally perform phagocytosis on microorganisms, so the activity of leukocidins reduces body defenses.

15.26 Are there any toxins that are directly responsible for the symptoms of disease?

Among the bacteria, the organisms of diphtheria, tetanus, and botulism are well-known for their exotoxins. The **diphtheria** exotoxin damages cells of the respiratory tract and causes pseudomembranes to form in the tract, leading to suffocation. The **tetanus** exotoxin encourages spasms of muscles by interfering with the activities taking place in the synapse. The **botulism** toxin is also active in the synapse, where it prevents the release of acetylcholine and other neurotransmitters, thereby causing paralysis in the muscles.

15.27 Which type of microorganisms produce exotoxins and which produce endotoxins?

Exotoxins are produced primarily by Gram-positive bacteria, while endotoxins are produced by Gram-negative bacteria. Viruses do not produce toxins; the toxins of fungi and protozoa have not been studied in depth.

15.28 What sort of effects do endotoxins have on the body?

Endotoxins cause nonspecific effects on the body. These effects can include fever, hemorrhage, and collapse of the cardiovascular system. This collapse is often referred to as **endotoxin shock.** It can occur when Gram-negative bacteria break down in the blood and release their endotoxins, often as a result of antibiotic therapy.

TYPES OF DISEASES

15.29 What is the difference between localized infections and systemic infections?

The pattern of infection in humans may vary according to where the infection occurs. A **localized infection** remains at a specific body site, usually the first site of infection. A **systemic infection**, by contrast, is one that spreads to alternative sites in the body such as deeper organs or tissues. A staphylococcal infection for instance, may remain as a localized infection on the skin, or it may become a systemic infection if it occurs in the blood, lungs, meninges, or other distant sites in the body.

15.30 Can an infection be due to several microorganisms at one time?

The general pattern is that a single microorganism is involved in a single disease. However, there are times when multiple pathogens cause infection in the body at the same time. Such an infection is known as a **mixed infection**.

15.31 What is the difference between a primary infection and a secondary infection?

A **primary infection** is established in a previously healthy body, while a **secondary infection** develops in the body after it has already been infected by a different microorganism. For instance, certain types of pneumonia (e.g., mycoplasmal pneumonia) are due to microorganisms acquired from another individual. These are primary infections. Other types of pneumonia (e.g., pneumococcal pneumonia) occur in the body when the body is already fighting an infection due to influenza virus, HIV, staphylococci, or other microorganisms.

15.32 How are acute and chronic infections distinguished from one another?

Acute infections are those that come on rapidly, are accompanied by severe symptoms, and progress quickly to a climax. **Chronic infections** occur slowly, generally are accompanied by mild to severe symptoms, and persist over a long period of time, with an extended convalescence. An example of an acute disease is measles, while a chronic disease is typified by infectious mononucleosis.

15.33 What is the difference between signs and symptoms of disease?

A **sign** of disease is a measurable evidence of disease noted by an observer, while a **symptom** is a subjective measure of disease as noted by the patient. In typhoid fever, a sign of disease is rose-colored spots on the body, while a symptom is the overwhelming malaise encountered by the patient.

15.34 Are all diseases accompanied by obvious signs and symptoms?

Not all diseases can be accompanied by apparent signs and symptoms. Sometimes a disease is **asymptomatic**. This term implies that no signs or symptoms are associated with a disease. A disease may also be **subclinical,** meaning that there are no clinical signs of infection.

15.35 How can diseases be transmitted among various individuals?

In order for transmission among individuals to occur, the pathogenic microorganisms must leave the body through a **portal of exit**. Transmission can occur in the form of respiratory secretions expelled from the respiratory tract, or microorganisms can exit in the feces or urine, or they may be removed when blood is ingested by mosquitoes, ticks, or other arthropods. Skin contact, including contact made during sexual intercourse, is another mechanism for transport to the next individual.

15.36 Which public health professionals are involved in the study of disease?

The discipline of **epidemiology** is concerned with the spread of infectious disease in the community. The public health professional in this field is an **epidemiologist**. An epidemiologist conducts surveillance studies and is responsible for detecting disease in a population and charting its course.

15.37 Which is the principal public health agency in the United States?

In the United States, the government agency that is concerned with epidemiology is the **Centers for Disease Control and Prevention (CDC)** in Atlanta, Georgia. Epidemiologists at the CDC publish the *Morbidity and Mortality Weekly Report,* a publication that summarizes the diseases present in the community and their significance.

15.38 How can disease outbreaks be classified to distinguish them from one another?

Outbreaks of disease can be classified as endemics, epidemics, and pandemics. An **endemic** (or endemic disease) refers to a disease present in a human population, but remaining within that population over a relatively long period of time. By contrast, an **epidemic** (or epidemic disease) breaks out of a population and spreads rapidly over a relatively short period of time. A **pandemic** is an epidemic occurring on a worldwide scale. The AIDS pandemic is an example of a pandemic of current times.

15.39 How are microorganisms maintained in nature when epidemics or pandemics are not taking place?

In the time between epidemics and pandemics, microorganisms are maintained in humans who are reservoirs or carriers. A **reservoir** is a human or animal that maintains the pathogen without showing any signs or symptoms of disease. Soil, water, plants and inanimate objects can also serve as reservoirs. **Carriers,** by contrast, are individuals who have had the disease and recovered but continue to maintain the pathogen and spread it to others. A carrier may be outwardly healthy but continue to spread the disease, nevertheless.

15.40 Are there any different types of carriers that can be present in a human population?

Carriers of pathogens may be of various types depending upon which stage of disease they are in. An **incubation carrier**, for example, spreads the pathogen while incubating the disease before any signs or symptoms emerge. Persons who carry HIV are incubation carriers. Another is the **convalescent carrier**, one who has had the disease and is now in the carrier stage. Hepatitis A is often transmitted by convalescent carriers. A **chronic carrier** is one who has recovered from the disease but suffers occasional mild relapses. One who has tuberculosis typifies a chronic carrier. A **contact carrier** is an individual who has picked up microorganisms from one source and carried them to another source. People in the food industry can be contact carriers.

15.41 Are animals able to transmit infectious disease among humans?

Animals that transmit human diseases are referred to as **vectors**. A vector can be a **biological vector** if it is infected by the pathogen and transmits the pathogen before it dies, or it may be a **mechanical vector** if it transmits the pathogen without becoming infected itself. Mosquitoes that transmit the yellow fever virus are biological vectors, while flies that transmit microorganisms among food are mechanical vectors.

15.42 Are all microbial diseases communicable?

A **communicable disease** is one that can be transmitted from one host to another and most infectious diseases fall into this category. **Contagious diseases** are communicable diseases that are particularly easy to transmit. There are some infectious diseases that are **noncommunicable**. These diseases are acquired directly from the environment or from organisms already in the body. Tetanus is an example of a noncommunicable disease; pneumococcal pneumonia is a second. People with these diseases do not normally transmit the disease to other individuals.

15.43 What is the difference between horizontal and vertical spread of a disease?

Horizontal spread of disease refers to spread among different members of the human population, one individual to the next. **Vertical spread** of disease refers to transmission among generations; that, is from parent to offspring by transmission across the placenta or during the birth process.

15.44 What are some general methods for transmitting pathogens among humans?

Pathogens may be transmitted among humans by direct or indirect methods. **Direct methods** involve some sort of human contact by which the microorganism exits from one person and enters the second. **Indirect transmission** may involve an inanimate object known as a fomite or some type of food, or some soil, water, or airborne particle. For example, the bits of mucus and saliva expelled in a sneeze (droplet nuclei) can be an indirect method for transmission of pathogens among humans.

Review Questions

Multiple Choice. Select the letter of the item that correctly completes each of the following statements.

1. A commensalism is a relationship between the normal flora and the human body in which (*a*) both populations benefit from the other (*b*) one population benefits and there is no damage to the other population (*c*) one population kills the other population (*d*) the normal flora is destroyed.

2. The enzyme hyaluronidase assists the spread of microorganisms by (*a*) forming clots around the body's defense mechanisms (*b*) breaking down blood clots that form (*c*) breaking down hyaluronic acid, which binds cells together (*d*) resisting phagocytosis.

3. Those infections that come on rapidly and are accompanied by severe symptoms are described as (*a*) chronic (*b*) opportunistic (*c*) acute (*d*) asymptomatic.

4. A convalescent carrier is one (*a*) who spreads disease while in the carrier state (*b*) has never had the disease but is carrying the organism (*c*) has recovered from the disease but still has mild relapses (*d*) is spreading the disease during the incubation stage.

5. An example of a noncommunicable disease in the human body is (*a*) diphtheria (*b*) whooping cough (pertussis) (*c*) salmonellosis (*d*) tetanus

6. Animals that transmit human diseases are referred to as (*a*) carriers (*b*) reservoirs (*c*) vectors (*d*) incubators.

7. Binding to the tissue is encouraged if bacteria possess (*a*) capsules (*b*) pili (*c*) flagella (*d*) granules.

8. The organisms that cause human disease in patients infected with HIV are described as (*a*) exogenous (*b*) pandemic (*c*) opportunistic (*d*) symbions.

9. Microbial poisons that interfere with the physiology of body cells and tissues are described as (*a*) toxins (*b*) bacteriocins (*c*) antibiotics (*d*) suppresins.

10. A mixed infection is one in which (*a*) several different types of poisons are produced by the microorganisms (*b*) several different kinds of antibodies are produced by the body (*c*) multiple different microorganisms are causing infection (*d*) multiple portals of entry have been used for penetration.

11. Which of the following are incapable of producing toxins in the body (*a*) Gram-positive bacteria (*b*) Gram-negative bacteria (*c*) bacterial rods (*d*) viruses.

12. All the following locations in the human body have a normal flora except (*a*) the skin and mucous membrane (*b*) the respiratory tract (*c*) the external portions of the eye (*d*) the liver.

13. The degree of pathogenicity of a species of microorganisms is defined as its (*a*) toxemia (*b*) intoxication (*c*) endemic potential (*d*) virulence.

14. An example of a microorganism having multiple portals of entry is the organism that causes (*a*) malaria (*b*) AIDS (*c*) tuberculosis (*d*) plague.

15. Leukocidins are microbial toxins that (*a*) have a destructive effect on red blood cells (*b*) permit penetration of microorganisms through the tissues (*c*) destroy white blood cells (*d*) cause spasms of the muscles.

16. A secondary infection is one that develops (*a*) in the body already infected by a different organism (*b*) at a secondary site in the body tissues (*c*) when symptoms are at their highest level (*d*) that is not communicable from one person to the next.

17. A pandemic is an outbreak of disease occurring on the (*a*) local level (*b*) worldwide level (*c*) urban level (*d*) only within a certain race of people.

18. Disease that passes from parent to offspring across the placenta is said to be transmitted by a (*a*) horizontal spread (*b*) oblique spread (*c*) vertical spread (*d*) circular spread.

19. Biological vectors are those that spread the microorganisms of disease and (*a*) themselves become infected (*b*) themselves remain uninfected (*c*) spread fungal diseases only (*d*) spread diseases to various points of entry.

20. One who is responsible for detecting disease in a population and charting its course is a (*a*) mycologist (*b*) bacteriologist (*c*) epidemiologist (*d*) urologist.

Completion. Add the word or words that correctly complete each of the following statements.

1. The condition in which pathogenic microorganisms penetrate the host defenses and multiply in its tissues is called ____________ .

2. A typical organism in the resident population of the human skin is the bacterium ____________ .

3. The inoculum required to establish disease in the host is called the ____________ .

4. The diphtheria toxin causes disease by damaging cells of the ____________ .

5. Those infections that occur slowly and are accompanied by mild to severe symptoms are referred to as ____________ .

6. A disease in which there are no signs or symptoms is said to be ____________ .

7. An example of a noncommunicable disease taking place in the respiratory tract is ____________ .

8. The biological vector that transmits yellow fever viruses is the ____________ .

9. For the human immunodeficiency virus, the infectious dose is ____________ .

10. When potentially infectious microorganisms exist in the body but have not yet invaded the tissue, the body is said to be ____________ .

11. An example of a body tissue in which there is no normal flora is the ____________ .

12. An organism that is highly virulent has a high potential for causing ____________ .

13. The microbial enzyme that digests collagen and encourages passage through the tissues is called ____________ .

14. Diseases due to the presence of large amounts of microbial toxins are known as ____________ .

15. The government agency concerned with epidemiology in the United States is the ____________ .

16. One who maintains pathogens in the tissues without showing any symptoms of disease is called a ____________ .

17. Those communicable diseases that are transmitted with particular ease are said to be ____________ .

18. Microbial toxins called hemolysins are known for their ability to destroy ____________ .

19. Within the normal flora of the mouth, the plaque around the teeth contains anaerobic species of ____________ .

20. Those organisms that cause human disease in patients infected with HIV are commonly described as ____________ .

21. Clams, oysters, and other shellfish are known for their ability to concentrate the viruses of ____________ .

22. The presence of microbial toxins in the tissues is referred to as ____________ .

23. Paralysis develops in the muscles when the release of acetylcholine is prevented by the toxin of ____________ .

24. An infection spreading to alternative sites of the body such as deeper organs is said to be ____________ .

25. Endotoxin shock is often due to endotoxins released by bacteria that are ____________ .

26. Subjective measures of disease as noted by the patient is referred to as ____________ .

27. A disease remaining within a human population over a long period of time without breaking out is described as ____________ .

28. A mechanical vector may transmit microorganisms among samples of ____________ .

29. Indirect transmission of microorganisms may be caused by inanimate objects known as ____________ .

30. Those toxins secreted by microorganisms when they infect tissues are described as ____________ .

True/False. In the space to the left of each of the following statements, enter a "T" if the statement is true in its entirety or an "F" if the statement or any part of it is false.

___ **1.** Infection is the same as disease.

___ **2.** All microorganisms are able to produce exotoxins.

___ **3.** Opportunistic microorganisms are not normally considered pathogenic.

___ **4.** Exotoxins are produced primarily by Gram-positive bacteria, while endotoxins are produced by many Gram-negative bacteria.

___ **5.** The CDC publishes the *Morbidity and Mortality Weekly Report*, a publication that summarizes the diseases present in the community and their significance.

___ **6.** A pandemic is a disease present in a human population but remaining in that population over a relatively long period of time without breaking out.

___ **7.** Chronic carriers of disease are those who have recovered from the disease but suffer occasional mild relapses and are capable of spreading the pathogen to others.

___ **8.** The enzyme hyaluronidase digests the collagen fibers in connective tissues and thereby destroys the white blood cells.

___ **9.** The resident population of the skin consists of microorganisms that generally do not grow there.

__ **10.** The botulism toxin acts in the synapse, where it prevents the release of neurotransmitters.

__ **11.** Acute infections occur slowly and are generally accompanied by mild symptoms and persist over a long period of time.

__ **12.** An example of a noncommunicable infectious disease is tetanus.

__ **13.** The horizontal spread of diseases refers to spread among different members of the human population.

__ **14.** Leukocidins are types of exotoxins that destroy white blood cells.

__ **15.** An example of the normal flora that can cause disease in the body is *Candida albicans*.

__ **16.** All microorganisms have a single portal of entry to the human body.

__ **17.** For certain microorganisms, such as those of typhoid fever, the infectious dose is very high.

__ **18.** The effect of the diphtheria exotoxin on the human body occurs in cells of the respiratory tract.

__ **19.** A primary disease is one that develops in the human body after infection by a different microorganism has already taken place.

__ **20.** An epidemiologist is a public health professional who is responsible for detecting disease in a population and charting the course of that disease.

Chapter 16

Host Resistance and the Immune System

OBJECTIVES

When confronted with infectious disease, the human body responds with various kinds of resistance. A primary form of resistance is the immune system, which is a major topic of this chapter. The objectives of the chapter are to:

1. Describe a number of mechanical, nonspecific defensive mechanisms that operate in response to disease.

2. Appreciate the importance of phagocytosis as a defensive mechanism in the body.

3. Recognize the difference between innate and acquired immunity in the body and describe the various forms of acquired immunity.

4. Conceptualize the significance of antigens and discuss how they stimulate a response by the immune system.

5. Explain the significance of two types of cells of the immune system and indicate how they originate in the body.

6. Describe the importance of antibodies in antibody-mediated immunity and provide characteristics of various kinds of antibodies.

7. Summarize the significance of T cells in the immune response and show how they function in cell-mediated immunity.

8. Describe various types of vaccines used for stimulating acquired immunity in the body.

THEORY AND PROBLEMS

16.1 How many lines of resistance exist in a host organism confronted by infectious disease?

The natural resistance present in the body can usually confine microorganisms to the surface of the body and prevent entry into the innermost organs. Occasionally, however, microorganisms possess unusual qualities that add to their pathogenicity and permit them to penetrate the surface barriers. Under these conditions, the human body employs three lines of defense against disease: surface defenses, phagocytic defenses, and specific defenses centered in the immune system. The surface defenses of the human body may be subdivided as mechanical defenses, chemical defenses, and microbial defenses.

16.2 Describe the various mechanical defenses available to the host organism?

The most essential **mechanical defense** mechanism available to the body is the intact skin surface and the mucous membranes that line many of the systems, such as the respiratory tract, gastrointestinal tract, urogenital tract, and conjunctiva of the eye. One of the first stages of disease is breaching these barriers by means of cuts, puncture wounds, or other trauma. In addition, mucous membranes can be weakened by tobacco smoke, air pollutants, and toxic gases.

16.3 Are there any factors that augment the mechanical defenses of the body?

There are several factors that augment the mechanical defenses. Among these are the mucus and cilia present in the respiratory tract. The sticky **mucus** traps particles that enter the tract and the **cilia** move the mucus along into the throat where it is swallowed. Coughing and sneezing also enhance the mechanical defenses by expelling microorganisms from the respiratory tract. Nasal **hairs** add to the mechanical defenses, and body hair prevents microorganisms from reaching the skin. **Urine** flow reduces the microbial population of the urinary tract, and **tears** flush the eyes constantly. Moreover, the release of dead skin cells removes many microorganisms from the skin surface.

16.4 Which types of chemical defenses are part of the surface defense displayed by the body?

The body synthesizes various chemicals to assist its surface defense. Among these are **lysozyme**, an enzyme in the tears and saliva that destroys the cell walls of Gram-positive bacteria. Many **fatty acids** in the sweat and earwax also have antimicrobial activities. The **hydrochloric acid** in the stomach kills bacteria in this organ, and the vagina contains a high acid content that also protects against microbial contamination. **Bile** from the gall bladder kills microorganisms in the gastrointestinal tract, and **enzymes** from the duodenum kill others.

16.5 How can microbial defense be used to inhibit the development of pathogenic organisms?

In many parts of the body, a normal flora can be found. The microorganisms in the normal flora compete with pathogenic microorganisms for the nutrients in the environment and available space. Since microorganisms of the normal flora are well-established, they generally can overcome the presence of pathogens that happen to enter the body accidentally. The normal flora microorganisms also produce substances that inhibit pathogens. For example, lactobacilli in the vagina produce acid, which inhibits the growth of pathogens. Microorganisms of the skin also produce acid.

PHAGOCYTOSIS

16.6 What is the basis for the phagocytic defenses of the body?

Phagocytic defenses of the body are centered in **phagocytosis**, a process performed by white blood cells known as phagocytic cells. These cells engulf microorganisms as shown in Figure 16-1, and combine their enzymes with the microorganisms. Enzymes from the lysosomes of the phagocyte digest the microbes quickly. Such enzymes as proteases, lipases, and peptidoglycanases are active in the phagocyte. The debris is then discharged at the conclusion of phagocytosis.

16.7 Which are the major cells that perform phagocytosis in the body?

Among the major cells involved in phagocytosis are the neutrophils. **Neutrophils** are granulocytic leukocytes. They have a multilobe nucleus and represent about 60 to 65 percent of the circulating white blood cells. Neutrophils are also called **polymorphonuclear leukocytes**. They are among the first white blood cells to arrive at an infection site, and they are part of the pus that accumulates.

16.8 Are there any other phagocytic cells in the bloodstream?

Another important phagocyte is the monocyte. **Monocytes** have a single, large horseshoe-shaped nucleus. They represent about 8 percent of the circulating white blood cells.

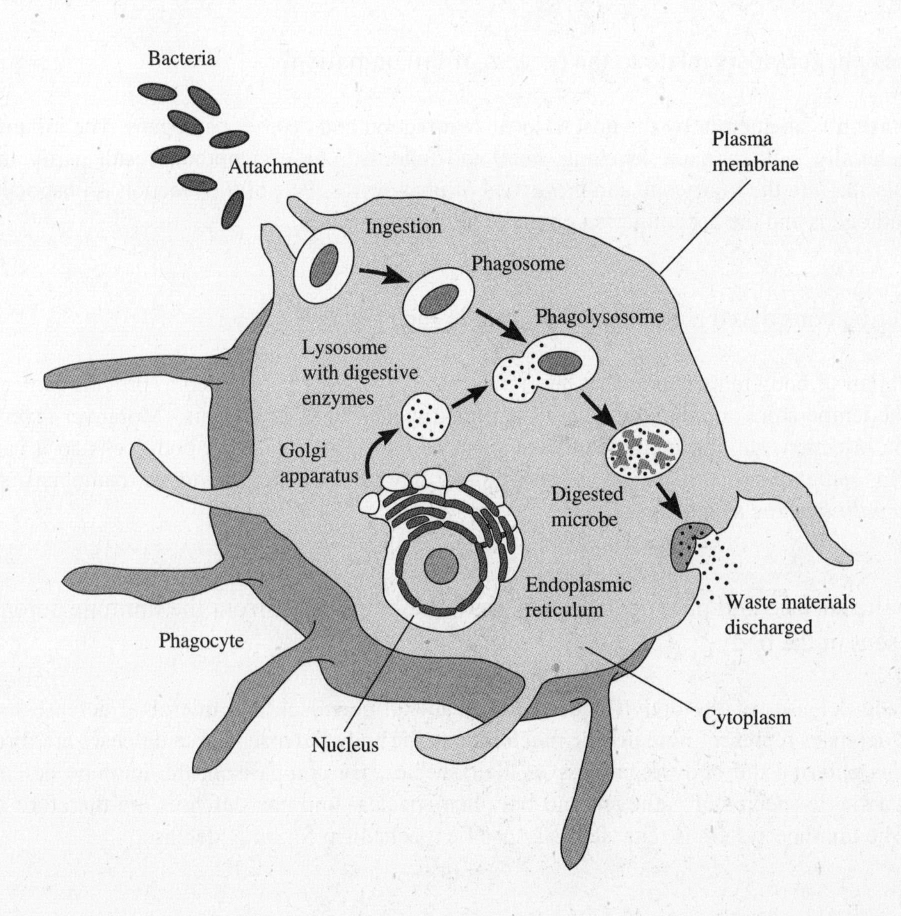

Fig. 16-1 An illustration of the process of phagocytosis, which provides nonspecific resistance to disease. The process begins with the attachment of the phagocyte to the infectious bacteria and concludes with discharge of waste materials. Enzymes are contributed to the digestion of the microbe by the lysosome.

16.9 Are any phagocytes found in the tissues but not in the bloodstream?

Among the tissues, specialized phagocytes called **macrophages** can be found. Macrophages occur in the liver, spleen, bone marrow, and within many of the other body tissues and organs. Their responsibility is to perform phagocytosis and begin the immune response, as we shall discuss shortly.

16.10 Which kinds of macrophages may be found in the tissues?

Macrophages can be classified as fixed macrophages and wandering macrophages. **Fixed macrophages** remain in one place and engulf potential pathogens as the blood and lymph flow by. **Wandering macrophages** migrate to sites where pathogens might be encountered. They accumulate at an infection site after penetration of the body's skin defenses has taken place. The fixed macrophages, wandering macrophages, monocytes, and other phagocytes make up the **reticuloendothelial system** of the body. The more contemporary name for this system is the **mononuclear phagocytic system**.

16.11 How does phagocytosis relate to the process of inflammation?

Inflammation is an attempt by the host to localize infection and destroy pathogens. The inflammatory response generally includes pain, swelling, heat, and redness. These symptoms result partly from the escape of plasma into the injury site and the arrival of phagocytes. Part of the reaction is phagocytosis of invading pathogens and the accumulation of pus.

16.12 Can fever be considered a defensive measure in the host?

The elevation of body temperature that takes place during a **fever** is believed to provide protection by elevating the temperature and thereby increasing the efficiency of phagocytosis. Moreover fever speeds blood to the infection site and increases the efficiency of metabolism in the body cells so it can resist infection. In some cases, the elevated temperature may retard the growth of pathogens such as *Treponema pallidum,* the agent of syphilis.

16.13 How do the surface and phagocytic defenses of the body differ from the immune defenses also present in the body?

The surface defenses of the body (including mechanical, chemical, and microbial defenses) and the phagocytic defenses represent **nonspecific mechanisms** for body defense. These defenses are directed at all pathogens entering the body regardless of their species. By comparison, the immune defenses are directed at a single species of pathogen and only that species. Immune defenses are therefore **specific defenses**. The immune system is regarded as a specific mechanism for body defense.

TYPES OF IMMUNITY

16.14 Which are the two major types of immunity that exist in the body?

The human body has two basic types of immunity: innate immunity and acquired immunity. Each yields specific host resistance when the body is confronted with infectious microorganisms.

16.15 What are the features of innate immunity in the body?

Innate immunity results from the anatomical, physiological, biochemical, and genetic makeup of an individual. This immunity develops independent of any previous experience with a specific pathogen. For example, humans are innately immune to the viruses that cause canine distemper because canine distemper viruses cannot attach to human cells since human cells lack the appropriate receptor sites. In addition, dogs do not suffer from AIDS because cells of the dog do not possess the receptor sites for HIV. Innate immunity is inborn.

16.16 How does acquired immunity develop in an individual?

Acquired immunity is that type of immunity resulting from exposure to pathogens or other foreign substances. In general, acquired immunity develops as a result of a highly specific response to the invading organism or substance.

16.17 Are there various forms of acquired immunity?

There are two recognized forms of acquired immunity: natural acquired immunity and artificial acquired immunity. **Natural acquired immunity** occurs during the natural course of events, while **artificial acquired immunity** results from the deliberate introduction of some foreign substance to stimulate an immune response.

16.18 Are there any subdivisions of natural acquired immunity?

Natural acquired immunity can be active or passive. **Naturally acquired active immunity** comes about when a person is exposed to a pathogen and responds by forming antibodies that attack and neutralize the pathogens. Naturally acquired passive immunity comes about when antibodies are transmitted from mother to child via the placenta and umbilical cord. The antibodies that pass to the developing fetus remain with the child for about six months after it is born.

16.19 Are there any subdivisions of artificially acquired immunity?

Artificially acquired immunity can also be active or passive. **Artificially acquired active immunity** comes about after a person has been injected with a vaccine or other immune-stimulating agent and has produced antibodies. The antibodies remain with the individual and neutralize the pathogens should they enter the body at a later date. **Vaccines** are composed of inactive viruses, dead bacteria, fragments of viruses or bacteria, or genetically engineered chemical components of microorganisms. The second type of artificially acquired immunity is **artificially acquired passive immunity**. This immunity comes about from the injection of antibodies into an individual. The antibodies provide immediate defense against disease, but the antibodies do not remain in the system for more than several days or weeks. **Gamma globulin** is a preparation of antibodies used to induce this form of immunity.

THE IMMUNE SYSTEM

16.20 What are antigens and how do they relate to immunity?

Antigens are chemical substances that stimulate a specific immune response resulting in antibodies or activated cells. Antigens are usually foreign to the host organism, but in certain cases, they can be the host's own chemical substances that have been interpreted as being foreign. The word **immunogen** is synonymous with antigen.

16.21 What are antigens composed of?

Antigens are large molecules of protein or polysaccharide. In rare instances, lipids or nucleic acids can function as antigens. Such things as the proteins in bacterial flagella and the polysaccharides in bacterial capsules contain antigens.

16.22 Can small molecules act as antigens to stimulate the immune system?

Small molecules function poorly as antigens because they are not attracted to macrophages, and macrophages must first engulf antigens to set off the immune response. However, there are small molecules such as penicillin molecules that unite with proteins in the body and form a large molecule that is antigenic. Small molecules such as penicillin molecules are known as **haptens**.

16.23 Does the antigen do the actual stimulation of the immune system?

The immune response is not directed at the entire antigen molecule. Instead, the immune response is directed toward small chemical molecules on the antigen molecule known as **epitopes,** or **antigenic determinants**. A single antigen molecule may have numerous different epitopes, each capable of stimulating an immune response.

16.24 How does the processing of antigens occur?

Before an antigen can stimulate an immune response it must be processed. This processing is performed by monocytes in the circulation and macrophages among the tissue cells. Both of these cells are granulated phagocytes. They engulf the antigen molecules and break them down into their epitopes. Then the epitopes are displayed on the membrane of the cell at the conclusion of the processing step.

16.25 Are there any other molecules at the cell surface that function in the immune response?

Epitopes are displayed at the membrane of the monocytes and macrophages together with a set of molecules called **major histocompatibility (MHC) molecules**. These molecules are unique for an individual, and they identify the macrophages and monocytes as belonging to that individual. The major histocompatibility molecules are encoded by major histocompatibility genes in the chromosomes of the person's cell.

16.26 Which cells of the immune system are responsible for the immune response?

The human immune system has two major arms. One arm is centered in a set of cells called **B lymphocytes**, while the other arm is centered in **T lymphocytes**. The B lymphocytes function in a form of immunity known as **antibody-mediated immunity**, while the T lymphocytes function in **cell-mediated immunity**. B lymphocytes are also known as B cells, while T lymphocytes are also known as T cells.

16.27 What is the origin of T cells in the human body?

Like all blood cells, T cells originate from stem cells in the bone marrow. In the fetal stage, the immature cells enter the circulation and pass through an organ known as the thymus gland shown in Figure 16-2. Located in the region of the neck, the thymus gland modifies the cells, and mature T cells emerge from the gland. The T cells (T lymphocytes) then move through the circulation and colonize the lymph nodes, spleen, tonsils, adenoids, and other tissues of the lymphatic system.

16.28 Are there different kinds of T cells in the immune system?

As they mature in the thymus gland, different clones of T cells are marked with different surface receptor molecules that will respond to different antigens. Therefore, there are numerous different colonies of T cells designed to respond to different antigens that may enter the body. T cells will respond to antigens from fungi, protozoa, virus-infected cell, cancer cells, and transplanted cells.

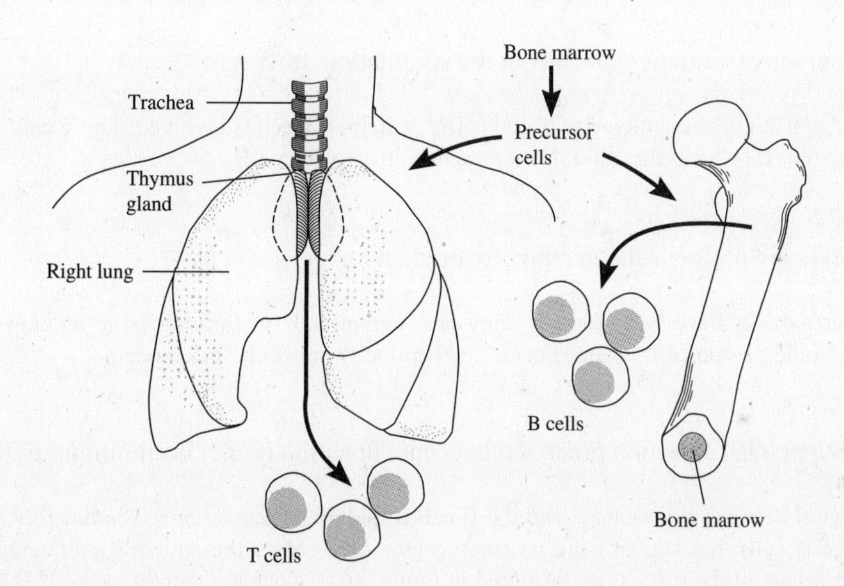

Fig. 16-2 The immune system is based on the activity of two cells, the B cells and the T cells. Precursor cells are modified possibly in the bone marrow to form B cells. Other precursor cells are modified in the thymus gland to yield T cells. Both B cell and T cells then colonize the lymphatic system.

16.29 How are B cells formed by the body?

B cells (B lymphocytes) also originate in the stem cells of the bone marrow, but the site of maturation in humans is not yet identified. Research evidence points to the possibility that the cells may mature in the liver, bone marrow, or tissues along the gastrointestinal tract. In the embryonic chick, the maturation occurs in an organ called the bursa of Fabricius. "B" lymphocytes are so-called due to the B in bursa.

16.30 Are there different kinds of B cells in the human immune system?

When the B cells are mature, they join the T cells at the lymph nodes, spleen, and other lymphatic tissues. Different clones of B cells carry different surface markers. These surface markers are antibody molecules. They recognize and unite with various antigens to stimulate the process of **antibody-mediated immunity**. B cells respond to antigens from bacteria, viruses, and large molecules not normally found in the body.

ANTIBODY-MEDIATED IMMUNITY

16.31 What is the nature of the recognition process in antibody-mediated immunity?

When macrophages and monocytes enter the lymphoid tissues, they display epitopes on their surface, and the B lymphocytes respond. The response includes a recognition of the MHC molecules and a union between the receptor sites and the epitope molecules. Once the recognition has taken place, the B cells differentiate and mature into plasma cells. **Plasma cells** produce and secrete large amounts of antibodies.

16.32 Is there any involvement of T cells in the stimulation of B cells?

During B cell stimulation, a type of T cell called a **helper T cell** is involved. The T cell helps bind the macrophage and B cell together, and its secretions help activate the B cells.

16.33 How active are plasma cells in antibody production?

Once plasma cells have been formed, they are "committed" to forming large amounts of antibody molecules. Some plasma cells produce over 2,000 antibody molecules per second.

16.34 What is the clonal selection process that occurs in antibody-mediated immunity?

The reaction between macrophage and the B cell is a highly selective one. The macrophage must find the clone of B cells that contains the proper receptor sites before the antibody producing process can begin. This aspect of the process is illustrated in Figure 16-3. Once the correct clone of B cells has been found, it is "selected" from the remaining B cells for conversion to plasma cells and antibody production. This is the **clonal selection process**.

16.35 What are antibodies composed of?

Antibodies are protein molecules of the globulin type. They are subdivided into five different types depending on their makeup and function. All basic antibody molecules consist of four chains of amino acids: two heavy chains each of about 450 amino acids, and two light chains each of about 200 chains. The chains are bound together by sulfur to sulfur covalent bonds.

16.36 Are antibodies known by any other name?

Because of their classification as globulin proteins, and due to the fact that they function in the immune response, antibodies are also known as **immunoglobulins**. The designation Ig is often used for an antibody molecule.

16.37 Which is the most common antibody molecule in the blood?

Of the five kinds of antibodies, **IgG** represents about 80 percent of the antibody in the body at any one time. This antibody is the most important long-acting antibody in the specific immune response. It consists of a single monomer of four amino acid chains, and it is able to pass across the placenta from mother to child in natural passive immunity.

16.38 What part of this antibody reacts with the antigen molecule?

Each antibody molecule has a **constant region** and a **variable region**. The variable region is the place where an antibody molecule unites specifically with the antigen that stimulated its production. The epitope of the antigen fits perfectly with the antibody-combining site.

16.39 Which is the second kind of antibody molecule?

The second important kind of antibody molecule is **IgM**. This antibody is huge. It consists of five monomer units, and because of its size, it cannot leave the bloodstream, nor can it pass the placenta. IgM represents about 5 to 10 percent of the antibody total in the blood. It is the first antibody to arrive at the scene to neutralize the antigen and it is the principal component of the primary antibody response.

16.40 What are the characteristics of the third antibody, IgA?

IgA is known as the secretory antibody. It is secreted into the cavities of the body such as the gastrointestinal tract, the respiratory tract, and onto the body surface. The antibody consists of two monomer units connected by a secretory piece known as the J chain, which is displayed in Figure 16-4. IgA is secreted in the mother's milk and is passed to the child who nurses.

16.41 What is the function of the fourth antibody IgD?

IgD represents only about 1 to 3 percent of the antibody in the serum. This antibody serves as a receptor site at the surface of B cells. The epitope displayed by the macrophage arriving in the lymphoid tissue unites with the antibody to stimulate the immune response. IgD consists of a single monomer unit.

16.42 Which is the fifth antibody of the immune response, and what is its function?

The fifth antibody is referred to as **IgE**. This antibody represents less than 1 percent of the antibody in the blood; it functions in many allergic reactions. The molecule attaches to granulated cells such as basophils and induces the cells to lose their granules when a succeeding antibody-antigen reaction occurs. This release leads to the allergic response.

16.43 What are some ways by which antibodies neutralize microorganisms and impart immunity to the body?

Antibodies neutralize microorganisms in a variety of ways. For example, they unite with the surfaces of viruses and prevent the viruses from attaching to their host cells during the replication process. In addition, antibodies unite with bacterial flagella and prevent them from functioning, and they unite with pili and prevent bacterial attachment to the tissues. In some cases antibodies encourage phagocytosis to occur, and in other cases they set off a series of reactions known as the **complement system**. The complement system products result in holes and perforations in the microbial surface leading to loss of cytoplasm and disintegration of the cells. Antibodies also cause microorganisms to clump together in large masses easy for phagocytosis.

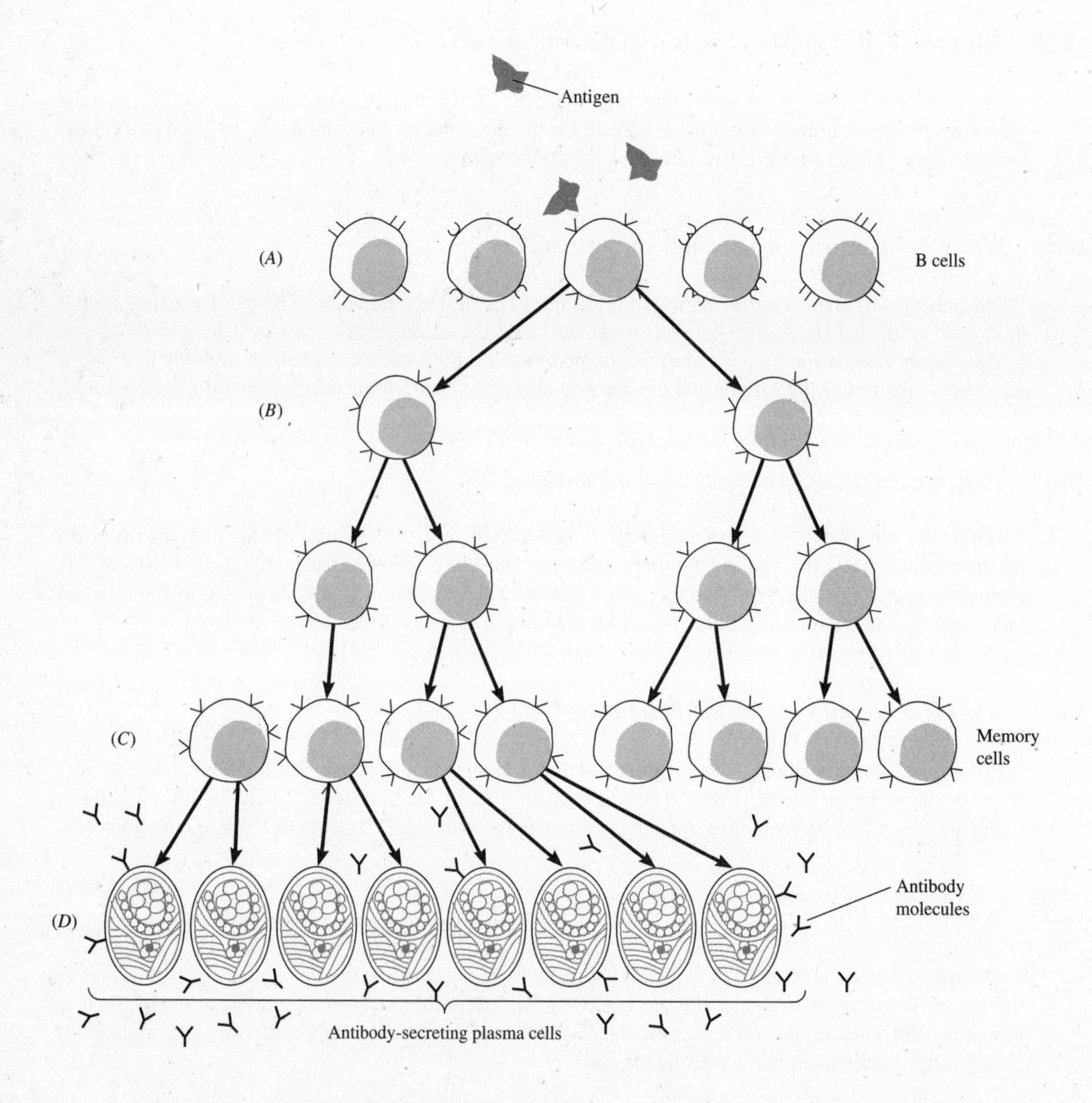

Fig. 16-3 Antibody-mediated immunity. (*A*) An antigen molecule reacts with a specific receptor site at the surface of B cells in the lymphatic system. B cells having different receptor sites are not involved. (B) The correct B cell is selected out and multiplies to form a colony (clone) of B cells. (*C*) The multiplication of B cells continues and some memory cells are produced. (*D*) The B cells convert to plasma cells, and the plasma cells begin the secretion of antibody molecules. The antibodies react with antigens to bring about an immune response.

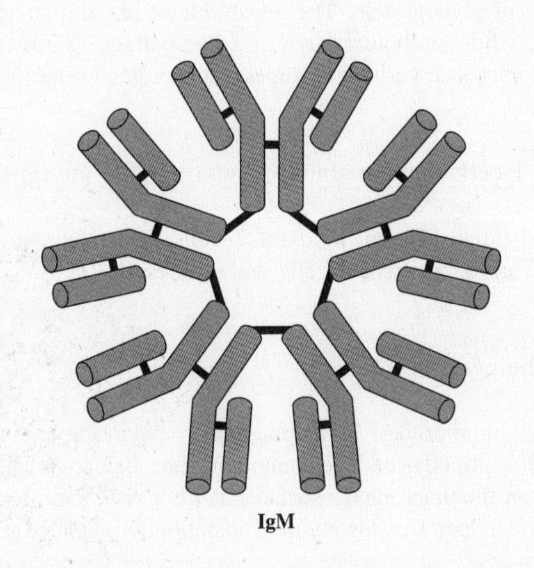

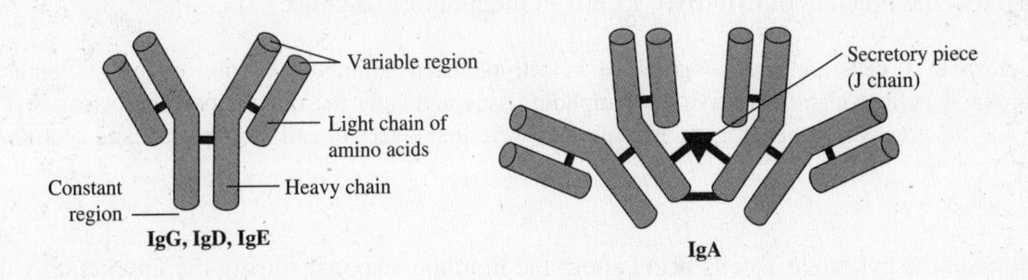

Fig. 16-4 The structure of five different types of antibody molecules. Note that IgG, IgD, and IgE consist of single monomers, while IgM has five monomers, and IgA has two monomers.

16.44 What is the function of T cells in the immune response?

The B cells bring about immunity through the activity of antibodies, but T cells bring about an immune response by direct intervention with antigens. The T cells themselves are involved in the immune response, so this type of immunity is known as **cell-mediated immunity**.

CELL-MEDIATED IMMUNITY

16.45 Do T cells have receptors on their membrane surfaces similar to those on the B cell?

B cells have receptor sites, which are antibody molecules. By comparison, T cells also have receptor sites, but the sites are molecules of glycoprotein. The glycoproteins are similar to antibodies, but they contain carbohydrate molecules, while antibodies have no carbohydrate. The receptor sites on T cells have variable regions so as to interact with variable epitopes from antigen molecules.

16.46 Are there different types of T cells that are able to function in the immune response?

Immunologists have identified three different kinds of T cells in the human immune system. The different types are called helper T cells, cytotoxic T cells, and suppresser T cells.

16.47 What is the function of the helper T cell?

The **helper T cells** assist the immune process by helping other cells in the immune system to achieve an efficient immune response. In antibody-mediated immunity, the helper T cell helps recognize the MHC molecule and the epitope on the macrophage surface, and it assists the interaction between the B cell and the macrophage. Thus, the helper T cell is involved in antibody-mediated immunity.

16.48 What is the function of cytotoxic T cells in the immune response?

Cytotoxic T cells are directly involved in cell-mediated immunity. Certain epitopes stimulate the cytotoxic T cells, which then leave the lymphoid tissue and enter the circulation. The cytotoxic T cells move to the area where the antigens and epitopes were first detected, and they interact and neutralize the antigens at this site.

16.49 How do the cytotoxic T cells bring about the immune response during the interaction with antigens?

Cytotoxic T cells unite specifically with cells containing antigens. The T cells secrete a number of substances including the enzyme **perforin**. Perforin and other enzymes damage the membranes of the cells and the cell cytoplasm leaks into the environment and the cells die, as Figure 16-5 demonstrates.

16.50 Which cells are attacked by cytotoxic T cells?

Cytotoxic T cells unite with various kinds of microorganisms and other foreign cells in the body. For example, they unite with fungal and protozoal cells, and immunity to these microorganisms is based largely on the activity of cytotoxic T cells. The cells also react with and neutralize virus-infected body cells, as well as tumor cells, and cells in transplanted tissue. By comparison, antibodies unite generally with bacteria, viruses, foreign molecules in blood, and haptens.

16.51 What is the function of the suppresser T cells?

The **suppresser T cells** guard against the overproduction of antibodies and the overactivity of cytotoxic T cells. Although the mechanisms are uncertain, T cells are known to dampen the immune response by both antibody-mediated and cell-mediated immunity. If these responses were not suppressed, then the immune system could cause substantial tissue damage in the body cells and tissues.

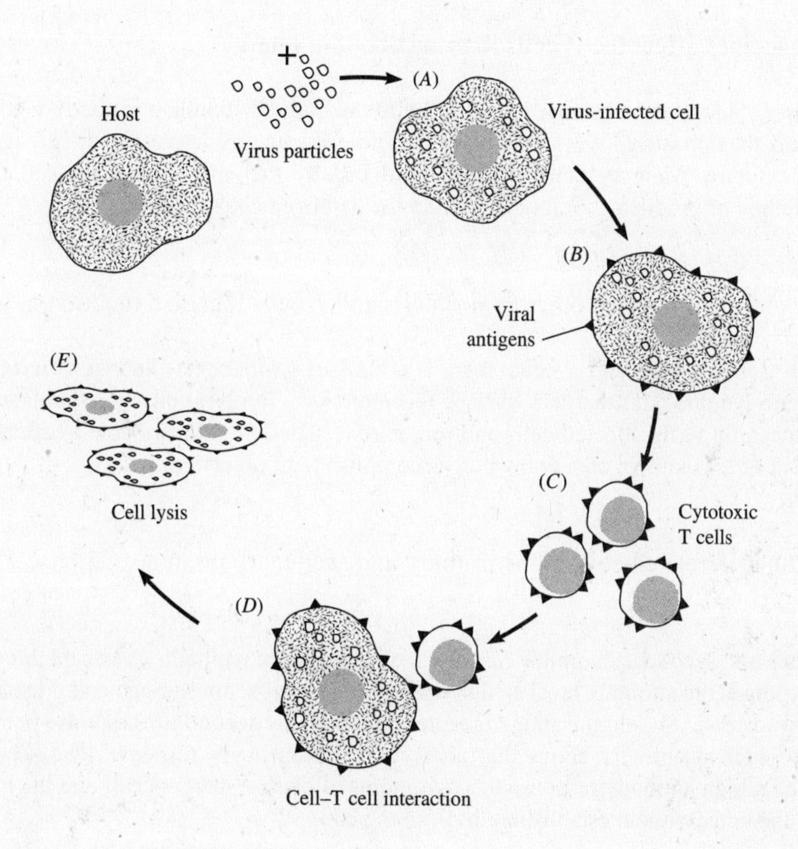

Fig. 16-5 Cell-mediated immunity. A host cell is invaded by virus particles. (*A*) The virus-infected cell accumulates the virus particles in its cytoplasm. (*B*) Antigens from the virus are displayed at the surface of the infected cells. Some of the infected cells are phagocytized and the antigens are transported to the immune system. (*C*) At the immune system, the antigens stimulate the production of cytotoxic T cells. The T cells have molecules at their surface that complement the viral antigens. (*D*) The cytotoxic T cells arrive back at the infection site and interact specifically with the virus-infected cells. (*E*) The infected cells undergo destruction (lysis) to complete the immune process.

16.52 Is there any memory factor in both the B cells and T cells?

Once the immune response has been completed, a series of **memory cells** are produced by both B cells and T cells. The memory B cells remain in the lymphoid tissue. Upon entry of the antigen to the body, they begin producing large quantities of antibody. The memory T cells, by comparison, are distributed throughout the body tissues and organs. When antigens enter the body, the cells bring about an immediate immune response through CMI. Thus, the body "remembers" the original stimulation by the immune system and the continuing immunity to infection and disease is related to the memory B and T cells.

16.53 How do vaccines affect the B cells to bring about immunity?

In a vaccines, a harmless substance is injected into the body to stimulate antibody-mediated immunity. The antibodies thus produced will circulate in the bloodstream and interact with pathogens should they appear in the future. Memory cells are also produced by the immune system and they provide an additional method of producing antibodies should the pathogen enter the body.

16.54 Are there any lymphocytes other than B cells and T cells that also function in immunity?

In addition to the B cells and T cells, there is a class of lymphocytes known as **natural killer (NK) cells**. These are lymphocytes that lack surface receptor sites. The NK cells are stimulated by antigens in the body, especially virus-infected cells and tumor cells. The NK cells are able to attack and neutralize these cells, but little is known about how they accomplish their objective.

16.55 What is the difference between the primary and secondary immune responses?

The primary and secondary immune responses develop when antibodies enter the bloodstream. In the **primary response**, the antibody level is high enough to neutralize the antigen and it usually brings about recovery from disease. Should the antigen reenter the body, the **secondary response** occurs. In this case, the antibody level reaches far above that achieved in the primary response. The secondary response results in such a high antibody response that symptoms of disease rarely occur, and the microorganism is driven from the body without establishing itself in the body.

16.56 How do the primary and secondary responses relate to vaccinations?

In the first injection of a vaccine, the antibody production is similar to that in a primary response, that is, a moderately high level of antibodies is produced in the body. When a booster, or second injection of vaccine is given, the antibody level reaches an extraordinary level and produces a much higher level of protection, as in the secondary response. For this reason, **booster injections** of vaccine are recommended at regular intervals. A modern injection gun shown in Figure 16-6 is commonly used.

16.57 What prevents the immune system from reacting against the body cells?

Humans enjoy a concept called **immunological tolerance**. This means they have learned to tolerate themselves and not develop an immune response to themselves. Although this tolerance is not fully explained, it apparently develops during the fetal stage when cells that might respond to body tissue are paralyzed. So long as this paralysis continues through life, the body will recognize itself, and it will respond only to foreign substances seen as **nonself.** In some individuals the tolerance for oneself breaks down and the person develops an **autoimmune disease** where they produce immune substances against their own cells. Rheumatoid arthritis and systemic lupus erythematosis are examples of autoimmune diseases.

16.58 When do the immune processes first become active in the body?

Although the newborn has the ability to respond to some antigens, the full immune response is not present until a child is about six months of age. Before that time, the child receives antibodies from its mother's circulation. These antibodies eventually wear off and are replaced by the child's antibodies. For the first six months of its life, a child is therefore dependent on its mother's antibodies.

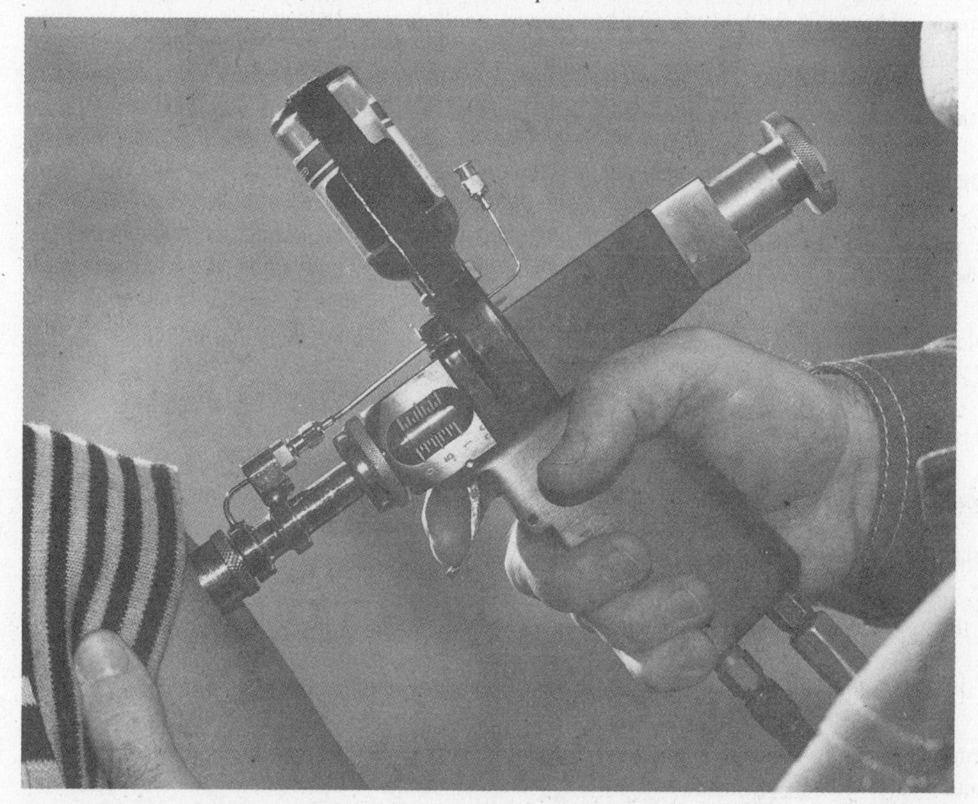

Fig. 16-6 A modern air-driven injector gun used to deliver vaccines to a recipient.

16.59 Do antibodies have a life span in the body circulation?

Antibodies, as noted, are protein molecules. They will either leave the body in secretions or be broken down by enzymes that degrade proteins in the body. The average life span of most antibody molecules is roughly 25 days or so. Continued antigen stimulation is necessary to continue the antibody response, and as the antigen disappears the stimulation disappears. The antibodies produced as a result of the initial stimulation remain in the bloodstream for a long period of time, with about half the quantity disappearing after 25 days and half the remaining disappearing after another 25 days, and so forth. In many cases, the antibody level is sufficient to sustain an individual for the remainder of his or her life, especially after recovery from disease. This is why one remains immune to a disease after suffering a bout.

Review Questions

True/False. For each of the following statements, mark the letter "T" next to the statement if the statement is true. If the statement is false, change the underlined word to make the statement true.

___ **1.** The intact skin is part of the immune defense mechanism available to the body.

___ **2.** The enzyme in tears and saliva that destroys the cells walls of Gram-positive bacteria is lysosome.

___ **3.** Antibody molecules are composed of polysaccharides of the globulin type.

___ **4.** The immune response of the body is directed toward small chemical molecules on the antigen that are known as epitopes.

___ **5.** During the stimulation of B cells, a type of T cell called the suppresser T cell is involved.

___ **6.** Antibodies are secreted in the body primarily by cells referred to as serum cells.

___ **7.** When a person forms antibodies as a result of an injection of a vaccine, the immunity that develops is said to be naturally acquired passive immunity.

___ **8.** The molecules that identify the macrophages of an individual as belonging to that individual are known as MHC molecules.

___ **9.** The antibody identified as IgE functions in many allergic reactions.

___ **10.** The primary immune response occurs when an antigen enters the body for the first time.

___ **11.** The foundation cells of the immune response are found in the spleen and heart.

___ **12.** Among the major cells involved in phagocytosis in the body are the lymphocytes.

___ **13.** Artificial active immunity comes about after the injection of antibodies from another individual.

___ **14.** The immunity resulting from the anatomical, physiological, biochemical, and genetic makeup of an individual is known as innate immunity.

___ **15.** Small molecules acting as antigens to stimulate the immune system are known as immunogens.

___ **16.** Antibody-mediated immunity is dependent upon the activity of T lymphocytes.

___ **17.** The term immunoglobulin is an alternate expression for an antigen.

___ **18.** The largest antibody molecule, consisting of five monomer units, is IgD.

___ **19.** Cytotoxic T cells help destroy antigen-containing cells by secreting the enzyme pepsin.

___ **20.** When vaccines are injected into the body, they induce the T cells to become active.

___ **21.** The fact that humans do not develop an immune response to themselves embodies the concept of immunological tolerance.

___ **22.** The average life span of most antibody molecules is about 25 years.

___ **23.** In a newborn child, the full immune response is not present until the child is about one year old.

___ **24.** The process of CMI refers to the activity of B cells in the human body.

___ **25.** On the surface of B cells, the antibody that serves as a receptor site is IgE.

__ **26.** During the clonal selection process, B cells are chosen for conversion to <u>cytotoxic T cells</u>.

__ **27.** During their formation, T cells are modified in the <u>thyroid</u> gland.

__ **28.** The major histocompatibility molecules exist on the <u>surface</u> of an individual's cells.

__ **29.** Naturally acquired active immunity comes about when a person is exposed to a <u>vaccine</u> and responds by forming antibodies.

__ **30.** The immune system is one of the <u>specific</u> mechanisms for body defense.

Multiple Choice. Select the letter of the item that correctly completes each of the following statements.

1. All the following represent nonspecific mechanisms of body defense except (*a*) mucus that traps particles in the respiratory tract (*b*) stomach acid (*c*) IgM (*d*) phagocytosis.

2. Fever accomplishes all the following in the body except (*a*) it speeds blood to the infection site (*b*) it increases the efficiency of metabolism in body cells (*c*) it increases the efficiency of phagocytosis (*d*) it stimulates the recognition of antigens by B cells.

3. An example of innate immunity is (*a*) the immune response to a vaccine by an individual (*b*) the lack of receptor sites for HIV on dog cells (*c*) the reactions of cell-mediated immunity (*d*) the injection of antibodies to the bloodstream.

4. Which of the following is true of antibodies (*a*) they are also known as immunoglobulins (*b*) they are produced by T cells (*c*) they are composed of protein (*d*) they exist in five different types.

5. The antibody secreted into the cavities of the body such as the gastrointestinal tract is (*a*) IgD (*b*) IgA (*c*) IgG (*d*) IgE.

6. Monocytes and neutrophils are important cells participating in (*a*) antibody production (*b*) perforin production (*c*) passive immunity (*d*) phagocytosis.

7. All the following are true of lysozyme except (*a*) it is found in the tears and saliva (*b*) it is an enzyme (*c*) it is a type of antibody (*d*) it destroys the cell walls of Gram-positive bacteria.

8. Those phagocytes found in the tissues but not in the bloodstream are (*a*) plasma cells (*b*) macrophages (*c*) neutrophils (*d*) basophils.

9. The opposite to innate immunity is (*a*) passive immunity (*b*) acquired immunity (*c*) T cell immunity (*d*) phagocytosis.

10. Small molecules function poorly as antigens (*a*) because they are attracted to macrophages (*b*) because they possess no epitopes (*c*) because they do not enter the bloodstream (*d*) because they are part of the nonspecific mechanism of defense.

11. T cells originate from stem cells located in the (*a*) liver (*b*) bone marrow (*c*) thyroid gland (*d*) gastrointestinal tract.

12. Among the organs of the body that are rich in mature T cells and B cells are the (*a*) brain and spinal cord (*b*) liver and gall bladder (*c*) small and large intestine (*d*) spleen and lymph nodes.

13. All basic antibody molecules consist of (*a*) four polysaccharides (*b*) four chains of amino acids (*c*) four enzyme molecules (*d*) four ATP molecules.

14. In every antibody molecule, one can locate a (*a*) high-energy region (*b*) constant and variable regions (*c*) ATP-activating area (*d*) epitope.

15. The concept of immunological tolerance explains (*a*) why people do not develop immune responses to themselves (*b*) why individuals mount immune responses to small molecules (*c*) why B cells and T cells are different (*d*) why the maturation site for B cells cannot be located.

16. For the first six months of its life, a child depends on antibodies (*a*) received in food (*b*) obtained from its mother (*c*) acquired from the environment (*d*) produced from its immune system cells.

17. Cytotoxic T cells are known for their ability to unite with (*a*) bacteria and viruses (*b*) small molecules known as epitopes (*c*) cells of fungi and protozoa (*d*) human tissue cells.

18. The complement system is a series of reactions set off as a result of (*a*) phagocytosis (*b*) T cell activity (*c*) antibody activity (*d*) large foreign molecule activity.

19. IgE is well known for its participation in the (*a*) phagocytic reaction (*b*) passive immunity reaction (*c*) allergic reaction (*d*) reticuloendothelial reaction.

20. The site of B cell maturation in the human body (*a*) is the thymus (*b*) is the brain and spinal cord (*c*) is not yet identified (*d*) is the tonsils and adenoids.

Completion. Add the word or words that correctly complete each of the following statements.

1. Those immune system cells that mature in the thymus gland are the ___________.

2. The macrophages, monocytes, and other phagocytes make up a system of the body known in the contemporary literature as the ___________.

3. The type of immunity acquired when a child receives antibodies from its mother is ___________.

4. During the immune response, there is a recognition of epitopes and molecules known as ___________.

5. All antibodies are composed of organic compounds called ___________.

6. The region of the antibody molecule where an antigen-antibody reaction takes place is the ___________.

7. The antibody secreted in mother's milk and passed to the child who nurses is ___________.

8. When cytotoxic T cells unite with cells containing antigens, the T cell enzymes damage the cells at the ___________.

9. T cells are known to react with tumor cells, transplanted tissue cells, and body cells infected with ___________.

10. Antibodies are known to stimulate a series of reactions that result in perforations in microbial surfaces resulting from activity of the ___________.

Chapter 17

Immune Tests and Disorders

OBJECTIVES

Several types of laboratory-based immune tests are available for detecting microbial diseases. This chapter will survey many of them. The immune system is also responsible for a number of disorders occurring in the body. The chapter will summarize some of these disorders. The objectives of the chapter are to:

1. Describe the basis for serological immune tests and explain the basis for agglutination and precipitation tests.

2. Explain methods by which antibodies can be separated in the laboratory prior to performing a serological test.

3. Summarize the basic theory of the complement fixation test, immunofluorescence, and ELISA tests.

4. State two examples of immunodeficiency diseases and explain the basis for each.

5. Recognize the characteristics of a hypersensitivity reaction and define and explain four different types of hypersensitivity reactions.

6. Define the important cells, mediators, and symptoms associated with anaphylaxis and allergy reactions.

7. Discuss why transfusion reactions typify cytotoxic hypersensitivity and discuss the immunological basis for taking or rejecting certain blood transfusions.

8. Describe the characteristic features of a delayed hypersensitivity and state several examples of this hypersensitivity in the body.

THEORY AND PROBLEMS

17.1 Why are immune tests valuable in microbiology?

Many disease conditions can be detected by confirming the presence of disease antigens or antibodies produced by the immune system. The immune tests are designed for these purposes. They help the physician know what disease is taking place in the patient, and how far along that disease has progressed.

SEROLOGICAL TESTS

17.2 Are immune tests the same as serological tests?

Immune tests are also known as **serological tests**. This is because the test material generally used in the procedure is a sample of serum. **Serum** is the cell-free component of blood. It normally contains the antigens or antibodies detected in the serological tests.

17.3 Are serological tests qualitative or quantitative?

Serological tests can be either qualitative or quantitative. Qualitative serologic tests reveal the presence of antigens or antibodies. Quantitative serologic tests measure the concentration of antigens or antibodies. A quantitative test helps monitor the progress of a disease by indicating whether the antigen or antibody level rises or falls.

17.4 How many kinds of serological tests are there?

Six basic types of serological tests are used in medical microbiology. They are agglutination tests, precipitation tests, complement fixation tests, immunofluorescence tests, radioimmunoassays, and enzyme-linked immunosorbent assays.

17.5 What is the basis for agglutination tests?

Agglutination tests are serological tests in which antigen molecules are attached to large particles and combine with a sample of the patient's serum. If the serum contains the complementary antibody, it will unite with the antigen on the large particle (such as a cell or latex bead), and a visible clump or particles will appear to the observer. Figure 17-1 shows how this reaction appears when viewed with the unaided eye.

17.6 Are all agglutination tests direct tests?

Many agglutination tests are **direct tests** when the antigen is naturally part of a large molecule such as an erythrocyte or a bacteria. For example, when typhoid bacilli are mixed with typhoid antibodies, the antibodies attack the antigens on the surface of bacilli and cause the bacilli to form large visible clumps. An agglutination test can also be an **indirect test**. In this case, antigens or antibodies are artificially adsorbed onto the surface of a latex bead. When mixed with an appropriate antigen or antibody, the beads will enter into the reaction and form the visible agglutination.

17.7 What are some examples of agglutination tests in serology?

Agglutination tests are used to detect typhoid fever and strep throat. They are also the basis for typing blood and testing blood prior to transfusion. Agglutination tests can also be used to detect *Salmonella* species and for numerous other viral and bacterial agents.

17.8 How can an agglutination test be used in quantitative manner?

To utilize the agglutination test for a quantitative procedure, a sample of patient's serum (containing antibodies) is diluted many times in a series of tubes. Then a standard amount of antigen is added to each tube. The dilution of the tube displaying the last evidence of agglutination reaction represents the **titer** of antibody in the patient's serum. The titer may be expressed as 1:32, 1:100, or 1:640.

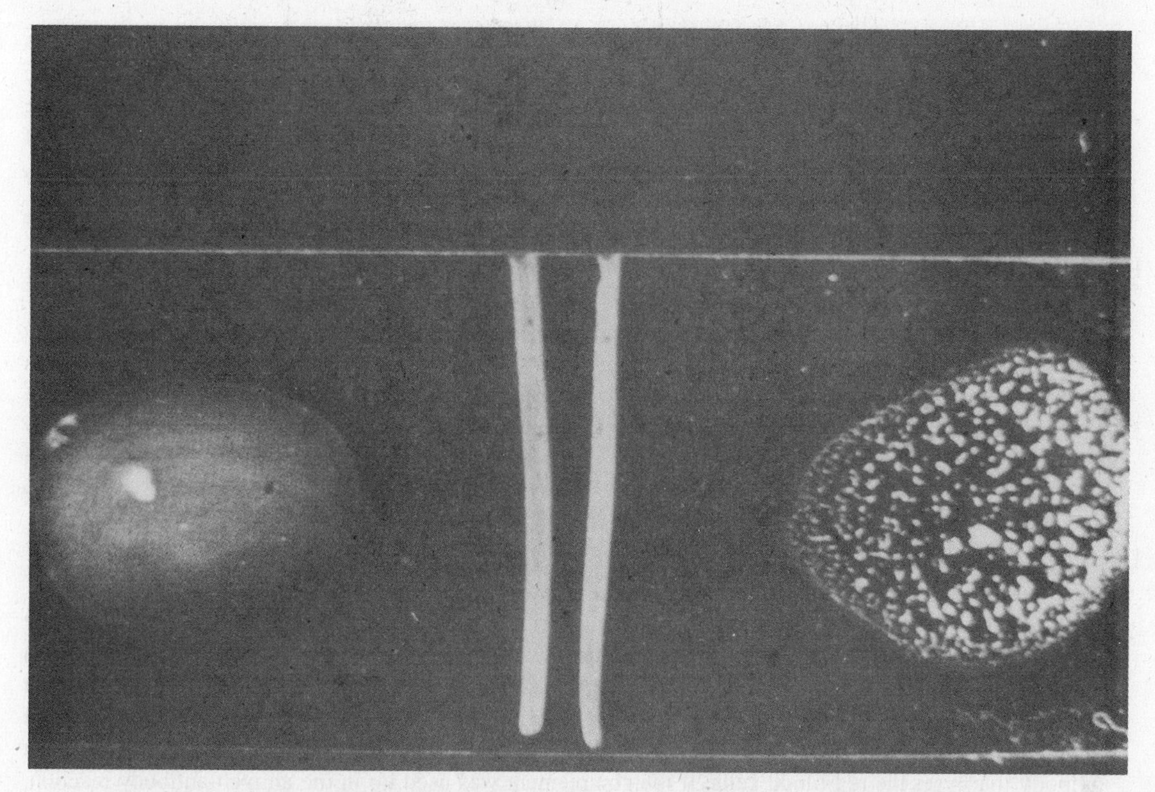

Fig. 17-1 A photograph of the agglutination test. On the left, agglutination has not occurred because the complementary antigens and antibodies are not present in the mixture. On the right, the large particles represent the result of agglutination. In this case the complementary antigens and antibodies were present. The cells clumping are red blood cells.

17.9 Is there any difference between agglutination tests and precipitation tests?

The basic difference between an agglutination test and a precipitation test is that in a **precipitation test** there are no cells or large particles to enter the reaction. Instead, antigen and antibody molecules react with one another and form aggregates and networks that precipitate and become visible within the test tube.

17.10 Which tests utilize the precipitation test in immunology?

The precipitation reaction can be performed in a liquid or in a gel. When performed in liquid, a solution of antigen is placed in the tube and a solution of antibody is layered carefully on top of it. The antigen and antibody molecules migrate to one another and form a precipitation ring at the interface of the two solutions. This is the **double-diffusion Oudin test.** When a similar test is performed in a gel, the antigen and antibody solutions are placed in wells hollowed out of the gel. The reactants move out of the well by diffusion and form a precipitation ring where they unite. If the antigens and antibodies do not match, no precipitation evidence is seen in the tube or the gel. This is the **Ouchterlony test.**

17.11 Can a series of antibodies in a mixture be separated before the precipitation test is performed?

A procedure called **electrophoresis** is used to separate the antibodies in an antibody mixture. This procedure utilizes an electric current to distinguish antibody molecules according to their size. Once the antibodies have been separated, antigen solutions can be added alongside the antibody mixture, and the various antibodies and antigens will migrate toward one another. Where they unite, visible bands of precipitate will appear.

17.12 What is the basic theory that underlines the complement fixation test?

The **complement fixation test** is a complex immunological procedure in which the antigen-antibody complex unites with and "fixes" the complement system. The complement system is a series of proteins in the blood that "completes" the antibody-antigen reaction on the surface of a microorganism. By determining whether the complement has been fixed it is possible to learn whether the antigen-antibody molecules react with one another.

17.13 How is the complement fixation test performed in the serological laboratory?

To begin the complement fixation test, a serum specimen is collected and the assumption is made that it contains a certain antibody. The serum is combined with its complementary antigen, and if the antigen complements the antibody in the serum an antigen-antibody complex will form. Now, a measured amount of complement is added. If the antigen-antibody reaction has occurred, it will fix (react with) the complement. Now an indicator solution is added. The solution consists of red blood cells and an antibody that lyses the red blood cells. If the complement was used up in the antigen-antibody reaction, none will be available to lyse the red blood cells in the presence of its antibody. The red blood cells will remain whole. If, however, the reaction did not occur in the first mixture, then the complement is still available. It will unite with the red blood cells, and the lysing antibody will bring about the destruction of the red blood cells. Therefore, by observing whether whole red blood cells are present in the tube, one may determine whether an antibody-antigen reaction occurred during the first part of the test and whether the patient possesses the antibody suspected in the serum.

17.14 Is the complement fixation blood test widely used in diagnostic microbiology?

The complement fixation test is difficult to perform, and it requires many different variations and much time. However, it is a very sensitive test, and it can be used to detect a wide variety of bacteria, viruses, fungi, and other microorganisms. By varying the antigen used in the first part of the test, numerous different antibodies can be detected in this procedure.

17.15 What is the underlying theory of the immunofluorescence test?

The **immunofluorescence test** is another contemporary procedure in serological microbiology. In this test, antibodies are tagged with fluorescent dyes that make objects visible when the antibodies react with the objects. In a direct immunofluorescence test, the fluorescent-tagged antibodies are combined with suspected microorganisms on a slide. If the antibodies locate their corresponding antigens on the microorganisms, they unite with the antigens, and the fluorescent dye will accumulate. When the slide is subjected to ultraviolet light, the dye will fluoresce and the microorganism can be identified.

17.16 Can the immunofluorescence test be used to identify antibodies as well as antigens?

The immunofluorescence test can be used in an indirect way to detect an antibody present in a serum sample. To perform this test, an antibody must be used that will bind to another antibody. This first antibody is prepared by injecting human antibodies into an animal such as a rabbit. The animal will treat the antibody as an antigen and form antibodies against it. The antibodies so formed are called human anti-antibodies. These anti-antibodies are then tagged with fluorescent dyes and combined with a serum sample. If a reaction between antibodies in the serum and the tagged anti-antibodies takes place, then the serum antibody can be identified since the anti-antibody is a known entity.

17.17 What is the basis of the radioimmunoassay test and how does it compare to the immunofluorescence test?

In the **radioimmunoassay test** an antibody is tagged with a radioactive molecule rather than a fluorescent molecule. The radioactive antibody is then combined with a suspected antigen-bearing microorganism. If a reaction occurs between the antibody and antigens on the surface of the microorganism, the radioactivity will accumulate on the surface. The radioactivity can be determined by a sensing device. Its presence indicates that the corresponding antibody and antigen have united and that the suspected microorganism is present.

17.18 Describe the basis for the enzyme-linked immunosorbent assay (ELISA).

In the **ELISA test,** the antibodies utilized are linked to an enzyme rather to a fluorescent dye or radioactive substance. When the enzyme unites with its substrate, it breaks down the substrate and produces a color change visible to the observer or in a sensing device. An ELISA test can be used in direct or indirect antibody tests.

17.19 How is the direct ELISA test performed in the laboratory?

In the direct ELISA test, an enzyme-linked antibody is combined with microorganisms suspected of containing the complementary antigen. After the reactants have been mixed, the substrate material is added. If the antibody has united with the antigen on the surface of the microorganism, the enzyme molecules gather and when the substrate is added, they will break down the substrate to produce a color reaction. The presence of the color reaction indicates that the suspected antigen and suspected microorganism were present.

17.20 How is the indirect ELISA test performed in diagnostic procedures?

An indirect ELISA test is performed to determine whether a certain antibody is present in a patient's serum. The test is illustrated in Figure 17-2. Plastic beads coated with antigen are combined with the serum sample, and if the complementary antibodies are present, they react with the antigen on the bead surface. The beads are then incubated with anti-antibodies linked to an enzyme. These antibodies will combine with the suspected antibodies if they are present on the bead surface and the enzyme will accumulate. Next the substrate is added, and a color change indicates a positive test for the suspected antibodies. If a color change fails to occur, the test implies that the suspected antibodies were not present in the patient's serum sample. A diagnosis can be conducted in this way.

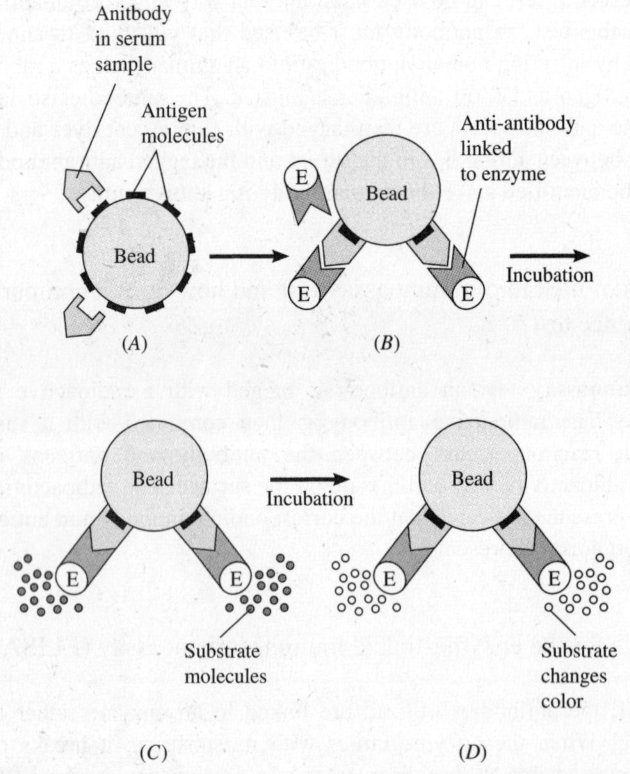

Fig. 17-2 The indirect ELISA test. (*A*) A plastic bead is coated with antigen molecules and combined with antibody molecules in a serum sample. A reaction takes place. (*B*) Anti-antibody molecules linked to enzyme molecules are added to the beads. A reaction takes place and the enzyme molecules accumulate. (*C*) Substrate molecules are added to the mixture. (*D*) Because the enzyme molecules have accumulated, there is a reaction with the substrate molecules, and the substrate changes color. The color change indicates that the suspected antibodies were present in the original serum sample.

IMMUNE DISORDERS

17.21 What is the nature of immune disorders?

There are several disorders traced to deficiencies of the immune system or dysfunction of the system. Immune system deficiencies develop because something is lacking in the immune system, while the dysfunctions come about because actions of the immune system are deleterious to the body.

17.22 What are two examples of immunodeficiency diseases?

An example of an immunodeficiency is **DiGeorge's syndrome**. In persons with this syndrome, the thymus fails to develop, and T cells fail to form. Without T cells, there is high susceptibility to disease by fungi and protozoa, and to virus-infected cells. The person cannot mount cell-mediated immunity. Another immunodeficiency disease is **Bruton's agammaglobulinemia**. People with this disease fail to form B cells. Without B cells they are unable to manufacture antibodies. Susceptibility to bacterial infections is usually acute in these individuals.

17.23 Is Bruton's agammaglobulinemia related to the genes of an individual?

Bruton's agammaglobulinemia appears to be related to a deficiency on the X chromosome, one of the two sex chromosomes. Therefore, it is a sex-linked disease. It manifests itself in males most commonly, and it is carried by the female.

17.24 Are any immunodeficiency diseases related to the stem cells in humans?

Both the B and T cells are derived ultimately from cells in the bone marrow called stem cells. When the stem cells are missing or defective, the individual fails to form B cells and T cells. This condition is known as **severe combined immunodeficiency disease (SCID)**. Confinement to a sterilized plastic bubble is usually required for these individuals because they are unable to mount immune defenses against any antigens.

17.25 Are there any immunodeficiency diseases that affect phagocytes?

There is an immunodeficiency disease of the phagocytes called **Chediak-Higashi disease**. In this disease, the lysosomes of phagocytes fail to unite with microorganisms after they have been engulfed. Therefore the ability to destroy microorganisms by phagocytosis is severely reduced. Bacterial and viral infections tend to be common in these patients.

17.26 Can an immunodeficiency disease be due to an agent outside the body?

The situation of **acquired immune deficiency disease (AIDS)** illustrates how an agent outside the body can destroy the immune system. The human immunodeficiency virus (HIV) destroys the T cells and reduces the body's ability to respond to protozoa, fungi, cancer cells, and other foreign cells in the body. With cell-mediated immunity (CMI) at a minimal, these microorganisms often cause severe illness and death in a patient.

17.27 What is hypersensitivity, and how many types of hypersensitivity reactions can occur in the body?

The hypersensitivity reactions represent the second type of immune disorder. **Hypersensitivities** are immune reactions traced to dysfunction of the immune system. Hypersensitivity implies an overly-aggressive sensitivity an antigen entering the body. The immune response ordinarily would neutralize the antigen, but in the case of hypersensitivity, the immune reaction results in damage to the body.

17.28 How are the hypersensitivity reactions classified in the body?

Hypersensitivity reactions are of four major types: type I, type II, type III, and type IV hypersensitivity. Each type of hypersensitivity is distinctly different from the others, and each depends upon a separate process.

17.29 What are the characteristics of type I hypersensitivity?

Type I hypersensitivity is also known as **immediate hypersensitivity** because reaction takes place almost immediately in the body. It is also known as anaphylaxis if it involves the entire body, or it is known as allergy if it involves a limited area of the body.

17.30 Which antibody functions in type I immediate hypersensitivity reactions?

The antibody that functions in type I hypersensitivity reactions is IgE. This antibody is produced by plasma cells and B cells on stimulation by antigens such as penicillin molecules, mold spores, certain foods, pollen grains, and dust particles. The antibody is released into the bloodstream as it is produced.

17.31 What happens to the IgE produced during a type I immediate hypersensitivity?

Once the IgE is released to the bloodstream, it circulates and fixes itself to the surface of mast cells and basophils. **Mast cells** are connective tissue cells located throughout the body. **Basophils** are circulating white blood cells found within the blood vessels. The IgE molecules attach themselves to these cells by their common domains. The body usually experiences no reactions from the attachment of IgE to basophils and mast cells, so as long as this is the first time that the antibodies have attached to the cells. As antibodies build up on the surface of cells, however, a person becomes increasingly sensitive to the antigen.

17.32 When does a reaction eventually occur in the basophils and mast cells?

At some time in the future the person is exposed to the same antigen once again. Now the antigens combine with IgE molecules on the surface of the cells. This induces the cells to release their granules, which release physiologically active substances such as histamine, serotonin, leukotrienes, and other substances. These substances are known as **mediators**.

17.33 What affect do the histamine and other mediators have on the body?

The mediators such as histamine released from the mast cells and basophils cause the smooth muscles of the body to contract. These contractions can result in dilation of the bronchial tubes, cramps of the gastrointestinal tract, and constriction of the blood vessels. These contractions can lead to hives, respiratory distress, gastrointestinal cramps, swellings of the skin, and other symptoms of allergic reaction. The reaction is summarized in Figure 17-3.

17.34 What is the difference between anaphylaxis and allergy?

When mediators such as histamine are released throughout the body and contractions occur throughout the body, then the result is a life-threatening condition called **anaphylaxis**. Closure of the bronchial tubes can lead to suffocation and death within minutes. When the reaction happens locally, then the reaction is known as allergy. **Allergy** is usually not a life-threatening condition.

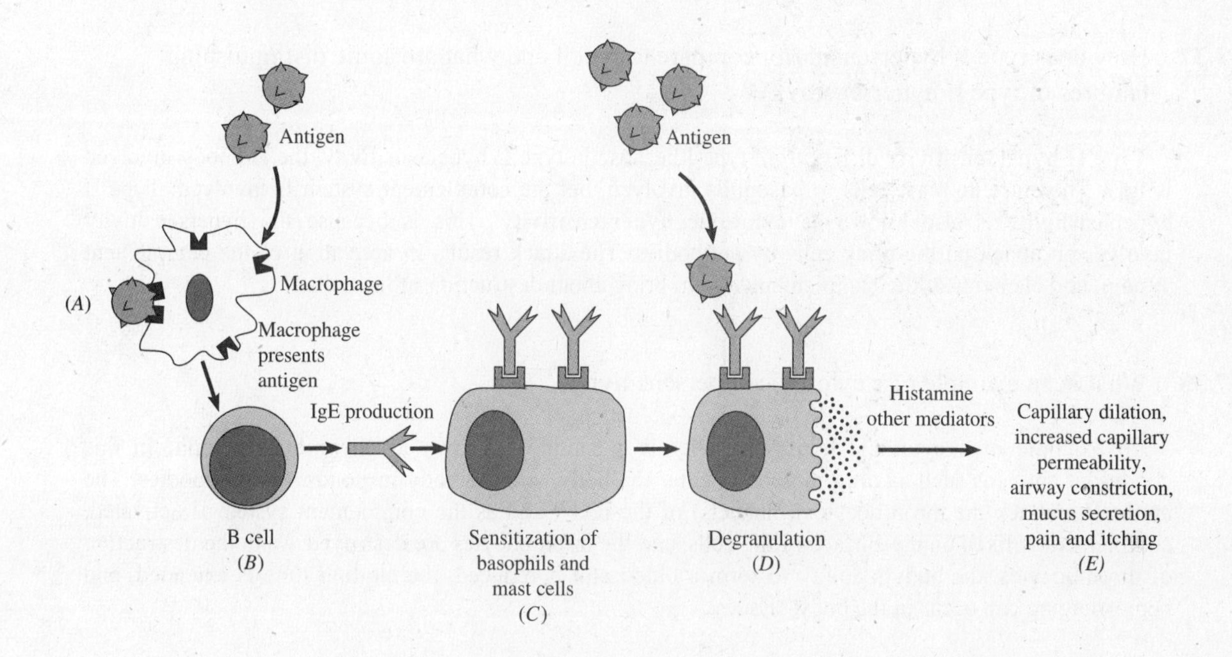

Fig. 17-3 The allergy and anaphylaxis reactions. (*A*) On entering the body, antigens are engulfed by macrophages and presented to B cells. (*B*) The B cells produce IgE, which enters the circulation. (*C*) The IgE molecules attach by the constant end to the surface of basophils and mast cells and sensitize them. (*D*) At a later date, antigens reenter the body and react with the IgE molecules on the cell surface. The cell undergoes degranulation, and releases various kinds of mediators. (*E*) The mediators induce the contraction of smooth muscles and bring about many of the symptoms of allergy reaction (localized) or anaphylaxis (whole body).

17.35 What can be done for the person who is suffering from anaphylaxis or allergy?

Several things can be done for the person in the midst of anaphylaxis or allergy. For example, injections of epinephrine (adrenaline) will stabilize the mast cells and basophils to prevent further release of granules. Epinephrine will also stimulate the heart and relieve some of the other symptoms. An antihistamine can be administered to neutralize the histamine being released. A smooth muscle relaxer can be used to offset the contractions of the smooth muscle, and cortisone can be used to reduce swelling.

17.36 How do "allergy shots" prevent the development of allergies during the allergy season?

To prevent a person from developing allergies, allergy shots can be given. An allergy injection consists of a very dilute solution of the antigen. The antigens in the solution gradually eliminate the sensitized mast cells and basophils so that when the allergy season occurs, the cells are much less abundant in the body, and the allergy reaction is much less severe. Also, the antigens in the allergy injection induce the formation of antibodies that block the activity of antigens with IgE. These antibodies are called **blocking antibodies**.

17.37 How does type II hypersensitivity compare to type I and what are some distinguishing features of type II hypersensitivity?

Type II hypersensitivity differs from type I because in type II hypersensitivity, the antibody involved is IgG. There are no mast cells or basophils involved, but the complement system is involved. Type II hypersensitivity is also known as **cytotoxic hypersensitivity**. This is because the hypersensitivity involves an attack on the body cells by antibodies. The attack results in activation of the complement system, and elements of the complement system bring about destruction of the cells.

17.38 What is an example of a cytotoxic hypersensitivity?

An example of a cytotoxic hypersensitivity is a condition known as **thrombocytopenia.** In this condition, antigens such as drug molecules enter the body, and the body responds with antibodies. The antibodies attack the thrombocytes (platelets) of the body, and as the complement system is activated, complement is fixed on the surface of the cells, and the thrombocytes are destroyed. With the destruction of thrombocytes, the body's ability to form a blood clot is reduced, the clotting time is extended, and hemorrhaging can occur in the body tissues.

17.39 Are there other examples of cytotoxic hypersensitivity?

There is another condition known as **agranulocytosis**. In this condition, the body forms antibodies against antigens that enter the body. The antibodies unite with neutrophils, and they activate the complement system. With the destruction of neutrophils, the ability to perform phagocytosis in the body is reduced. This removes an important defense mechanism of the body?

17.40 Why is the transfusion reaction an example of cytotoxic hypersensitivity?

The **transfusion reaction** typifies a cytotoxic hypersensitivity because antibodies attack human cells, activate the complement system, and destroy them. For example, a person who has blood type B possesses B antigens on the red blood cells and anti-A antibodies in the serum. By contrast, a person who has type A blood has A antigens on the red blood cells, and anti-B antibodies in the serum. If type B blood were to be donated to the person with type A blood, then the red blood cells entering the recipient would have B antigens on their surface. The recipient has anti-B antibodies in the serum. These antibodies would react with the incoming antigens of the donor's blood, and the reaction would set off the complement system leading to the destruction of red blood cells and clumping reactions probably leading to death. Therefore, type B blood cannot be transfused into a person with type A blood because a type II cytotoxic hypersensitivity is possible.

17.41 Suppose a type AB individual wishes to donate to a recipient having type O blood. Could this transfusion be made?

A type AB individual has both A and B antigens on the surface of the blood cells, but no anti-A or anti-B antibodies in the serum. The recipient in this situation is type O, and has no antigens on the surface of the red blood cells, but both anti-A and anti-B antibodies exist in the serum. If the transfusion were to be made, the A antigens of the incoming cells would react with the anti-A antibodies, and the B antigens of the incoming cells would react with the anti-B antibodies. A cytotoxic hypersensitivity would take place, and destruction of the red blood cells and severe clumping and death would follow.

17.42 Suppose the situation were to be reversed and the donor had type O blood and the recipient had type AB blood. Could this transfusion work?

In this situation, the donor has neither A nor B antigens on the red blood cells. The recipient has neither anti-A nor anti-B antibodies in the serum. Therefore, the incoming red blood cells have no antigens, and the recipient's serum has no corresponding antibodies. Therefore the transfusion could be successfully made.

17.43 Is it common practice to transfuse different types of blood to different individuals?

There are many blood groups that must be taken into account to determine if two blood types are compatible. What has been described exists only for the ABO system, and it should not be taken to mean that blood types can be mixed indiscriminately. However, in a dire emergency, it might be possible to mix blood types so long as the antigens on the donor's red blood cells and the antibodies in the recipients serum are considered. It is extremely important to avoid mixing antigens and their compatible antibodies.

17.44 How does the Rh antigen become important in cytotoxic hypersensitivity?

The Rh antigen (or Rh factor) is another antigen present on the red blood cells of the great majority of individuals, perhaps in excess of 85 percent. Those possessing the antigen are said to have "positive" blood such as type A-positive or O-positive. Cytotoxic hypersensitivities due to this antigen are not considered important in adults, but in newborns, the hypersensitivity can result in a condition called hemolytic disease of the newborn.

17.45 How does hemolytic disease of the newborn come about?

When an Rh positive male (e.g., blood type A+) marries an Rh negative female (e.g., A-), there is a 75 percent probability 3:1 chance that the offspring will be Rh positive (e.g., A+). During the pregnancy the danger is minimal, but at birth some of the child's blood may enter the blood of the mother. Should this happen, the mother's immune system will respond to the Rh antigens and produce Rh antibodies. However, these Rh antibodies will not affect the child since it has already been born.

17.46 When can the Rh antibodies pose a danger to a child?

If a succeeding conception results in another Rh positive child (e.g., A+), then the Rh antibodies may pass across the placenta (along with other antibodies of the mother) and enter the blood of the child. If they do so, the antibodies will attack the Rh antigens on the surface of the red blood cells of the child. This attack will result in a destruction of the red blood cells, a cytotoxic hypersensitivity, and possibly death in the unborn child. This is **hemolytic disease of the newborn**.

17.47 Has hemolytic disease in the newborn been known by any other names?

In some anatomy and physiology books, hemolytic disease of the newborn is also known as **erythroblastosis fetalis**. This is because young red blood cells called erythroblasts appear in the child's blood to replace the erythrocytes being destroyed in the child's blood by the cytotoxic hypersensitivity. The disease is also known as **Rh disease**. It is described in Figure 17-4.

17.48 Can anything be done to prevent the development of hemolytic disease of the newborn?

To prevent hemolytic disease of the newborn, it is important that the mother's immune system be prevented from forming Rh antibodies. This can be accomplished by giving the woman an injection of RhoGAM shortly after the birth of the first child. RhoGAM contains Rh antibodies. These antibodies unite with Rh antigens coming from the child's blood and neutralize the Rh antigens. When the Rh antigens are neutralized, they will not be available to stimulate the woman's immune system, and she will not produce Rh antibodies. Therefore, during the second pregnancy there will not be any Rh antibodies to interfere with the blood of the child. She must also be given an injection of RhoGAM after the second birth and every time she gives birth to an Rh positive child.

17.49 What are the characteristic signs of a type III hypersensitivity?

Type III hypersensitivity reactions involve the formation of immune complexes. **Immune complexes** develop when dissolved antibody molecules react with dissolved antigen molecules. The antibody and antigen molecules combine in such a fashion that a large lattice forms. This lattice is composed of interlocking antigen and antibody molecules that form a mass so large that it is visible. The mass is the immune complex. It sets off the complement system and the complement system brings about destruction of the local tissue.

17.50 What are some organs of the body that can be affected by immune complexes?

Immune complexes can form in various regions of the body including the skin, joints, and kidneys. When the complexes form in the skin, a condition called **systemic lupus erythematosis** can result. This condition is characterized by a butterfly-shaped rash on the skin and tissue damage in blood-rich organs such as the kidneys and spleen. In the joints, the immune complexes can lead to a condition called **rheumatoid arthritis**.

17.51 Are there any bacterial diseases associated with immune complexes?

One disease associated with immune complexes is **bacterial endocarditis**. In this condition, streptococcal antigens and streptococcal antibodies form immune complexes, which fix themselves to the valves of the heart. Destruction of the valve tissue results in poor blood flow through the heart, a symptom of endocarditis. In other cases, the immune complexes form in the kidney tissues and result in **glomerulonephritis**. Poor urine formation and excretion are symptoms of this disease.

17.52 What sort of hypersensitivity is exhibited in serum sickness?

Serum sickness is also characterized by a type III immune complex hypersensitivity. Serum sickness occurs when antiserum is injected into humans to prevent or treat a disease such as tetanus. In this situation, the antiserum is usually produced in an animal, and antigens present in the serum stimulate the formation of antibodies in the body. When the antigens unite with the antigens immune complexes are generated.

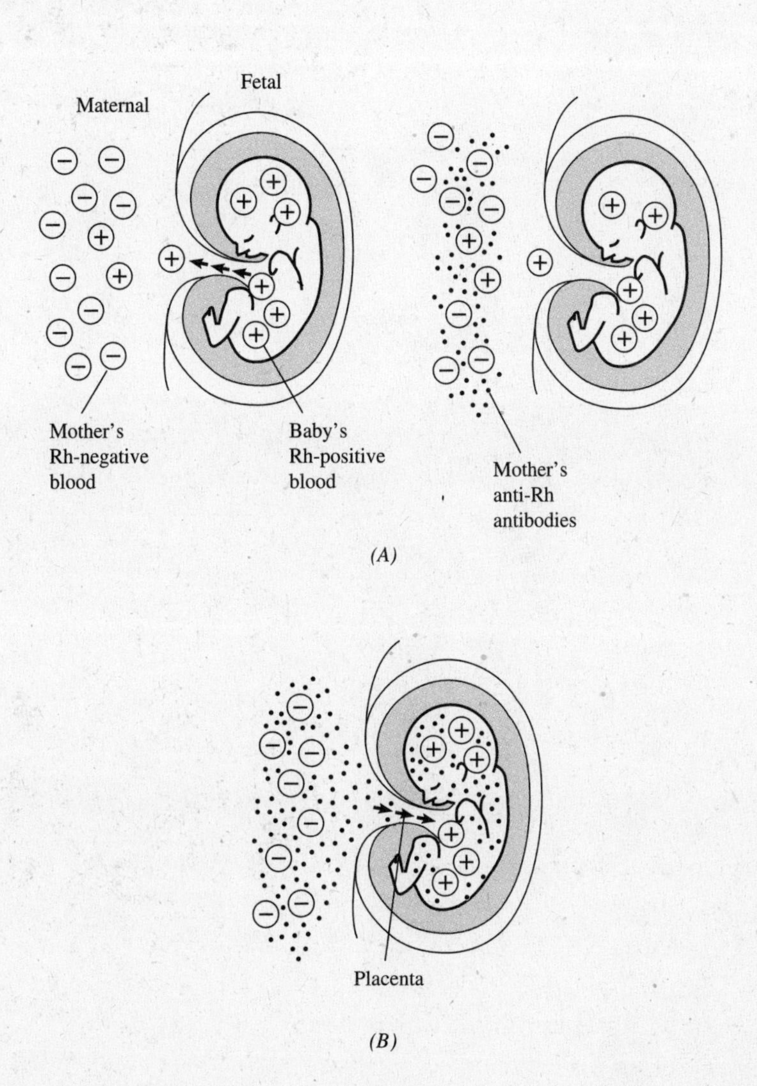

Fig. 17-4 Hemolytic disease of the newborn. (*A*) A woman who is Rh negative is pregnant with a child who is Rh positive. At birth, some of the child's red blood cells carry the Rh antigen into the mother's bloodstream. The mother's immune system responds by producing anti-Rh antibodies. There is no damage to the child, however, because it has already been born. (*B*) At a later time, the woman may become pregnant with another Rh positive child. During the fetal development, the anti-Rh antibodies may pass the placenta barrier and enter the bloodstream of the fetus, where they attack the Rh antigens on the red blood cells. This attack leads to destruction of red blood cells and symptoms of disease. The possibility of this happening can be prevented by administering Rh antibodies to the woman immediately after the birth of the first child.

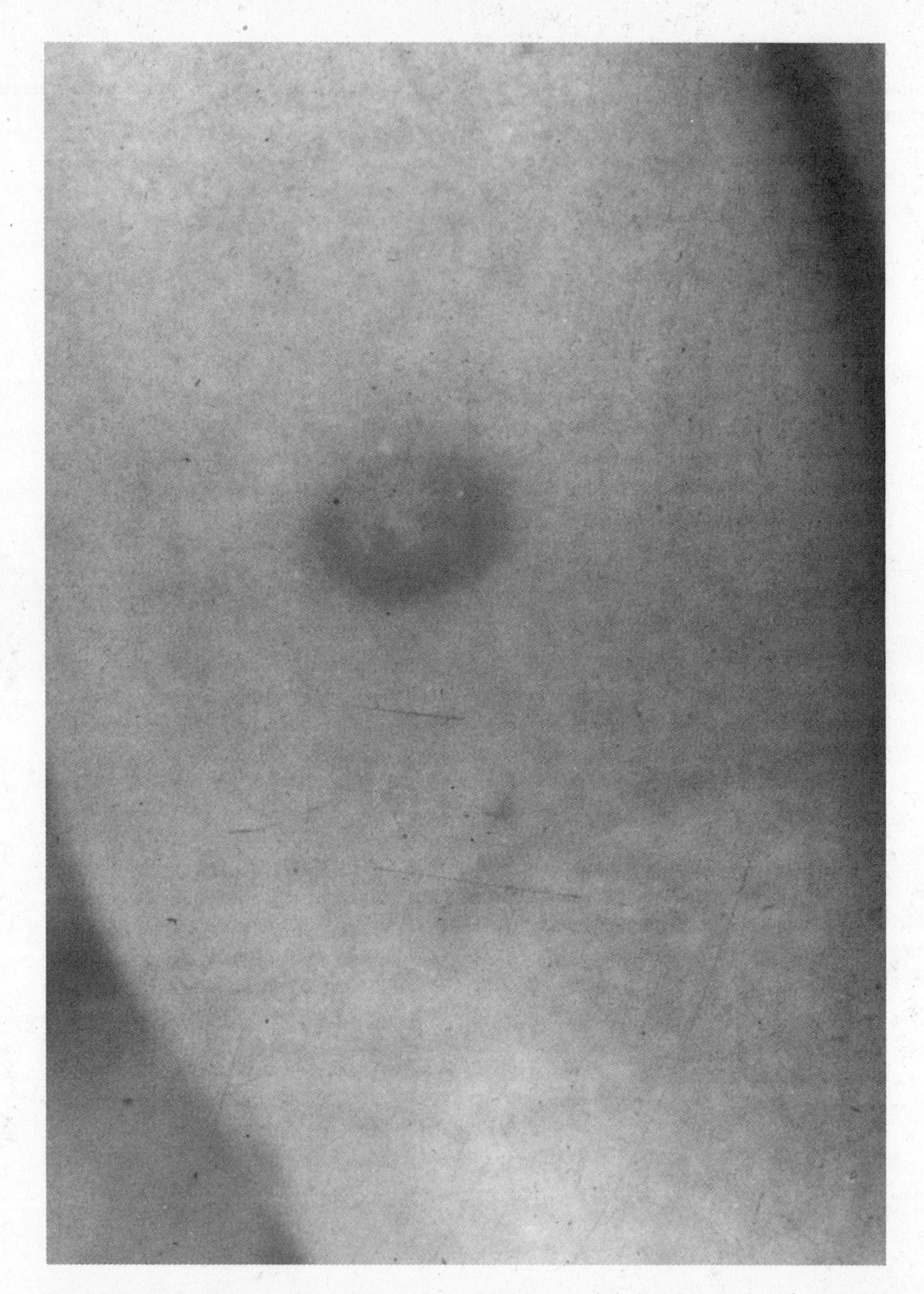

Fig. 17-5 A photograph of a positive tuberculin skin test showing the results of a type IV delayed hypersensitivity. This individual has had a previous exposure to antigens from tubercle bacilli.

17.53 What are the characteristic features of type IV hypersensitivity?

Type IV hypersensitivity is also known as **delayed hypersensitivity**. This is because the hypersensitivity reaction occurs some hours or days after union between the antigens and the immune system products. In the case of type IV hypersensitivity, there are no antibodies. Instead, the reaction involves cell-mediated immunity. T cells are involved in this reaction. Figure 17-5 presents an example of a delayed hypersensitivity as shown by the tuberculin skin reaction. The reaction is explained below.

17.54 Can the type IV hypersensitivity reaction be used in diagnostic medicine?

The **tuberculin skin test** is an example of a type IV delayed hypersensitivity. When a person has been exposed to tubercle bacilli or has had a tuberculosis immunization, the body responds with sensitized T cells that distribute themselves throughout the body, including the skin. In the tuberculin skin test, a preparation of tuberculosis known as PPD (purified protein derivative) is injected superficially into the skin. The sensitized T cells react with the antigens and the site of the injection becomes reddened and thickened within several hours. In two to three days, the reaction reaches its peak, and it is clear that the person has been exposed to tuberculosis antigens at some time in the past. The influx of macrophages at the injection site contributes to the tissue damage.

17.55 What is another example of a type IV hypersensitivity reaction?

Another example of a type IV hypersensitivity is **contact hypersensitivity**. This reaction develops as a result of reactions with antigens in poison ivy plants. T cells are produced by the body in response to those antigens and the cells react with antigens. The reaction results in raised, itching welts on the skin that are characteristic of **poison ivy**.

17.56 Are there any other examples of contact hypersensitivity?

Contact hypersensitivities are also manifested by allergic reactions that individuals show to metals such as nickel in certain jewelry, or to chromium salts in certain leather products or to dyes in clothing, or to aluminum salts in deodorants, or to other materials. In many cases, the reaction between antigens and T cells is manifested by a raised thickening of the skin known as **induration.** Formaldehyde and formalin products can also produce a type IV hypersensitivity reaction.

Review Questions

Multiple Choice. Select the letter of the item that correctly completes each of the following statements.

1. Serological tests are used in diagnostic microbiology to detect (*a*) the presence of antigens or antibodies in blood (*b*) whether bacterial flagella are present or absent (*c*) whether the serum of an individual contains red blood cells (*d*) whether protozoa and fungi have invaded the tissues of the body.

2. An example of an agglutination test used in serology is that in which (*a*) the dissolved cells break open during lysis (*b*) viruses multiply within bacteria (*c*) a visible clump appears to the observer (*d*) the body's B cells diminish in number.

3. The titer of an individual represents the (*a*) amount of antigen present in the body tissues (*b*) amount of antibody in the patients serum (*c*) number of flagella present on a bacterial cell (*d*) number of spores formed by a fungus.

4. In order to separate the antibodies in an antibody mixture, the laboratory technologist may use a procedure called (*a*) transfusion (*b*) complement fixation (*c*) electrophoresis (*d*) gene amplification.

5. At the conclusion of the complement fixation test, one observes whether (*a*) the red blood cells have been destroyed or not (*b*) whether visible masses of precipitate have formed or not (*c*) whether bacteria have mutated to resistant forms or not (*d*) whether a contact hypersensitivity has developed.

6. The immunofluorescence test can be used to identify (*a*) protein molecules and polysaccharide molecules (*b*) lipid molecules and nucleic acid molecules (*c*) antibody molecules and antigen molecules (*d*) cytoplasmic molecules and cell wall molecules.

7. In the ELISA test, an essential requirement is a reaction brought about by (*a*) a fungus and a protozoan (*b*) the substance complement (*c*) a radioactive isotope (*d*) an enzyme

8. At the conclusion of the ELISA test, (*a*) radioactivity is produced (*b*) a clumping reaction is seen (*c*) cells undergo lysis (*d*) a color change takes place.

9. Tests for typhoid fever and strep throat as well as blood typing utilize a serological test known as (*a*) precipitation (*b*) electrophoresis (*c*) agglutination (*d*) complement fixation.

10. Serological tests can be used for (*a*) distinguishing prokaryotes from eukaryotes (*b*) detecting the presence of infectious disease (*c*) forming sodium chloride assays in blood (*d*) detecting whether a toxemia is in progress.

11. DiGeorge's syndrome is one in which (*a*) the B cells fail to form (*b*) the stem cells fail to form (*c*) the T cells fail to form (*d*) agammaglobulinemia occurs.

12. Type I hypersensitivity is also known as (*a*) anaphylaxis (*b*) agglutination (*c*) the transfusion reaction (*d*) contact hypersensitivity.

13. Hemolytic disease of the offspring may develop if a child results from the marriage of (*a*) an Rh positive male and an Rh positive female (*b*) an Rh positive male to an Rh negative female (*c*) an Rh negative male to an Rh positive female (*d*) an Rh negative male to an Rh negative female.

14. Bacterial endocarditis and glomerulonephritis both are associated with the formation of (*a*) the complement system (*b*) radioactive particles in blood (*c*) immune complexes (*d*) the ELISA reaction.

15. The tuberculin skin test is an example of a (*a*) type IV delayed hypersensitivity (*b*) allergy reaction (*c*) precipitation reaction (*d*) serum sickness.

16. The antibody IgE and the mast cells and basophils are all associated with (*a*) contact hypersensitivity (*b*) immunofluorescence tests (*c*) electrophoresis (*d*) immediate hypersensitivity.

17. In the condition called thrombocytopenia, there is a reduction of the body's (*a*) level of antibodies (*b*) platelets (*c*) ability to agglutinate typhoid antibodies (*d*) ability to develop a contact hypersensitivity.

18. A recipient individual who has type A blood may in an emergency receive a transfusion of blood from a donor whose blood type is (*a*) either A or B (*b*) either B or O (*c*) A, B, or O (*d*) either A or O.

19. An individual whose blood type is B may in an emergency donate blood to a donor whose blood type is (*a*) B or O (*b*) AB or A (*c*) A or O (*d*) AB or B.

20. To prevent hemolytic disease of the newborn from occurring in a successive pregnancy, and Rh negative woman giving birth to an Rh positive child will receive shortly after birth an injection of (*a*) Rh antigens (*b*) A and B factors (*c*) Rh antibodies (*d*) type O blood.

Matching. In Column A, place the letter of the serological test from Column B that best fits the characteristic.

Column A	Column B
____ 1. Requires red blood cells	(*a*) Agglutination test
____ 2. Color change at test conclusion	(*b*) Precipitation test
____ 3. Performed in a gel	(*c*) Complement fixation test
____ 4. Used to detect *Salmonella*	(*d*) Immunofluorescence test
____ 5. Requires an indicator solution	(*e*) Radioimmunoassay
____ 6. Uses ultraviolet light	(*f*) Enzyme-linked immunosorbent assay
____ 7. Radioactivity detected	
____ 8. Two distinct parts in test	
____ 9. Ouchterlony test	
____ 10. Used to determine titer	

True/False. For each of the following, enter the letter "T" if the statement is true in its entirety or "F" if any part of the statement is wrong.

____ **1.** The agglutination tests can only be used as direct tests for qualitative determinations.

____ **2.** The electrophoresis procedure utilizes an electric current to distinguish antibody molecules according to their size and separate them.

____ **3.** If the corresponding antigen and antibody are present at the beginning of the complement fixation test, the red blood cells will undergo lysis at the conclusion of the test.

____ **4.** The ELISA test is preferred to the radioimmunoassay because the ELISA test utilizes radioactivity for the detection of antigens and antibodies.

____ **5.** In the immunofluorescence test, antibodies are tagged with fluorescent dyes and combined with suspected microorganisms on a slide to perform a direct test.

____ **6.** The complement fixation test is a relatively simple procedure used in serological microbiology.

____ **7.** In an agglutination test, a visible clump of particles appears to the observer.

____ **8.** In the precipitation test, cells and large particles enter the reaction and clump together.

____ **9.** The double-diffusion Oudin test utilizes the precipitation reaction as its basis.

____ **10.** The immunofluorescence test can be used to detect antibodies as well as antigens.

____ **11.** In the condition known as Bruton's agammaglobulinemia, the body fails to form T cells.

___ **12.** Chediak-Higashi disease, an immunodeficiency disease, affects the phagocytes of the body.

___ **13.** An allergy reaction is one that involves the entire body, while an anaphylaxis involves only a limited area of the body.

___ **14.** The purpose of allergy shots is to eliminate sensitive lymphocytes in the body before the allergy season occurs.

___ **15.** Thrombocytopenia is a type of hypersensitivity in which IgE attacks mast cells and basophils.

___ **16.** A potentially deadly transfusion reaction can occur if an individual with type A blood donates blood to an individual with type AB blood.

___ **17.** When an Rh-positive male marries an Rh-negative female and they decide to have a child, there is a 100 percent probability that the offspring will be Rh-positive.

___ **18.** Immune complex formation is an essential feature in the development of system lupus erythematosis.

___ **19.** The tuberculin skin test typifies a type IV delayed hypersensitivity and is used to determine whether the person has had a previous exposure to tubercle bacilli or has had a tuberculosis immunization.

___ **20.** A cytotoxic hypersensitivity is typified by hay fever or other common allergies that involve histamine release in the body.

Chapter 18

Microbial Diseases of the Skin and Eyes

OBJECTIVES

Skin diseases are among the most numerous in humans because the skin is constantly exposed to the environment. This chapter will consider the variety of microbial diseases of the skin and eyes. The objectives of the chapter are to:

1. Become familiar with the variety of diseases due to species of staphylococci.

2. Learn the different types of streptococci and the diseases for which they are responsible.

3. Describe the symptoms of various pox diseases caused by viruses.

4. Summarize the skin diseases due to herpesviruses.

5. Explain the nature of the immunization preparations available for chickenpox, measles, and rubella.

6. Define the symptoms of fungal diseases of the skin known as tinea diseases.

7. Recognize the various diseases due to *Candida albicans*.

8. Summarize the microbial diseases of the eye, including pinkeye, gonococcal ophthalmia, trachoma, and keratitis.

THEORY AND PROBLEMS

BACTERIAL DISEASES

18.1 Why are staphylococci so often associated with skin infections?

Staphylococci are Gram-positive spheres about 0.5 to 1.0 μm in diameter. They form irregular grapelike clusters and frequently come in contact with the skin. Over the eons of time, staphylococci have adapted to the conditions on the skin and they have evolved to become inhabitants of the skin.

18.2 Which species of staphylococci are involved in skin diseases?

There are two main species of staphylococci: *Staphylococcus aureus* and *Staphylococci epidermidis*. *Staphylococcus aureus* is **coagulase-positive** and forms colonies that are golden yellow. Its coagulase causes fibrin clots to develop in tissue and protect the organism from phagocytosis. The organism also produces toxins that interfere with physiological processes. For example, **leukocidin** destroys leukocytes, **exfoliative toxin** causes scalding of the skin, and **enterotoxin** induces peristaltic contractions of the gastrointestinal tract. *Staphylococcus epidermidis* is a coagulase-negative organism that lives on the skin and is generally not pathogenic. However, when the skin barrier is penetrated such as by a cut, wound, or medical procedure, then this species may cause skin disease.

18.3 What types of skin diseases are caused by staphylococci?

There are numerous types of staphylococcal skin diseases. Infection of the hair follicle results in **folliculitis,** which is also called **pimples.** Infection at the base of the eyelash results in a **sty**. A pus-filled lesion due to staphylococci is called a **boil**, or **abscess**. The term **furuncle** is also used. An enlarged lesion, especially on the neck, is known as a **carbuncle**. The carbuncle is hard, round, and deep and involves the tissues beneath the skin.

18.4 Can staphylococci be involved in impetigo?

Impetigo refers to a highly contagious, pus-producing skin infection possibly due to staphylococci. Impetigo is common in children but not adults. It is transmitted by skin contact as well as by inanimate objects such as utensils and toys. Streptococci may also cause this skin disease.

18.5 What is unique about the scalded skin syndrome caused by staphylococci?

The **scalded skin syndrome** is accompanied by large, soft vesicles occurring over the entire body. The vesicles peel to leave large, scalded-looking regions. The condition is brought about by a toxin produced by staphylococci. Vigorous antibiotic therapy is usually needed to resolve the condition, and high fever and extensive blood invasion of staphylococci accompanies the disease.

18.6 Which diseases are associated with staphylococcal toxins?

One of the well-known staphylococcal conditions associated with a toxin is the **toxic shock syndrome**. The disease came to prominence in the late 1970s when an outbreak was linked to high absorbency tampons. Toxic shock syndrome is accompanied by high fever, a sudden drop in blood pressure (shock), and a rash on the skin that resembles sunburn. Another symptom is the skin peeling especially on the hands, as shown in Figure 18-1. Blood transfusions can relieve the symptoms, and most cases are now associated with materials that remain in a wound to absorb fluids. Another toxin-associated condition is **staphylococcal food poisoning,** which is caused by an enterotoxin produced by staphylococci in food (Chapter 21).

18.7 Can toxins be produced by streptococci as well as staphylococci?

Streptococci are Gram-positive cocci occurring in chains. Certain strains are able to produce toxins, particularly the **erythrogenic toxin**. Strains producing this toxin generally belong to the species *Streptococcus pyogenes*. When this species enters the bloodstream, its toxin is responsible for the reddening of the skin characteristic of **scarlet fever.**

18.8 Can streptococci proliferate in the blood even though they produce no erythrogenic toxin?

There are strains of *Streptococcus pyogenes* that produce no erythrogenic toxin, but cause serious disease. **Strep throat** is an example of such a disease which can be a serious problem of the blood and respiratory tract, even though there are no skin symptoms. Transmission can occur by respiratory droplets, and therapy with penicillin is usually recommended.

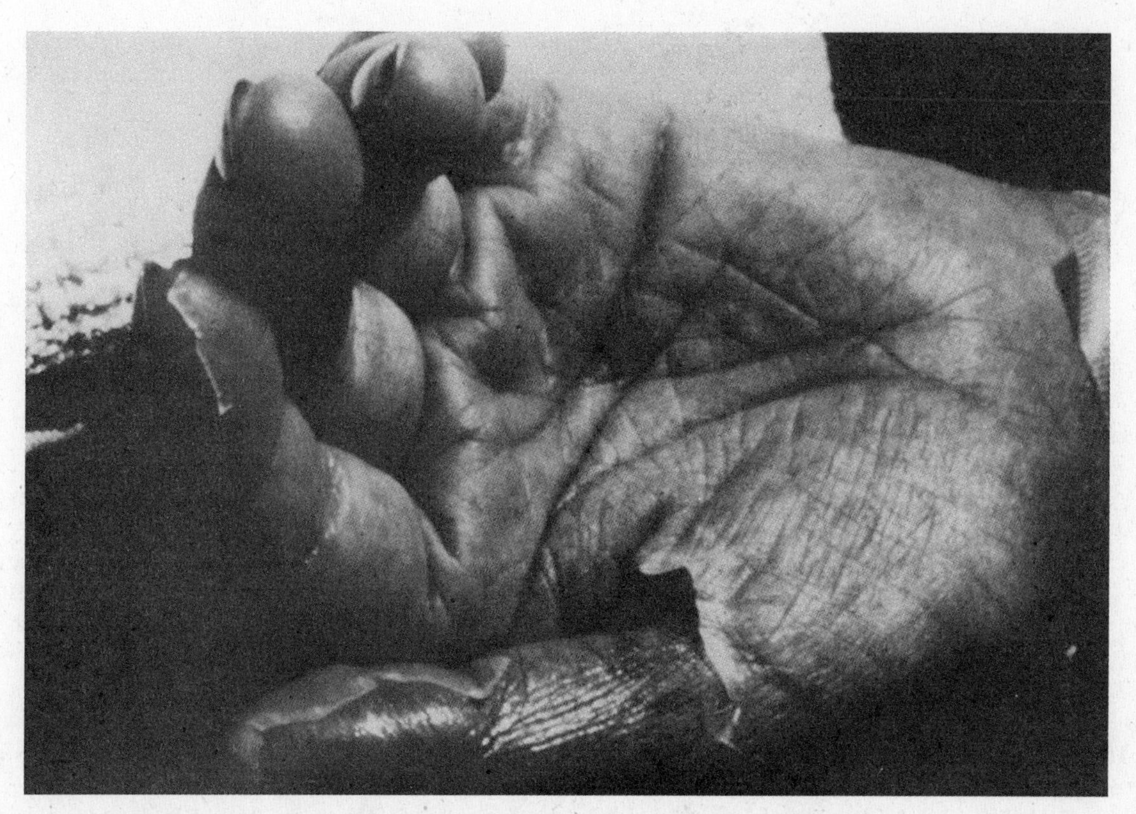

Fig. 18-1 The peeling skin often associated with toxic shock syndrome.

18.9 Do streptococci produce hemolysins?

Streptococci are among the most well-known producers of **hemolysins**, a series of toxins that damage red blood cells. If the damage to the red blood cells results in complete dissolution of the cells, the streptococci are known as **beta-hemolytic streptococci**; if the red blood cells are partially destroyed, the streptococci are called **alpha-hemolytic streptococci**; and if no destruction of red blood cells occurs because there are no toxins, the streptococci are designated **gamma-hemolytic streptococci**.

18.10 Is there an alternate way to differentiate streptococci?

In addition to the hemolytic designations, streptococci can be classified in groups A, B, C, D, and so on to O. These designations refer to the antigens present in their cell walls. Group A streptococci are the most pathogenic. They include the bacteria responsible for skin infections.

18.11 What group do the "flesh-eating" streptococci belong to?

The so-called "flesh-eating" streptococci belong to group A. These streptococci infect the fascia which underlies the skin and causes a condition called **necrotizing fasciitis**. Skin and fascia destruction accompany this disease, and entry to the deep skin tissues is often a result of a wound.

18.12 Can streptococci produce toxins when they invade the bloodstream?

Like the staphylococci, streptococci can also produce toxins that may bring on **toxic strep syndrome**. As in toxic shock syndrome, there is blood pressure drop, fever, and a skin rash. The infection can be serious, and prompt therapy with antibiotics and fluid infusions are recommended.

18.13 Is erysipelas an infection of the skin?

A streptococcal infection called **erysipelas** affects the dermis of the skin. Red patches with thickening and swelling at the margins characterize erysipelas. The infection is usually due to streptococcal-derived toxins. In classical literature, erysipelas was known as St. Anthony's fire.

18.14 Are there any long-range effects of streptococcal infections of the skin?

When a person has had streptococcal sore throat or streptococcal blood disease, the body produces antibodies that react with streptococcal antigens. Sometimes the reaction between antigen and antibody results in immune complexes that fix themselves to the valves of the heart and the joints. The immune complexes activate the complement system, and tissue damage ensues. The result in the heart valves is referred to as **endocarditis**, while in the joints it is known as **rheumatoid arthritis**. If it occurs in both locations, the condition is called **rheumatic heart disease**.

18.15 Which is the most common skin disease found in humans?

Probably the most prevalent disease of the skin is acne, also known as **acne vulgaris**. In this condition, various microorganisms grow within the secretions of sebaceous glands and inflame the surrounding tissue to form "blackheads." The species *Propionibacterium acnes* is among the most frequent causes of acne. Acne scars and pustules result from the bacterial infection.

18.16 Are there any means to control skin diseases such as acne?

The effects of acne can be minimized by frequently using ointments to cleanse the skin and reduce the possibility of infection. Some physicians recommend tetracycline ointments to eliminate the bacterial infection. Another drug called Accutane inhibits sebum production, but intestinal bleeding can accompany its use.

18.17 Are infections of burnt tissue considered skin infections?

When the skin is burned, the tissue provides conditions that attract bacterial infection, especially by the Gram-negative rod *Pseudomonas aeruginosa*. This organism produces a fluorescent green pigment in the burnt tissue, and infection can be recognized by the presence of this pigment. Left untreated, the bacterium produces powerful toxins that erode the skin and can lead to death. Other bacteria sometimes causing infections in burnt tissue, include *Serratia marcescens* and *Providencia* species.

VIRAL DISEASES

18.18 Is warts considered an infectious disease of the skin?

Warts can be considered infectious diseases of the skin, because most cases are caused by viruses. The principal virus involved is the **papilloma virus**. The viruses can be transmitted by contact, especially sexual contact, and they often cause warts in the genital region. The tissue multiplies in a benign fashion as compared to tumor tissues, and the warts can be physically removed with chemicals or surgery.

18.19 How does the disease molluscum contagiosum compare to warts?

Molluscum contagiosum is caused by a poxvirus. The virus of molluscum contagiosum has a boxlike appearance, whereas the papilloma virus of warts is icosahedral. The skin lesions of molluscum contagiosum are pink and express a milky-white fluid when they burst. Warts by comparison are usually thin-walled vessels that express clear fluid.

18.20 Is the disease smallpox accompanied by skin lesions?

Smallpox begins with pink pimples called **macules** on the skin surface. The macules progress to red lesions called **papules**, which increase in size and become fluid-filled bodies called **vesicles**, as the photograph in Figure 18-2 depicts. The vesicles develop pus and burst to become pustules. These pustules occur throughout the body surface and are usually so severe that they lead to death. Secondary bacterial invasion is also possible. The virus that causes smallpox is a DNA-containing poxvirus that has a boxlike appearance.

18.21 Is smallpox still a serious threat in the world?

Smallpox has not occurred on the Earth since 1977. That year, the worldwide campaign to eliminate smallpox came to an end with the last reported case of smallpox in nature. No additional cases have been detected since that historic year, although as of this writing smallpox viruses continue to be maintained in certain laboratories.

18.22 Is it possible to immunize against smallpox?

Since 1798 and the work of Edward Jenner, it has been possible to immunize an individual against smallpox by injecting the viruses of **cowpox**, also known as **vaccinia**. The cowpox viruses stimulate the immune system to produce antibodies that unite with the cowpox viruses as well as the smallpox viruses. So-protected, the individual does not develop smallpox when exposed to the virus.

18.23 Does cowpox still exist in the world today?

Cowpox still exists in the world. The virus that causes it is a cubelike poxvirus containing DNA. In the disease, mild skin lesions form and there is some inflammation of the lymph nodes. The disease is rarely serious, however, and it appears to be most common in cattle.

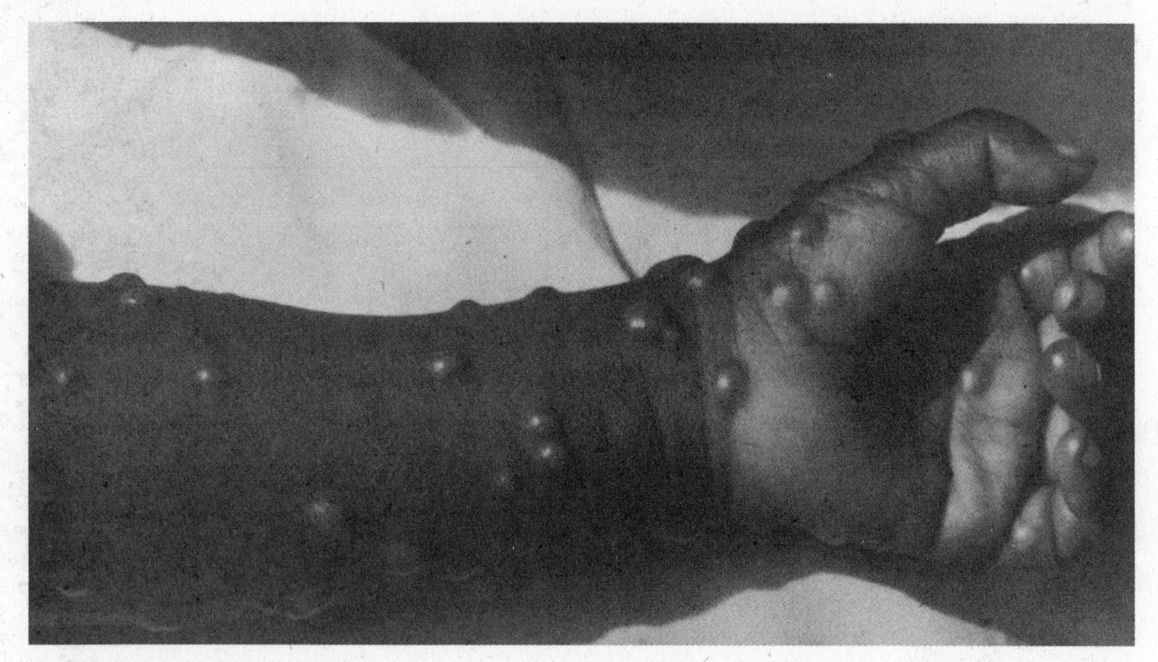

Fig. 18-2 The pus-filled vesicles associated with smallpox. The vesicles burst to become pustules that form scars (pox) on healing.

18.24 Can chickenpox be considered a microbial disease of the skin?

Chickenpox is a viral disease caused by a herpesvirus of the family Herpesviridae. The disease is transmitted by respiratory droplets, and the viruses migrate through the bloodstream and localize in the skin. Here they cause tissue destruction and cause the skin to develop lesions that resemble teardrops. Highly infectious fluid fills the lesions, and disease transmission can be easily effected by contact with the fluid.

18.25 Do the lesions of chickenpox appear simultaneously throughout the body?

The lesions of chickenpox do not occur simultaneously, but in crops over a period of days. After an incubation period of about 14 days, the first crop of chickenpox lesions develops on the scalp and trunk. New crops then occur on the face and limbs and occasionally in the mouth and throat. One of the distinguishing features of chickenpox is the sequence of lesion formation.

18.26 Do chickenpox viruses remain in the body after the disease has resolved?

One of the characteristics of herpesviruses is the ability to remain in the body tissues. Chickenpox viruses are known to remain for years in nerve cells within the skin tissues.

18.27 Is shingles related to chickenpox?

Shingles is a disease also known as **herpes zoster**. It is relatively certain that shingles results from a reactivation of a chickenpox virus, especially those viruses latent in the dorsal root ganglia. The viruses affect the sensory nerves in the skin and induce the formation of localized vesicles. The vesicles are commonly found near the waist and are extremely sensitive to the touch, often yielding severe pain. Individuals with immune disorders are particularly sensitive to the development of shingles.

18.28 Are there any drugs available for the treatment of chickenpox or shingles?

In recent years, the drug **acyclovir** has been prescribed for use in limiting the effects of chickenpox and shingles. It is also possible to relieve the symptoms, but aspirin is not used because of the possibility of developing Reye's syndrome.

18.29 Can a child be immunized against chickenpox?

A chickenpox vaccine was approved for general use in 1994. The vaccine consists of attenuated viruses, that is, weakened chickenpox viruses. The vaccine is injected at a very young age and is believed to induce lifetime immunity, but this has not yet been proven.

18.30 Are there any other herpesviruses capable of causing skin infections?

Perhaps the most well-known herpesvirus is that of **herpes simplex.** The herpes simplex is a DNA icosahedral enveloped virus of the Herpesviridae family. When it infects the skin, the virus may be the cause of **canker sores, cold sores,** or **fever blisters**. Traumas such as exposure to the sun or emotional upsets commonly trigger this skin infection. Type 1 herpes virus is usually involved. The virus may also cause an infection of the eye known as **herpetic keratitis** as well as infection of the pharynx known as **gingivostomatitis**.

18.31 Is the same virus responsible for genital herpes as for other herpes infections?

It is believed that type 2 herpes simplex virus is responsible for outbreaks of **genital herpes**. In genital herpes, thin-walled vesicles form on the external or internal genital organs. Transmission is usually effected by sexual contact, and recurrent attacks of genital herpes are common.

18.32 Is any drug available to treat genital herpes?

The drug **acyclovir** is commercially available and effective for genital herpes. The drug does not eliminate the disease, but it increases the time between recurrences of disease and it reduces the severity of the disease. Acyclovir works by interfering with viral DNA synthesis in the cytoplasm of the host cell.

18.33 Can herpes simplex affect any of the deeper organs?

The herpes simplex virus is capable of infecting the cells of the nervous system. Thus, it may cause **herpes encephalitis**, a brain inflammation. Acyclovir may be suitable for treating this disease.

18.34 Can the herpes simplex virus pass to the newborn if a pregnant woman has an outbreak of herpes simplex?

The herpes simplex virus is one of those microbial agents capable of passing across the placenta and infecting the fetus. In this situation, it may cause infection of the nervous system in the offspring and a condition known as **neonatal herpes**. Herpes simplex represents the H in the **TORCH** group of diseases, a number of microbial diseases that pass across the placenta and affect the fetus. Other diseases in this group include toxoplasmosis (T), rubella (R), and cytomegalovirus (C).

18.35 Is measles a true skin disease?

Strictly speaking, **measles** is not a disease of the skin. It is a respiratory disorder caused by an RNA virus that is transmitted by respiratory droplets. However, since a skin rash is a typical sign of measles, the disease is considered a microbial skin infection. The measles rash appears as a blush on the skin. It begins at the forehead and spreads to the upper appendages, body trunk, and lower appendages. The rash is more diffuse than the vesicles in chickenpox. Figure 18-3 compares the rashes. Measles is also known as **rubeola** because of its red rash.

18.36 Are there any characteristic signs early in the development of measles?

One of the useful diagnostic signs for measles is the appearance of **Koplik spots** along the oral mucosa near the molars. Koplik spots are small red patches having white specks at the center. They appear two or three days before the fever appears and several days before the skin rash becomes apparent.

18.37 Are there any complications to cases of measles?

Cases of measles can be complicated by infections of the middle ear or respiratory lung infection. In some cases, **measles encephalitis**, a brain inflammation, occurs in patients, and in some individuals there develops a condition of the nervous system known as subacute sclerosing panencephalitis.

18.38 Can measles be treated or prevented?

There is no effective treatment currently available to lessen the symptoms of measles. However, the disease can be prevented by an injection of attenuated measles viruses contained in the MMR vaccine. Although several vaccines used in the past have only yielded limited immunity, the current vaccine is believed to bring about lifetime immunity.

18.39 Is there any difference between measles and German measles?

About the only similarity between measles and German measles is that both are viral diseases accompanied by rash. **German measles**, also known as **rubella**, is caused by a completely different virus than the measles virus. The virus is an RNA-containing virus of the Togaviridae family. It is transmitted by respiratory droplets and causes a mild skin rash with mild fever.

18.40 Are there any serious side effects to rubella?

Rubella can be a serious problem if it occurs in a pregnant woman. In this situation, the virus passes across the placenta and infects the developing fetus. Depending on how early in the pregnancy the virus is contracted, the damage can be mild to serious and can occur in the fetal eyes, ears, and heart.

18.41 Is rubella included in the TORCH group of diseases?

The TORCH diseases are a group of infectious diseases transmitted during pregnancy from mother to child. The R in the TORCH group represents rubella.

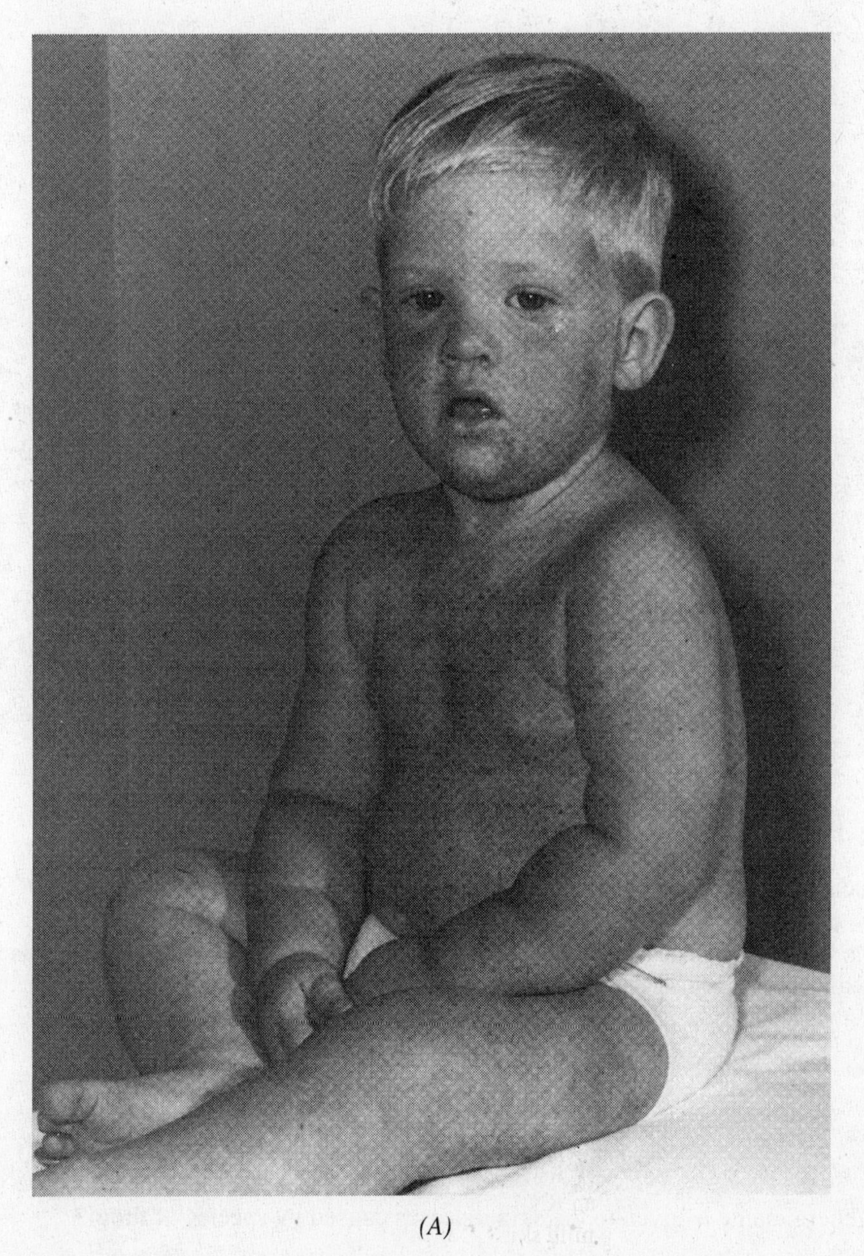

(A)

Fig. 18-3 The skin rashes associated with measles (*A*) and chickenpox (*B*). The measles rash is more like a blush, while the chickenpox appear as discrete vesicles.

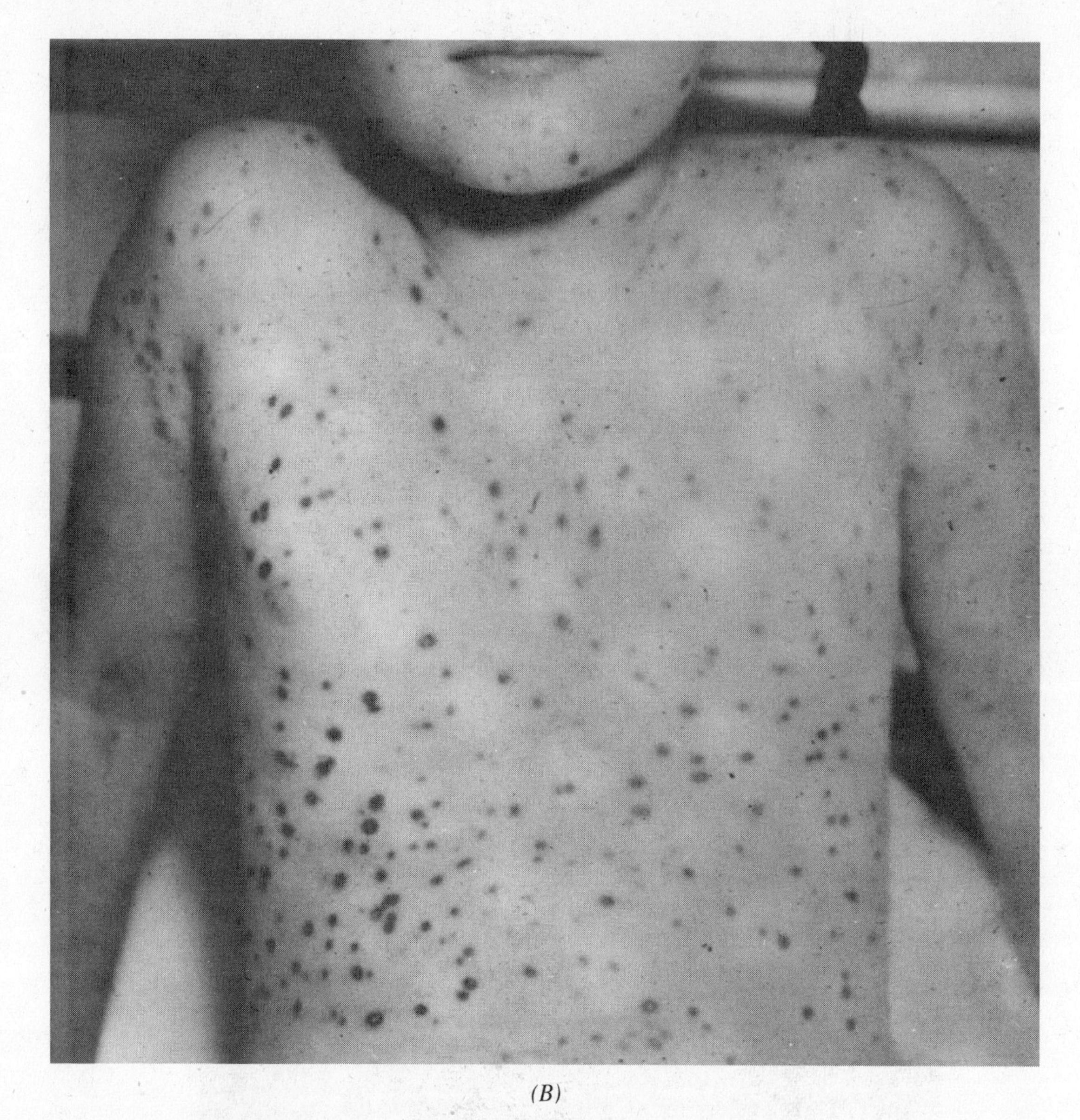

(B)

Fig. 18-3 *(cont.)*

18.42 Is a vaccine available for the immunization against rubella?

A rubella vaccine has been available since the late 1960s. The vaccine consists of attenuated viruses. It is combined with the measles and mumps vaccines in the **MMR vaccine** and is administered to children in their first year.

OTHER SKIN DISEASES

18.43 What general name is given to the skin diseases caused by species of fungi?

The skin diseases of fungal origin are known as **dermatomycoses**. The fungi causing these diseases are referred to as **dermatophytes**. The general word for any fungal infection is a **mycosis**.

18.44 What are two general types of fungal skin diseases?

Fungal disease can be cutaneous mycoses or subcutaneous mycoses. The **cutaneous mycoses** occur in the hair, nails, and outer layer of the skin, while the **subcutaneous mycoses** occur in the deeper skin layers after the skin has been pierced.

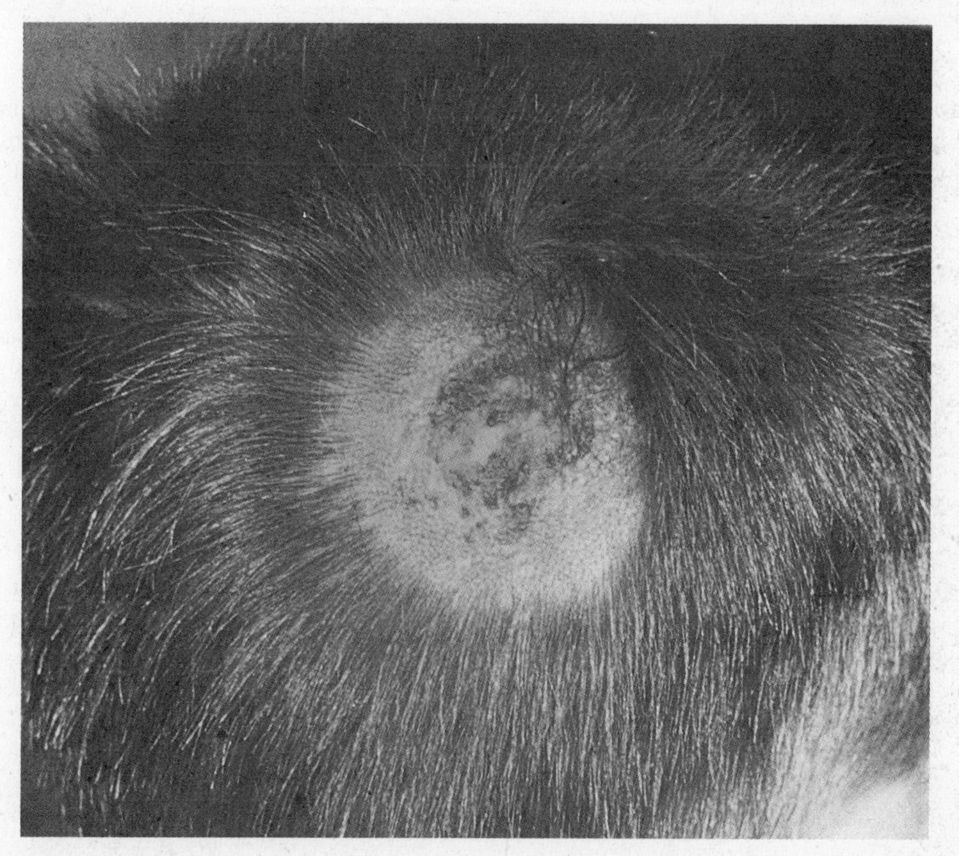

Fig. 18-4 The characteristic appearance of ringworm of the scalp.

18.45 Which type of skin disease is athlete's foot?

Athlete's foot is a cutaneous mycosis. It is caused by species of *Microsporum, Trichophyton,* and *Epidermophyton.* The fungi invade the skin in the webs of the toes causing dry, scaly lesions. Where the environment is moist, the lesions fill with fluid, and eventually the skin cracks and peels, which opens the skin to secondary bacterial infection.

18.46 Is athlete's foot the same as ringworm?

All the fungal skin infections are collectively known as **tinea diseases**. One of the other tinea diseases is tinea capitis, a fungal disease of the head known as **ringworm**. The fungi grow in the scalp and cause a circular ring of lesions, once believed to reflect the presence of a worm. Figure 18-4 shows the characteristic sign of the disease.

18.47 How can the tinea diseases be acquired?

There are many different tinea diseases and many different sources from which they can be acquired. For example, tinea pedis (ringworm of the scalp) is transmitted by infected combs, hats, and surfaces, such as the back rest of a seat in a movie theater. Tinea pedis (athlete's foot) can be obtained from fungal fragments left on the shower room floor or on towels. Tinea diseases are generally acquired by some form of contact with an infected surface.

18.48 What are other types of tinea diseases?

Other tinea diseases include **tinea corpora**, which is ringworm of the body; **tinea cruris**, or ringworm of the pubic area; **tinea unguium**, or ringworm of the nails; and **tinea barbae**, which is ringworm of the face and ears.

18.49 How are the tinea diseases diagnosed and treated?

Diagnosis of the tinea diseases usually depends on microscopic observation of fungi from scrapings of the infected area. Cultivation of the dermatophytes is performed to further identify them. Some species glow when they are illuminated by ultraviolet light. Treatment usually consists of applying a chemical such as **Whitfield's ointment** (undecylenic acid) to the affected area. **Griseofulvin** can also be used.

18.50 Can *Candida albicans* cause infections of the skin?

Strictly speaking, *Candida albicans* causes diseases of the mucous membranes. This disease often manifests itself in the oral cavity as an overgrowth of milky, white *Candida albicans* on the tongue. This condition is called **thrush**. *Candida albicans* also causes infection of the mucus membranes of the vagina, a condition known as **"yeast disease"** or **vaginitis**. The infection is accompanied by discomfort, and sometimes a white, cheesy discharge. Irritation of the vaginal canal is common.

18.51 What are the technical names for *Candida albicans* infections?

Although known as thrush, vaginitis, and other names, the technical name for diseases caused by *Candida albicans* is **candidiasis**.

18.52 What leads to the development of *Candida albicans* infections in the body?

Candida albicans is usually a member of the normal flora of the mouth, vaginal tract, intestines, and other organs of the body. When excessive antibiotic treatment is rendered, the bacteria in the mouth or intestines, disappear, and *Candida albicans* overgrows the region. A similar condition occurs in the vagina, when antibiotics kill the lactobacilli that normally produce acid in this organ. Without the acid, the *Candida albicans* flourishes. *Candida albicans* can also cause infection of the skin called **onchyosis**. People whose hands are in water for extended periods of time may be susceptible to this disease.

18.53 Which drugs are commonly used for the treatment of candidiasis?

Several drugs are available for treating candidiasis including **nystatin** and the **imidazole** antibiotics. Imidazole antibiotics include miconazole, clotrimazole, ketoconazole, and itraconazole. They are available by prescription and in pharmacy products such as Monistat and Gyne-Lotrimin.

18.54 Why is candidiasis an important disease in patients with HIV infection?

The normal body defenses available against candidiasis are centered in the T lymphocytes. In patients with HIV infection, these lymphocytes are reduced in number and, consequently the body defenses are diminished. *Candida albicans* flourishes in these patients.

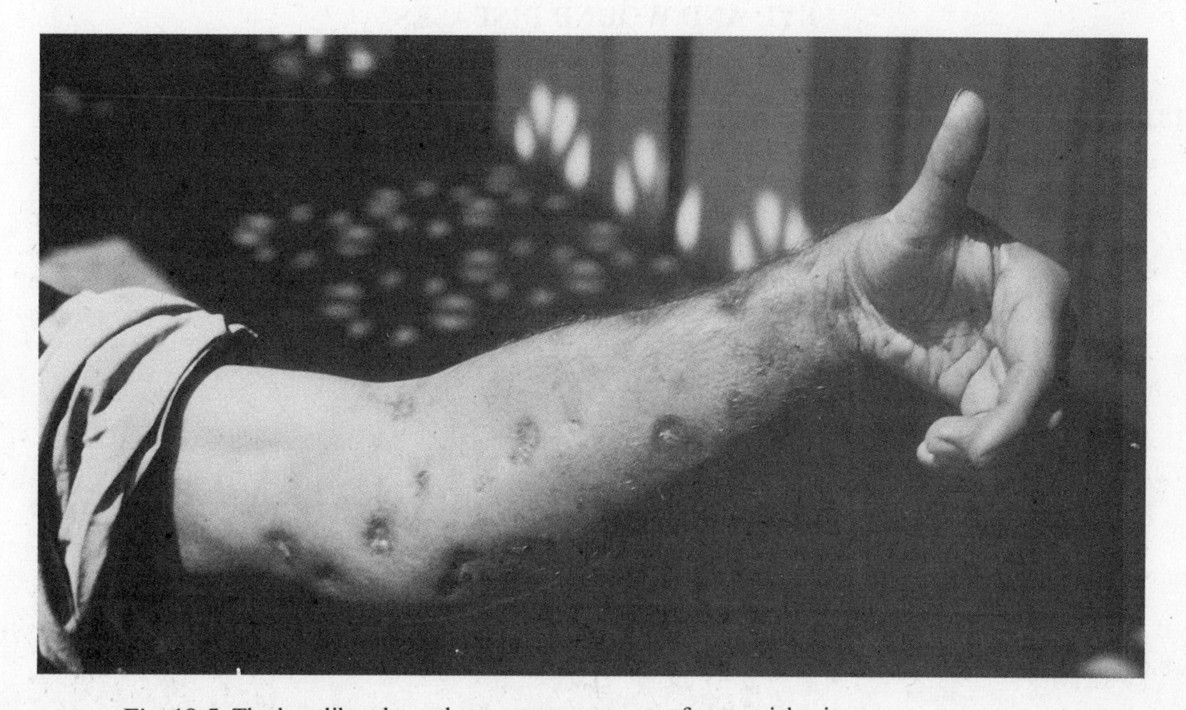

Fig. 18-5 The knotlike ulcers that accompany cases of sporotrichosis.

18.55 What type of infection does the fungus *Sporothrix schenckii* cause?

Sporothrix schenckii causes a subcutaneous mycosis called **sporotrichosis**. Acquired from sphagnum moss and the thorns of roses and barberry bushes, sporotrichosis manifests itself as lesions at the site of the wound and infection of the deeper subcutaneous tissue. Small ulcers often occur at the infection site, and they become pustulated and granulomatous as they spread. Often they appear as knots under the skin, as Figure 18-5 displays.

18.56 What is Madura foot?

Madura foot is a skin disease due to a number of different funguslike bacteria belonging to the genus *Nocardia* and *Actinomyces*. In Madura foot, the bacteria enter the skin during wounds, and cause chronic granular lesions. The lesions are pus-filled and may cause severe distortion of the foot, possibly requiring amputation.

18.57 Are there any multicellular parasites that cause skin diseases?

Most multicellular parasites such as worms invade the deeper tissues of the body, but there are at least two parasites that remain in the skin tissue. One is called *Schistosoma*. This flatworm causes **swimmer's itch**, a condition developing when the flatworm burrows into the skin, and the body develops an immune response. Another parasite causing skin disease is *Dracunculus medinesis*. This parasite is a roundworm. It reaches the skin in contaminated water and lives within the skin tissues, often causing skin lesions and a condition called **dracunculiasis.**

EYE AND WOUND DISEASES

18.58 Are there any serious bacterial diseases of the eye?

Bacterial eye infections generally arise from exposure to contaminated air or from a wound of the eye. Sometimes, exposure to bacteria during birth takes place. An example is a baby born to a woman with an active case of gonorrhea. In this case, *Neisseria gonorrhoeae* reaches the baby's eye and causes a serious infection called **gonococcal ophthalmia** that can lead to blindness. Much pus is present, and ulceration of the cornea can develop. Silver nitrate and antibiotics are used in the newborn's eye to prevent this condition.

18.59 Are newborns susceptible to infection by any other bacteria during the birth process?

Another sexually transmitted disease called chlamydia can also be transferred to the newborn as an eye infection called **chlamydial ophthalmia**. The causative organism is *Chlamydia trachomatis*. Symptoms in the newborn are similar to those occurring in gonorrhea.

18.60 Which bacteria are responsible for conjunctivitis in the eye?

Conjunctivitis is also known as **pinkeye**. It is accompanied by swelling and encrusting of the eye, reddening of the sclera, and a discharge that often drips down the cheek. Possible causes of conjunctivitis include *Haemophilus aegyptius*, a small Gram-negative rod; various staphylococci and streptococci; and species of *Pseudomonas*, a long, slender Gram-negative rod.

18.61 Is conjunctivitis able to pass from person to person.

Conjunctivitis is an extremely contagious disease, which is easily passed among children. It is sometimes spread in respiratory droplets expelled into the air. Also, the bacteria are picked up from the surface of the face onto the hands and spread to the next child's hands or face, from which infection takes place. The disease is treated with various antibiotics and is not considered a fatal disease.

18.62 Why is trachoma considered an important disease in humans?

Trachoma is an eye disease that often leads to blindness. It is caused by a chlamydia called *Chlamydia trachomatis*, which infects the cornea and causes permanent scarring. Hundreds of millions throughout the world have the disease, and the potential for blindness is acute. Treatment with tetracycline is important to prevent the blindness from developing.

18.63 Are there any important viral diseases of the eye?

Viruses cause two important diseases of the eye. One is due to an adenovirus, which causes a disease called **adenoviral keratoconjunctivitis**. This disease is accompanied by inflammation of the conjunctiva and severe pain in the eye, with swelling. The second disease is due to the herpes simplex virus. This virus causes a cornea infection called **herpetic keratitis**, a corneal infection with ulcers and inflammation. Both of these infections can be acquired from instruments used in optometric examinations. Adenoviruses are also acquired from contaminated water in swimming pools.

18.64 Which protozoan may be responsible for disease of the eye?

Although protozoal disease of the eye is relatively rare, there is one genus of protozoa called *Acanthamoeba* that can cause eye infection. The organism most often involved is *Acanthamoeba castellani*. It causes a corneal infection called **keratitis**. Most cases are accompanied by mild inflammation, but some include severe pain and damage to the cornea. Recent cases have involved contact lenses, and contaminated cleaning solutions used for these lenses.

18.65 Are there any multicellular parasites that infect the eye?

A roundworm called *Loa loa* is capable of infecting the eye tissues and causing **loiasis**. The roundworm lives in the cutaneous tissues of the eye and can often be seen on the surface of the cornea and conjunctiva. Various drugs are available to eliminate the worm. Deer flies are important in its transmission.

18.66 Are bite infections considered skin diseases?

Wounds inflicted by animal bites can be considered skin infections because they are usually accompanied by manifestations on the skin.

18.67 What is an example of a bacterial wound infection?

One example of a bacterial wound infection is **cat scratch fever**. Cat scratch fever is transmitted by the bite or scratch of a cat. After some days, a pus-filled lesion occurs at the scratch location, and swollen lymph nodes are experienced on the side of the body where the bacteria entered. Sore throat and headache may also occur.

18.68 Which bacterium is responsible for cat scratch fever?

At the present time, the bacterium responsible for cat scratch fever is not identified with absolute certainty. Two leading candidates are *Afipia felis,* a Gram-negative rod and *Rochalimeaea henselii*, a species of rickettsia.

18.69 Can the bite of a rat transmit a disease?

The bite of a rat may result in either of two forms of **rat bite disease**. One type is caused by *Streptobacillus moniliformis*, a Gram-positive chain of bacilli. This organism is transmitted in the saliva of a rat, and disease is accompanied by headache, fever, and a skin rash beginning at the site of the bite. The second type of rat bite fever is caused by *Spirillum minor*, a spiral bacterium that moves by polar flagella. An open ulcer mark occurs at the site of entry, and fever and rash generally follow. Antibiotics are used to treat both types of rat bite fever.

Review Questions

Completion. Add the word or words that correctly complete each of the following statements.

1. Boils and carbuncles are usually due to infection by members of the genus ____________.

2. Those streptococci that have a partially destructive effect on red blood cells are classified as ____________.

3. The infection onchyosis is caused by *Candida albicans* and occurs on the ____________.

4. Cases of chickenpox or shingles may be treated with the drug ____________.

5. Burnt tissue is particularly conducive to infection by the Gram-negative rod ____________.

6. Molluscum contagiosum is accompanied by skin lesions and is caused by a ____________.

7. Since 1977, public health officials have not detected a single case of ____________.

8. An early diagnostic sign for the presence of measles is the appearance of ____________.

9. An alternative expression for German measles is ____________.

10. The most pathogenic streptococci belong to the group ____________.

11. The virus that causes chickenpox belongs to the family ____________.

12. The TORCH diseases are those transmitted from a woman when she is ____________.

13. The tinea diseases are due to fungi and can be treated with the drug ____________.

14. Serious eye infections can occur in the newborn if the mother has an active case of ____________.

15. The bacterial disease trachoma attacks the ____________.

16. A condition of the nervous system called subacute sclerosing panencephalitis is associated most commonly in individuals who have previously had ____________.

17. Type 2 herpes simplex virus is most often responsible for cases of ____________.

18. The toxin produced by staphylococci that causes scalding of the skin is referred to as ____________.

19. Those strains of *Streptococcus pyogenes* that produce the erythrogenic toxin are responsible for cases of ____________.

20. Most cases of warts are due to ____________.

21. Chickenpox has a distinctive incubation period of ____________.

22. The vaccine used for immunization to measles and to other childhood viral diseases is known as the ____________.

23. Ringworm and athlete's foot are related to one another and are collectively known as ____________.

24. Sporotrichosis is caused by a microorganism belonging to the group ____________.

25. The "yeast disease" is also known as vaginitis and is most commonly due to infection by ____________.

26. Reactivation of the chickenpox virus is usually responsible for outbreaks of herpes zoster, also known as ____________.

27. The drugs clotrimazole and ketoconazole are used primarily for infections caused by ____________.

28. The bacterium *Haemophilus aegyptius* is responsible for infections of an eye structure known as the ____________.

29. The protozoan *Acanthamoeba* is often responsible for infections that occur in the ____________.

30. Madura foot is a skin infection of the tissues of the foot due to members of the genus *Actinomyces* and ____________.

Multiple Choice. Select the letter of the item that correctly completes each of the following statements.

1. Cutaneous mycoses are the collective name for (*a*) viral diseases of the eye (*b*) protozoal diseases of the skin (*c*) bacterial diseases of the blood (*d*) fungal diseases of the hair, nails, and outer skin layers.

2. All the following apply to the disease rubella except (*a*) it is caused by an RNA-containing icosahedral virus (*b*) it is transmitted by respiratory droplets (*c*) it is also known as German measles (*d*) the skin rash is usually severe.

3. The viruses of the following diseases are known to remain in the body tissues for long periods of time (*a*) trachoma and ringworm (*b*) herpes simplex and chickenpox (*c*) smallpox and boils (*d*) scarlet fever and warts.

4. The work of Edward Jenner is notable because it led to an immunization for (*a*) mumps (*b*) athlete's foot (*c*) smallpox (*d*) measles.

5. Rat bite disease may result from infection by (*a*) *Pseudomonas* or *Candida* (*b*) *Spirillum* or *Streptobacillus* (*c*) *Sporothrix* or *Staphylococcus* (*d*) *Chlamydia* or *Afipia*.

6. Excessive antibiotic use in the body may result in infections due to (*a*) *Staphylococcus aureus* (*b*) tinea corpora (*c*) *Candida albicans* (*d*) *Acanthamoeba*.

7. If the organisms of chlamydia or gonorrhea are transmitted to a baby during birth, the baby may develop a serious disease of the (*a*) lungs (*b*) outer skin layers (*c*) liver (*d*) eyes.

8. Streptococci that damage the red blood cells and cause them to dissolve are designated (*a*) gamma-hemolytic streptococci (*b*) group D streptococci (*c*) herpes streptococci (*d*) beta-hemolytic streptococci.

9. All the following characteristics apply to the disease chickenpox except (*a*) the lesions occur in crops (*b*) the disease is transmitted by respiratory droplets (*c*) the disease is caused by the same virus that causes herpes simplex (*d*) the viruses localize in the skin.

10. Cases of herpes encephalitis and genital herpes may be treated with the drug (*a*) acyclovir (*b*) nystatin (*c*) penicillin (*d*) clotrimazole.

11. Cases of toxic shock syndrome and food poisoning may both be related to (*a*) *Haemophilus influenzae* (*b*) *Staphylococcus aureus* (*c*) *Pseudomonas aeruginosa* (*d*) herpes simplex viruses.

12. Endocarditis may be described as a (*a*) disease of the heart valves due to a species of *Streptococcus* (*b*) infection of the endocardium due to pox viruses (*c*) disease of the endometrium due to *Candida albicans* (*d*) one of the TORCH diseases transmitted by pregnant women.

13. The campaign to eradicate smallpox was based in large measure on (*a*) eliminating all carriers of smallpox viruses (*b*) eliminating the bacteria that cause smallpox (*c*) discovering novel antibiotics for use in smallpox victims (*d*) widespread usage of the vaccine containing cowpox viruses.

14. The herpes simplex virus is capable of causing all the following conditions with the exception of (*a*) a brain inflammation called herpes encephalitis (*b*) neonatal herpes (*c*) thin-walled vesicles on the external or internal genital organs (*d*) the lesions of chickenpox.

15. Which of the following is most descriptive of the measles rash (*a*) it consists of pustules (*b*) it appears as a blush on the skin (*c*) it occurs in crops over a long period of time (*d*) it rarely occurs.

16. Most of the tinea diseases are acquired by (*a*) contact with an object containing the fungus (*b*) droplets exhaled from the respiratory tract (*c*) food contamination by water containing fecal material (*d*) direct contact with someone with the disease.

17. The thorns of barberry bushes and rose bushes are both known to transmit the fungus that causes (*a*) sporotrichosis (*b*) swimmer's itch (*c*) tinea barbae (*d*) Madura foot.

18. Many cases of conjunctivitis, also known as pinkeye, are caused by a (*a*) Gram-positive staphylococcus (*b*) Gram-negative staphylococcus (*c*) Gram-positive rod (*d*) Gram-negative rod.

19. The bacterium that causes cat scratch fever causes at the site of infection (*a*) a pus-filled lesion (*b*) a dry lesion (*c*) a staphylococcal infection (*d*) a gangrenous wound.

20. Swimmer's itch is usually caused by an immune response to the presence of a (*a*) Gram-positive coccus in chains (*b*) icosahedral virus (*c*) box-shaped virus (*d*) flatworm.

True/False. For each of the following, enter the letter "T" if the statement is true in its entirety. Enter an "F" if the statement or any part of it is wrong.

___ **1.** Coagulase-positive staphylococci pose a threat to disease, because coagulase causes fibrin clots to develop in the tissue.

___ **2.** Impetigo is a highly contagious, pus-producing skin infection common in children but not in adults.

___ **3.** Toxic shock syndrome is due to numerous species of streptococci, which are Gram-positive cocci occurring in chains.

___ **4.** The hemolysins are a series of toxins, which are produced by streptococci and damage red blood cells.

___ **5.** The disease of the heart valves known as endocarditis is caused by streptococci and is also known as St. Anthony's fire.

___ **6.** The species *Propionibacterium acnes* is among the most frequent causes of acne.

___ **7.** The tissue of burn victims is commonly contaminated by papilloma viruses, which produce a fluorescent green pigment in the tissue.

___ **8.** Molluscum contagiosum is a pox virus disease accompanied by skin lesions that are pink and express a milky-white fluid when they burst.

___ **9.** Approximately 10 cases of smallpox occur in the world each year.

___ **10.** In general terms, the disease cowpox is more dangerous to human health than is smallpox.

___ **11.** Chickenpox viruses have the ability to remain in the body's nerve cells for many years and reemerge to cause shingles.

___ **12.** Aspirin is not recommended for treating chickenpox because of the possibility of developing Grave's syndrome.

___ **13.** The virus that causes herpes simplex is an RNA helical virus having an envelope and belonging to the Herpesviridae family.

___ **14.** The drug acyclovir is recommended for the treatment of genital herpes.

___ **15.** Measles is technically a respiratory disorder accompanied by a skin rash and caused by a DNA virus transmitted in food.

___ **16.** The MMR vaccine provides immunization to measles, mumps, and rubeola.

___ **17.** Dermatophytes are fungi that cause diseases of the skin.

___ **18.** The fungal disease sporotrichosis is accompanied by lesions at the wound site and infection of the deeper subcutaneous tissue.

___ **19.** The fungus Candida albicans may be responsible for infections in the oral cavity, vagina, skin, or intestines.

___ **20.** The organisms that cause gonorrhea and tinea pedis may be transmitted to the newborn during birth and may cause serious infection of the eye.

Chapter 19

Microbial Diseases of the Nervous System

OBJECTIVES

The microbial diseases of the nervous system are among the most dangerous known because they affect a vital organ system in the body. For this reason, the diseases of the nervous system are among the best known. The objectives of this chapter will be to:

1. Explain some of the characteristics and some of the different forms of bacterial meningitis.

2. Recognize the effects on the nervous system of the toxins produced in cases of tetanus and botulism.

3. Consider some of the important characteristics of leprosy and point out why this is a disease of the nervous system. Discuss the viral disease rabies, with emphasis on the immunization available to prevent the development of this disease.

4. Outline the components and uses of the two polio vaccines that have led to substantial reduction of the incidence of the disease.

5. Describe the characteristics of prions and their place in human disease.

6. Identify the important diseases of the nervous system due to fungi and protozoa, with emphasis on sleeping sickness.

THEORY AND PROBLEMS

19.1 Which microorganisms cause diseases of the nervous system?

There are many different kinds of organisms that cause diseases of the nerve cells, nerves, and other aspects of the nervous system such as the meninges. Bacteria, viruses, protozoa, and fungi all cause diseases. In many cases, the etiological agents enter the body through the respiratory tract or through the gastrointestinal tract or broken skin.

19.2 What are the characteristic signs of meningitis due to microbial infection?

The meninges are the three coverings over the spinal cord and brain. In most cases meningitis is accompanied by severe headache and inflammation. Coma is present in some situations, and the mortality rates vary with the type of microorganism involved.

BACTERIAL DISEASES

19.3 Which type of meningitis is caused by *Neisseria meningitidis*?

Neisseria meningitides is a Gram-negative, small diplococcus. It is responsible for **meningococcal meningitis**, which affects about 3000 Americans annually.

19.4 How is meningococcal meningitis transmitted?

Meningococcal meningitis is generally transmitted by airborne droplets. The bacteria inhabit the upper respiratory tract causing influenzalike symptoms, before passing into the bloodstream. From the bloodstream, they pass to the meninges, and meningitis follows. Most symptoms of meningitis are believed due to toxins produced by the organisms.

19.5 What are the symptoms of meningococcal meningitis during the blood phase?

When meningococci enter the blood, they cause severe malaise and toxemia, a syndrome known as **meningococcemia**. Endotoxin shock accompanies this condition, and many patients die during the blood phase of the disease.

19.6 What are some characteristics of meningeal infection?

The person suffering infection of the meninges usually displays a stiff and arched neck. Headaches are extremely painful, and skin spots often occur on the body surface. In addition, there is inflammation of the adrenal glands accompanied by severe hemorrhaging. This condition is called the **Waterhouse-Friderichsen syndrome**.

19.7 How are diagnosis and treatment for meningococcal meningitis carried out?

The diagnostic procedure for meningococcal meningitis usually involves taking a spinal tap of the patient. The cerebrospinal fluid is then examined for the presence of Gram-negative diplococci. Aggressive therapy with penicillin, rifampin, and other antibiotics is the preferred treatment.

19.8 Are any forms of meningitis less severe than meningococcal meningitis?

Another form of meningitis is caused by *Haemophilus influenzae* type b. This disease is often called **Hib disease**. It also affects the meninges, but most cases occur in children between the ages of two and five. In some cases, the disease may be fatal, but the mortality is generally lower than for meningococcal meningitis. Some mental retardation is observed in recoverers.

19.9 Is there any way of preventing *Haemophilus* meningitis?

A vaccine is available known as the **Hib vaccine** which consists of polysaccharides from the etiological agent. It is available to children and is often combined with the DPT or DTaP injection.

19.10 Which drugs are routinely used for *Haemophilus* meningitis treatment?

The drug **rifampin** is routinely used for the treatment of *Haemophilus* meningitis. This drug can be administered if exposure has taken place and it will preclude the development of the disease. Prophylactic therapy such as this is particularly important if the child has not been immunized against *Haemophilus influenzae* type b.

19.11 Are there any other bacteria that can cause meningitis?

Meningitis can be caused by various species of streptococci and staphylococci. For example, *Streptococcus pneumoniae* is a common cause of meningitis when it infects the blood after establishing itself in the nasopharynx. Staphylococcal meningitis can be due to *Staphylococcus aureus* which enters the body through the respiratory tract or by a serious skin wound. Both of these diseases can result in fatality, so they must be treated aggressively.

19.12 What causes listeriosis, and what are its characteristic symptoms?

Listeriosis is a caused by a small Gram-positive rod called *Listeria monocytogenes*. It is primarily a blood disease, with fever, malaise, and a general ill feeling. In recent years, public health officials have noted that cases of listeriosis are often accompanied by inflammation of the meninges and the development of **listeric meningitis**.

19.13 How is listeric meningitis obtained and in which group of people is it particularly dangerous?

Cases of listeric meningitis have been related to unpasteurized milk products such as cheeses. In this situation, the bacteria enter the bloodstream through the gastrointestinal tract before causing meningitis. The disease is particularly dangerous in pregnant women because the disease may cause miscarriage or pass to the fetus and cause damage of the fetal tissues.

19.14 Why is tetanus considered a disease of the nervous system?

Tetanus is a disease usually acquired by puncture of the skin during a wound with a piece of glass or other pointed object. The etiological agent grows in the dead tissue and produces a number of toxins, which have their affect on this tissue. These toxins prevent the destruction of acetylcholine in the synapse and encourage nerve impulses to pass into the muscles, where they cause continual muscle contractions symptomatic of tetanus. Since the primary affect is in the nervous system, tetanus is considered a disease of this system.

19.15 Which organism causes tetanus?

Tetanus is caused by a Gram-positive, sporeforming anaerobic rod called *Clostridium tetani*. This organism exists as the spore in the human and animal intestine and passes to the soil in the feces. When spores enter the dead tissue of a wound they germinate to form the bacilli that produce the toxins.

19.16 What treatment can be given to the person who is suffering from tetanus?

Tetanus is primarily due to a toxin affecting the nervous system called a neurotoxin. Antibiotics are of minimal value against neurotoxins, and antibodies must be administered to neutralize the toxin. The antibodies are known as antitoxins.

19.17 Can tetanus be prevented?

A preparation of tetanus toxin is chemically altered to produce the tetanus toxoid. This **tetanus toxoid** is then injected into the body in the DPT injection, whereupon it induces the immune system to produce protective antibodies. A tetanus injection should be administered every 10 years or sooner to provide a protective state of immunity.

19.18 Do any *Clostridium* species cause disease of the nervous system?

A *Clostridium* species called *Clostridium botulinum* causes a highly fatal disease called **botulism**. *Clostridium botulinum* is an anaerobic, sporeforming Gram-positive rod. It exists as spores in the soil, which it enters from the intestines of animals and humans.

19.19 How does the agent of botulism enter the body?

Spores of *Clostridium botulinum* cling to foods harvested from the soil. When the foods are prepared and placed in unsterilized vacuum-sealed containers, the spores germinate to vegetative cells in the anaerobic environment of the can and produce a highly lethal exotoxin. The exotoxin is ingested in the food.

19.20 Why is botulism considered a disease of the nervous system?

Although botulism begins with entry of the exotoxins into the gastrointestinal tract, the disease is more correctly a disease of the nervous system. This is because the botulism toxin molecules pass into the bloodstream and localize in the synapses where the dendrites of nerve cells come together with the axons of other nerve cells or with muscle cells at the neuromuscular junction. In the synapse, the botulism toxin is absorbed into end brushes of the axons, where it prohibits the release of neurotransmitters such as acetylcholine from the terminal endings. Without neurotransmitters, nerve impulses cannot pass from one nerve cell to another or from the nerve cell to the muscle cell.

19.21 What symptoms arise from the effect of botulism toxin taking place in the synapse?

As a result of the neurotransmitter inhibition, a progressive form of paralysis takes place in the body. Numbness of the extremities is experienced, the facial muscles enter a state of flaccid paralysis, speech is slurred, vision is impaired, and when paralysis of the intercostal muscles and diaphragm takes place, respiratory distress is experienced. Paralysis of the breathing muscles eventually leads to death.

19.22 Is there any treatment available for botulism?

When a person displays the symptoms of botulism, the physician may administer a preparation of antitoxin to neutralize the toxins. Antitoxins are antibody molecules produced in persons whose immune systems have been stimulated by the botulism toxin. Since botulism is not characterized by a bacterial infection, it is not a transmissible disease.

19.23 What precautions can consumers take to avoid the botulism toxin in their foods?

The botulism toxin is a high molecular weight protein that is susceptible to heat. Therefore, if food is heated to boiling before consumption, the toxin will be destroyed if it is present. Most cases of botulism are related to foods consumed without heating or with mild heating.

19.24 Why is leprosy considered a disease of the nervous system?

Although **leprosy** is usually considered a skin disease, the bacilli affect the peripheral nerves. With the disruption of the nerve endings, a patient becomes insensitive to environmental stimuli. This is the basis for leprosy.

19.25 How does the peripheral nerve destruction lead to the symptoms of leprosy?

With the destruction of the nerve endings, the patient is highly susceptible to injury of the skin. For example, burns, injuries, or wounds are not felt. Often these wounds become infected easily because the pain endings have been destroyed. Also, there is some destruction to the bone tissue because the nerve endings servicing these tissues have been lost. In some cases, the bones become distorted and the patient experiences the "claw hand" shown in Figure 19-1. Leprosy is not considered a highly fatal disease unless secondary infection causes severe complications.

19.26 What are the causative agent and incubation period of leprosy?

Leprosy is caused by *Mycobacterium leprae*, an acid-fast rod. The organism is transmitted by multiple skin contacts or by droplets coughed from the superficial layers of the respiratory tract. Leprosy has an extremely long incubation period of between three and six years. This incubation period is among the longest for any bacterial disease.

19.27 What is the alternative name for leprosy and how is it treated?

Because of the historic view of leprosy, the term "leprosy" is considered a pejorative term. Therefore, the disease is preferably known as **Hansen's disease**, after Gerhard Hansen, who studied the disease in the late 1800s. Many modern antibiotics and chemotherapeutic agent can be used for the treatment of leprosy. Among the most useful drugs is dapsone, chemically known as diaminodiphenylsulfone. Other drugs used against leprosy include rifampin and clofazimine.

19.28 Are there different forms of leprosy (Hansen's disease)?

The form of leprosy (Hansen's disease) that affects the skin and causes the person to lose sensation is called **tuberculoid leprosy.** A second form, called **lepromatous leprosy,** involves the formation of skin nodules, which enlarge and disfigure the skin. Diagnosis is assisted by the recovery of acid-fast rods in either of these two situations.

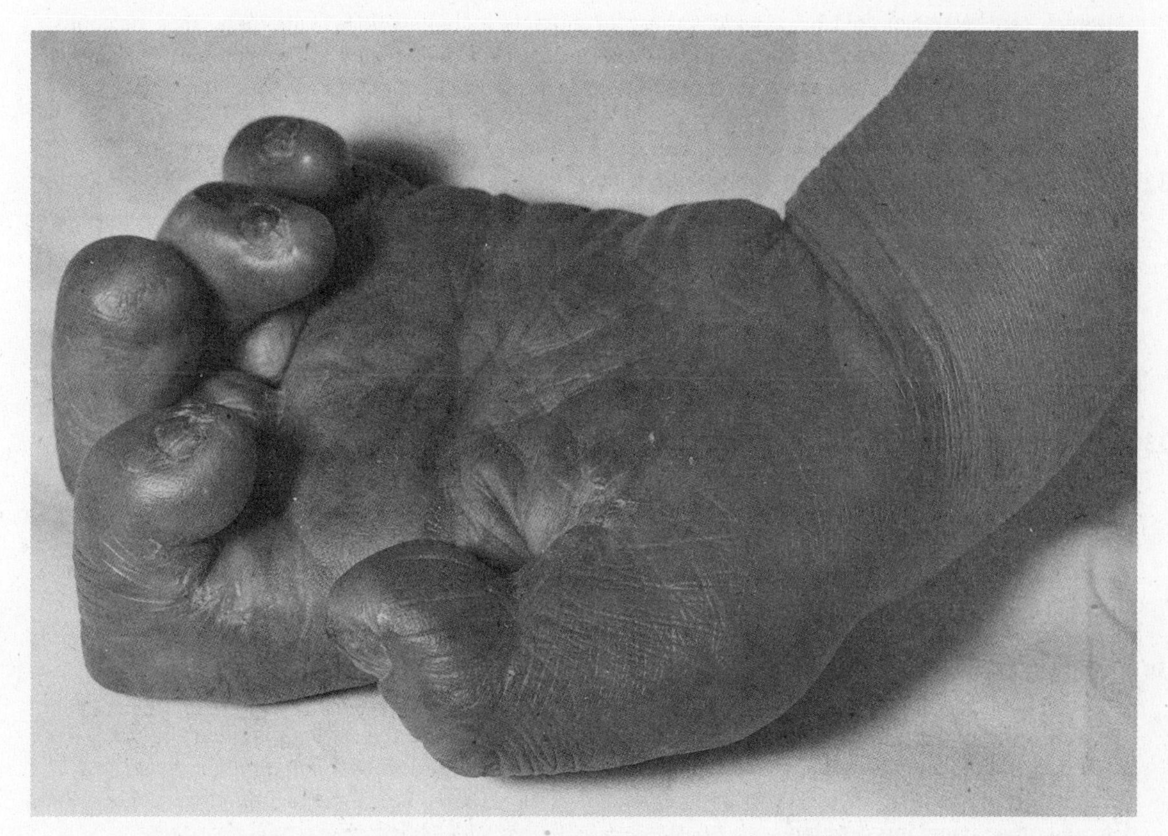

Fig. 19-1 The "claw hand," a result of the destruction of peripheral nerves occurring during cases of leprosy.

VIRAL DISEASES

19.29 Which is the most serious viral disease of the nervous system?

Without doubt, the most serious disease of the nervous system is **rabies**, a disease due to a helical, bullet-shaped RNA virus. In this disease, the viruses affect the brain and the fatality rate approaches 100 percent. With the destruction of the nerve tissue, a person experiences severe paralysis as well as spasms of the pharyngeal muscles when a person attempts to drink fluid or thinks of fluid. This condition is called **hydrophobia**, meaning fear of water.

19.30 Approximately how many cases of rabies occur annually in the United States?

Actual cases of rabies in the United States are very rare. Less than 10 usually occur in a single year, and in some years there are no cases. However, over 25,000 people receive rabies vaccinations each year because they have been exposed to the virus through the bite of an animal.

19.31 Which animals are involved in the transmission of rabies?

Rabies can be transmitted by a wide variety of animals including dogs, cats, bats, rats, raccoons, horses, wolves, and numerous other species. If a bite has taken place, then the person must assume to have been inoculated and the rabies vaccine should be given.

19.32 Where are rabies vaccinations given and how long do they last?

For many decades, rabies immunizations consisted of up to two dozen injections given in the abdominal fat, a very painful experience. Since 1980, however, rabies injections have been given in the arm, and the series numbers three or four injections. The material used is attenuated rabies viruses cultivated in human tissue cells. The immunity is believed to last a lifetime.

19.33 Are rabies immunizations given only after a bite has been received?

Anyone who wishes protection against rabies may receive the vaccine. Animal workers and people who work with animals should consider **preexposure immunizations**, and persons who have been bitten by an animal must receive **postexposure immunizations**.

19.34 What is the normal incubation period for rabies?

The incubation period for rabies varies according to how close to the nervous system the virus has entered the body and how much virus has been inoculated. The period of incubation can be as few as several days or as long as several months. Generally, there is enough time to immunize against the disease before symptoms develop.

19.35 Why is polio considered a disease of the nervous system?

Polio is an abbreviated term for poliomyelitis. The disease is caused by a very small, icosahedral RNA virus. This disease is considered with nervous system diseases because the primary effects occur in the meninges and in the stem at the base of the brain known as the medulla oblongata. In mild cases of polio, the virus infects the lymph nodes of the neck and ileum, and causes respiratory symptoms and some malaise, but no further effects. In serious forms of the disease, the virus affects the motor nerve cells of the upper spinal cord and induces paralysis, possibly even respiratory failure. In its most severe case, the virus causes permanent paralysis when it affects the brain stem.

19.36 How is the polio virus transmitted?

The virus of polio is generally transmitted by contaminated food and water because the virus is present in the gastrointestinal tract. Gastrointestinal symptoms such as nausea, cramps, and diarrhea are often early signs of the disease.

19.37 Which two vaccines are available for preventing polio?

The two vaccines available are the Salk vaccine and the Sabin vaccine. The **Salk vaccine** ("dead viruses") is prepared with viruses inactivated by chemicals; the **Sabin vaccine** ("live viruses") is prepared with viruses weakened by multiple passages through laboratory culture cells. The Salk vaccine must be injected in the arm, while the Sabin vaccine can be taken orally in a sugar cube or drops of liquid. Use of the vaccines has played a large role in the reduction of the incidence of polio in the United States, as displayed in Figure 19-2.

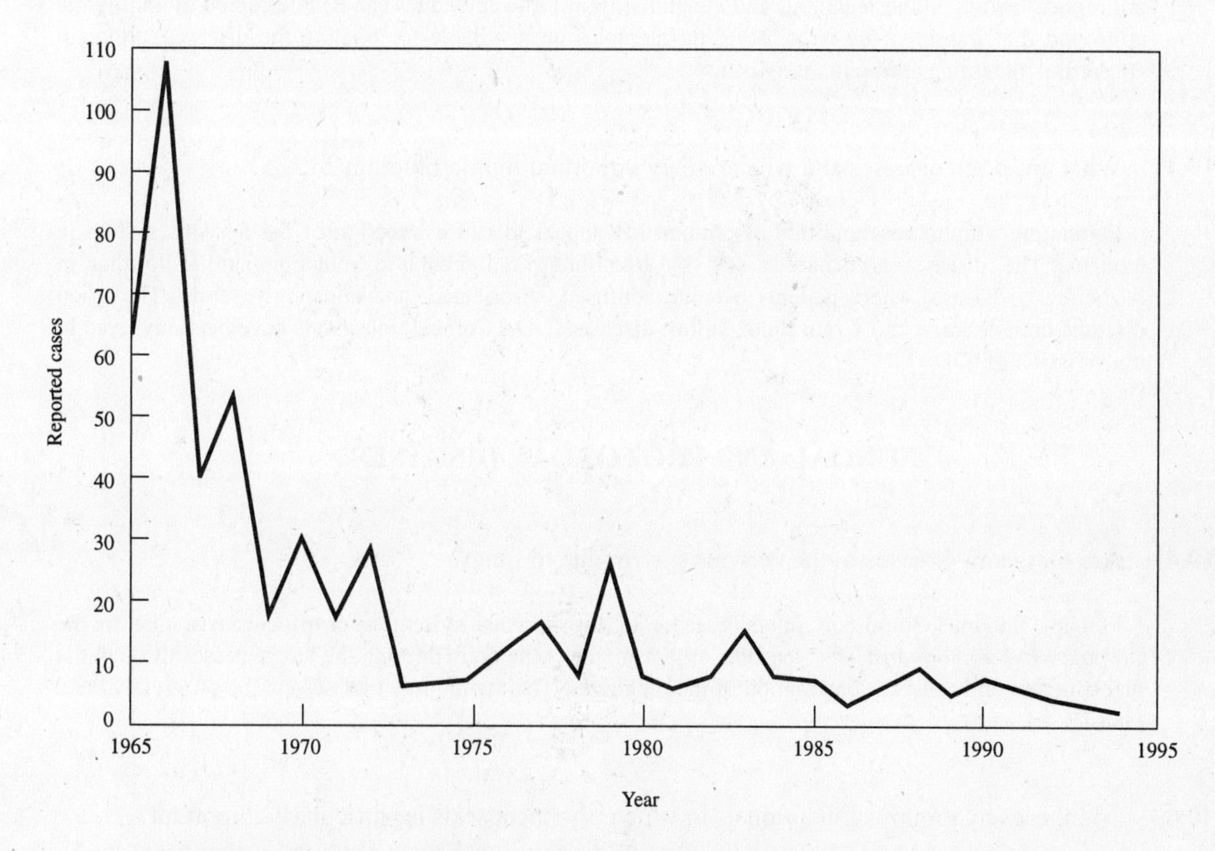

Fig. 19-2 Reported cases of polio in the United States in the period from 1965 to 1994. The dropoff of cases is due largely to the use of polio vaccines.

19.38 Which is the preferred vaccine for preventing polio?

There is no general agreement as to whether the Salk or Sabin vaccine is the preferred. At present, in the United States, the Sabin vaccine is most widely used; in other parts of the world Salk vaccine is preferred. There are advantages and disadvantages to using each vaccine. For example, the Sabin vaccine gives a stronger immune response, but there is some risk of a mild poliolike disease in the recipient. The Salk vaccine gives a weaker immune response but carries less risk of infection.

19.39 What are some of the characteristic signs of viral encephalitis?

Viral encephalitis is a disease of the brain tissue. Usually it is accompanied by headache, fever, mental distress, and in some cases permanent damage to the nervous system. Most cases of viral encephalitis are transmitted by arthropods such as mosquitoes and ticks. Brain damage, deafness, and other neurological problems vary with the various forms of encephalitis. Various RNA viruses may be responsible for viral encephalitis.

19.40 Which are some of the types of viral encephalitis that are possible?

Among the major types of viral encephalitis are St. Louis encephalitis (SLE), LaCrosse encephalitis, Japanese B encephalitis, Eastern equine encephalitis (EEE), and Western equine encephalitis (WEE), both of horses. Many of these forms exist in birds and other animals in nature and are transmitted by arthropods feeding in these animals and the human skin. The epidemics can be interrupted by killing the arthropod that transmits the virus. No drug therapies are available for treating the diseases, although antiserum containing antibodies may be used.

19.41 What are prion diseases and why are they important in microbiology?

Prions are ultramicroscopic bits of protein that appear to cause infection of the nervous system in humans. The diseases are characterized by tremblings and mental dysfunction similar to that in Alzheimer's disease, where patients become confused, disoriented, and apparently senile. The prion diseases include **kuru** and **Creutzfeldt-Jakob disease** (CJD). Corneal transplants have been involved in transmission of CJD.

FUNGAL AND PROTOZOAL DISEASES

19.42 Are there any diseases of the nervous system due to fungi?

Perhaps the most important fungal disease of the nervous system is **cryptococcosis**, due to the *Cryptococcus neoformans*. This yeastlike organism enters the body through the lungs, passes through the bloodstream, and causes inflammation of the meninges. The meningitis it causes can be progressive and fatal.

19.43 Are there any groups of individuals in which cryptococcosis is particularly important?

Cryptococcosis is an opportunistic diseases seen in patients with HIV infection and AIDS. The fungi, which resemble yeasts, invade the brain and spinal cord from the lungs and cause fatal illness as a consequence to AIDS. Other immune-suppressed individuals such as those taking therapy for transplants and for cancer are also susceptible to the effects of cryptococcosis.

19.44 Are there any drugs used for the treatment of cryptococcosis and how is it acquired?

The drug amphoteracin B is often used to treat cryptococcosis. However, this is a very dangerous drug and must be used under careful supervision. *Cryptococcus neoformans* is believed to be transmitted by pigeon droppings when wind gusts pass through the dried droppings. Therefore, the disease often happens in urban areas where pigeons congregate. The fungal cells can be seen in the lung tissues before the disease progresses to the nervous system. Figure 19-3 shows how they appear.

19.45 Why is sleeping sickness considered a disease of the nervous system?

Sleeping sickness is a disease of the nervous system because the causative protozoa that cause it multiply in the tissues of the central nervous system and cause severe headache, tremors, and progressive paralysis. In many cases, the patient has convulsions and develops a coma. This is the "sleeping" of sleeping sickness.

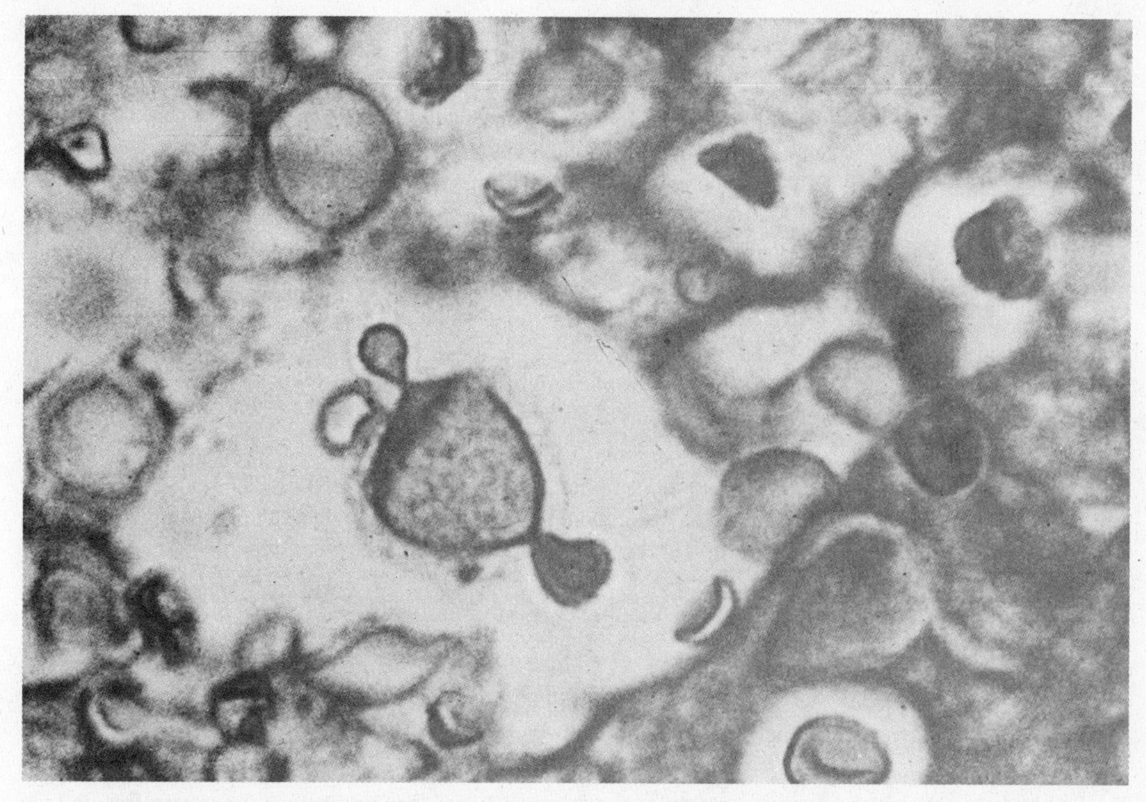

Fig. 19-3 A photomicrograph of *Cryptococcus neoformans* as seen in the lung tissue. The fungi are the darker stained, somewhat circular bodies among the lung cells.

19.46 Which organisms cause sleeping sickness and how are they transmitted?

There are two kinds of sleeping sickness: African sleeping sickness and South American sleeping sickness. **African sleeping sickness** is caused by a flagellated protozoan called *Trypanosoma bruceii*, while **South American sleeping sickness** is caused by *T. cruzi*, which is also a flagellate. African sleeping sickness is transmitted by the tsetse fly, while South American sleeping sickness is transmitted by the triatomid bug.

19.47 Are there any major differences between the two types of sleeping sickness?

Both forms of sleeping sickness affect the nervous system, but the South American form is also accompanied by infection of the heart muscle. Irregular heart beats and fluid accumulation around the heart often lead to permanent damage to the heart in this disease. South American sleeping sickness is also called **Chagas' disease** for the investigator who studied it early in this century.

19.48 Which drugs can be used to treat sleeping sickness?

There are several drugs for treating sleeping sickness including suramin and pentamidine isethionate.

19.49 Do any other diseases of protozoal origin affect the nervous system?

Another nervous system disease of protozoal origin is encephalitis due to an amoeba called *Naegleria fowleri*. This disease is rare in the United States. The responsible protozoa apparently enter the body through the nasal mucosa and spread by the olfactory nerve to the brain. Here they cause brain inflammation and tissue destruction that is fatal in most cases.

Review Questions

Multiple Choice. Select the letter of the item that correctly completes each of the following statements.

1. Which of the following sets of characteristics applies to meningococcal meningitis? (*a*) it is generally transmitted by contaminated food (*b*) it may be accompanied by inflammation of the adrenal glands (*c*) it is caused by a Gram-positive diplococcus (*d*) there is no blood phase of the disease.

2. The effects of the botulism toxin are observed in the (*a*) gastrointestinal system (*b*) respiratory system (*c*) urogenital system (*d*) nervous system.

3. The disease polio is transmitted by contaminated food and water and is caused by a (*a*) bacteria (*b*) protozoan (*c*) virus (*d*) fungus.

4. The incubation period for rabies is (*a*) a few hours (*b*) always less than two days (*c*) between five and ten days (*d*) highly variable.

5. Ultramicroscopic bits of protein that appear to cause infection of the nervous system are called (*a*) inclusion bodies (*b*) prions (*c*) metachromatic granules (*d*) viroids.

6. All the following apply to *Cryptococcus neoformans* except (*a*) it causes a disease of the nervous system (*b*) it infects individuals who have AIDS (*c*) it can be transmitted by pigeon droppings (*d*) it is a type of protozoan.

7. Viruses are involved in all the following diseases of the nervous system except (*a*) sleeping sickness (*b*) rabies (*c*) St. Louis encephalitis (*d*) Japanese B encephalitis.

8. The treatment available for patients with botulism consists of (*a*) penicillin antibiotics (*b*) tetracycline antibiotics (*c*) antitoxins (*d*) toxoids.

9. Although leprosy is usually considered a skin disease, it is more correctly a disease of the (*a*) central nervous system (*b*) urogenital tract (*c*) sense organs (*d*) peripheral nerves.

10. Species of *Trypanosoma* and *Naegleria* are both (*a*) transmitted by tsetse flies (*b*) treated with penicillin antibiotics (*c*) types of protozoa (*d*) causes of sleeping sickness.

11. Amphoteracin B is most often used to treat cases of (*a*) polio (*b*) botulism (*c*) rabies (*d*) cryptococcosis.

12. All the following organisms are capable of causing meningitis except (*a*) *Haemophilus influenzae* type b (*b*) *Mycobacterium leprae* (*c*) *Streptococcus pneumoniae* (*d*) *Staphylococcus aureus* .

13. An alternative expression and the preferred term for leprosy is (*a*) Hansen's disease (*b*) sleeping sickness (*c*) Pasteur's illness (*d*) Pfeiffer's disease.

14. All the following are transmissible diseases of humans except (*a*) meningitis (*b*) polio (*c*) leprosy (*d*) tetanus.

15. The Hib vaccine is used to induce protection against (*a*) sleeping sickness (*b*) botulism (*c*) rabies (*d*) meningitis.

16. Most cases of tetanus are acquired by (*a*) bites of arthropods (*b*) puncture of the skin (*c*) respiratory droplets (*d*) consuming unpasteurized milk.

17. Which of the following describes the organism that causes listeriosis: (*a*) an anaerobic member of the genus *Clostridium* (*b*) a fungus (*c*) a small Gram-positive rod (*d*) an icosahedral virus.

18. A person has the following symptoms: slurred speech, impaired vision, and paralysis of the diaphragm and intercostal muscles. The disease most likely present is (*a*) encephalitis (*b*) polio (*c*) botulism (*d*) leprosy.

19. The bacterium that causes leprosy is (*a*) acid-fast (*b*) anaerobic (*c*) transmitted by arthropods (*d*) a small Gram-negative diplococcus.

20. Both tetanus and botulism are caused by (*a*) protozoa (*b*) prions (*c*) *Cryptococcus neoformans* (*d*) a species of *Clostridium.*

21. A patient is unable to feel the sensations of pain when a flame is applied to the skin of the fingers. The patient is most likely suffering from (*a*) Chagas' disease (*b*) rabies (*c*) listeric meningitis (*d*) leprosy.

22. The Waterhouse-Friderichsen syndrome is most often associated with cases of (*a*) cryptococcosis (*b*) Eastern equine encephalitis (*c*) kuru (*d*) meningococcal meningitis.

23. The Salk vaccine used to protect against rabies consists of (*a*) inactivated viruses (*b*) attenuated viruses (*c*) genetically engineered viruses (*d*) viral proteins.

24. Most cases of encephalitis are transmitted by (*a*) wound punctures in the skin (*b*) arthropods (*c*) contaminated water (*d*) animal bites.

25. A valuable way of preventing the transmission of botulism is (*a*) avoiding arthropods (*b*) treating animal bites promptly (*c*) heating foods well before consuming them (*d*) avoiding contact with one who has the disease.

Matching. Match the characteristic in Column B to the disease in Column A by placing the correct letter next to the disease name.

Column A	Column B
_____ 1. Polio	(*a*) Caused by a *Trypanosoma* species
_____ 2. Encephalitis	(*b*) Prion disease
_____ 3. Meningococcal meningitis	(*c*) Accompanied by cryptococcosis
_____ 4. Cryptococcosis	(*d*) South American sleeping sickness
_____ 5. Kuru	(*e*) One type is LaCrosse
_____ 6. *Haemophilus* meningitis	(*f*) Related to a *Mycobacterium* species
_____ 7. Sleeping sickness	(*g*) Caused by a fungus
_____ 8. Botulism	(*h*) Transmitted by bats, dogs, and raccoons
_____ 9. Chagas' disease	(*i*) Due to an ultramicroscopic prion

____ 10. Listeric meningitis	(*j*) Immunization with Sabin vaccine
____ 11. Rabies	(*k*) Develops after a skin puncture
____ 12. AIDS	(*l*) Prevention with Hib vaccine
____ 13. Tetanus	(*m*) Related to unpasteurized dairy products
____ 14. Leprosy	(*n*) Due to a *Neisseria* species
____ 15. Creutzfeldt-Jakob disease	(*o*) Toxin present in food

True/False. Enter the letter "T" if the statement is true in its entirety. Enter the letter "F" if the statement or any part of the statement is false.

___ **1.** Cases of listeric meningitis are particularly dangerous in pregnant women because the disease may cause miscarriage or the disease may pass to the fetus and cause damage.

___ **2.** To prevent tetanus in individuals an injection of tetanus toxoid is injected into the body in the DPT injection.

___ **3.** Polio is a disease of the nervous system where the primary effects occur in the meninges and in the cerebrum.

___ **4.** The fungi that cause cryptococcosis resembles a yeast and invades the brain and spinal cord after causing infection of the gastrointestinal tract.

___ **5.** When Neisseria meningitidis enters the bloodstream, it causes a severe malaise and toxemia, which are part of the syndrome called meningococcemia.

___ **6.** The drug penicillin is routinely used for the treatment of Haemophilus meningitis, and is administered prophylactically to prevent development of the disease.

___ **7.** The tetanus toxin causes disease in the individuals by preventing the release of acetylcholine in the synapse.

___ **8.** Individuals who suffer an animal bite receive rabies injections in the arm.

___ **9.** Most cases of polio are transmitted by infected arthropods, which bite unimmunized individuals.

_ **10.** Trembling and mental dysfunction similar to that in Alzheimer's disease may be characteristic signs of Creutzfeldt-Jakob disease, which is believed to be due to a prion.

_ **11.** Both African and South American sleeping sickness are due to flagellated fungi that are transmitted by arthropods.

_ **12.** One of the symptoms of rabies is hydrophobia, which is accompanied by spasms of the pharyngeal muscles.

_ **13.** The organism that causes botulism is an anaerobic, sporeforming rod that contaminates food and food products.

_ **14.** Cases of meningitis may be due to numerous bacteria, including species of *Listeria, Streptococcus,* and *Naegleria.*

_ **15.** Most cases of cryptococcosis are due to microorganisms transmitted from pigeon droppings by wind gusts.

__ **16.** Rabies is transmitted to humans by dogs and cats, but few other species of animals

__ **17.** The incubation period for leprosy is a short two to three weeks, and the disease is caused by an acid-fast rod.

__ **18.** The Hib vaccine is currently used in medicine to provide protection against possible cases of polio.

__ **19.** Cases of rabies in the United States are very rare, and less than ten cases usually occur in a single year.

__ **20.** Mosquitoes and ticks are among the arthropods, which are able to transmit cases of viral encephalitis.

Chapter 20

Microbial Diseases of the Respiratory System

OBJECTIVES

The respiratory tract is the site of numerous microbial diseases including pneumonia, influenza, and other well-known problems. Transmission of these diseases is relatively easy because the respiratory tract is open to the environment. The objectives of this chapter are to:

1. Identify the variety of diseases and symptoms associated with streptococcal infection of the respiratory tract.

2. Summarize the symptoms occurring in pertussis and indicate which vaccine is available for its prevention.

3. Summarize some important characteristics relating to tuberculosis such as its etiologic agent, symptoms, treatments, diagnosis, incidence, and immunization.

4. Discuss the various bacterial species responsible for pneumonia, and compare the characteristic symptoms of the different types.

5. Outline the importance of chlamydia and rickettsiae in the occurrence of respiratory disease.

6. Specify some of the characteristics associated with influenza and influenza viruses and compare these to the characteristics of the common cold.

7. Outline the various respiratory diseases due to fungi, including valley fever, histoplasmosis, and blastomycosis.

8. Describe the importance of opportunistic protozoa such as *Pneumocystis carinii* in the occurrence of respiratory disease associated with AIDS.

THEORY AND PROBLEMS

20.1 Why are respiratory tract diseases so common in humans?

Infections of the respiratory tract are very common in humans primarily because they are transmitted by droplets of mucus and saliva. Since people crowd together often, passage of microorganisms occurs easily. Also, many of the infections take place at the outer epithelium of the respiratory tract. Since the blood may not be involved, the immune system response is less vigorous than it might otherwise be.

BACTERIAL DISEASES

20.2 Which streptococci cause infections of the respiratory system?

There are three major groups of **streptococci**: the alpha-hemolytic streptococci, the beta-hemolytic streptococci, and the gamma-hemolytic streptococci. Alpha-hemolytic streptococci digest blood partially and cause blood to assume an olive green color. The bacteria in this group are involved in tooth decay, intestinal disorders, and other mild streptococcal diseases. There is only one species of streptococci classified as beta-hemolytic. It is *Streptococcus pyogenes*. This Gram-positive chain of cocci is responsible for numerous infections of the skin, blood, and meninges. In addition, it is the etiologic agent of **strep throat**, also known as **septic sore throat**. Gamma-hemolytic streptococci are nonpathogenic.

20.3 How can the physician know whether a respiratory infection is a strep throat?

Traditionally, the diagnosis of strep throat has consisted of taking a throat swab and smearing the throat swab on blood agar. Beta-hemolytic streptococci dissolve blood cells, and if the infection is strep throat, the blood in the blood agar will disappear. In more contemporary medicine, antibody tests are used in which streptococcal antibodies are bound to latex particles, and a throat swab is placed with the particles. If streptococcal antigens are present, the particles will undergo agglutination.

20.4 What are some characteristic signs of strep throat?

In most cases of strep throat, the tissues of the pharynx become bright red, and pus is seen on the tonsils and walls of the pharynx. The tongue often becomes bright red, and the papillae often become upright, giving the condition known as "strawberry tongue." The cough is substantial, the fever is high, and the cervical lymph nodes enlarge and become tender. Infection of the middle ear may also occur because the Eustachian tube connects the pharynx to the middle ear and permits passage of the bacteria to this region. Middle ear infection is known as **otitis media**.

20.5 Why is treatment of strep throat important and why should it be immediate?

Treatment for streptococcal sore throat is usually conducted with penicillin or a suitable alternative if the patient is allergic to penicillin. Treatment should be prompt and aggressive, because if the streptococci pass into the bloodstream, they cause severe blood symptoms and rapid temperature rise. In addition, streptococcal antigens react with streptococcal antibodies and fix themselves to the tissue of the heart valves, causing a condition known as **bacterial endocarditis.** Endocarditis can lead to degeneration of the heart valve, especially if there is an underlying condition such as a mitral valve prolapse.

20.6 Is a strep throat related to scarlet fever?

Scarlet fever and strep throat are both caused by *Streptococcus pyogenes*. The difference is that scarlet fever is due to a toxin-producing strain that penetrates to the bloodstream. The toxin, called **erythrogenic toxin**, damages the capillaries and allows blood to leak into the skin tissues, causing a pink-red skin rash and a high fever. The rash may also be due to a hypersensitivity reaction taking place in the blood. Strawberry tongue accompanies scarlet fever, and the skin often undergoes desquamation, or peeling. This condition is similar to staphylococcal scalded skin.

20.7 Which organism is the etiologic agent of pertussis?

Pertussis is caused by a small, Gram-negative rod named *Bordetella pertussis*. Transmitted by respiratory droplets, the organism lives in the epithelial lining of the respiratory tract and produces several toxins that help it adhere to the cilia of the epithelial cells. Pertussis was once prevalent in the United States, but the incidence has been reduced by the use of public health surveillance methods and by the administration of vaccines. In recent years, there has been a slight resurgence of the disease. Figure 20-1 shows the incidence of pertussis over a 30-year period.

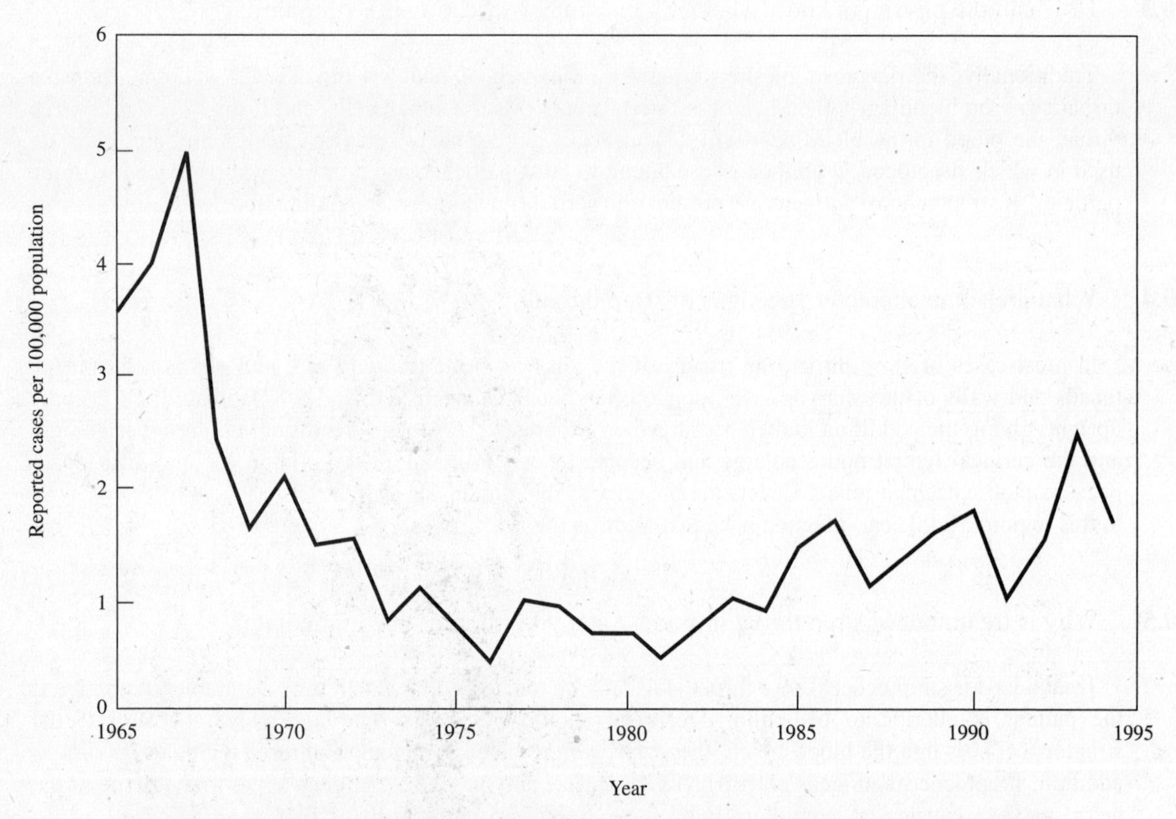

Fig. 20-1 A graph showing the incidence of pertussis in the United States between 1965 and 1994. Since the 1980s, the incidence of pertussis has risen, but with the introduction of the new vaccine, it is expected to decline.

20.8 What are some of the characteristic signs of pertussis?

Typical cases of pertussis occur in three stages. The first stage is accompanied by fever, vomiting, and mild cough. In the second stage, there are paroxysms of staccatolike coughing in which strings of mucus accumulate in the air passageways. The paroxysms of cough end in a high-pitched "whoop" as air is drawn over the narrow passageways. For this reason, pertussis is also known as **whooping cough**. In the third stage, the coughing continues for several weeks during convalescence. Secondary infections with other bacteria can occur during this period.

20.9 How is pertussis diagnosed and treated?

The diagnosis of pertussis generally consists of isolating bacteria from the respiratory tract and cultivating on Bordet-Gengou medium. Erythromycin is effective against the disease. However, it is very difficult to eliminate symptoms completely, even though the bacteria have been killed.

20.10 Which vaccine for pertussis is effective for preventing the disease?

The vaccine against pertussis has traditionally consisted of killed bacteria in the **DPT** vaccine, where P stands for pertussis. The more recent vaccine consists of a cell-free preparation of polysaccharides and is known as the acellular pertussis vaccine. Combined with the diphtheria and tetanus immunizing agents, the triple vaccine is known as **DTaP**. Immunizations begin at the age of about two months.

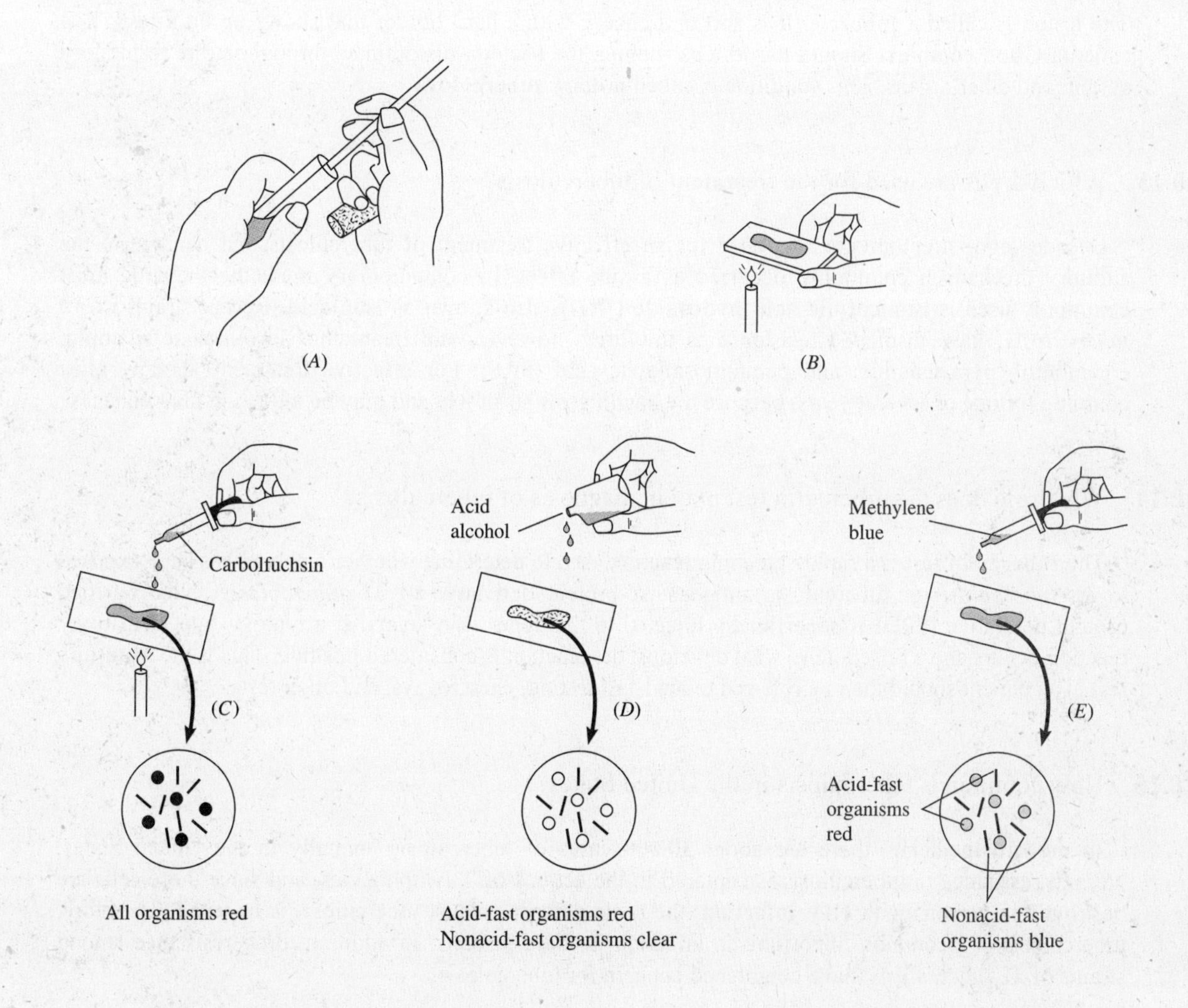

Fig. 20-2 The acid-fast technique. Bacteria are obtained from the culture tube (*A*) and placed on a slide and heat-fixed (*B*). The smear is covered with carbolfuchsin (*C*) and all organisms stain. Acid alcohol is added (*D*), but acid-fast organisms remain stained. A counterstain is added (*E*) to stain the nonacid-fast organisms. The acid-fast organisms retain the original stain which is red.

20.11 Which bacterium is responsible for tuberculosis and what are the characteristics of the organism?

Tuberculosis is caused by a slender rod known as *Mycobacterium tuberculosis*. This slow-growing rod is neither Gram-positive nor Gram-negative; rather, it is **acid-fast**. When stained with carbolfuchsin, steam must be used to force the stain through the bacterial cell wall. Once stained, however, the organism cannot be decolorized even if dilute acid-alcohol is used. This acid-fast characteristic is due to the presence of a very thick cell wall containing mycolic acid. Figure 20-2 illustrates how the technique is performed.

20.12 Why is tuberculosis sometimes known as consumption?

Tuberculosis is sometimes called **consumption** because the lungs are "consumed" by the unrelenting growth and multiplication of bacteria. No toxins are produced by the bacteria, but as the bacteria multiply, they cause an acute inflammation with the development of lesions containing many dead cells. The lesion is called a **tubercle**. It is soft and cheesy with a hard border that shows up on x-rays; it is called a **Ghon complex**. Should the lesions rupture, the bacteria disseminate throughout the respiratory system and other organs. This condition is called **miliary tuberculosis**.

20.13 Which drugs are used for the treatment of tuberculosis?

Decades ago, streptomycin was used for an effective treatment of tuberculosis, but damage to the auditory mechanism commonly occurred as a side effect. In contemporary medicine the drug most commonly used is **isonicotinic acid hydrazide (INH)**, also known as **isoniazid.** Recent strains of *M. tuberculosis* have displayed resistance to this drug, however, and treatments now include rifampin, ethambutol, pyrazinamide, and paraminosalicylic acid (PAS). For effective treatment, therapy must continue for one or several years, because the bacilli grow so slowly and may be hidden in macrophages.

20.14 What role does the tuberculin test play in diagnosis of tuberculosis?

The **tuberculin test** is a rapid-screening reaction used to determine whether a person has been exposed to *M. tuberculosis* or tuberculosis antigens. A protein derivative of *M. tuberculosis* called purified protein derivative (PPD) is superficially injected to the upper skin layers. If a type IV hypersensitivity reaction occurs and a raised itchy weal develops, the reaction is considered positive. This is the **Mantoux test.** The patient should then be referred to acid-fast testing, chest x-rays, and other tests.

20.15 How common is tuberculosis in the United States?

In modern medicine, there are about 30,000 cases of tuberculosis annually in the United States. Normal resistance to tuberculosis is centered in the actions of T lymphocytes, and since these cells are destroyed in persons with **HIV infection**, the basic defense against tuberculosis is lessened. Therefore, tuberculosis is a common opportunistic infection in these persons. In addition, drug resistance among strains of *M. tuberculosis* have heightened concern for tuberculosis.

20.16 Is there any immunization possible against tuberculosis?

The vaccine against tuberculosis is called **BCG**, for **bacille Calmette Guerin**. The vaccine consists of a strain of *Mycobacterium* that causes tuberculosis in cows. The strain is attenuated and is not believed to cause disease in humans. Though used in other parts of the world, the vaccine is not widely used in the United States because methods for diagnosis and treatment are established.

20.17 Are other species of *Mycobacterium* capable of causing tuberculosislike infections?

Two strains of mycobacterium are important causes of tuberculosislike infections. *Mycobacterium bovis* causes **bovine tuberculosis**, and the organism can be transmitted by raw milk. Pasteurization is important in preventing the transmission because a tuberculosislike disease of the lungs can develop in humans. *Mycobacterium avium-intracellulare* causes an infection in the lungs called **mycobacteriosis,** also known as **MAI disease** after the organism. Persons with HIV infection and AIDS are susceptible to this disease.

20.18 Which bacterium is responsible for cases of diphtheria, and how common is this disease in the United States?

Diphtheria is a rare disease in the United States, accounting for about a dozen cases annually. It is caused by *Corynebacterium diphtheriae*, a club-shaped, Gram-positive rod. The organisms contain phosphate-laden **metachromatic granules** that stain with methylene blue and become red, a characteristic sign of the organism.

20.19 What are some symptoms associated with diphtheria?

On entering the upper respiratory tract, *C. diphtheria* produces exotoxins that interfere with protein synthesis in the epithelial cells and cause the cells to deteriorate. Damaged cells, bacilli, white blood cells, and other extraneous material build up and form a **pseudomembrane** in the respiratory tract, blocking the airway and possibly inducing suffocation. Figure 20-3 displays this symptom. Other organs such as the heart, nervous system, and kidneys may also be affected. Treatment with diphtheria antitoxin and antibiotics are effective.

20.20 What does the vaccine for diphtheria consist of?

The D in the **DPT vaccine** stands for diphtheria. The material used is a chemically treated diphtheria toxin. The chemical causes the toxin to become a toxoid that induces immunity without causing any toxic effects on the body tissues.

20.21 Which bacterial species are responsible for bacterial pneumonias?

The term **pneumonia** applies to inflammation of the lung tissue due to chemical involvement or microbial infection. Among the bacteria, several species may be the etiologic agents of bacterial pneumonia: *Streptococcus pneumoniae, Staphylococcus aureus, Klebsiella pneumoniae,* and *Haemophilus influenzae.* A pneumonia can be **lobar pneumonia** if it infects an entire lobe of the lung, **double pneumonia** involves two lobes of the lung, **bronchial pneumonia** involves the bronchi and bronchioles, and **walking pneumonia** refers to a mild case of pneumonia.

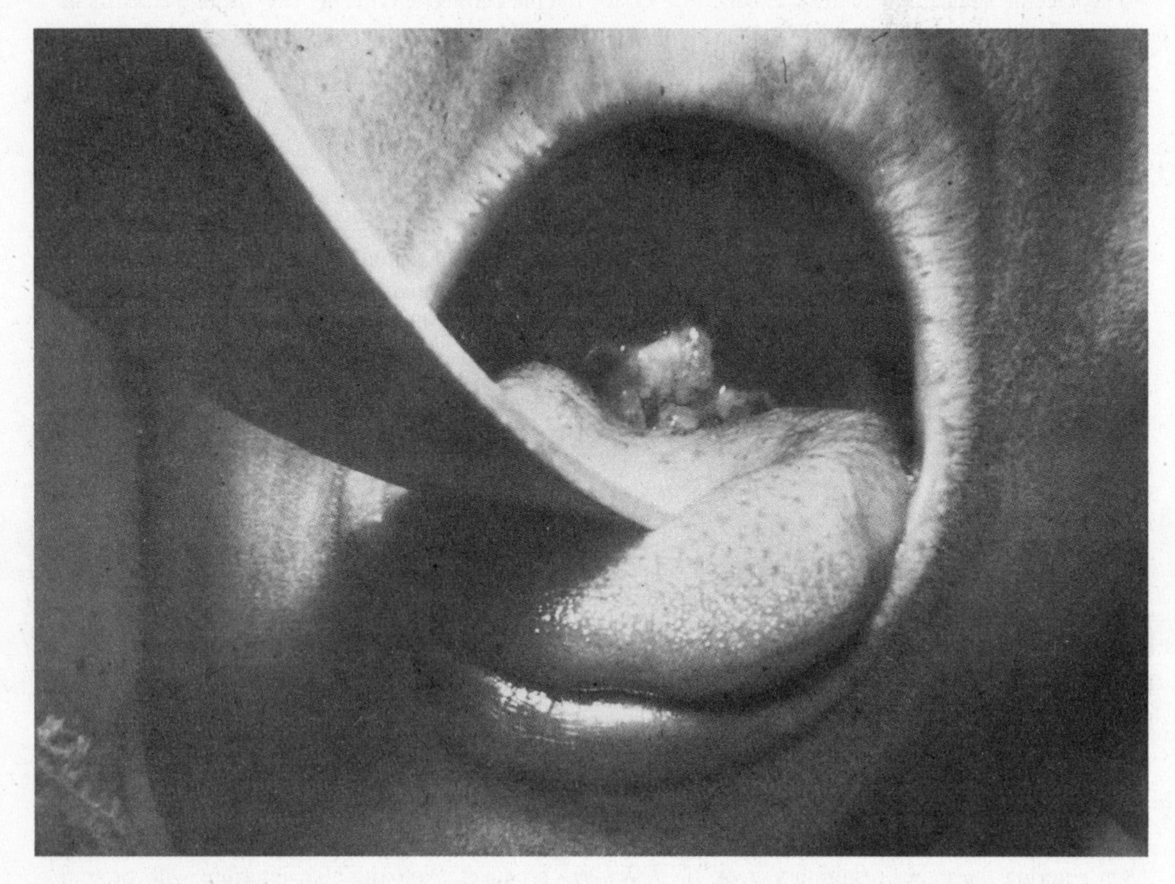

Fig. 20-3 A view into the pharynx showing the pseudomembrane of an individual with diphtheria.

20.22 Which is the most common cause of bacterial pneumonia?

Probably the most common type of adult pneumonia is **pneumococcal pneumonia** due to *Streptococcus pneumoniae*, colloquially known as the pneumococcus. Patients experience violent chills, high fevers, severe chest pain, cough, and rust-colored sputum. Treatment with penicillin or a related antibiotic is an effective way of treating the disease, but without antibiotic therapy, mortality rates tend to be high. In many cases, the streptococci have been present in the upper respiratory tract of the patient, and disease develops when the immune system is compromised or the body tissues are traumatized such as during a case of influenza.

20.23 Is there any vaccine for pneumococcal pneumonia?

For those individuals susceptible to pneumococcal pneumonia, a vaccine containing polysaccharides may be administered. There are close to 100 strains of pneumococcus capable of causing pneumonia, and the vaccine contains polysaccharides from about 25 types.

20.24 What is meant by an atypical pneumonia?

Atypical pneumonias are caused by microorganisms that are different than the "typical" bacterial species because they have unusual properties or are difficult to cultivate from infected patients. Atypical pneumonias may be caused by species of chlamydiae, viruses, or mycoplasmas.

20.25 What is the etiologic agent of mycoplasmal pneumonia?

Mycoplasmal pneumonia is caused by *Mycoplasma pneumoniae,* a submicroscopic bacterium that has no cell wall. The organisms cannot be seen with the usual light microscope, and therefore, they have no Gram reaction. Transmitted by airborne droplets, the mycoplasmas can be cultivated in the laboratory, where their colonies have a "fried egg" appearance.

20.26 What are the characteristic kinds of mycoplasmal pneumonia?

Mycoplasmal pneumonia is generally a mild infection and is often referred to as **walking pneumonia** because most patients remain ambulatory. The disease is usually accompanied by low grade fever and mild cough, and treatment with erythromycin is usually recommended. Mycoplasmal pneumonia is referred to as **primary pneumonia** because it is transmissible, whereas secondary pneumonias develop from organisms already in the body.

20.27 What sort of disease is Legionnaires' disease?

Legionnaires' disease is a type of pneumonia caused by bacteria infecting the lung tissue. Patients experience high fever, chills, severe headaches, and pain in the chest. Shock and kidney failure may be the causes of death.

20.28 How did Legionnaires' disease get its name and what is its etiologic agent?

Legionnaires' disease acquired its name because an outbreak occurred during a convention of the American Legion in Philadelphia in 1976. The etiologic agent is *Legionella pneumophila*, a Gram-negative rod. The organism lives in water and is airborne by breezes and wind gusts, then is inhaled into the respiratory tract. Erythromycin is useful for treating the disease. The organism is also responsible for a closely related disease called **Pontiac fever**.

20.29 Which respiratory diseases are known to be caused by species of *Chlamydia*?

Species of *Chlamydia* are known to cause two respiratory diseases. *Chlamydia pneumoniae* causes **chlamydial pneumonia**, a transmissible pneumonia that resembles mycoplasmal pneumonia. *Chlamydia psittaci* causes **psittacosis**, a respiratory disease in parrots, parakeets, and other birds such as ducks, pigeons, and sea gulls. The disease, also known as **ornithosis**, is accompanied by influenzalike symptoms. Both chlamydial diseases are treated with tetracycline.

20.30 What are some of the typical properties of chlamydiae?

The chlamydiae involved in respiratory diseases are extremely tiny bacteria, invisible with the light microscope. They live within cells and have a complex reproductive cycle that involves the formation of granulelike inclusions called **elementary bodies**. The elementary bodies become **reticulate bodies** before reverting to elementary bodies once again. Then they exit the host cell, as Figure 20-4 indicates.

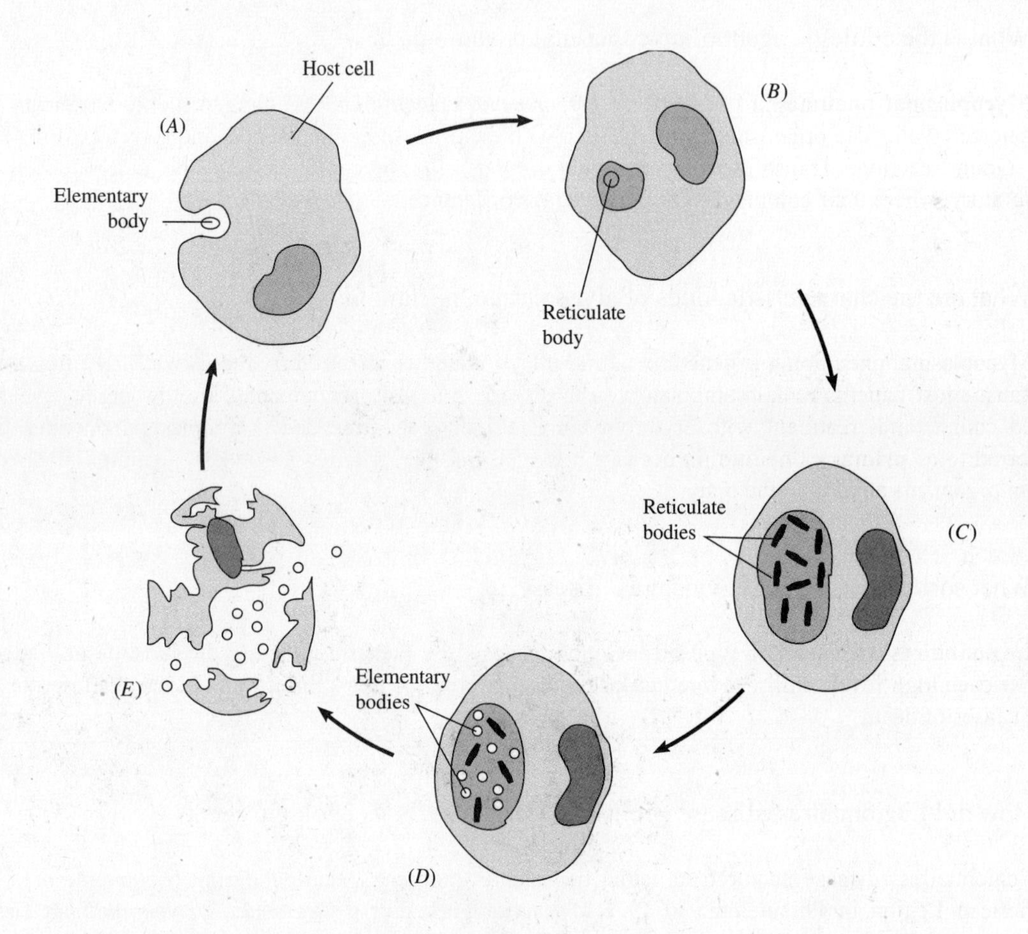

Fig. 20-4 The life cycle of chlamydiae (*A*). The chlamydia enters its host cell in the form of an elementary body (*B*). It undergoes a transformation to a reticulate body (*C*). The reticulate body enlarges, divides, and becomes numerous reticulate bodies (*D*). The reticulate bodies revert to many elementary bodies (*E*). The host cell undergoes lysis, and the elementary bodies are released. These are new chlamydiae.

20.31 Are there any rickettsiae involved in respiratory diseases?

A rickettsia named *Coxiella burnetii* causes an influenzalike disease called **Q fever**. Transmitted by respiratory droplets, as well as ticks and raw milk, Q fever manifests itself as fever, chest pains, severe headache, and other evidence of respiratory involvement. The disease also resembles mycoplasmal pneumonia in its symptoms.

VIRAL DISEASES

20.32 Which virus causes the common cold?

There is no one cause of the **common cold**. Instead, a number of different viruses can cause this condition, among them the rhinoviruses, the coronaviruses, and adenoviruses. **Rhinoviruses** are the most commonly encountered causes of colds, causing the large majority of headcolds. **Coronaviruses** cause bronchitis, mild pneumonias, and some cases of gastroenteritis. **Adenoviruses** generally cause infections of the middle respiratory tract as well as eye infections and some cases of viral meningitis.

20.33 Which are the common symptoms of the common colds?

Common colds generally have a brief incubation period of 3 to 4 days. Symptoms include headache, cough, sore throat, excessive sneezing and congestion, and low to mild fever.

20.34 Why is there no vaccine for the common cold?

Because so many different viruses are capable of causing common colds, it is unlikely that a vaccine will be developed. There are hundreds of strains of rhinoviruses, coronaviruses, and adenoviruses that cause the common cold syndrome. Most treatments consist of treating the symptoms until they disappear.

20.35 Which virus is responsible for cases of influenza?

Influenza, or "the flu," is caused by an RNA virus occurring in eight segments. Each segment is bounded by protein, and the eight segments are enclosed in an envelope containing spikes. Two antigens called hemagglutinin (H) and neuraminidase (N) are found in the spikes. The H and N antigens vary among influenza viruses and account for the many influenza strains in our society.

20.36 What is antigenic shift as it applies to influenza viruses?

The changes occurring in the H and N antigens are examples of an **antigenic shift**. Antigenic shift results in varying strains in influenza viruses and is responsible for the yearly outbreaks of influenza. For example, the body may develop antibodies against a certain strain in one year, only to have a new strain occur the following year. Since antibodies produced the previous year are ineffective against this strain of the virus, the patient suffers another attack of disease.

20.37 How are influenza viruses named?

Viruses are named according to which H and N antigens are present: for example H3 and N2. In addition, influenza viruses are classified as types A, B, or C. Thus, an influenza virus may be designated A(H3N3).

20.38 Are there any characteristic signs of infection with influenza virus?

The incubation period for influenza is about two days. The patient then suffers high fever and sore muscles, with coughing, sore throat, and intestinal upset. During this time, fluid accumulation in the lung may lead to bacteria pneumonia, which may be a cause of death. Antibiotics are then given to preclude this possibility. The antiviral drug **amantadine** is also useful to reduce influenza symptoms.

20.39 Is it possible to immunize against influenza?

It is possible to receive an injection of influenza vaccine, but it must be given annually because different strains emerge each year. Persons who are susceptible to influenza, such as those with compromised immune systems or damaged lungs, are particularly encouraged to receive the vaccine.

20.40 What is the major difference between influenza and parainfluenza?

Parainfluenza viruses contain RNA and generally multiply in the mucous membranes of the nose and throat as compared to influenza viruses multiplying in the lower respiratory tract. Parainfluenza viruses often cause **croup**, defined as hoarse coughing. The larynx becomes swollen, and the cough takes on the characteristics of a high-pitched bark. There are four different parainfluenza viruses as opposed to multiple strains of influenza viruses.

20.41 What is the RS virus and what type of disease does it cause?

The term RS stands for **respiratory syncytial virus**, an RNA virus. This virus causes culture cells to cling together in large, functionless masses. In a child or infant, the RS virus causes **viral pneumonia**, which can have a high fatality rate. In adults, the RS virus replicates primarily in the upper respiratory tract and is less dangerous.

20.42 Does the hantavirus cause respiratory symptoms?

In the 1990s, there have been several outbreaks of infection by the hantavirus. A notable epidemic occurred in the southwest United States in 1994. Known as the "four corners disease," the disease is accompanied by viral pneumonia and extensive hemorrhaging. Rodents are usually associated with the disease, and hantavirus infections occur sporadically.

FUNGAL AND PROTOZOAL DISEASES

20.43 Which is the etiologic agent of valley fever?

Valley fever is a respiratory disease associated with the fungus *Coccidioides immitis*. The fungal spores are found in the southwest United States and are blown about by the wind and transferred to the lungs. The disease they cause is known as valley fever, because it occurs in the San Joaquin Valley of California. The technical name is **coccidioidal mycosis**.

20.44 What are some characteristic signs of coccidioidal mycosis?

Patients with coccidioidal mycosis experience fever, coughing, and weight loss accompanied by sharp chest pains. Most patients recover, but in a few cases a progressive disease disseminates in the body. The disease resembles tuberculosis and is often mistaken for this disease. A characteristic sign of the organism is a body called the **spherule**, which contains endospores. This observation aids in diagnosis.

20.45 What type of disease is caused by *Histoplasma capsulatum*?

Histoplasma capsulatum is a fungus of the soil that is responsible for **histoplasmosis**, also known as **Darling's disease**. In river valleys such as the Mississippi and Ohio valleys, the spores are inhaled to the lungs where they cause lesions seen on x-ray examination. Figure 20-5 shows the organism in its typical appearance in the body tissue.

20.46 Can histoplasmosis be a serious disease in certain individuals?

Although histoplasmosis is generally a mild disease, it can be serious in people with suppressed immune systems, such as persons with AIDS. In these individuals, the organism spreads to other organs such as the lymph nodes, liver, and spleen, and the disease may be fatal. A drug known as **amphotericin B** can be used for treatment. Air containing dried chicken feces and dried bat feces contains *Histoplasma capsulatum.*

20.47 Is blastomycosis a possible disease of the respiratory tract?

Although usually a skin disease, **blastomycosis** can occur in the lungs. In this case, it is accompanied by mild respiratory symptoms, although in persons with compromised immune systems, the disease may disseminate. As a lung infection, the disease is usually acquired by inhaling spores.

20.48 Which is the etiologic agent of blastomycosis?

Blastomycosis is caused by the fungus *Blastomyces dermatitidis*. This organism is found widely in the Mississippi valley and is a soilborne organism. If skin contact is made with this organism in the soil, skin ulcers may result. Amphotericin B is used for severe infections.

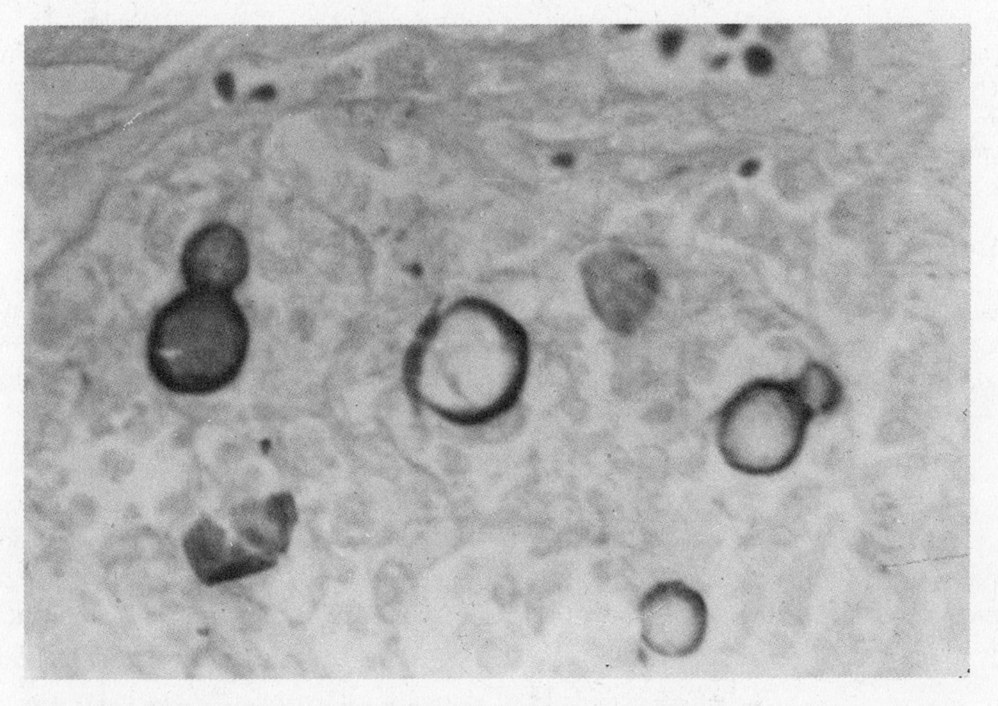

Fig. 20-5 *Histoplasma capsulatum* seen in lung tissue. The organism appears in its yeast form, and the cells cling together to resemble a figure 8.

20.49 Are there any opportunistic fungi that may be involved in respiratory disease?

In addition to the established pathogens, several fungi can cause disease when given the opportunity to infect the tissues. These fungi are known as opportunistic fungi. They can cause disease in persons undergoing transplant therapy with immune-suppressing drugs, or in AIDS patients, or when a large number of spores enter the body. Examples of the opportunistic fungi include *Aspergillus fumigatus, Mucor* species, and *Rhizopus* species.

20.50 Are there any protozoa that can cause diseases of the respiratory tract?

One of the most severe diseases of the respiratory tract is caused by the protozoan *Pneumocystis carinii*. This organism invades the lung tissues in AIDS patients, and is responsible for over 50 percent of the deaths associated with AIDS. The protozoan takes up all the air spaces, prevents the exchange of gases, and generally leads to death by consolidation and suffocation. Many of the alveolar membranes also rupture in this disease. The disease is commonly known as ***Pneumocystis carinii* pneumonia** or *Pneumocystis* pneumonia.

20.51 Which treatments are available for *Pneumocystis* pneumonia?

For established cases of *Pneumocystis* pneumonia, the patients are treated with the drug **pentamidine isethionate**. Other drugs are also available. For persons with HIV infections, aerosols of pentamidine are often used to prevent the development of *Pneumocystis* pneumonia, since *Pneumocystis carinii* is a common inhabitant of the lung.

20.52 Why is *Pneumocystis carinii* sometimes considered a fungus?

There is no universal agreement about whether *Pneumocystis carinii* is a protozoan or a fungus. Analysis of its RNA indicates a relationship to the fungi, but the activities of the organism's life cycle are more like that of a protozoan. Until a definitive decision has been made to change classifications, *Pneumocystis carinii* is considered a protozoan.

20.53 Are there any multicellular parasites involved in lung disease?

One multicellular parasite, the **lung fluke**, causes disease in many parts of the world, particularly Asia. The etiologic agent is *Paragonimus westermani*. The organisms are taken into the body in crab meat or crayfish meat, whereupon the parasites pass through the tissues and localize in the lungs. Infected patients suffer chronic cough, difficult breathing, and in some cases blood in the sputum.

Review Questions

Completion. Add the missing word or words in each of the following to complete the thought.

1. Tuberculosis is caused by the ___________ rod known as *Mycobacterium tuberculosis.*

2. Mycoplasma pneumoniae is the submicroscopic bacterium that has no ____________ and is known to be the cause of mycoplasmal pneumonia.

3. The drug ____________ is used for treating fungal infections of the lungs such as histoplasmosis.

4. Those streptococci that dissolve ____________ and cause strep throats are known as beta-hemolytic streptococci.

5. The DPT vaccine and the DTaP vaccine provide protection against diphtheria, tetanus, and ____________.

6. The drug isoniazid is widely used for the treatment of ____________ and is also known as isonicotinic acid hydrazide.

7. The lung fluke known as ____________ is usually taken into the body when one consumes shellfish.

8. Although *Pneumocystis carinii* is generally considered a ____________, there is some evidence that it may be a fungus.

9. The "four corners disease" is now known to be caused by hantaviruses, which are usually spread by ____________.

10. The phenomenon of ____________ results in varying strains of influenza viruses and is responsible for yearly outbreaks of influenza.

11. It is unlikely that a ____________ will be developed for the common cold because so many different viruses are able to cause this condition.

12. In Legionnaire's disease, bacteria cause pneumonia when they infect the ____________ tissue.

13. Pneumococcal pneumonia is a serious bacterial pneumonia caused by ____________ pneumoniae.

14. The etiologic agent of diphtheria produces ____________ that interfere with protein synthesis taking place in the epithelial cells.

15. A rickettsia named ____________ causes an influenzalike disease known as Q fever.

16. The respiratory disease psittacosis affects parrots, parakeets, and other birds and is also known as ____________.

17. MAI disease is known to be caused by a species of ____________.

18. The bacteria that cause pertussis are spread among individuals by respiratory ____________.

19. The antiviral drug ____________ may be used to reduce the symptoms of influenza.

20. The respiratory disease known as ____________ is associated with the fungus Coccidioides immitis.

21. Histoplasmosis, also known as ____________, occurs in river valleys such as the Mississippi and Ohio valleys.

22. AIDS patients having ____________ pneumonia are commonly treated with the drug pentamidine isethionate.

23. The erythrogenic toxin damages the capillaries of the body, causing blood to leak into the skin tissues, resulting in the pink-red rash of ___________.

24. The ___________ test is a form of tuberculin test in which PPD is injected to the upper skin layers.

25. The material used to immunize against ___________ in the DPT vaccine consists of chemically treated toxin.

Multiple Choice. Select the letter of the item that correctly completes each of the following statements.

1. The organism that causes pertussis is a (*a*) fungus (*b*) large Gram-positive rod (*c*) small Gram-negative rod (*d*) virus.

2. All the following apply to the disease tuberculosis except (*a*) immunization is rendered by the DPT vaccine (*b*) the organism that causes it is acid-fast (*c*) the disease is sometimes called consumption (*d*) disseminating tuberculosis is called miliary tuberculosis.

3. Pseudomembranes that form in the respiratory tract and block the airways and induce suffocation are a characteristic sign of (*a*) scarlet fever (*b*) blastomycosis (*c*) diphtheria (*d*) common colds.

4. Vaccines are available for all the following diseases except (*a*) pneumococcal pneumonia (*b*) pertussis (*c*) diphtheria (*d*) toxoplasmosis.

5. Chlamydial diseases are commonly treated with the antibiotic (*a*) penicillin (*b*) tetracycline (*c*) amantadine (*d*) amphoteracin B.

6. Coronaviruses, rhinoviruses, and adenoviruses are all possible causes of (*a*) viral encephalitis (*b*) viral pneumonia (*c*) valley fever (*d*) common colds.

7. The respiratory syncytial virus is a possible cause of (*a*) tuberculosis (*b*) viral pneumonia (*c*) diphtheria (*d*) strep throat.

8. Valley fever is a fungal disease commonly found in the (*a*) Mississippi and Ohio valleys (*b*) Rocky Mountain valleys (*c*) southwest United States (*d*) valleys of the northeast.

9. Air containing the dried feces of chickens or bats may be a possible mode of transmission of (*a*) tuberculosis (*b*) histoplasmosis (*c*) *Pneumocystis* pneumonia (*d*) common colds.

10. The BCG preparation consists of a strain of bacteria intended to immunize against (*a*) scarlet fever (*b*) tuberculosis (*c*) influenza (*d*) blastomycosis.

11. Which of the following applies to scarlet fever: (*a*) it is caused by *Corynebacterium* species (*b*) it is a viral disease (*c*) it is a fungal disease (*d*) it is caused by *Streptococcus pyogenes.*

12. All the following are true of tuberculosis except (*a*) penicillin is used in therapy (*b*) the Mantoux test is used as a rapid screening test (*c*) the lesions of the lung are called tubercles (*d*) normal resistance is centered in the actions of T lymphocytes.

13. The number of cases of diphtheria occurring in the United States annually is (*a*) over 1,000 (*b*) over 10,000 (*c*) very low (*d*) unusually high.

14. Walking pneumonia generally refers to a (*a*) mild case of pneumonia (*b*) atypical case of pneumonia (*c*) type of fungal pneumonia only (*d*) pneumonia caused by chlamydiae.

15. The etiologic agent of Legionnaire's disease (*a*) exists in contaminated food (*b*) is transmitted by pigeon droppings (*c*) is usually found in bat caves (*d*) lives in water.

16. The elementary body is a characteristic granulelike inclusion formed by (*a*) all rickettsiae (*b*) all chlamydiae (*c*) all acid-fast bacilli (*d*) Gram-negative rods only.

17. The letters H and N refer to antigens associated with (*a*) rhinoviruses (*b*) the fungus that causes histoplasmosis (*c*) hantaviruses (*d*) influenza viruses.

18. Pneumonia caused by *Pneumocystis carinii* is an important cause of death in individuals (*a*) who live in mosquito-infested lands (*b*) who drink contaminated water (*c*) who have AIDS (*d*) who work in hospitals and laboratories.

19. Hoarse coughing, also known as croup, is a characteristic sign of infection by (*a*) *Coccidioides immitis* (*b*) parainfluenza viruses (*c*) RS viruses (*d*) lung flukes.

20. Darling's disease is an alternative expression for a (*a*) viral disease of the nose (*b*) fungal disease of the lungs (*c*) protozoal disease of the upper respiratory tract (*d*) bacterial disease of the pharynx.

21. Metachromatic granules stain with methylene blue and indicate the presence of the agent that causes (*a*) viral pneumonia (*b*) valley fever (*c*) scarlet fever (*d*) diphtheria.

True/False. For each of the following statements, mark the letter "T" next to the statement if the statement is true. If the statement is false, change the underlined word to make the statement true.

___ **1.** Those streptococci that are considered nonpathogenic are referred to as beta-hemolytic streptococci.

___ **2.** Endocarditis is a disease of the heart muscles generally related to streptococcal antibodies.

___ **3.** A condition known as strawberry tongue is a characteristic sign of pertussis.

___ **4.** The acid-fast characteristic of Mycobacterium is due to the presence of a very thick cell wall.

___ **5.** The organism that causes tuberculosis is capable of producing numerous toxins.

___ **6.** Vaccination against tuberculosis is possible with a strain of attenuated bacilli known as DPT.

___ **7.** The presence of a pseudomembrane in the respiratory tract is a characteristic sign of the disease tuberculosis.

___ **8.** Probably the most common type of adult pneumonia in the United States is mycoplasmal.

___ **9.** The primary infection associated with Legionnaire's disease takes place in the lung tissue.

__ **10.** Elementary bodies are granulelike inclusions associated with the presence of rickettsiae.

__ **11.** Pontiac fever is a respiratory infection most closely associated with Q fever.

__ **12.** The designation A(H3N2) is most likely associated with the respiratory syncytial virus.

__ **13.** Valley fever, which occurs in the San Joaquin Valley of California, is caused by spores derived from a bacterium.

__ **14.** The drug penicillin is widely used for the treatment of diseases caused by fungi.

__ **15.** Over 50 percent of the deaths associated with influenza are caused by lung infection due to *Pneumocystis carinii.*

Chapter 21

Microbial Diseases of the Digestive System

OBJECTIVES

The microbial diseases of the digestive system are related to foods and food products. These foods may contain infectious microorganisms or the toxins they produce. In either case, disease will be established in the body, as this chapter indicates. The objectives of the chapter are to:

1. Recognize the useful purposes served by microorganisms normally found in the gastrointestinal tract.

2. Identify some important characteristics of periodontal disease and dental caries, including the names of the responsible organisms.

3. Describe staphylococcal food poisoning as an example of a foodborne intoxication of the body and point out alternative organisms that may be involved.

4. Summarize the modes of transmission and symptoms of salmonellosis as a typical type of foodborne infection.

5. Compare the etiologic agents, characteristic signs, and treatments of such intestinal diseases as typhoid fever, bacillary dysentery, and cholera.

6. Specify the intestinal diseases caused by lesser known bacteria such as *Campylobacter, Yersinia,* and *Escherichia coli.*

7. Explain the involvement of bacteria in the development of peptic ulcers.

8. Summarize the various types of viral hepatitis that can affect the human liver with emphasis on the symptoms and immunizations for the types.

9. Describe the important characteristics of protozoal diseases of the intestine such as giardiasis, amoebiasis, and cryptosporidiosis.

10. Define the importance of tapeworms, flukes, and roundworms in the development of diseases of the digestive system.

THEORY AND PROBLEMS

21.1 Which part of the digestive system is most heavily populated by microorganisms?

Certain parts of the digestive system such as the stomach and first part of the small intestine have relatively small populations of microorganisms due in part to the acidity and enzymes present in these regions. However, the mouth contains numerous species of microorganisms, and the latter part of the small intestine and large intestine contain huge populations. It has been calculated, for example, that a gram of feces contains up to 100 billion bacteria.

21.2 Do microorganisms of the gastrointestinal tract serve any useful purpose?

Intestinal microorganisms serve several important uses. They produce various growth factors such as vitamin K. They digest otherwise indigestible materials and make the nutrients available to body cells. And they hold pathogenic microorganisms in check by using up nutrients the pathogens need and producing chemical factors that inhibit their growth.

BACTERIAL DISEASES

21.3 What role do microorganisms play in the development of dental caries?

In the mouth, microorganisms accumulate in the **dental plaque**, a mass of organic material that forms near the tooth enamel. Within the plaque, oral bacteria produce large amounts of acid from the fermentation of carbohydrates such as sucrose. The acid causes decalcification of the enamel, and as the enamel breaks down, a cavity develops.

21.4 What causes cariogenic bacteria to cling to the plaque?

Cariogenic bacteria are those that cause dental caries. These bacteria produce a gummy polysaccharide called **dextran** which permits bacteria to adhere. Dextran results from the metabolism of the monosaccharides fructose and glucose, which result from the hydrolysis of sucrose.

21.5 Which are the most important cariogenic microorganisms?

The most important cariogenic bacteria are species of streptococcus such as *Streptococcus mutans*. Other cariogenic bacteria include *S. salivarius* and *S. sanguis*, as well as several filamentous forms and various species of *Eikenella* and *Lactobacillus*.

21.6 Is periodontal disease the same as dental caries?

Dental caries is a disease of the tooth enamel and tooth itself, while **periodontal disease** is a combination of gum inflammation and erosion of the periodontal ligaments and bone that support the teeth. During periodontal disease, the gums recede and sometimes undergo necrosis, while the teeth loosen as the surrounding bone and ligaments erode (Figure 21-1). In some cases, the cementum that supports and protects the roots of the teeth is also broken down. In its mildest form, periodontal disease is known as gingivitis; while in its severest form it is called **acute necrotizing ulcerative gingivitis** (ANUG), also known as **trench mouth**.

21.7 Which organisms are responsible for periodontal diseases?

A number of Gram-negative rods may be involved in periodontal disease, including species of *Bacteroides, Eikenella,* and *Eubacterium.* A long, spearlike anaerobic rod called *Fusobacterium* may also be involved, and species of *Treponema* as well as *Actinomyces* have been related to the condition.

21.8 Which viral disease is of particular importance in the oral cavity?

A paramyxovirus causes an infection of the salivary glands called **mumps**. The virus infects the parotid gland or glands and causes swelling of the tissues, and taught skin overlying the glands. The disease is also known as **epidemic parotitis**; cases of it in the United States are rare because of vaccinations with attenuated virus in the MMR (measles-mumps-rubella) preparation.

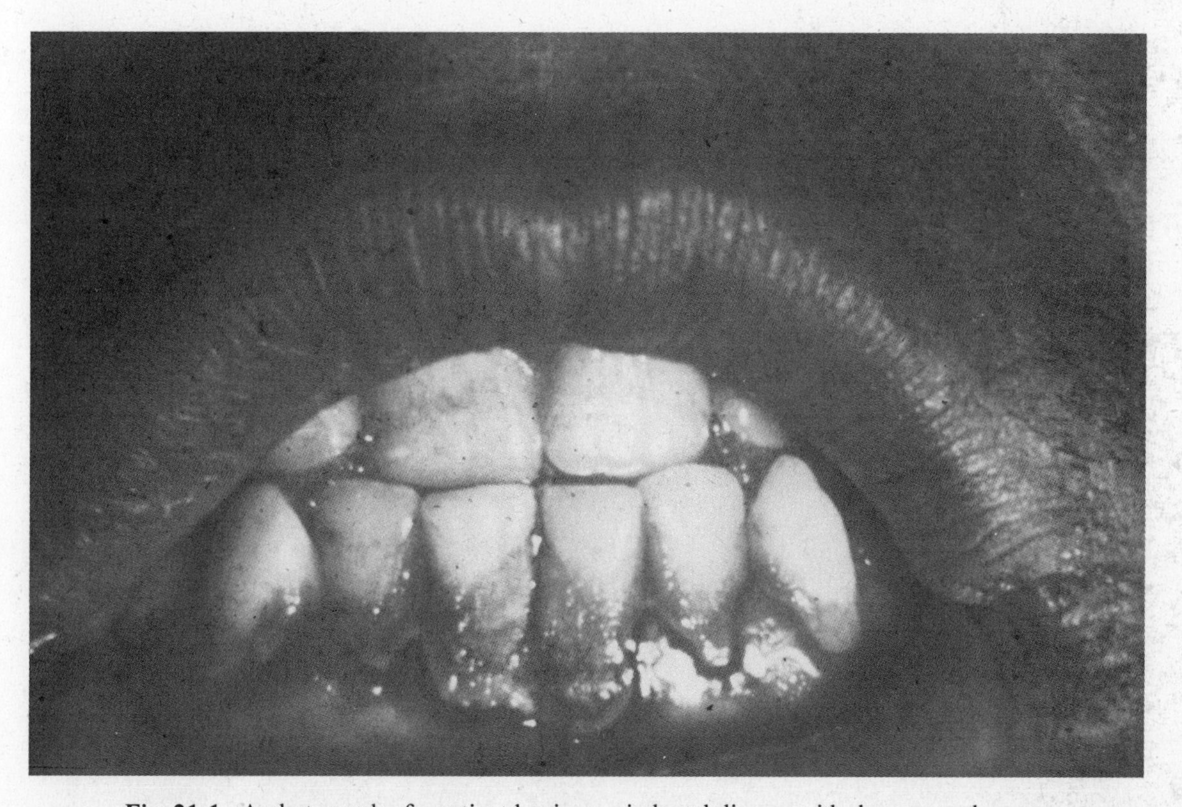

Fig. 21-1 A photograph of a patient having periodontal disease with damage to the gum tissues and erosion of the teeth.

21.9 In which age groups does mumps normally occur?

Mumps is normally a disease in children under the age of ten. However, the virus may affect adult populations, and in post-pubic males the disease is often accompanied by inflammation of the testes, a condition known as **orchitis**. In rare cases, sterility results.

21.10 What is the difference between infections and intoxications of the digestive system?

Microbial diseases of the digestive system may be separated into two major categories: intoxications and infections. **Intoxications** are caused by toxins ingested in food after pathogenic bacteria have grown in the food. **Infections**, by contrast, develop when bacteria grow within the digestive tract itself. Normally, the infection takes longer to develop than an intoxication (the incubation period is longer), and the effects of the infection tend to be longer-lasting than for intoxication.

21.11 What is an example of a foodborne intoxication in the human body?

One of the most common intoxications occurring in the body is **staphylococcal food poisoning**. This disease is caused by ingestion of an enterotoxin produced by *Staphylococcus aureus* when it has contaminated food. The staphylococci are resistant to heat, drying, and salt, factors that permit growth in foods. The organisms deposit their toxins during prolonged refrigeration, and if the food is eaten without sufficient heating the toxins will be consumed into the digestive tract.

Fig. 21-2 A diagnostic method for *Staphylococcus aureus*. The bacterium is cultivated on a plate of nutrient medium, and drops of bacteriophages are placed at various areas to see which one interacts with the staphylococcus. Clear areas reflect the places where the particular bacteriophage multiplied in the staphylococci and destroyed the cells. The strain of *S. aureus* is deduced by this method.

21.12 What are some of the symptoms of staphylococcal food poisoning?

When staphylococcal toxins enter the body, they are absorbed into the mucosa where they cause tissue damage and induce abdominal pain, nausea, vomiting, and diarrhea (but little fever). The incubation period is a short few hours, and the symptoms last for up to eight hours. Recovery is usually complete, and additional episodes may occur at a later date because immunity is not generated. In the laboratory, *S. aureus* can be identified by placing various bacteriophages on a plate of *S. aureus* to see where multiplication of the phages occurs, as Figure 21-2 shows.

21.13 Are any bacteria other than *Staphylococcus aureus* capable of causing food poisoning?

The Gram-positive, sporeforming rod *Clostridium perfringens* is another cause of food poisoning. This organism's toxin is produced under anaerobic conditions such as in undercooked meats and in gravies kept warm for a long period of time. The characteristic sign of **clostridial food poisoning** is diarrhea, and the incubation is generally longer than for staphylococcal food poisoning. The etiological agent is also the cause of gas gangrene. Another possible cause of food poisoning is *Bacillus cereus*, an aerobic Gram-positive, sporeforming rod. Vomiting and diarrhea characterize food poisoning by this organism, and rice and dairy products are commonly involved. Botulism, caused by *Clostridium botulinum*, is a type of foodborne intoxication, but its symptoms occur in the nervous system. The disease is discussed in Chapter 19.

21.14 Why is it incorrect to use the term *Salmonella* food poisoning?

Members of the genus *Salmonella* cause a foodborne "infection" in humans rather than a foodborne poisoning. The disease is correctly called **salmonellosis**. It is characterized by extensive growth of bacteria in the intestine, as compared to staphylococcal food poisoning, in which the toxin is ingested.

21.15 Why are varieties of *Salmonella* called serological types?

In contemporary microbiology there are hundreds of different varieties of *Salmonella* capable of causing salmonellosis. The varieties are referred to as **serological types** rather than species because they are distinguished by the type of antibodies they elicit in the body. The serological types are named in the same way as species are named, and the term species is commonly used with *Salmonella*, even though the term is not technically correct.

21.16 Are any animals particularly susceptible to the effects of *Salmonella*?

Various species (serological types) of *Salmonella* are present in chicken, turkey, and poultry products as well as eggs from these birds. Therefore, salmonellosis is often related to poultry products inadequately cooked and to eggs that are consumed raw or only partially cooked. *Salmonella* species are also known to be present in reptiles such as lizards and pet turtles.

21.17 Are there any symptoms that distinguish salmonellosis?

Salmonellosis is accompanied by abdominal pain, fever, and diarrhea, often with blood or mucus in the stool. The incubation period is about 24 hours after the ingestion of food and invasion of the intestinal mucosa generally takes place. The mortality rate of patients is very low, and large numbers of

Salmonella cells can be detected in the feces during the diagnostic procedure. Endotoxins may be related to the fever. In some cases, a more serious condition called **enterocolitis** may occur. In this case, intestinal symptoms last for several weeks and are moderately severe. Prevention of the disease depends on sanitary handling of foods and food products and proper cooking of poultry and egg products.

21.18 Which *Salmonella* type is responsible for typhoid fever?

Typhoid fever is an extremely serious infection of the gastrointestinal tract caused by *Salmonella typhi*. The disease is transmitted by contaminated food and water, often from a carrier individual, that is, an individual who has had the disease and has recovered, but continues to shed the bacilli. *Salmonella typhi* is a Gram-negative rod.

21.19 What are the characteristic signs of typhoid fever?

Typhoid fever begins with a long incubation period of about two weeks. The bacteria enter the blood from the intestine and invade many tissues causing intense spiking fever and unrelenting headache. Characteristic "rose spots" appear on the trunk and remain for several days, and many patients become delirious and suffer internal hemorrhaging.

21.20 Are there any drugs used to treat typhoid fever?

The drug of choice for treating typhoid fever is chloramphenicol, also known as Chloromycetin. Recoverers usually become chronic carriers and harbor the salmonellae in the gall bladder. Gall bladder removal may be necessary. Ceftriaxone and other cephalosporin antibiotics are also used.

21.21 Which bacteria are responsible for bacillary dysentery?

Bacillary dysentery is also known as **shigellosis**. It is caused by several species of *Shigella*, including *S. sonnei, S. flexneri, S. boydii,* and *S. dysenteriae.* The etiologic agent is found in contaminated food and water, and the disease is spread by ingesting these.

21.22 What are the common signs of bacillary dysentery (shigellosis)?

Bacillary dysentery (shigellosis) is due to a powerful toxin called the **shiga toxin**, which is produced in the intestine. The toxin causes tissue damage in the mucosa resulting in severe diarrhea containing blood and mucus. Many cases of disease are accompanied by severe dehydration, and fluid and electrolyte balances are upset in this life-threatening illness. The disease has an incubation period of about four days, and symptoms last for a week.

21.23 What treatments are used for patients suffering from bacillary dysentery?

Fluid and electrolyte replacements are essential for treating sufferers of shigellosis. Several antibiotics are used in combination because many resistant strains of *Shigella* have evolved in recent years. Recoverers are often carriers and can spread the disease by the food or water they contact.

21.24 Which organism is responsible for the disease cholera?

The etiologic agent of **cholera** is *Vibrio cholerae*, a comma-shaped Gram-negative rod. The organism has two biotypes and is particularly prevalent in Asia where it causes numerous epidemics.

21.25 What are the characteristic signs of cholera?

The bacilli of cholera produce an enterotoxin that binds to the mucosa of the intestine and encourages the release of large quantities of water containing mucus. The outstanding characteristic of the disease is a sudden loss of massive amounts of fluids and electrolytes, sometimes reaching many liters per day. The rapid fluid loss results in severe shock, dehydration, and in many cases, death.

21.26 What treatments are required for cholera patients?

The replacement of lost fluids and electrolytes is essential for recovery from cholera. This can be accomplished by intravenous infusions or by special salt tablets that encourage the retention of water. Antibiotics can be used, but their use is secondary to fluid and electrolyte replacement. Recoverers often become carriers and infect others when they contaminate food or water.

21.27 Is there any other species of *Vibrio* that is pathogenic in humans?

A second species of *Vibrio* called *Vibrio parahemolyticus* causes a form of gastroenteritis in humans. The organism is common in eastern Asia, and it is often associated with shrimp, crabs, and other shellfish. Proper cooking of shellfish interrupts spread of the disease. Vomiting, diarrhea, and abdominal pain accompany the infection.

21.28 Can *Escherichia coli* be the cause of intestinal infection in humans?

Escherichia coli has been used for many decades as highly valuable research tool in biochemistry and biotechnology. Certain strains, however can cause illness in humans. In tropical regions of the world, *E. coli* strains exist that are enteroinvasive and enterotoxic. These strains are common causes of **traveler's diarrhea** which tourists contract during visits to warm climates. Since the *E. coli* strain is present in water and foods, the disease can be avoided by boiling or chemically treating water before use, and by eating well-heated foods.

21.29 Are there any strains of *E. coli* that are enterohemorrhagic?

In recent years, an enterohemorrhagic strain of *E. coli* has emerged in the United States. This strain is known as *E. coli* 0157:H7. This strain of *E. coli* is transmitted in food. It produces shigalike toxins that cause hemorrhaging in the colon, a condition accompanied by intestinal bleeding. Hemorrhaging can also occur in the kidneys, which can lead to kidney failure. Foods, particularly meats, are involved in transmission of the bacteria, and thorough cooking is encouraged to avoid disease.

21.30 What type of disease does *Campylobacter jejuni* cause?

Campylobacter jejuni is a curved, Gram-negative rod that causes a form of gastroenteritis called **campylobacteriosis**. The *Campylobacter* is transmitted in raw and unpasteurized milk and in foods such as poultry. Infections are accompanied by diarrhea, abdominal pain, and foul-smelling feces. Recovery normally occurs without complications after several days of symptoms.

21.31 Do species of *Yersinia* cause intestinal infections?

A species of *Yersinia* called *Yersinia enterocolitica* is a cause of gastroenteritis in humans. The organism is a Gram-negative rod which displays bipolar staining, a condition in which the stain gathers at the poles of the cells and causes the organism to resemble a safety pin. The disease is referred to as **yersiniosis** and is accompanied by sharp abdominal pain, diarrhea, and fever. Most cases resolve without complications.

21.32 Which bacteria are involved in ulcers of the gastrointestinal tract?

Recent research has indicated that the bacterium *Helicobacter pylori* is responsible for a condition known as **peptic ulcer**, also called **peptic disease syndrome**. In this condition, gastric and duodenal ulcers occur in the patient, and relief can be obtained by using a number of antibiotics. The etiologic agent is a Gram-negative curved rod similar to *Campylobacter jejuni.*

21.33 How can *Helicobacter pylori* grow in the highly acidic environment of the stomach?

Apparently, *H. pylori can* survive in the highly acidic stomach environment by producing the enzyme urease. This enzyme digests urea and produces ammonia. The ammonia forms ammonium hydroxide which neutralizes the acid of the stomach and creates a less acidic environment in which the bacteria grow. The method by which *H. pylori* is spread among humans has not yet been established.

VIRAL DISEASES

21.34 Which viruses are responsible for viral gastroenteritis?

Viral gastroenteritis is a general term for viral disease of the intestines. The viruses that cause viral gastroenteritis are given the general term **enteroviruses**. Among the enteroviruses are rotavirus, perhaps the most common cause of viral gastroenteritis, especially in children. Another possible cause is Norwalk virus, a virus that occurs in all age groups. In both cases, the infection is accompanied by diarrhea and vomiting. Symptoms are generally mild, and the infections are usually self-resolving. Rehydration is administered in severe cases.

21.35 What sort of infections do Coxsackie viruses and echoviruses cause?

Both Coxsackie viruses and echoviruses can cause mild to serious gastrointestinal infections, with damage to the cells. Echoviruses can also infect the skin tissues and cause a rash called **exanthem**, while Coxsackie viruses are also involved in cases of meningitis, and a disease of the heart muscle known as **myocarditis**.

21.36 Which viruses are responsible for hepatitis A and hepatitis B?

Both hepatitis A and hepatitis B are diseases of the human liver. **Hepatitis A** is caused by a very resistant RNA-containing virus, while **hepatitis B** is caused by a very fragile DNA-containing virus. The hepatitis A virus is usually transferred by contaminated food and water (since the virus remains active in the environment outside the body), while the hepatitis B virus is usually transmitted directly from blood to blood or semen to blood (because the virus cannot remain active outside the body environment).

21.37 What are the general symptoms associated with hepatitis?

Most cases of hepatitis are accompanied by malaise, nausea, diarrhea, anorexia, and jaundice. The skin and the whites of the eyes become yellow and the urine is dark. The liver is tender and enlarged, and pain is felt in the upper left quadrant of the abdomen. The incubation period can be variable, with hepatitis A having an incubation time of about one month, and hepatitis B having an incubation time of about three months. Treatment for all forms of hepatitis consists of avoiding liver irritants such as alcohol and fats. Antibody preparations may also be available for administration.

21.38 Is any immunization available for hepatitis A and hepatitis B?

A vaccine for hepatitis B has been available since 1987 in the form of genetically engineered protein fragments. The vaccine is given in three doses and is believed to yield lifetime immunity against hepatitis B. For hepatitis A, a vaccine consisting of attenuated viruses was licensed in 1995.

21.39 Which other types of hepatitis exist besides hepatitis A and hepatitis B?

There are many different kinds of hepatitis in addition to hepatitis A and B. At this writing, viruses have been identified for hepatitis C, hepatitis D, and hepatitis E. The diseases caused by these viruses are collectively known as **non-A non-B (NANB) hepatitis**. Other forms of hepatitis also exist. These diseases have not been studied as well as hepatitis A and B and no vaccines are available. Laboratory tests are available for certain forms of NANB, but not for other forms.

21.40 Are there any long-range effects of hepatitis?

Certain forms of hepatitis, such as hepatitis B, result in chronic liver disease. Hepatitis B has also been related to liver cancer (hepatocarcinoma). Immunization against this form of hepatitis is therefore important.

OTHER MICROBIAL DISEASES

21.41 Which fungi are associated with gastrointestinal disease in humans?

There are no fungi that multiply within the gastrointestinal environment and thereby cause infection. However, there are several fungi that grow in foods and produce toxins, called **mycotoxins**. Ingested in food, the mycotoxins can cause human disease.

21.42 Which fungi can produce mycotoxins in foods?

One fungus that produces mycotoxins is *Aspergillus flavus*. The mycotoxins, known as **aflatoxins**, apparently affect the liver and are believed to induce liver cancer. *Claviceps purpurea* produces a mycotoxin called **ergot**. Rye and wheat plants may contain this toxin, which causes fever, hallucinations, and convulsions in humans. A poisonous mycotoxin is also produced by the **mushroom** *Amanita*. This toxin affects liver cells, causing jaundice and vomiting, and the toxin can be lethal if ingested in quantity.

21.43 What type of gastrointestinal disease is caused by the protozoan *Giardia lamblia*?

Giardia lamblia (also known as *Giardia intestinalis*) is a flagellated protozoan taken into the body in contaminated food and water. The organism is displayed in Figure 21-3. It causes an infection called **giardiasis**, which is accompanied by malaise, nausea, weight loss, and diarrhea. There is no invasion of the intestinal cells, but the protozoa grow on the surface of the intestine, and they shed cysts in the feces as they grow. These cysts pass to the next individual and spread the disease.

21.44 How is giardiasis normally acquired?

Giardia lamblia commonly exists in contaminated water supplies near the mountains. Therefore, backpackers, hikers, and campers should be aware of its presence. The organism infects numerous animals and is shed into the water by their feces. Chlorination does not destroy the cysts, and water should be boiled before consumed or bottled water should be used. Drugs such as metronidazole and quinacrine can be used to treat severe cases.

21.45 Why is *Entamoeba histolytica* a dangerous pathogen in the human intestine?

Entamoeba histolytica is a type of amoeba that invades the intestinal wall and causes **amoebiasis**, also called **amoebic dysentery**. The organism uses its proteolytic enzymes to invade the intestinal wall, cause deep ulcers, and perforate the wall of the intestine. Abscesses often spread to the liver and lung and bacterial infection of the ulcers can lead to fatality.

21.46 How does *Entamoeba histolytica* avoid the stomach acid on its way to the intestine?

Entamoeba histolytica has a **cyst** form that is resistant to acid and permits passage through the stomach environment. Cysts are obtained in contaminated food and water. In the intestine, the cysts rupture and release the feeding form of the amoeba, known as the **trophozoite**. The trophozoite causes the ulcers, and cysts are continually passed from the intestine in the feces. Diagnosis consists of observing cysts in stool samples. Metronidazole has been used for treatment.

21.47 Are any ciliated protozoa able to cause intestinal infection?

One ciliated protozoan, *Balantidium coli*, causes **balantidiasis**, a gastrointestinal illness accompanied by dysentery. There is extensive diarrhea and some invasion of the tissue of the intestinal wall. The organism is very large, and it can be observed in specimens of feces. Pigs are a reservoir in nature, and pork products should be thoroughly cooked to avoid the protozoan. Water is also a means of transmission. The protozoan is pictured in Figure 21-4.

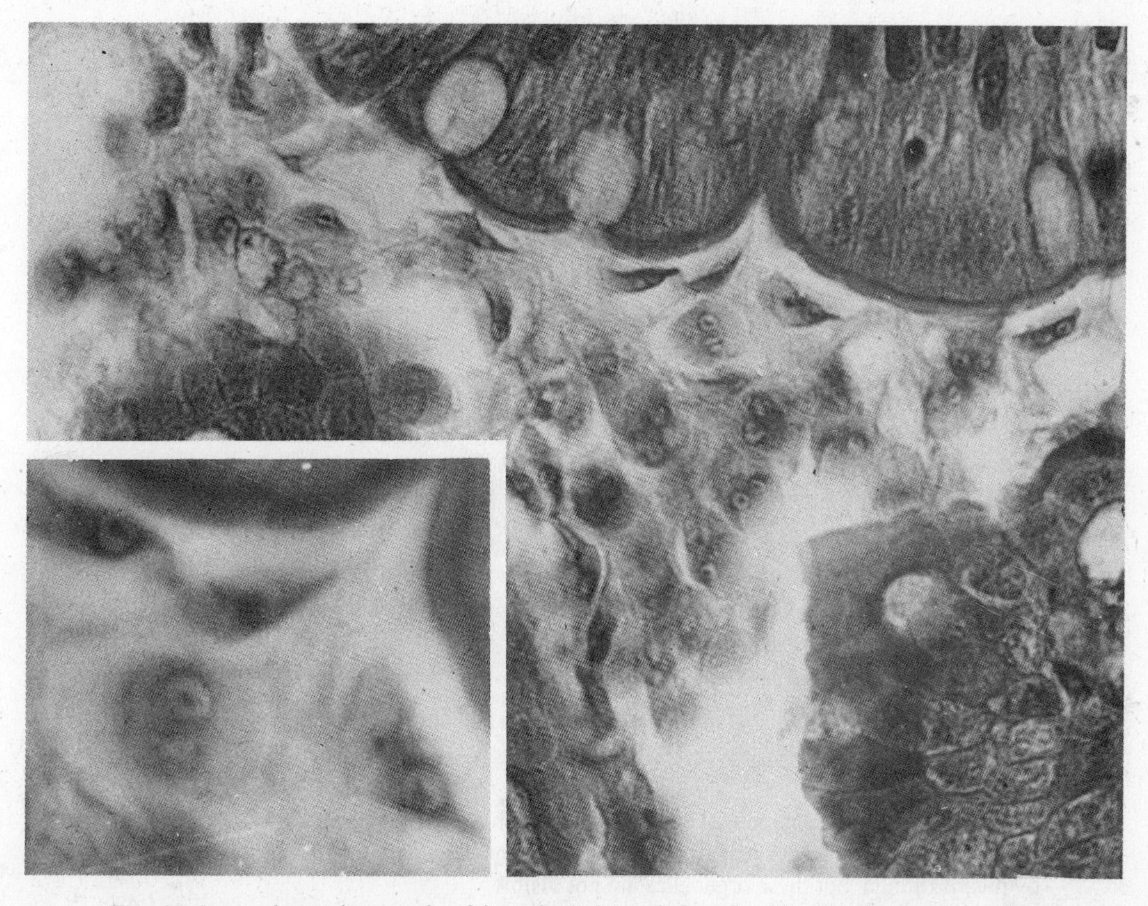

Fig. 21-3 A photomicrograph of intestinal tissue infected with *Giardia lamblia*. The protozoa can be seen as bodies having two large nuclei, which appear as eyes. A closeup is shown in the lower left.

21.48 Which protozoan causes severe intestinal illness in AIDS patients and people with HIV infection?

The protozoan *Cryptosporidium parvum* is an important opportunistic organism in patients with AIDS or HIV infection. This complex organism multiplies rapidly when the immune system is suppressed, and it causes cryptosporidiosis, a choleralike diarrhea that results in severe dehydration and emaciation. The diarrhea can be life threatening.

21.49 Are there any effective treatments for cryptosporidiosis.

No effective drug treatment for cryptosporidiosis is currently available. Therefore, any treatment must include rehydration of the tissues with fluids and electrolytes. In those with functional immune systems, the disease is much milder and is of short duration. Recent outbreaks have called attention to the fact that normal municipal water treatment systems may be unable to detect and remove *Cryptosporidium* from waters.

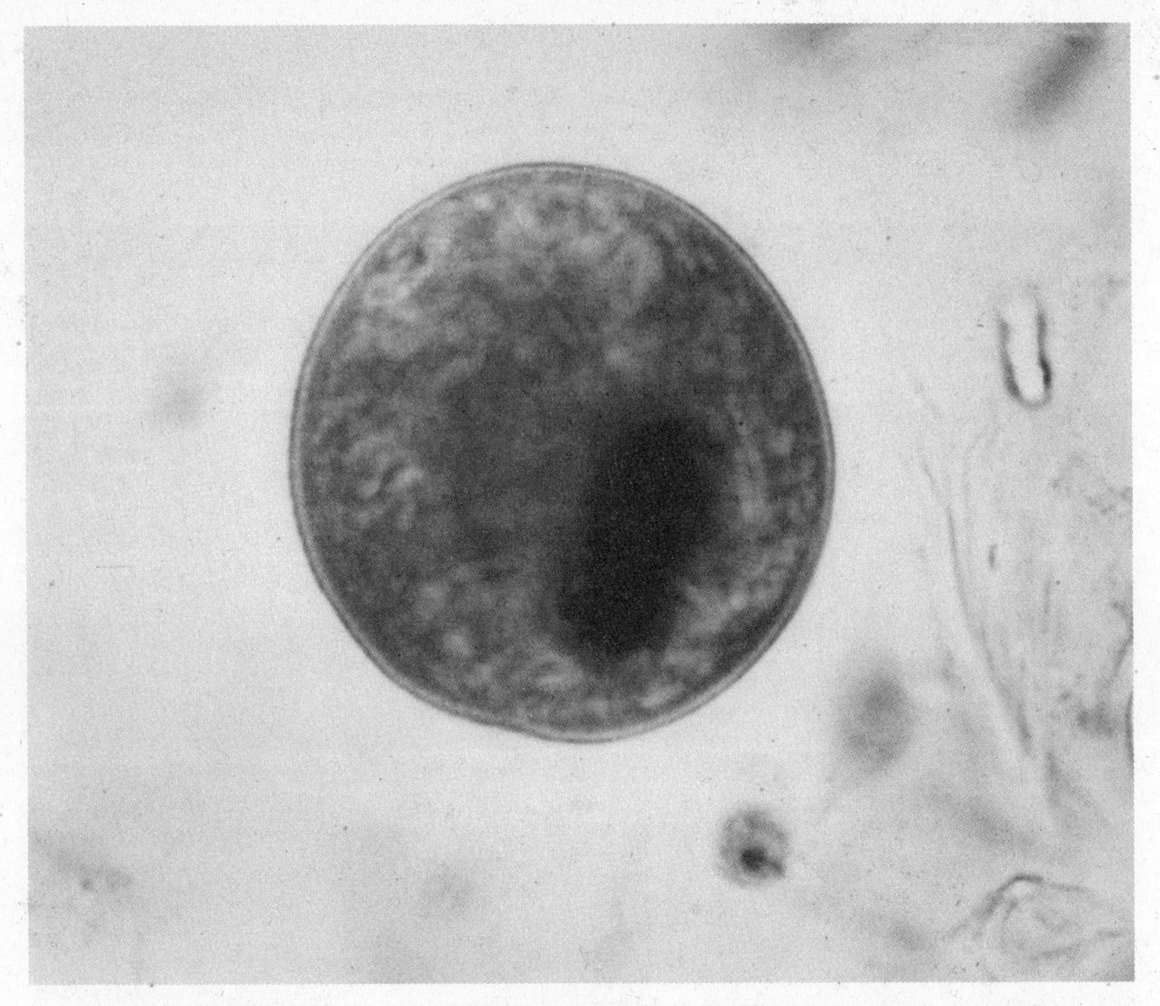

Fig. 21-4 A photomicrograph taken with the oil immersion lens of *Balantidium coli*. This large protozoan has a very prominent nucleus shown as the dark area in the cell. It moves by means of cilia, but these organelles are not visible.

21.50 Which tapeworms can cause infection of the intestine?

Tapeworms are types of **flatworms** having numerous segments called proglottids. The tapeworm attaches by a sucker device to the infected tissue and proceeds to feed on that tissue, thereby bringing on disease. There are several tapeworms capable of causing human infection of the intestine. All worms such as these are considered multicellular parasites. The **beef tapeworm** is *Taenia saginata;* while the **pork tapeworm** is *Taenia solium;* the **fish tapeworm** is *Diphyllobothrium latum;* the **dog tapeworm** is *Echinococcus granulosis*; and the **dwarf tapeworm** is *Hymenolepis nana.*

21.51 What is the general mode of transmission for tapeworms involved in human disease?

Tapeworm cysts or eggs generally enter the human gastrointestinal tract through contaminated food or water. Infection can also occur by contact with dogs, fish, or grain products, depending on the tapeworm. Some of the tapeworms, such as the pork and beef tapeworm, remain as worms in the intestine, but others such as the dog tapeworm migrate to the liver and other organs where they cause infection by multiplying and creating heavy worm burdens. Proper cooking of foods and avoiding animal contacts allows one to avoid the tapeworm.

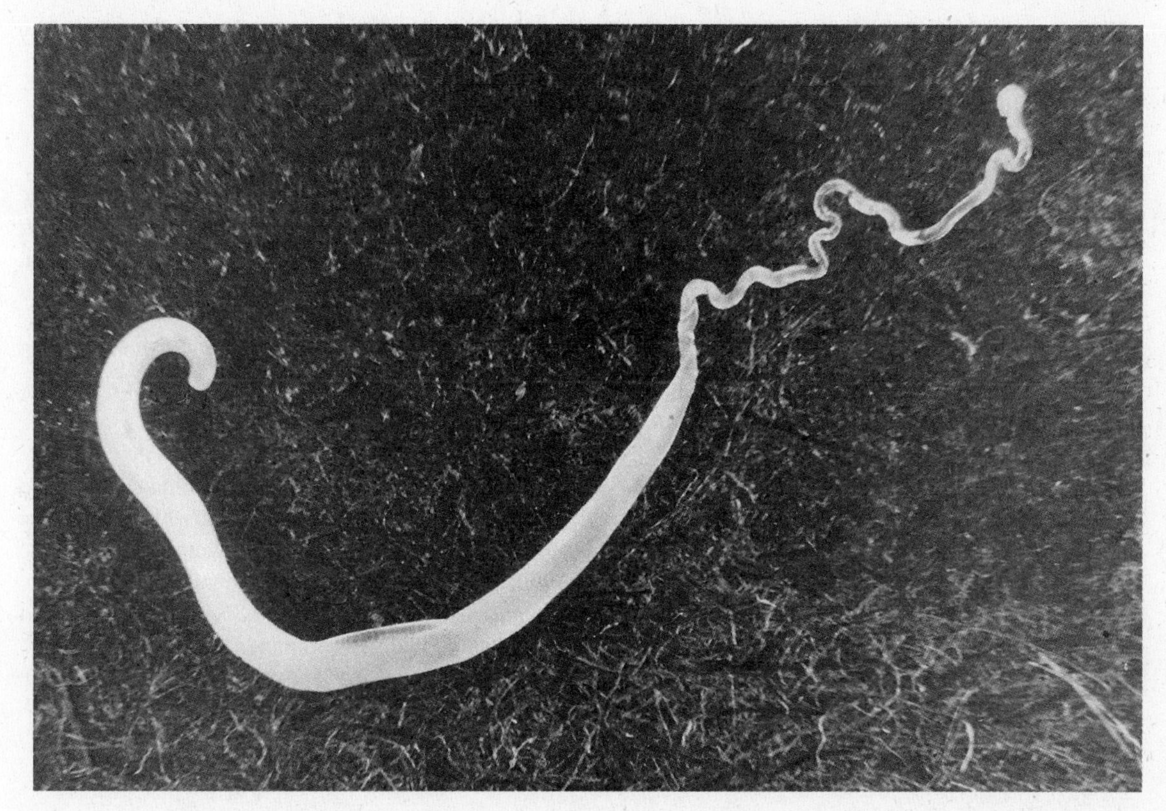

Fig. 21-5 A photomicrograph of the whipworm *Trichuris trichiura*, an infectious roundworm. The whiplike appearance of this worm is visible.

21.52 Which flukes are capable of causing gastrointestinal disease in humans?

Flukes are types of flatworms that are leaflike in appearance. Human pathogens include the **sheep liver fluke**, *Fasciola hepatica*; the **pork intestinal fluke**, *Fasciolopsis buski;* and the **Chinese liver fluke**, *Clonorchis sinensis*.

21.53 What is the general life cycle for the flukes?

In their life cycles, these flukes commonly use intermediary hosts such as the **snail**. The fluke's life cycle includes immature forms called the cercaria and the metacercaria. Humans acquire the flukes from vegetation or a fish or crustacean. The flukes live on the tissue of the intestine or liver and often cause blockages that lead to consolidation and heavy worm burdens. Eggs of the flukes are released in the feces to complete the fluke's life cycle. Care in the cooking and consumption of food helps one avoid fluke diseases.

21.54 Which roundworms are able to cause human disease?

Roundworms are so-named because they are round on the long axis like an earthworm. An important roundworm involved in human disease is *Trichinella spiralis,* the cause of **trichinosis**. Larvae of the roundworm are ingested in poorly cooked pork. The worms then develop into adults in the human intestine. Larvae of the worms pass through the blood and invade the liver, heart, lungs, and other tissues in humans. Cysts form in the skeletal muscles, especially in those of the eye. Trichinosis is avoided by thoroughly cooking pork products.

21.55 Which other roundworms are able to cause human disease?

Various roundworms are related to human disease including the **hookworms**, *Ancylostoma duodenalis* (the Old World hookworm) and *Necator americanus* (the New World hookworm); the **roundworm**, *Ascaris lumbricoides;* the **pinworm**, *Enterobius vermicularis*; and the **filarial worm**, *Strongiloides stercoralis*. Another is the **whipworm**, *Trichuris trichiura*; which Figure 21-5 depicts.

21.56 How do these roundworms cause human disease?

Roundworms produce eggs that invade the body either through the skin or by contaminated food and water. Hookworms and filarial worms penetrate the bare skin from contaminated soil, while pinworms, whipworms, and roundworms enter through contaminated food. In the intestine, the worms develop into adults and generally cause blockage of the intestine and intestinal distress. In some cases, migration to the lungs via the blood takes place and infection can also occur in these distant sites. Sanitary handling of food and avoiding contaminated soil by wearing shoes prevent infection in humans.

Review Questions

True/False. For each of the following statements, mark the letter "T" next to the statement if the statement is true. If the statement or any part of the statement is false, enter the letter "F."

___ **1.** Dental caries is identical to periodontal disease.

___ **2.** Staphylococcal food poisoning is an example of a foodborne intoxication of the body, while salmonellosis is an example of a foodborne infection.

___ **3.** The organism responsible for typhoid fever is *Salmonella typhi*, a Gram-positive type of coccus.

___ **4.** The characteristic signs of cholera include a sudden loss of massive amounts of fluids and electrolytes, as well as severe shock and dehydration.

___ **5.** *Escherichia coli* 0157:H7 is an enterohemorrhagic strain that is transmitted in food and causes an intestinal infection accompanied by bleeding.

___ **6.** The viruses of hepatitis A and hepatitis B are identical.

___ **7.** *Campylobacter jejuni* is a curved, Gram-negative rod that causes a form of gastroenteritis after the bacterium has been transmitted in raw, unpasteurized milk.

___ **8.** Vaccines are available for the prevention of both hepatitis A and hepatitis B.

___ **9.** Members of the genus *Penicillium* are well-known for their production of mycotoxins that affect the liver and are believed to induce liver cancer.

__ **10.** The protozoan *Balantidium coli* is a type of amoeba whose cyst form is resistant to stomach acid.

__ **11.** In patients with AIDS or HIV infection, the protozoan *Cryptosporidium parvum* multiplies rapidly and causes a choleralike diarrhea that results in severe dehydration.

__ **12.** Recent research has indicated that the bacterium *Taenia solium* is responsible for a condition known as peptic ulcer.

__ **13.** *Giardia lamblia* is a flagellated protozoan that causes severe infection of the lung tissues and other areas of the respiratory tract.

__ **14.** Among the tapeworms capable of causing human infection of the intestine are the fish tapeworm, *Diphyllobothrium latum*, and the dwarf tapeworm, *Hymenolepis nana*.

__ **15.** The disease trichinosis is caused by a fluke that is present in pork and can be killed by thorough cooking.

__ **16.** Among the characteristic signs of typhoid fever are delirium, internal hemorrhaging, and "rose spots" appearing on the trunk of the patient.

__ **17.** For patients suffering from shigellosis, it is essential that fluids and electrolytes be replaced because they have been lost in large amounts.

__ **18.** *Streptococcus mutans* is an important species of bacterium involved in dental caries.

__ **19.** Most of the bacteria involved in diseases of the human intestinal tract are Gram-positive.

__ **20.** Cases of cholera are generally associated with chicken, turkey, and poultry products because the responsible bacteria are present in these foods.

Multiple Choice. Select the word or phrase that best completes each of the following statements.

1. Which of the following is not a useful purpose served by microorganisms of the intestine? (*a*) they produce various growth factors (*b*) they hold pathogenic microorganisms in check (*c*) they digest otherwise indigestible materials (*d*) they breakdown aging red blood cells.

2. Cariogenic bacteria adhere to the surface of a tooth by using the gummy polysaccharide (*a*) sucrose (*b*) dextran (*c*) starch (*d*) polymerase.

3. All of the following characteristics apply to the disease mumps except (*a*) it is caused by Gram-negative rod (*b*) it involves the parotid gland or glands (*c*) it is also known as epidemic parotitis (*d*) it may be accompanied by inflammation of the testes if it occurs in post-pubic males.

4. Cases of staphylococcal food poisoning can be prevented by (*a*) being inoculated with the vaccine (*b*) taking large doses of antibiotics (*c*) thoroughly cooking leftover foods (*d*) taking injections of antitoxins.

5. All of the following are possible causes of bacterial food poisoning except (*a*) *Bacillus cereus* (*b*) *Clostridium botulinum* (*c*) *Clostridium perfringens* (*d*) *Corynebacterium xerosis*.

6. Infections caused by members of the genus *Salmonella* are correctly called foodborne "infections" because (*a*) there is extensive growth of bacteria in the intestine (*b*) the bacteria are infected with viruses while in the intestine (*c*) no symptoms are present during the infectious stage (*d*) the disease cannot be transmitted by contaminated food and food products.

7. Cases of salmonellosis are commonly associated with (*a*) poultry products and eggs (*b*) contaminated pork and lamb (*c*) consumption of fermented sausage (*d*) canned foods such as peppers and mushrooms.

8. Which of the following is associated with the disease typhoid fever? (*a*) it is caused by *Vibrio parahemolyticus* (*b*) treatment is rendered with chloramphenicol (*c*) the disease is caused by a virus (*d*) symptoms associated with the disease are usually very mild.

9. Bacillary dysentery is sometimes referred to by its alternative name of (*a*) trichinosis (*b*) gastroenteritis (*c*) shigellosis (*d*) giardiasis.

10. The bacterial disease cholera is accompanied by (*a*) peptic ulcers (*b*) rapid fluid loss from the intestine (*c*) infection of the heart muscles (*d*) rose spots.

11. Cases of traveler's diarrhea contracted by tourists visiting warm climates are commonly due to strains of (*a*) *Salmonella typhi* (*b*) *Taenia solium* (*c*) *Escherichia coli* (*d*) echovirus.

12. The peptic disease syndrome and peptic ulcers are now known to be due to the bacterium (*a*) *Shigella sonnei* (*b*) *Helicobacter pylori* (*c*) *Giardia lamblia* (*d*) *Enterobius vermicularis.*

13. All the following apply to hepatitis A except (*a*) it is caused by a DNA-containing virus (*b*) the virus is usually transmitted by contaminated food and water (*c*) the disease is usually accompanied by jaundice (*d*) the incubation time is about one month.

14. All the following are true of hepatitis B except (*a*) a vaccine is available for prevention (*b*) the incubation period is about three months (*c*) the disease is accompanied by jaundice (*d*) the virus that causes the disease is very resistant.

15. Mycotoxins are toxins associated with gastrointestinal disease and which are produced by (*a*) viruses (*b*) fungi (*c*) Gram-negative rods (*d*) Gram-positive rods.

16. All the following are protozoa associated with human intestinal disease except (*a*) *Balantidium coli* (*b*) *Entamoeba histolytica* (*c*) *Aspergillus flavus* (*d*) *Giardia lamblia.*

17. The disease cryptosporidiosis is particularly serious in (*a*) patients with hepatitis (*b*) AIDS patients (*c*) persons with dermatitis (*d*) children under the age of six.

18. The intermediary host commonly used by flukes is the (*a*) snail (*b*) bird (*c*) lobster (*d*) snake.

19. Poorly cooked pork may transmit the roundworm that causes (*a*) hepatitis C (*b*) bacillary dysentery (*c*) trichinosis (*d*) tapeworm disease.

20. All the following roundworms are able to cause human disease except (*a*) *Necator americanus* (*b*) *Trichuris trichiura* (*c*) *Taenia saginata* (*d*) *Ascaris lumbricoides.*

Completion. For each of the following, add the word that best completes the statement.

1. In its most severe form, periodontal disease is known as acute necrotizing ulcerative gingivitis, also known as ____________.

2. One of the causes of periodontal disease is a long, spearlike anaerobic rod belonging to the genus ____________.

3. In post-pubic males, mumps may be accompanied by the testes and a condition known as ____________.

4. *Staphylococcus aureus* is one of the most common causes of an intoxication referred to as ____________.

5. Instead of the term species, the varieties of *Salmonella* are correctly known as ____________.

6. Diagnostic tests for salmonellosis seek to detect *Salmonella* cells in the ____________.

7. Cases of typhoid fever can often be related to an individual who has recovered from the disease and is termed a ____________.

8. Most cases of shigellosis are due to organisms transmitted by contaminated food and contaminated ____________.

9. The etiologic agent of cholera is a comma-shaped bacillus that is Gram- ____________.

10. The gastroenteritis associated with *Vibrio parahemolyticus* is often related to the consumption of ____________.

11. The disease yersiniosis is caused by *Yersinia enterocolitica*, which is a Gram-negative rod that displays ____________.

12. During the 1990s, *Helicobacter pylori* has been identified as an important cause of ____________.

13. Norwalk virus and rotavirus are but two of the numerous viruses that cause ____________.

14. In addition to intestinal infections and meningitis, Coxsackie viruses are also known to cause a disease of the heart muscle called ____________.

15. The vaccine for hepatitis B contains protein fragments, while the vaccine for hepatitis A contains ____________.

16. The protozoan *Giardia lamblia* causes gastrointestinal disease in humans and moves about by means of its ____________.

17. One of the most poisonous mycotoxins is produced by mushrooms belonging to the genus ____________.

18. In order to survive the passage through the stomach and avoid the effects of stomach acid, many protozoa exist in a resting form known as the ____________.

19. Organisms of the genera *Taenia*, *Echinococcus,* and *Hymenolepis* are all types of worms known as ____________.

20. The hookworm, pinworm, whipworm, and filarial worm are all examples of a type of an intestinal worm known as a ____________.

Chapter 22

Microbial Diseases of the Blood and Viscera

OBJECTIVES

The diseases of the blood and viscera are usually accompanied by fever, and in some cases, a skin rash. Several of these diseases will be surveyed in this chapter, whose objectives are to:

1. Describe the different diseases related to species of *Streptococcus*.

2. Identify some important characteristics of anthrax and note the important similarities between plague and tularemia.

3. Discuss the etiologic agent, transmission, symptoms, and treatment for Lyme disease.

4. Summarize the important characteristics of rickettsiae and note the names and key concepts of several rickettsial diseases.

5. Compare yellow fever and dengue fever with respect to typical symptoms, mode of transmission, and etiologic agent.

6. Discuss the place of the Epstein-Barr virus in human pathology.

7. Specify the characteristics of such protozoal diseases as malaria, toxoplasmosis, and babesiosis.

THEORY AND PROBLEMS

BACTERIAL DISEASES

22.1 What is the general name given to "blood poisoning?"

Blood poisoning is an infection of the blood given the general name of **septicemia**. Various bacteria can cause septicemia including such Gram-positive bacteria as *Staphylococcus aureus* and *Streptococcus pneumoniae*. Septicemias can also be caused by Gram-negative organisms such as *Pseudomonas aeruginosa, Serratia marcescens, Klebsiella pneumoniae,* and *Enterobacter aerogenes*.

22.2 What are some typical symptoms of septicemias?

The symptoms of septicemias vary, but they usually include fever, shock, and red streaks on the skin arising from inflammation of the lymphatic vessels beneath the skin. These red streaks are called **lymphangitis**. The shock associated with septicemia is generally called **septic shock**.

22.3 Which bacterium is responsible for cases of childbed fever?

Childbed fever is also known as **puerperal fever**. It is usually due to a beta-hemolytic streptococcus known as *Streptococcus pyogenes*. Usually these bacteria are associated with infection of the uterus of the mother. During the birth process, the streptococci enter the blood of the mother causing septicemia, inflammation of the pelvic organs, and a serious disease that can be fatal. Penicillin is currently used to prevent the serious consequences of puerperal fever.

22.4 Which infection of the heart valves is due to streptococci?

The same organism that causes puerperal fever, *Streptococcus pyogenes*, causes a disease of the heart valves known as **endocarditis**. Endocarditis is an infection of the endocardium of the heart, the tissue of which the valves are composed. It usually results in severe fever, anemia, and heart murmur, as the heart valves are damaged and blood leaks backwards into the atrium. If untreated, endocarditis can result in heart valve disintegration and death in the patient.

22.5 Are there any predisposing factors to bacterial endocarditis?

A previous exposure to *Streptococcus pyogenes* predisposes one to bacterial endocarditis. For example, when a person has a strep throat due to *Streptococcus pyogenes*, the streptococci pass into the cardiovascular system and damage the mitral valve of the heart. This condition is referred to as **rheumatic heart disease**. In this condition, the heart valves are sensitive to future infection by streptococci, and endocarditis can be a result.

22.6 Which treatments are available for bacterial endocarditis?

Cases of bacterial endocarditis are usually treated aggressively with penicillin or other antibiotics for Gram-positive bacteria. Surgical valve replacement may be required if the valves are damaged beyond repair. Endocarditis may also be caused by *Staphylococcus aureus,* which is treated with penicillin antibiotics.

22.7 Which serious blood disease is caused by *Bacillus anthracis,* and what are the characteristics of that disease?

Bacillus anthracis is a Gram-positive, sporeforming rod. It is the cause of **anthrax**, a disease that affects a variety of animals as well as humans. The organism is usually transported from the soil, and it enters the bloodstream through a skin wound. It may also enter through the respiratory or digestive tracts by air or food, respectively. The disease is accompanied by severe septicemia, with hemorrhaging blood. A blackening of the spleen and other blood-rich organs characterizes the disease.

22.8 Why is anthrax also known as woolsorters' disease?

Anthrax is also called **woolsorters' disease** because the disease occurs in sheep, and the spores can be inhaled by people who shear sheep and work with wool. In this case, anthrax manifests itself as a form of respiratory pneumonia. Mortality rates can be very high, but antibiotic therapy with penicillin antibiotics lowers the mortality rate.

22.9 Do anthrax spores have any significance in modern warfare?

Because anthrax spores are extremely resistant to environmental changes, they remain alive for extraordinarily long periods of time in the atmosphere and in the soil. Because of this survival and the serious disease they cause, anthrax spores can be used in biological warfare. However, there are international treaties that attempt to control these uses.

22.10 Does plague still happen in the United States?

Plague has a rich and colorful history as one of the great classical diseases of all times. Several cases occur in the United States each year and in the world, and it occurs periodically in regions of Asia (especially India), Africa, and South America.

22.11 Which is the etiologic agent of plague?

Plague is caused by *Yersinia pestis*, a Gram-negative rod that stains at the poles and displays bipolar staining. The rod is transmitted by the rat flea, *Xenopsylla cheopis*. The disease occurs primarily in rats, but it also occurs in prairie dogs, chipmunks, ground squirrels, and other small rodents. Since rat fleas also bite humans, the disease can be transmitted from the rodent to its human host by the arthropod vector.

22.12 How many different kinds of plague are there?

There are three different kinds of plague: bubonic plague, septicemic plague, and pneumonic plague. **Bubonic plague** is generally the first to occur. It is accompanied by swellings called buboes that occur in the lymph nodes of the armpits, groin, and elsewhere. Hemorrhaging in the lymph nodes turns the skin black, and gives the disease its alternative name, Black Death. The organisms then move from the lymphatic system to the general circulation, where they cause **septicemic plague**. Hemorrhaging occurs throughout the body and the disease is usually fatal at this point. **Pneumonic plague** can develop if the bacteria invade the lungs, where they rapidly cause consolidation and death. Antibiotic therapy can be used to halt the spread of disease if used early and aggressively.

22.13 Can plague be spread by any means beside the rat flea?

In the pneumonic stage, plague can be spread by airborne droplets. For this reason, patients must be isolated. Tetracycline, streptomycin, and other drugs for Gram-negative bacteria can be used to treat the disease. A vaccine is available for research workers and medical personnel.

22.14 What is the relationship between plague and tularemia?

Tularemia is a blood disease caused by an organism closely related to the plague bacillus. The etiologic agent of tularemia is *Francisella tularensis*, a Gram-negative rod that displays bipolar staining. Tularemia is transmitted by numerous arthropods living in the fur of rodents such as rabbits. It is transmitted by ticks and deer flies, which bite humans and transfer the bacilli. A small ulcer usually occurs at the site of the infection. The local lymph nodes soon become enlarged and filled with pus. Septicemia, pneumonia, and abscesses follow, but very often there are no distinctive symptoms that identify this disease.

22.15 Which other ways can tularemia be transmitted?

Tularemia is capable of being acquired by several means other than by arthropod bites. For example, the bacilli can be inhaled such as during a skinning of infected animals. In this case, they will cause an infection of the lungs. The bacilli can also be consumed in contaminated meat, such as rabbit meat. An intestinal infection will follow. Minor cuts, abrasions, or bruises may also be responsible for transmission. Several antibiotics such as gentamicin and streptomycin can be used for therapy.

22.16 What are the alternate names for brucellosis and which organism causes this disease?

Brucellosis is also known as **undulant fever**. A characteristic sign of the disease is periods of high and low fever in an undulating pattern. The disease is also called **Malta fever** due to an outbreak occurring on the island of Malta in the late 1800s. It is known as **Bang's disease** for the investigator who studied it in the early 1900s, and it is sometimes called **contagious abortion**, since the disease in animals is often accompanied by abortion of the animal fetus and sterility in the female animal. The disease is caused by *Brucella abortis*, a small Gram-negative rod occurring in cows. *Brucella mellitensis* occurs in sheep and goats, *B. suis* occurs in pigs, and *B. canis* is found in dogs.

22.17 How is brucellosis acquired by humans?

Brucella species enter human system by contaminated dairy products such as milk. Contaminated meat is also a method of transfer. In humans, the primary manifestations occur in the circulatory system with an accompanying septicemia. High and low cycles of fever are evident. Chronic aches and pains and nervousness often accompany the disease. Antibiotic therapy is usually successful, but the disease may linger for many months.

22.18 Which human disease is caused by *Borrelia recurrentis*?

Borrelia recurrentis is the cause of **epidemic relapsing fever**. This disease is transmitted by ticks and is accompanied by high fever, recovery, and numerous relapses. Each relapse is accompanied by the appearance of a slightly different strain of bacteria, which is able to escape the body's defenses.

22.19 What is the shape of the organism that causes relapsing fever, and how is the disease treated?

The organism *Borrelia recurrentis* is a large spirochete. The spirochete can be stained easily and observed under the microscope. Cases of relapsing fever may be treated with tetracycline and, since cases are transmitted by lice and ticks, the best preventative is control of these arthropods.

22.20 Which other contemporary disease is caused by a species of *Borrelia?*

Borrelia burgdorferi, shown in Figure 22-1, is the etiologic agent of **Lyme disease**, also known as **Lyme borreliosis**. This disease is transmitted by ticks belonging to the genus *Ixodes*. The ticks are deer ticks, although dog and horse ticks are also known to transmit the disease. At the current time, Lyme disease has been observed in almost all regions of the United States.

22.21 What are the typical symptoms of Lyme disease?

Lyme disease occurs in several stages, the first symptoms usually being a rash at the site of the tick bite. The rash is known as **erythema chronicum migrans (ECM)**. It is an expanding rash sometimes known as a bull's-eye rash. Weeks or months later, the patient experiences arthritis especially at the large joints such as at the elbow and knee. Flulike symptoms also occur. Later symptoms can include neurological defects such as facial paralysis, meningitis, and heart defects such as irregular heartbeats.

22.22 Which treatments are available for Lyme disease?

Several antibiotics can be used to limit the symptoms of Lyme disease and treat the disease successfully. Tetracycline drugs such as doxycycline and penicillin antibiotics such as amoxicillin are used. They are most effective when used early in the course of disease. The diagnostic tests for the disease include antibody tests as well as observation of characteristic signs and symptoms.

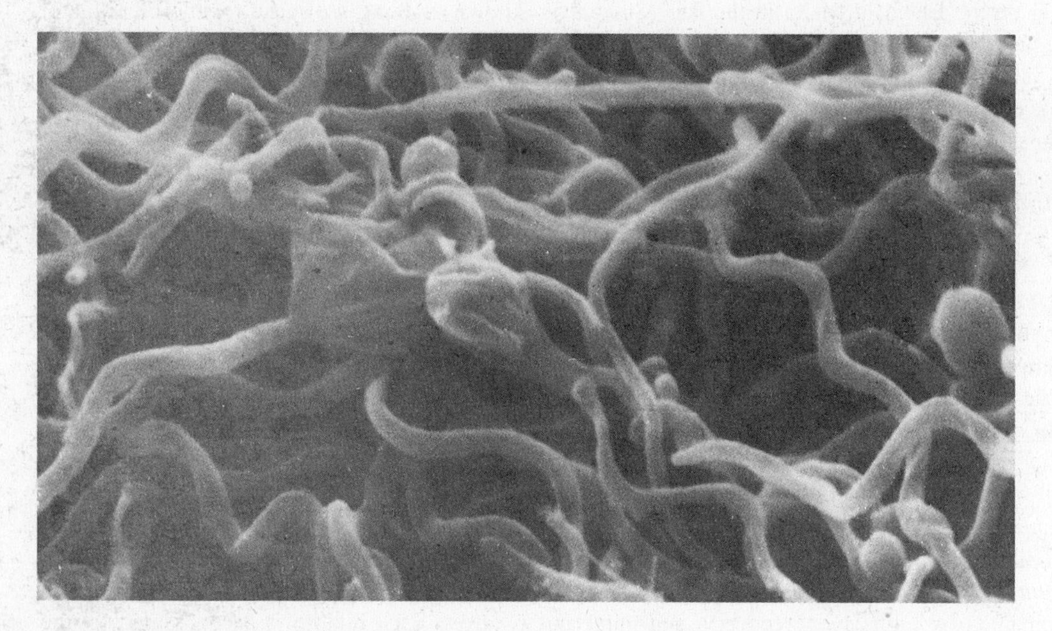

Fig. 22-1 The spirochete *Borrelia burgdorferi*, which causes Lyme disease.

22.23 Is there any vaccine available to protect against Lyme disease?

At the present time, dogs can be immunized against Lyme disease, but no vaccine is available for humans. Without a vaccine, control of Lyme disease depends in large part on avoiding ticks.

22.24 Which bacterium is the cause of gas gangrene and why is the organism unique?

Gas gangrene is caused by *Clostridium perfringens*. This organism is an anaerobic, sporeforming, Gram-positive rod. It grows within the tissue of the muscle and produces enormous amounts of gas. The gas swells the tissue, pulling the muscle cells away from their blood supply. The muscle cells soon begin to die from oxygen starvation.

22.25 Do the gas gangrene bacilli produce any toxins during time of disease?

The clostridia of gas gangrene produce numerous toxins that encourage the death of muscle cells and produce anaerobic conditions favorable for additional growth. The toxins also enter the bloodstream and spread to other tissues. Moreover, the clostridia produce enzymes that degrade the collagen and other proteins of the body. If a woman is pregnant, the clostridia can invade the wall of the uterus.

22.26 What treatments are available for cases of gas gangrene?

Cases of gas gangrene are difficult to treat. Treatment often involves removal of the necrotic tissue, a procedure called **debridement**. In many cases, amputation is necessary to prevent the spread of the gangrene. Patients can also be placed in a high-pressure chamber in order to force oxygen into the tissue and induce an aerobic environment.. The chamber is called a **hyperbaric chamber**. Penicillin is used to kill the bacteria and prompt treatment is necessary to prevent further damage.

RICKETTSIAL DISEASES

22.27 What is the general description of rickettsiae?

Rickettsiae are types of small bacteria, barely visible with the light microscope. The coverings of rickettsiae have the characteristics of Gram-negative bacteria, and the rickettsiae are cultivated, with exception, in living tissue cultures. These cultures include fertilized eggs, tissue cultures, and live animals. The rickettsiae are transmitted primarily by arthropod vectors such as ticks, fleas, and lice, displayed in Figure 22-2. Most infections caused by rickettsiae involve the endothelial cells of the cardiovascular system.

22.28 What are some general characteristics of diseases caused by rickettsiae?

Rickettsial diseases are generally accompanied by extremely high fevers and skin rashes called **maculopapular rashes**. These rashes appear initially as spots, and rickettsial diseases are often called spotted fevers. Most rickettsial diseases are treated with tetracycline or chloramphenicol. Prolonged treatment is generally advised.

22.29 What is the difference between epidemic typhus and endemic typhus?

Typhus is a rickettsial disease accompanied by high fever and a skin rash. **Epidemic typhus** is caused by *Rickettsia prowazekii*, which is transmitted by the human body louse. The disease occurs in epidemic proportions because lice are natural parasites of humans. **Endemic typhus**, by contrast, is caused by *Rickettsia typhi*, which is transmitted by rat fleas. The disease occurs primarily in rats, squirrels, and other rodents, and only occasionally in humans. The symptoms of endemic typhus (murine typhus) are generally less severe than those of epidemic typhus.

22.30 Which is the third form of typhus other then epidemic and endemic typhus?

A form of typhus, called **scrub typhus**, is transmitted by mites. The etiologic agent is *Rickettsia tsutsugamushi*. Fever and rash accompany the disease, which is found primarily in Asia and Australia.

Fig. 22-2 Three arthropods involved in the transmission of rickettsial disease. (*A*) A tick with its egg mass. (*B*) A louse. (*C*) A flea.

22.31 What are the characteristic signs of Rocky Mountain spotted fever?

Rocky Mountain spotted fever is named for the western United States region where it was prevalent in the early 1900s. The disease is caused by *Rickettsia rickettsii*, which is transmitted by the tick. The disease is accompanied by fever, and a maculopapular rash beginning on the ankles and wrists and proceeding toward the body trunk. Tick control is important to prevention of this disease.

22.32 Are any pox diseases due to rickettsiae?

A species of rickettsia called *Rickettsia akari* is known to form a disease called **rickettsialpox**. Transmitted by mites, rickettsialpox resembles chickenpox, with its mild skin lesions and fever. Mice are known to carry the mites that transmit the disease.

22.33 Which rickettsia is responsible for trench fever, and what are the symptoms of this disease?

Trench fever takes its name from its prevalence during World War I trenches. The disease is caused by the rickettsia *Rochalimaea quintana*, one of the few rickettsiae cultivated outside living tissue. The disease is transmitted by body lice and is accompanied by several days of high fever with severe leg pains (shinbone fever). Lice control is an important preventative.

22.34 Which disease is caused by a member of the genus *Ehrlichia*?

The rickettsia *Ehrlichia canis* causes a human disease called **ehrlichiosis**. Patients experience headache, fever, but no rash. The disease is transmitted by ticks and occurs in dogs. A closely related condition called **human granulocytic ehrlichiosis (HGE)** has also been recognized. The bacteria infect human granulocytes; the condition is due to *Ehrlichia chaffeensis*.

VIRAL DISEASES

22.35 How is yellow fever transmitted, and which organism causes this disease?

Yellow fever is transmitted by the mosquito *Aedes aegypti*. It is caused by an RNA virus belonging to the flavivirus group. The disease occurs in Central and South America as well as regions of Africa. Monkeys are a reservoir for the virus.

22.36 What are the typical symptoms associated with yellow fever?

Yellow fever is accompanied by viral infection of the bloodstream, a condition called **viremia**. The disease affects the liver where it causes jaundice, which is the yellowing of the skin due to bile pigments present in the bloodstream and deposited in the skin. The heart, kidney, spleen, and other organs are also damaged, and fatality rates are generally high.

22.37 Is there any vaccine available for yellow fever?

Although there is no treatment for yellow fever, two vaccines are available. One is administered in the skin, the other subcutaneously. Both establish effective levels of immunity against the disease.

22.38 What are some similarities between yellow fever and dengue fever?

Both yellow fever and **dengue fever** are caused by RNA viruses and both are blood disorders transmitted by the mosquito *Aedes aegypti*. Both diseases occur in tropical regions.

22.39 Which major differences distinguish yellow fever and dengue fever?

Dengue fever is accompanied by severe joint and bone pain and is often called **breakbone fever**. Also, dengue fever has a lower fatality rate than yellow fever, and there is no jaundice associated with the disease. Moreover, dengue fever often returns as a hemorrhagic disease, a pattern not seen with yellow fever.

22.40 What are the characteristic signs associated with infectious mononucleosis and which virus causes this disease?

Infectious mononucleosis is accompanied by mild fever, sore throat, swollen lymph nodes, and enlarged spleen. Most microbiologists agree that the disease is caused by **Epstein-Barr virus.** This virus is also known to cause Burkitt's lymphoma, a condition accompanied by cancerous tumors of the connective tissues of the jaw. The virus is an icosahedral DNA herpesvirus and is known to remain latent in the body tissues for long periods of time.

22.41 Which cells are the particular target for Epstein-Barr virus?

Epstein-Barr viruses are usually transmitted to the body by saliva transfer and they make their way to the bloodstream where they interact with the lymphocytes. The lymphocytes become granulated, vacuolated, and distorted, and they become characteristic **Downy cells**. Detection of an increased number of these cells in the bloodstream is a sign of infectious mononucleosis.

22.42 What treatments are available for infectious mononucleosis and which complications are associated with the disease?

There is little treatment for infectious mononucleosis other than bed rest. Antibiotics are used to prevent secondary bacterial infections, and physicians watch for complications such as an enlarged spleen. The presence of heterophile antibodies in the blood is a diagnostic sign for infectious mononucleosis.

22.43 Is Epstein-Barr virus also related to chronic fatigue syndrome?

Many physicians believe that Epstein-Barr virus is responsible for **chronic fatigue syndrome**, a condition accompanied by fever and persistent fatigue, together with other symptoms reminiscent of infectious mononucleosis. However, the relationship between the cause and effect have not been definitely established, and recent reports point to a different herpesvirus as the possible cause. Epstein-Barr virus is presumed to be the cause of **Epstein-Barr virus disease** which is apparently a very mild form of mononucleosis.

22.44 Which viruses are responsible for viral fevers occurring in humans?

A wide variety of viruses can cause viral fevers in humans. Among the viruses are a group of **filoviruses**. These viruses are filamentous, with branching, fish hook, or "U" shapes. They are RNA viruses and include Ebola virus and Marburg virus. The viruses cause severe skin and organ hemorrhages as well as extremely high fevers. A group of **bunyaviruses** also cause viral fevers. The viruses usually are icosahedral, and they include the LaCrosse virus, the California encephalitis virus, the Rift Valley fever virus, and the Haantan virus.

22.45 Which diseases do Coxsackie viruses cause?

Coxsackie viruses usually enter the body through water or food. They may cause gastroenteritis before passing into the bloodstream and moving to other organs. Coxsackie viruses cause an infection of the heart muscle called myocarditis, as well as a type of meningitis called aseptic meningitis (Chapter 21). Infections of the throat are also due to this virus, especially in children.

22.46 Why is the fifth disease so-named, and which virus causes this disease?

The fifth disease is the **fifth disease** in a series of diseases accompanied by body rashes (others include measles and scarlet fever). The fifth disease is also known as **erythema infectiosum**. It is caused by parvovirus strain B19. Therefore, it is also called **B19 disease**.

22.47 What are some characteristic signs of fifth disease?

In fifth disease, the viruses move through the bloodstream causing intense viremia. They cause the blood to leak into the skin, and infected children display a bright red rash on the cheeks of the face. The disease is therefore called "slapped-cheek disease." A low fever may be present, but the disease is not considered dangerous. Most cases are transmitted by respiratory droplets.

22.48 When did the AIDS epidemic begin?

The AIDS epidemic began in 1981, when physicians in Los Angeles and other cities noted an unusually large number of opportunistic microbial infections. Destruction of T lymphocytes of the immune system cells were associated with these infections, and it soon became obvious that an epidemic of disease was in progress. By 1984 the responsible virus had been identified, and in 1986, it was given the name **human immunodeficiency virus (HIV).**

22.49 How is the human immunodeficiency virus (HIV) transmitted?

HIV is a very fragile virus, and for this reason, it does not survive long periods of exposure outside the body. Most cases are transmitted directly from person to person via transfusions of blood or semen. The disease is associated with intravenous drug users who use contaminated needles and with individuals who perform anal intercourse, since bleeding is often associated with this practice. Heterosexual intercourse can also be a mode of transmission, especially if lesions occur on the reproductive organs.

22.50 How does penetration of T lymphocytes occur in infected individuals and what happens once HIV gets inside the cell?

In infected individuals, HIV infects T lymphocytes by combining its spike glycoproteins with the **CD4 receptor sites** of T lymphocytes. The nucleocapsid enters the cytoplasm of the T lymphocyte, and the viral enzyme **reverse transcriptase** synthesizes a DNA molecules using the RNA of HIV as a template (for this reason, the virus is called a **retrovirus**). The DNA molecule migrates to the cell nucleus and becomes part of a chromosome in the T lymphocyte nucleus. Figure 22-3 shows the process.

22.51 What happens once the DNA from the HIV particle enters the cell nucleus?

The DNA molecule, known as a **provirus**, assumes a relationship with the DNA of the T lymphocyte, and the provirus enters the state of **lysogeny**. From this point in the nucleus, the provirus encodes new HIV particles, which acquire their envelope by budding through the membrane of the T lymphocyte. The human body attempts to keep up with the mass of new viral particles, but eventually the newly emerging strains of HIV overwhelm the body defenses and the T lymphocyte count begins to drop. Normally, it is approximately 800 T lymphocytes per cubic millimeter of blood, but as the disease progresses, the count drops into the low hundreds and tens. This drop may occur as soon as six months after infection or as long as 12 years or longer after infection.

22.52 How is HIV infection defined and what are its symptoms?

While the T lymphocytes are infected, and so long as the T-lymphocyte level remains close to normal, the patient is said to have **HIV infection**. The patient occasionally will suffer swollen lymph nodes, mild prolonged fever, diarrhea, malaise, or other nonspecific symptoms. **AIDS** is the end stage of the disease. It is signaled by the appearance of **opportunistic infections** such as candidiasis, an excessively low T-lymphocyte count, or by a wasting syndrome or deterioration of the mental faculties.

22.53 How are opportunistic infections related to AIDS?

When a person has AIDS, an opportunistic infection is usually present. This infection may be pneumonia due to *Pneumocystis carinii*, diarrhea due to *Cryptosporidium*, encephalitis due to *Toxoplasma gondii*, severe eye infection and blindness due to cytomegalovirus, candidiasis of the mucous membranes and esophagus due to *Candida albicans*, meningitis due to *Cryptococcus neoformans*, or herpes simplex, tuberculosis, or cancer of the skin known as Kaposi's sarcoma. These are treatable with various drugs, but the AIDS patient is constantly fighting one or the other.

22.54 What was the extent of the AIDS epidemic in the United States as of June 1996?

As of June 1996, close to 600,000 cases of AIDS had been recognized in the United States and approximately 400,000 patients had died.

22.55 Which drugs are available to treat HIV infection and AIDS?

As of 1996, two types of drugs were available to inhibit the multiplication of HIV. One group is the **chain terminators**, such as **azidothymidine** (AZT), **dideoxycytidine** (ddC), and **dideoxyinosine** (ddI). These drugs interfere with the synthesis of the DNA molecule using the viral RNA as a template. They effectively interfere with the activity of the reverse transcriptase. The second group consists of **protease inhibitors**. They prevent the synthesis of the viral capsid at the conclusion of the replication cycle by interfering with one of the last steps in production of the capsid protein.

22.56 Which diagnostic tests are available to detect HIV infection?

Diagnostic tests for AIDS are usually antibody-based tests that determine the presence of antibodies against HIV. It takes about six weeks for the body to produce sufficient antibodies for a positive test, so a test performed before this period has elapsed can lead to a negative test (a false-negative). Other tests called antigen-based tests are designed to detect the virus itself. These tests use gene probes that unite with and signal the presence of the viral DNA (the provirus) if it is present in the T lymphocytes.

22.57 What prospects are there for an AIDS vaccine?

Thus far, **vaccines** are not available against HIV. There is question, for example, whether whole viruses or viral fragments are preferred for the vaccine. Two glycoproteins called **gp120** and **gp41** from the envelope spikes are being investigated as possible vaccines. Vaccine development is hampered however, since animal models are not available for testing, and it is difficult to find volunteers who would become antibody-positive and could suffer discrimination as a result of antibody presence. Nevertheless, candidate vaccines have been prepared with gp120 and gp41 molecules as well as with simian immunodeficiency virus (SIV), which infects primates, and viruses mutated so as to have no envelopes. Many candidate vaccines are now in the testing stage, and it is hoped that one will soon be available for the general population.

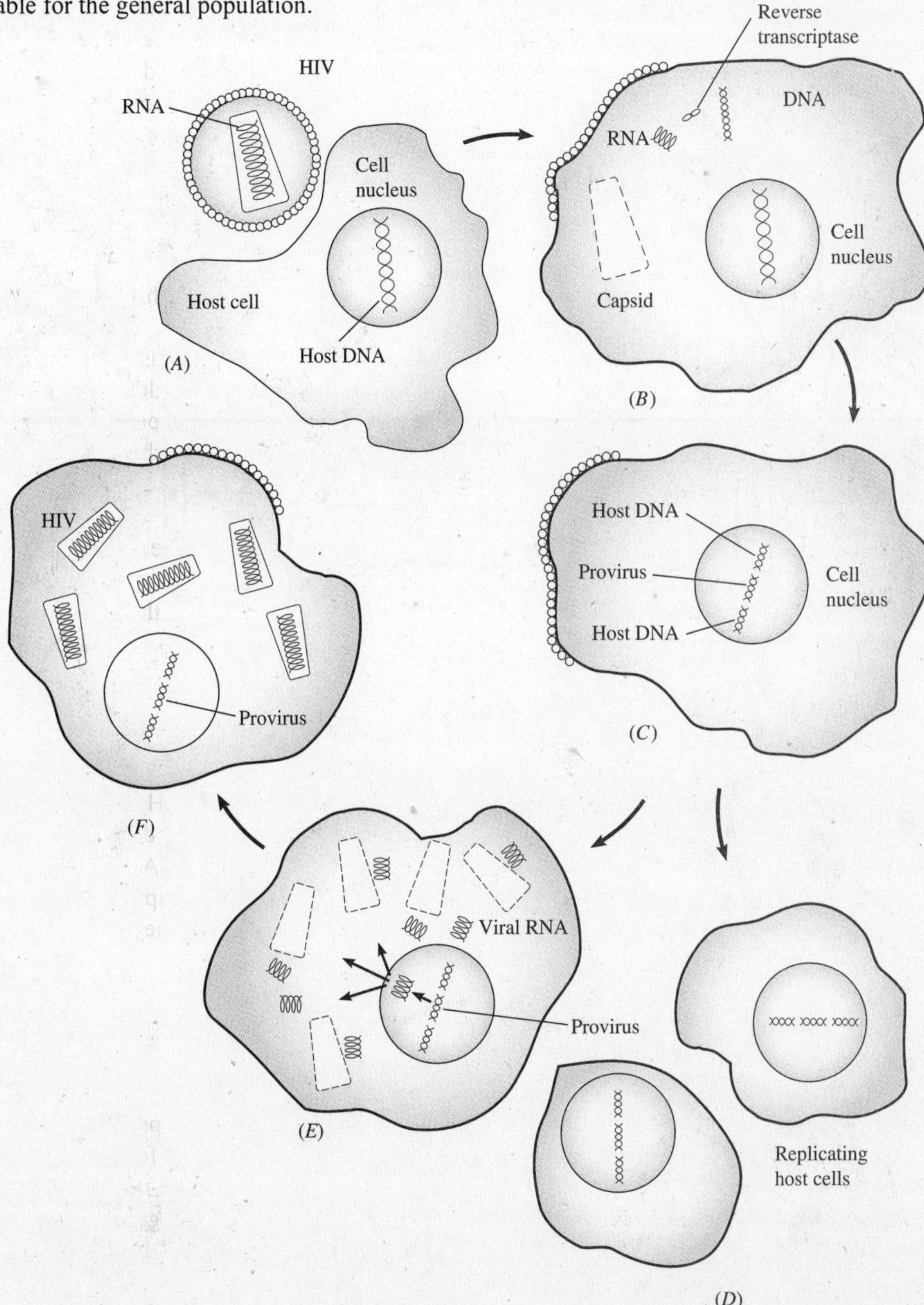

Figure 22-3 The replication cycle of the human immunodeficiency virus (HIV). (*A*) The viral envelope fuses with the host cell membrane, and the nucleocapsid enters the cytoplasm. (*B*) The capsid dissolves and reverse transcriptase uses the RNA as a model to synthesize DNA, which (*C*) incorporates into the cell's nucleus as a provirus. (*D*) The provirus replicates when host cell replicate. (*E*) It also encodes new viruses. (*F*) The latter bud through the membrane of the host cell and go on to infect new cells.

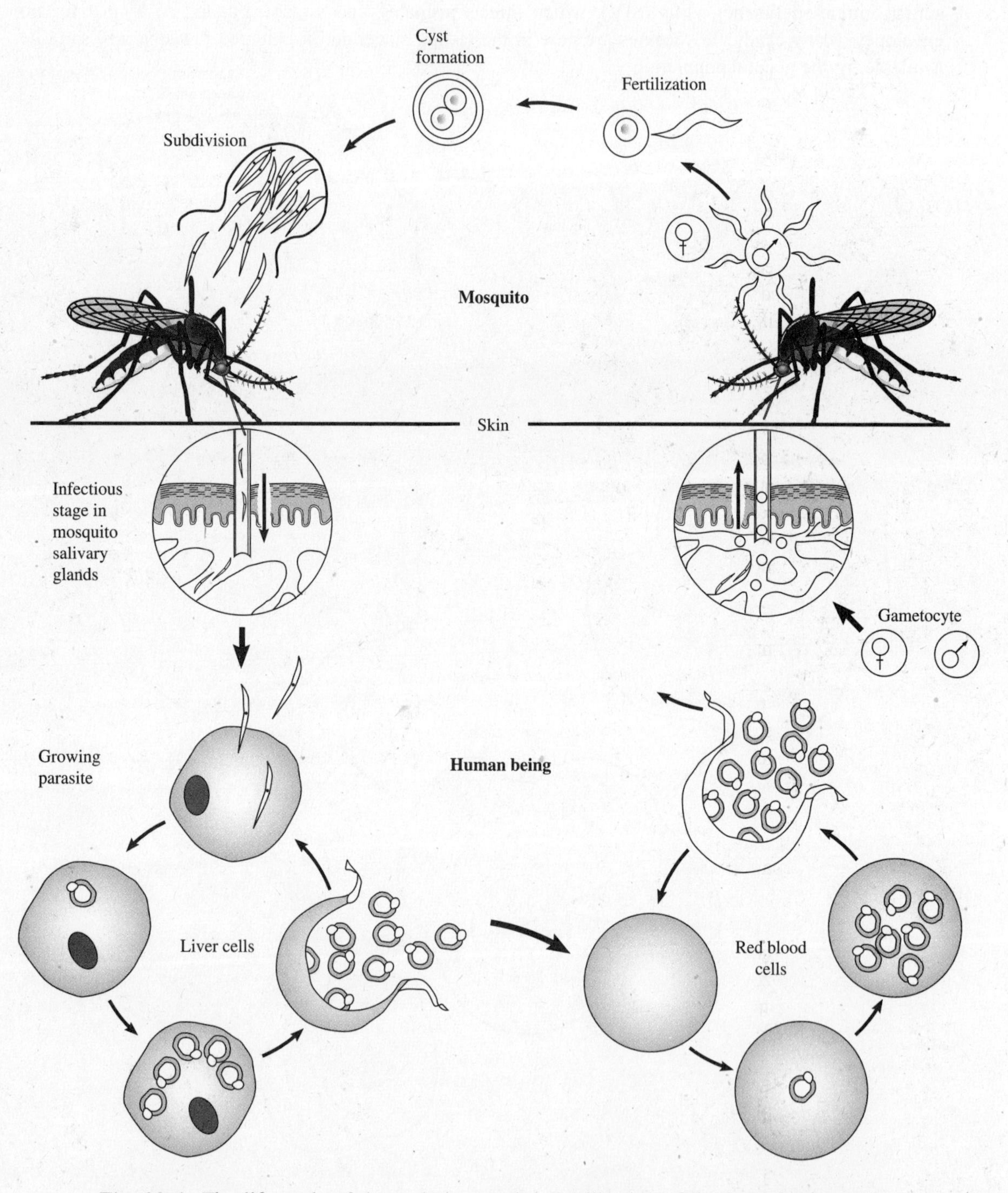

Fig. 22-4 The life cycle of the malaria parasite. Fertilization of the reproductive cells takes place in the mosquito, then the mosquito injects the parasites into the blood of a human victim. The parasite undergoes transformations in the liver cells, then emerges to infect the red blood cells where the reproductive cells (gametocytes) are formed. The massive destruction of red blood cells leads to death from the disease.

OTHER DISEASES

22.58 Which is the most widespread infectious disease of the blood due to a protozoan?

Probably the world's most important public health problem is **malaria**. Up to 300 million cases of this disease occur annually. The disease is caused by several species of the protozoan *Plasmodium* including *P. vivax, P. malariae, P. ovale,* and *P. falciparum.* The life cycle of the parasite is pictured in Figure 22-4.

22.59 Which cells are affected by malaria?

Malaria is a disease of the red blood cells. It is transmitted by mosquitoes of the genus *Anopheles.* After being injected into the bloodstream, the parasites penetrate the red blood cells and undergo a series of phases in their life cycles. In doing so, they destroy the red blood cells.

22.60 How are the symptoms of malaria explained?

The symptoms of malaria relate to the destruction of red blood cells. The patient suffers severe anemia, and alternating chills and spiking fevers. During the malaria attack, the parasites invade new red blood cells and repeat the cycle. Substantial obstructions of the blood vessels occur due to the damaged red blood cells. The disease is sometimes called **blackwater fever** due to the presence of hemoglobin pigments in the urine.

22.61 Which treatments are available for malaria?

The traditional treatment for malaria is quinine and its synthetic derivatives chloroquine and primaquine. In recent years, resistant strains of *Plasmodium* have emerged and the use of these drugs has been threatened. Mosquito control is a major method for preventing epidemics and vaccine development is ongoing.

22.62 Which protozoan is responsible for cases of toxoplasmosis, and how is this disease transmitted?

The organism that causes **toxoplasmosis** is the protozoan *Toxoplasma gondii.* Like the malaria parasite, this protozoan belongs to the class Sporozoa. It is present in domestic housecats and is usually transmitted by contact such as cleaning the litter box. The organism can also be transmitted in raw or poorly cooked beef.

22.63 In which two situations is toxoplasmosis a concern?

Toxoplasmosis is dangerous in the fetus. If a pregnant woman passes the organism to her child, the fetus may suffer congenital defects such as blindness, mental retardation, and movement disorders. Toxoplasmosis is also an opportunistic disease in persons with HIV infection. Here it occurs as a brain inflammation accompanied by epilepticlike disorders. Drugs such as pyrimethamine and sulfa derivatives are often used to treat the disease.

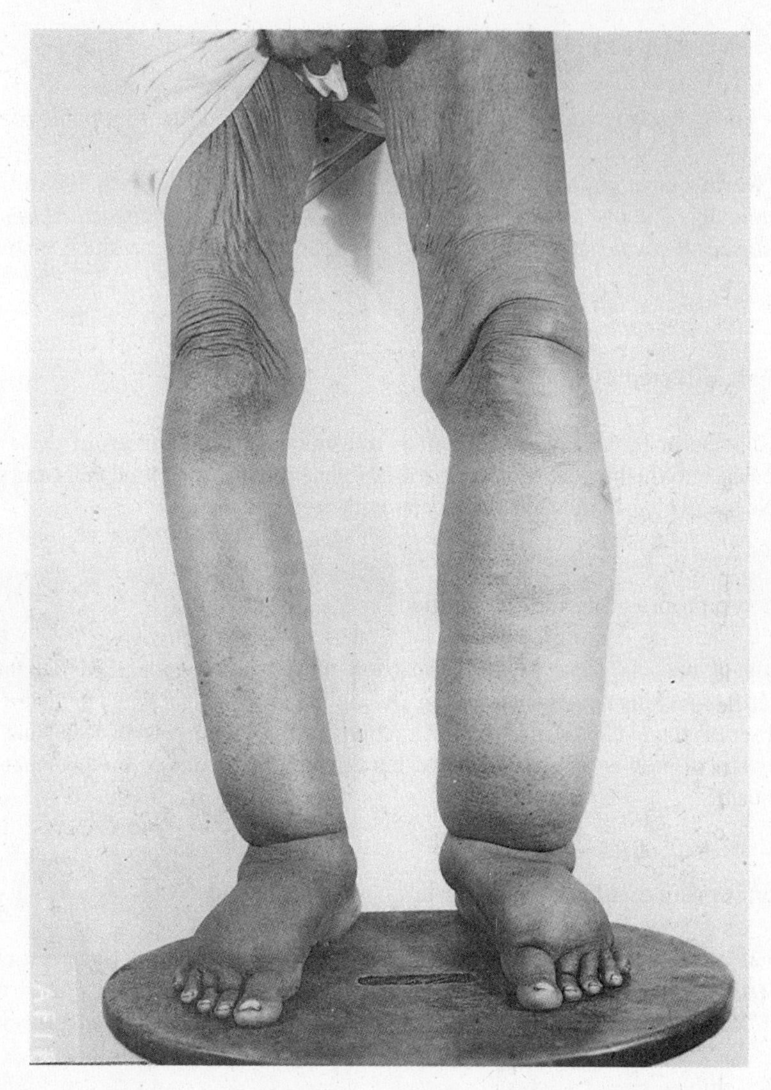

Fig. 22-5 Symptoms of elephantiasis, a disease of the lymph vessels due to a roundworm.

22.64 Which protozoan is the cause of babesiosis?

Babesiosis is caused by a protozoan known as *Babesia microti*. The protozoan is transmitted by infected ticks, and it causes a disease in the red blood cells. High fever and anemia accompany the disease, and the symptoms are similar to those of malaria but are much milder.

22.65 Which multicellular parasite causes severe blood and lymph vessel disease in humans?

The roundworm *Wuchereria bancrofti* is transmitted to the lymph vessels by several mosquitoes, including species of *Aedes* and *Anopheles*. The worm reproduces in the vessels and causes severe inflammation of the lymph ducts leading to edema and gross deformities called **elephantiasis**. Figure 22-5 shows how a patient with the disease may display its symptoms. Often, this condition occurs in the limbs of the body.

Review Questions

Multiple Choice. For each of the following, select the word or phrase that best completes the statement.

1. Endocarditis is an infection of the endocardium of the heart valves due to (*a*) *Yersinia pestis* (*b*) *Brucella abortis* (*c*) *Streptococcus pyogenes* (*d*) *Francisella tularensis.*

2. Most cases of anthrax develop after the etiologic agent enters the bloodstream (*a*) in food (*b*) through a skin wound (*c*) after an animal bite (*d*) after an arthropod bite.

3. The organism that causes plague is a (*a*) Gram-positive cocci in chains (*b*) Gram-positive rod (*c*) Gram-negative rod displaying bipolar staining (*d*) a protozoa.

4. The two means for spreading plague are (*a*) contaminated food and water (*b*) airborne droplets and a mosquito bite (*c*) rat fleas and airborne droplets (*d*) sexual contact and animal bites.

5. Periods of high and low fever in an undulating pattern is a characteristic sign of (*a*) plague (*b*) Lyme disease (*c*) mononucleosis (*d*) brucellosis.

6. All the following are true of epidemic relapsing fever except (*a*) it is caused by a large spirochete (*b*) it is accompanied by high fever, recovery, and numerous relapses (*c*) it may be treated with tetracycline (*d*) it is transmitted by skin contact.

7. The typical symptoms of Lyme disease include (*a*) erythema chronicum migrans (*b*) rose spots (*c*) severe meningitis (*d*) lesions of the respiratory tract.

8. All of the following are diseases caused by rickettsiae except (*a*) scrub typhus (*b*) Rocky Mountain spotted fever (*c*) trench fever (*d*) tularemia

9. Cases of human ehrlichiosis are transmitted by (*a*) rat fleas (*b*) ticks (*c*) mites (*d*) lice.

10. The etiologic agent of gas gangrene has all the following characteristics except (*a*) it is anaerobic (*b*) it is Gram-positive (*c*) it is a coccus (*d*) it is sporeforming.

11. Which of the following applies to the disease yellow fever? (*a*) it is caused by a virus (*b*) it is transmitted by the body louse (*c*) it occurs widely in North America (*d*) no vaccine is available for the disease.

12. The disease caused by *Rickettsia akari* resembles (*a*) plague (*b*) childbed fever (*c*) anthrax (*d*) chickenpox.

13. The rash accompanying many of the rickettsial diseases is described as (*a*) bulls-eye (*b*) maculopapular (*c*) petechial (*d*) rose spots.

14. Treatment in a hyperbaric chamber and removal of necrotic tissue by debridement are often necessary for cases of (*a*) Lyme disease (*b*) septic shock (*c*) gas gangrene (*d*) bubonic plague.

15. Most cases of brucellosis are acquired by humans after (*a*) a bite by a mosquito (*b*) consumption of contaminated dairy products (*c*) a bite by a tick (*d*) contact with an infected patient.

16. Cases of Lyme disease and relapsing fever are both caused by (*a*) spirochetes (*b*) viruses (*c*) protozoa (*d*) rickettsia.

17. All the following characteristics apply to Lyme disease except (*a*) it is caused by a species of *Borrelia* (*b*) treatment is administered with tetracycline drugs (*c*) arthritis often occurs in the large joints (*d*) the disease is transmitted by a mosquito.

18. Dengue fever is similar to yellow fever in that (*a*) both are transmitted by ticks (*b*) both are caused by RNA viruses (*c*) both affect the lungs (*d*) both occur in Arctic regions.

19. The Epstein-Barr virus is associated with the disease (*a*) fifth disease (*b*) yellow fever (*c*) infectious mononucleosis (*d*) brucellosis.

20. Coxsackie viruses are associated with all the following human conditions except (*a*) aseptic meningitis (*b*) myocarditis (*c*) gastroenteritis (*d*) pneumonia.

21. Parvovirus strain B19 is believed to be the cause of (*a*) chronic fatigue syndrome (*b*) Dengue fever (*c*) trench fever (*d*) fifth disease.

22. Downy cells are characteristic lymphocytes associated with the (*a*) LaCrosse virus (*b*) Ebola virus (*c*) Epstein-Barr virus (*d*) Marburg virus.

23. Anthrax spores are extremely resistant to environmental changes and therefore may be used in (*a*) biological warfare (*b*) viral research (*c*) biochemical mutations (*d*) industrial processes.

24. One of the characteristic signs of fifth disease is (*a*) intestinal lesions (*b*) a bright red rash on the cheeks of the face (*c*) swelling of the lymph nodes (*d*) a lesion at the site of entry.

25. Contaminated rabbit meat may be a possible mode of transmission for (*a*) malaria (*b*) tularemia (*c*) infectious mononucleosis (*d*) dengue fever.

26. Malaria is a disease of the human (*a*) astrocytes (*b*) connective tissues (*c*) red blood cells (*d*) histiocytes.

27. A woman who is pregnant should avoid a possible exposure to the organism of (*a*) toxoplasmosis (*b*) pneumonic plague (*c*) relapsing fever (*d*) rickettsialpox.

28. Quinine and its synthetic derivative chloroquine and primaquine are used most effectively for the treatment of (*a*) toxoplasmosis (*b*) fifth disease (*c*) malaria (*d*) Lyme disease.

29. The etiologic agents of toxoplasmosis and malaria both belong to the (*a*) Gram-negative rods (*b*) class Sporozoa (*c*) filovirus group (*d*) spirochetes.

30. Severe disease of the lymph vessels may be caused by the round worm (*a*) *Aedes aegypti* (*b*) *Yersinia pestis* (*c*) *Wuchereria bancrofti* (*d*) *Anopheles*.

True/False. For each of the following statements, mark the letter "T" next to the statement if the statement is true. If the statement is false, change the underlined word to make the statement true.

___ **1.** Cases of endocarditis are related to *Streptococcus pyogenes* which causes infection of the heart muscle.

___ **2.** *Bacillus anthracis* is a Gram-positive sporeforming coccus that causes a serious blood disease known as anthrax.

___ **3.** Plague is caused by *Yersinia pestis*, a rod that displays acid-fast staining and is transmitted by the rat flea.

___ **4.** The three forms of plague are bubonic plague occurring in the lymph nodes, septicemic plague occurring in the general circulation, and pneumonic plague which occurs in the lungs.

___ **5.** Because of the periods of high and low fever, brucellosis is also known as relapsing fever.

___ **6.** *Borrelia burgdorferi* , the agent of Lyme disease, is a bacterial rod that is difficult to see under the light microscope.

___ **7.** The rash occurring with most rickettsial diseases initially appear as spots and is known as a hemorrhagic rash.

___ **8.** The arthropod responsible for transmission of cases of Rocky Mountain spotted fever is the mite.

___ **9.** Yellow fever is due to a bacterium and is transmitted by the mosquito *Aedes aegypti*.

__ **10.** Because of the sever pains occurring in the joints and bones dengue fever is also known as break bone fever.

__ **11.** The Epstein-Barr virus is believed to be the cause of infectious mononucleosis and is usually transmitted by saliva transfer.

__ **12.** The Ebola and Marburg viruses are RNA viruses called filoviruses that are capable of causing severe skin and organ hemorrhages as well as extremely high fever.

__ **13.** A characteristic sign of malaria is the bright red rash occurring on the cheeks of the face.

__ **14.** Toxoplasmosis is due to the bacterium *Toxoplasma gondii*, which is transmitted from the domestic housecat.

__ **15.** Infected mosquitoes are believed to be the mode of transmission for the protozoan that causes babesiosis.

Matching. For each of the following diseases in Column A, match the characteristic from column B that applies most closely.

Column A	Column B
_____ 1. Relapsing fever	(*a*) Due to *Rickettsia prowazeckii*
_____ 2 . Plague	(*b*) Also called "woolsorters" disease
_____ 3. Yellow fever	(*c*) Transmitted by mites
_____ 4. Childbed fever	(*d*) Causes contagious abortion in animals
_____ 5. Epidemic typhus	(*e*) Accompanied by bull's-eye rash
_____ 6. Brucellosis	(*f*) Disease of the red blood cells
_____ 7. Fifth disease	(*g*) Transmitted by rabbit meat
_____ 8. Endocarditis	(*h*) Related to domestic housecat
_____ 9. Malaria	(*i*) Due to beta-hemolytic streptococci

____ 10. Lyme disease
____ 11. Toxoplasmosis
____ 12. Anthrax
____ 13. Scrub typhus
____ 14. Tularemia
____ 15. Gas gangrene

(*j*) Caused by an anaerobic sporeforming rod
(*k*) Transmitted by rat flea
(*l*) Due to *Borrelia recurrentis*
(*m*) Known as erythema infectiosum
(*n*) Streptococcal disease of heart valves
(*o*) Caused by an RNA virus

Chapter 23

Microbial Diseases of the Urogenital System

OBJECTIVES

The microbial diseases of the urogenital tract include the sexually transmitted diseases (STDs), which affect millions of Americans annually. The objectives of this chapter will be to:

1. Specify the microorganisms associated with infections of the urinary and reproductive tracts.

2. Outline the typical symptoms and available treatments for toxic shock syndrome and trichomoniasis.

3. Explain the manifestations and other important characteristics of gonorrhea, with reference to the complications associated with this disease.

4. Recognize the general course of infection associated with the three stages of syphilis.

5. Summarize the state of knowledge of chlamydia, indicating why this disease bears a striking resemblance to gonorrhea.

6. Recognize the causes and characteristics signs of lesser known STDs such as ureaplasmal urethritis, lymphogranuloma venereum, and granuloma inguinale.

7. Outline some of the microbiology associated with the various forms of herpes simplex, with reference to transmission and treatment.

THEORY AND PROBLEMS

23.1 Which factors predispose an individual to infections of the urinary tract?

Several factors may dispose one to urinary tract infections. These factors include diabetes mellitus, obstructions to the flow of urine (such as kidney stones and tumors), and the proximity of the anus to the female urethra. Because of the later, the urinary tract may become infected with intestinal bacteria such as *Escherichia coli*, species of *Proteus*, staphylococci, and species of *Pseudomonas. Candida albicans*, the opportunistic fungus, may also be transmitted from the intestinal tract.

23.2 What is the difference between urethritis, cystitis, and ureteritis?

Urinary tract infections may occur in the urethra, where they are called **urethritis**; if they occur in the bladder, they are referred to as **cystitis**; and if they develop in the ureters, they are called **ureteritis**.

Infection of the kidney is generally called **pyelonephritis**. In males, inflammation of the prostate gland, called **prostatitis**, often accompanies urinary tract infection.

URINARY TRACT DISEASES

23.3 Which bacterium is the cause of leptospirosis?

Leptospirosis is due to the spirochete *Leptospira interrogans*. The organism normally causes disease in domestic and wild animals, but it may be acquired by humans during contact with contaminated urine in soil or water. This is because infected animals shed the spirochete in their urine. Skin abrasions are commonly a method for transfer into the human body.

23.4 What are some characteristic symptoms of leptospirosis?

Leptospirosis is accompanied by muscle aches, fever, and chills that appear abruptly. There are no other specific symptoms that delineate the disease, although in some cases, the fever disappears then reappears. Some individuals experience liver infection known as **Weil's disease**. Disease of the kidney may result in death.

23.5 Which treatments are available for leptospirosis?

The agent of leptospirosis is susceptible to numerous antibiotics, including penicillin. However, antibiotic therapy should begin early in disease, which is why diagnosis must be performed promptly. Antibiotics can also be used in infected pets, and preventive vaccines are available for these animals.

23.6 Can *Streptococcus* species cause urinary tract infection?

It often happens that blood infection with beta-hemolytic streptococci, *Streptococcus pyogenes,* leads to inflammation in the kidney. This inflammation is known as **glomerulonephritis** (also called **Bright's disease**). The disease reflects an immune complex reaction resulting from activity of the immune system and a type III hypersensitivity. Fever and high blood pressure accompany the disease.

GENITAL TRACT DISEASES

23.7 Which bacterium is responsible for cases of bacterial vaginitis?

Bacterial vaginitis is an infection of the vagina, usually caused by organisms that multiply when the normal flora has been disturbed. Such things as excessive antibiotic treatment can destroy the normally present acid-producing lactobacilli. Other bacteria then multiply in the acid-free environment. One of the common causes of vaginitis is *Gardnerella vaginalis,* a Gram-negative rod. Anaerobic bacteria such as species of *Bacteroides* are sometimes involved as well.

23.8 What conditions accompany the development of toxic shock syndrome in the reproductive tract?

Toxic shock syndrome normally occurs from *Staphylococcus aureus* present in the vaginal flora. When these bacteria multiply profusely, such as in high-absorbency tampons, they secrete exotoxins, which are absorbed into the bloodstream. Here the toxins induce fever, a red rash on the trunk, and low blood pressure. The low blood pressure brings on shock. Reduced use of high-absorbency tampons or other devices used in the vagina for long periods of time reduces the possibility of TSS.

23.9 Which protozoan is responsible for most cases of sexually transmitted disease?

Trichomonas vaginalis is responsible for most cases of **trichomoniasis**, the primary sexually transmitted disease due to a protozoan. Other species that can cause the disease include *T. hominis* and *T. tenax*. All three are flagellated protozoa with undulating membranes.

23.10 What are the typical symptoms of trichomoniasis when they occur in humans?

The symptoms of trichomoniasis include internal discomfort, intense itching, and a whitish discharge from the reproductive tract. Also, there is a foul smell present. The symptoms occur more prominently in females than in males. Examination of the discharge reveals the trichomonads.

23.11 Which treatments are available for trichomoniasis?

Restoring the normal microbial population of the vagina is essential for resolving cases of trichomoniasis. The drug metronidazole (Flagyl) is also used, but it is contraindicated during pregnancy because it can induce abortion.

23.12 Which organism is responsible for cases of gonorrhea?

Gonorrhea is one of the most commonly reported sexually transmitted diseases in the United States, with over 600,000 cases reported annually. Although the incidence has improved over the past few years, as Figure 23-1 displays, the numbers remain very high. The disease is caused by *Neisseria gonorrhoeae,* a small Gram-negative diplococcus. The diplococcus resembles a pair of beans lying face-to-face. Most strains of the organism are considered fragile, but some strains can resist drying and exposure to the environment outside the body. The organisms have pili, which permit adherence to cells lining the urinary tract.

23.13 What are the characteristic signs of gonorrhea?

Gonorrhea is a disease of the urethra, the tube leading from the urinary bladder to the exterior. In males, the urethra is a long tube extending through the penis; in females the urethra is very short. Gonococci (the colloquial expression for *N. gonorrhoeae*) infect the urethral cells, but in females the infection is usually asymptomatic. In males, a whitish discharge is emitted from the urogenital tract, and there is internal pain and discomfort. Transmission occurs during sexual contact.

23.14 Are there other manifestations of gonorrhea besides the infection in the urethra?

Gonorrhea occurs in numerous parts of the body. For example, gonorrhea may occur in the pharynx when transmission occurs during oral sex. Infection of the rectum may also occur, with bleeding and constipation. In the female, infection occurs in the cervix, and many females develop **pelvic inflammatory disease** (PID) when infection spreads to other organs of the pelvic cavity. Sterility may develop if PID results in scarring of the Fallopian tubes. Joint infection accompanied by arthritis may also occur.

23.15 Is gonorrhea a danger in newborns?

During the birth process, newborns can acquire gonococci during passage through the birth canal. To preclude that possibility, the eyes of all newborns are treated with antibiotics or silver nitrate immediately after birth. Eye infection of the newborn is called **gonococcal ophthalmia**.

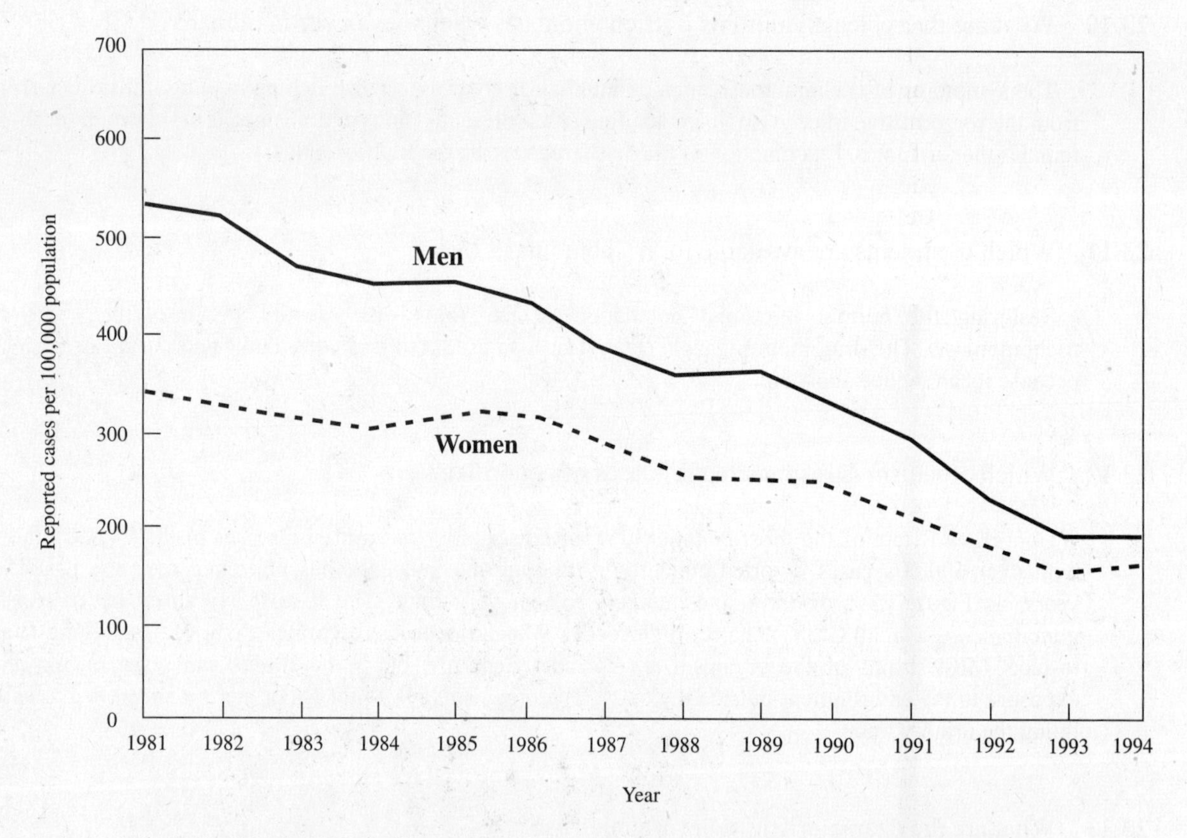

Fig. 23-1 The incidence of gonorrhea in the United States between 1981 and 1994. Early detection and treatment have accounted for the reduction in incidence, especially in men. Relative to other diseases, the incidence remains high.

23.16 What methods are available for the diagnosis and treatment of gonorrhea?

For the diagnosis of gonorrhea, laboratory cultures are secured, and gonococci are identified in the cultures. Sophisticated enzyme tests are available to confirm laboratory observations of gonococci. For the treatment of gonorrhea, penicillin has traditionally been recommended. In recent years, however, penicillin-resistant strains of *N. gonorrhoeae* have emerged, and alternative therapies have been sought. These include spectinomycin, tetracycline, and fluoroquinolone antibiotics.

23.17 Which bacterium is the etiologic agent of syphilis?

The etiologic agent of **syphilis** is the spirochete *Treponema pallidum.* The organism is motile and can be observed moving about under the darkfield microscope. It is extremely difficult to cultivate in the laboratory and direct observation is usually required for diagnosis. The organism is transmitted among humans by sexual contact such as during sexual intercourse.

23.18 What is the general course of development for cases of syphilis?

The development of syphilis is generally an involved series of events. The **primary stage** consists of a painless, hard lesion called a **chancre** at the site where spirochetes have entered the body (usually on the external or internal genital organs). After several weeks, the chancre disappears and a latent period ensues. The secondary stage appears many weeks or months or years later. The **secondary stage** is accompanied by a skin rash with pustular lesions and skin eruptions as the patient in Figure 23-2 shows. The hair on the head and eyebrows is often lost. Liver inflammation is common, and an influenzalike syndrome may appear. The person is highly contagious at this point. After some time, the lesions heal and another latent stage develops. A **tertiary stage** may appear years later. This stage is characterized by the formation of **gummas**, which are gummy, granular lesions forming in the brain and major blood vessels. The patient often becomes paralyzed and insane, and usually suffers permanent damage to the blood vessels. Death usually accompanies destruction of the heart and blood vessel tissues.

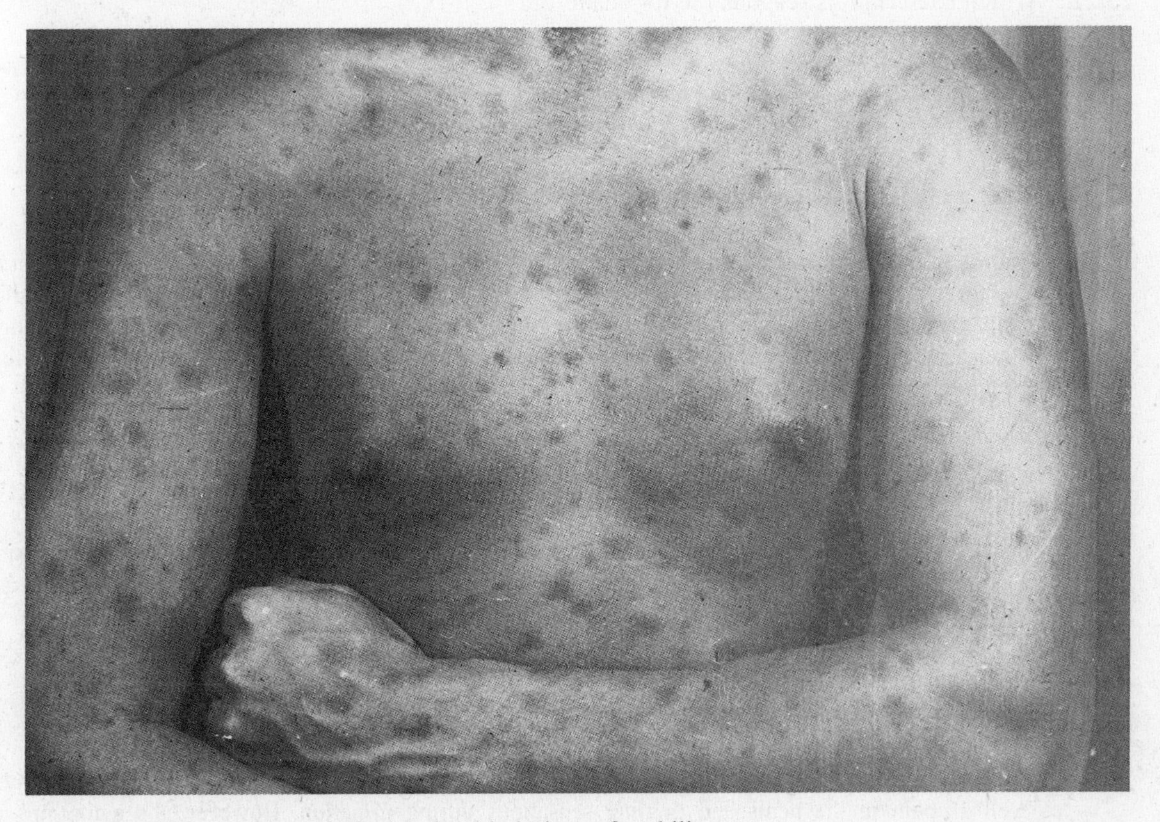

Fig. 23-2 A patient displaying the skin lesions of syphilis.

23.19 How are cases of syphilis diagnosed?

Cases of syphilis can be diagnosed by observing spirochetes from the chancre in the primary stage. Blood tests for syphilis antibodies are positive during the first latent period. During the secondary stage, spirochetes can be obtained from the body lesions and tissues. During the tertiary stage, the effects are generally due to reactions by the immune system, and diagnosis is based on the appearance of symptoms. The VDRL test is used to detect syphilis antibodies, and the complement fixation (Wasserman) test can also be employed, although it is not often used.

23.20 How can cases of syphilis be treated?

Cases of syphilis usually succumb to treatment with penicillin. Treatment is particularly valuable during the incubation, primary, latent, and secondary stages, but not in the tertiary stage. Tetracycline can be substituted for penicillin.

23.21 Can syphilis pass from a mother to child?

One of the dangers of syphilis is the possible development of **congenital syphilis**. In this instance, the spirochetes cross the placenta and enter the fetal blood from the mother's blood. Newborns show signs such as notched incisors (Hutchinson's teeth), a perforated palate, an aged looking face, and damage to the nose. Congenital syphilis may also result in stillbirth.

23.22 Which bacterium is responsible for chancroid?

Chancroid is a sexually transmitted disease caused by the Gram-negative rod *Haemophilus ducreyi*. The organism occurs in strands and is transmitted during sexual contact.

23.23 What are the characteristic signs of chancroid?

Chancroid, also known as **soft chancre**, is distinguished by soft, painful lesions often occurring on the genital organs several days after transmission. The lesions (chancres) bleed easily and are often accompanied by a burning sensation during and after urination. They may also develop on the tongue and lips if transmission has occurred to these organs. An enlarged mass of tissue called a **bubo** may form in the lymphatic tissue of the groin.

23.24 Are any drugs available for the treatment of chancroid?

Most cases of chancroid may be treated with antibiotics such as tetracycline and erythromycin. Lesions heal rapidly with therapy, but scars may remain as a result of the tissue destruction. The disease is usually diagnosed by isolating the etiologic agent from the soft chancres.

23.25 Which disease is most similar to the disease known as chlamydia?

The disease **chlamydia** bears a striking resemblance to gonorrhea. The symptoms include a discharge as well as pain in the penis and tingling sensations during urination. However, the symptoms are generally milder than in gonorrhea. Pelvic inflammatory disease can also result as a complication to chlamydia, as it does in gonorrhea. Chalmydia is sometimes referred to as **nongonococcal urethritis** to distinguish it from gonorrhea, which is called gonococcal urethritis. Cases of chlamydia can also be transmitted to the newborn during birth and can cause eye disease in the newborn.

23.26 Which organism is responsible for cases of chlamydia?

Chlamydia is caused by *Chlamydia trachomatis*, an extremely small bacterium belonging to the group **chlamydiae**. Chlamydiae measure about 0.25 μm in diameter and are beneath the resolving power of the light microscope. They are very difficult to cultivate in the laboratory and are also responsible for diseases such as trachoma and lymphogranuloma.

23.27 How are cases of chlamydia treated?

Cases of chlamydia respond to antibiotics such as tetracycline. Sulfa drugs are also useful, both for therapeutic and prophylactic purposes. Erythromycin is used in the eyes of newborns to prevent **chlamydial ophthalmia**.

23.28 Are any sexually transmitted diseases due to mycoplasmas?

The mycoplasma called *Mycoplasma hominis* causes an infection called **mycoplasmal urethritis**. The symptoms of this infection are similar to those in gonorrhea and chlamydia. Treatment is successful with erythromycin, but not with penicillin because mycoplasmas do not have cell walls.

23.29 Can an STD be caused by a ureaplasma?

A **ureaplasma** is a type of mycoplasma that metabolizes urea. A ureaplasma known as *Ureaplasma urealyticum* is responsible for a type of nongonococcal urethritis. The organism cannot be seen with a light microscope because of its small size (about 0.15 μm). **Ureaplasmal urethritis**, as the disease is known, often causes low sperm counts and poor sperm motility, conditions that lead to infertility. Tetracycline is used to control the infection.

23.30 What are the characteristic signs of lymphogranuloma venereum?

The characteristic signs of **lymphogranuloma venereum** are enlargement and pain of the lymph nodes to form pus-filled **buboes** (swollen lymph nodes). Swelling of the skin in the genital region is also known due to obstruction of the lymphatic vessels. Rectal infection can occur, and chronic infection of the rectum may result. Conjunctivitis, or infection of the conjunctiva of the eye, is also possible.

23.31 Which organism causes lymphogranuloma venereum and how is the disease transmitted?

Lymphogranuloma venereum is caused by *Chlamydia trachomatis*, a strain of the same organism that causes chlamydia. Transmission occurs most frequently during sexual contact. The disease is most prevalent in warmer regions of the world.

23.32 What is the preferred drug for treating lymphogranuloma venereum and how is diagnosis performed?

For the treatment of lymphogranuloma venereum, the drug of preference is tetracycline. Erythromycin may be substituted if the patient is allergic to tetracycline. Diagnosis is performed by locating inclusion bodies in pus obtained from infected lymph nodes.

23.33 Which microorganism is responsible for granuloma inguinale?

A small Gram-negative encapsulated rod is responsible for cases of **granuloma inguinale**. The organism is named *Calymmatobacterium granulomatis*.

23.34 What are the characteristic signs of granuloma inguinale and how is it treated?

The characteristic sign of granuloma inguinale is a painless, irregular ulcer near the genital organs. The disease is transmitted during sexual contact, and the etiologic agent can be transmitted to other body parts by finger contamination. Tetracycline and ampicillin are used for therapy.

23.35 How are Donovan bodies related to cases of granuloma inguinale?

Donovan bodies are accumulations of large mononuclear cells formed in tissues during cases of granuloma inguinale. Scrapings from the lesions reveal the bodies, which are used as a diagnostic sign.

23.36 Which is the most prevalent sexually transmitted disease related to a virus?

The most prevalent STD related to a virus is **herpes simplex**. Herpes simplex virus type I usually causes infections above the waist, while herpes simplex virus type II generally is the cause of most cases of genital herpes. The virus multiplies in the tissues of the genital region and induces the formation of small, fluid-filled vesicles, which are very painful. The vesicles heal without scarring in a period of about three weeks.

23.37 Can cases of herpes simplex reoccur in the future?

One of the characteristics of herpes simplex is the reactivation of the infection after a period of latency. These recurrences happen in a small percentage of those infected. The reinfection occurs from viruses remaining in the ganglia of the nerve tissues near the skin.

23.38 Is genital herpes transmitted exclusively by sexual intercourse?

Although most cases of genital herpes are transmitted by sexual intercourse, other practices such as oral sex can transmit the viruses. Although rarely transmitted by inanimate objects, such things as infected towels and the water in hot tubs may pass the virus among individuals.

23.39 Are there any special problems for women who have genital herpes?

Women who have genital herpes may find that the virus is transmitted to the newborn across the placenta. The H in the TORCH group of diseases stands for herpes simplex. Miscarriage may occur, or the newborn may be born with neurologic damage or skin lesions. Transmission may also occur during the birth process if the woman is infected, and symptoms may occur later in the newborn.

23.40 Is there any drug available for the treatment of genital herpes?

The drug **acyclovir** has been used with success for genital herpes. It is available in pharmacies as Zovirax. It does not eliminate genital herpes completely, but it lengthens the time between attacks and limits the severity of the disease during attacks.

23.41 Do herpes simplex viruses cause any diseases other than genital herpes?

Herpes simplex viruses cause a constellation of diseases including cold sores of the mouth (canker sores), lesions of the pharynx in children (gingivostomatitis), infections of the cornea and conjunctivitis (keratoconjunctivitis), infection of the lungs (herpes pneumonia), infection of the brain (herpes encephalitis), and herpes whitlow (herpes lesions on the fingers). These various diseases are spread among individuals by release of the virus from the area affected.

23.42 Which organisms are known to cause genital warts?

Recent studies have indicated that **genital warts** is a transmissible disease, usually a sexually transmitted disease. The etiologic agent is any of a number of human papilloma viruses. The incidence of genital warts has increased, and the disease is among the most common sexually transmitted diseases in American society.

23.43 What is an alternate expression for genital warts?

The disease genital warts is also known as **condyloma accuminata**. The term refers to the figlike appearance of the warts occurring in the disease. The warts can occur on any of the genital organs in males and females, both internally and externally.

23.44 Why is cytomegalovirus disease considered a sexually transmitted disease?

Cytomegaloviruses are released from the body from many different types of fluids including saliva, urine, blood, and semen. Therefore, they may be transferred to others by numerous means including during sexual intercourse. Cytomegaloviruses can pass from the woman who is pregnant to the unborn child. The fetus becomes infected in numerous organs and brain damage may occur. The C in the TORCH group refers to cytomegaloviruses.

23.45 What type of disease does *Candida albicans* cause in the reproductive tract?

Candida albicans is a yeastlike fungus that grows in the urogenital tract when the correct conditions develop. These conditions often involve loss of the acidity in the vaginal tract due to the destruction of lactobacilli with the overuse of antibiotics. Under these conditions, *Candida albicans* multiplies and causes the "**yeast disease**."

23.46 What are the characteristic signs of *Candida albicans* infections?

Characteristic signs of *Candida albicans* infections include severe itching, irritation, and a thick, yellow, cheesy discharge. A yeasty odor is also possible. Treatment with miconazole, clotrimazole, or nystatin is effective in retarding the infection.

23.47 Is AIDS considered a sexually transmitted disease?

AIDS is considered a disease of the lymphocytes, and thus it is discussed in Chapter 22. It bears noting, however, that the virus of AIDS is present in the semen and is transmitted during sexual intercourse. Transmission is enhanced if the receptive partner has lesions of the genital organs through which the virus can pass into the blood.

Review Questions

Multiple Choice. Select the letter of the item that correctly completes each of the following statements.

1. Most cases of leptospirosis are transmitted by (*a*) airborne spores (*b*) sexual contact (*c*) contaminated soil or water (*d*) contaminated vegetable crops.

2. Glomerulonephritis is a type of inflammation of the kidney usually due to (*a*) a Gram-negative rod (*b*) *Streptococcus pyogenes* (*c*) *Staphylococcus epidermidis* (*d*) fungi such as *Candida albicans.*

3. Fever, a red skin rash on the body trunk, and low blood pressure as well as use of high-absorbency tampons may indicate the presence of (*a*) condyloma accuminata (*b*) AIDS (*c*) granuloma inguinale (*d*) toxic shock syndrome.

4. Cases of trichomoniasis are due to a (*a*) protozoa (*b*) Gram-positive chain of cocci (*c*) Gram-negative chain of cocci (*d*) virus.

5. The organism that causes gonorrhea is described as a (*a*) small Gram-negative diplococcus (*b*) Gram-positive rod that has a capsule (*c*) Gram-positive rod that forms spores (*d*) Gram-positive rod that is anaerobic.

6. In addition to the urethra, the organism of gonorrhea may also infect the (*a*) muscles of the upper extremity (*b*) eyes (*c*) muscles of the lower extremity (*d*) liver.

7. Sterility may be the long-range result of gonorrhea if inflammation and scarring take place in the (*a*) urethra (*b*) labia majora (*c*) rectum (*d*) Fallopian tubes.

8. The etiologic agent of syphilis is (*a*) *Trichomonas vaginalis* (*b*) *Treponema pallidum* (*c*) *Ureaplasma urealyticum* (*d*) *Neisseria gonorrhoeae.*

9. The characteristic sign of the tertiary stage of syphilis is the (*a*) chancre (*b*) circulatory collapse (*c*) bubo (*d*) gumma.

10. Cases of syphilis may be diagnosed by observing (*a*) viruses from the chancre (*b*) tangled threads of hyphae in the blood (*c*) spirochetes from skin lesions (*d*) Gram-negative rods from the urethral discharge.

11. Most cases of syphilis can be treated with the antibiotic (*a*) chlotrimazole (*b*) gentamicin (*c*) chloramphenicol (*d*) penicillin.

12. *Haemophilus ducreyi* is best known as the bacterium that causes (*a*) genital warts (*b*) genital herpes (*c*) chancroid (*d*) Bright's disease.

13. The disease chlamydia closely resembles in symptoms the disease (*a*) syphilis (*b*) granuloma inguinale (*c*) gonorrhea (*d*) cytomegalovirus infection.

14. All the following characteristics apply to the chlamydiae except (*a*) they are difficult to cultivate in the laboratory (*b*) one type causes trachoma (*c*) they are among the larger species of bacteria (*d*) they respond to antibiotics such as tetracycline.

15. Which of the following descriptions applies to *Ureaplasma urealyticum*: (*a*) the organism is a mycoplasma (*b*) the organism is visible under the light microscope (*c*) the organism does not cause any known human disease (*d*) the organism does not respond to any antibiotics.

16. Swollen lymph nodes, also known as buboes, are characteristic signs of the infection (*a*) trichomoniasis (*b*) prostatitis (*c*) toxic shock syndrome (*d*) lymphogranuloma venereum.

17. The organism responsible for granuloma inguinale is described as a (*a*) large virus (*b*) small Gram-negative rod (*c*) large, Gram-positive rod (*d*) anaerobic sporeforming Gram-positive rod.

18. Accumulations of large mononuclear cells develop in cases of granuloma inguinale and are known as (*a*) Lipshutz bodies (*b*) councilman bodies (*c*) Negri bodies (*d*) Donovan bodies.

19. Cases of herpes simplex are often accompanied by a (*a*) period of latency (*b*) substantial red skin rash (*c*) growth of fungi in body regions (*d*) whitish discharge from the genital organs.

20. During periods of infection, the viruses of herpes simplex remain (*a*) in the intestinal tract (*b*) in the ganglia of nerves (*c*) in the salivary glands (*d*) on the skin surface.

21. Transmission of herpes simplex viruses can occur to the newborn (*a*) during the birth process (*b*) from contaminated formula (*c*) in airborne droplets from the mother (d) from infected toys.

22. The viruses involved in genital warts are believed to be (*a*) herpes viruses (*b*) *pox viruses* (*c*) papilloma viruses (*d*) polymer viruses .

23. The C in the TORCH group of diseases refers to (*a*) cytopathic viruses (*b*) cytomegaloviruses (*c*) complimentary viruses (*d*) concentrated viruses.

24. Excessive use of antibiotics may result in vaginal infection with (*a*) *Candida albicans* (*b*) human immunodeficiency virus (*c*) genital herpes virus (*d*) polyoma virus.

25. Which of the following would not be useful for treating yeast infections of the reproductive tract (*a*) nystatin (*b*) chlotrimazole (*c*) acyclovir (*d*) miconazole.

Completion. Add the word or words that correctly complete each of the following statements.

1. Infections occurring in the bladder are generally referred to as ____________.

2. The alternative name for leptospirosis when it occurs in the liver is ____________.

3. Most cases of leptospirosis occur in domestic and wild ____________.

4. Cases of toxic shock syndrome are often caused by excessive multiplication of the bacterium ____________.

5. The protozoan *Trichomonas vaginalis* has an undulating membrane and moves by means of ___________.

6. Gonorrhea is caused by *Neisseria gonorrhoeae,* a small diplococcus that is Gram-___________.

7. Most cases of gonorrhea occur in the tube known as the ___________.

8. The symptoms of gonorrhea are generally more apparent in ___________.

9. Should a child be born to a woman when she is suffering from gonorrhea, the newborn may suffer an eye infection known as ___________.

10. The organism that causes syphilis has the shape of a ___________.

11. Most of the obvious symptoms in patients suffering from syphilis occur in the ___________.

12. Most cases of syphilis can be treated with the drug ___________.

13. Soft, painful lesions occurring on the genital organs develop in chancroid, a disease caused by the bacterium ___________.

14. Cases of chlamydia bear a very close resemblance to cases of ___________.

15. For cases of mycoplasmal urethritis, the preferred drug is ___________.

16. The disease lymphogranuloma venereum is caused by a species of organism known as a ___________.

17. During cases of granuloma inguinale, the tissues contain Donovan bodies which are accumulations of large ___________.

18. The most prevalent sexually transmitted disease related to a virus is ___________.

19. A number of human papilloma viruses are believed responsible for cases of ___________.

20. The cytomegaloviruses are transmitted from mother to unborn child, and, therefore, cytomegaloviruses are included in a group of diseases known by the acronym ___________.

21. A yeastlike odor from the urogenital tract generally indicates infection with ___________.

22. Reinfection by herpes simplex viruses occurs from viruses remaining in the ganglia of the ___________.

23. Among the many herpes simplex infections is herpes lesions on the fingers, a condition known as ___________.

24. Infections of the urogenital tract by *Candida albicans* generally accompany the loss from the tract of ___________.

25. Zovirax is the trade name for a drug used against genital herpes and known as ___________.

True/False. For each of the following statements, mark the letter "T" next to the statement if the statement is true, or "F" if the statement or any part of the statement is false.

___**1.** Skin abrasions may account for the transfer of *Leptospira interrogans* into the body.

___**2.** Glomerulonephritis reflects a manifestation of an immune complex resulting from an activity of the immune system and a type III hypersensitivity.

___**3.** *Trichomonas vaginalis* is the only protozoan capable of causing trichomoniasis.

___**4.** The possibility of toxic shock syndrome can be lessened by avoiding devices that remain in the vaginal tract for long periods of time.

___ **5.** One of the possible long-range effects of cases of gonorrhea is sterility.

___ **6.** The primary stage of syphilis is accompanied by a skin rash with pustular lesions and skin eruptions, liver inflammation, and an influenzalike syndrome.

___ **7.** The diagnosis of syphilis in the tertiary stage depends upon isolation of the bacteria from characteristic lesions.

___ **8.** Most cases of chancroid may be treated with antibiotics.

___ **9.** The eyes of newborns are usually treated as a preventive measure to preclude gonococcal ophthalmia and chlamydial ophthalmia.

__ **10.** Type I herpes simplex virus is generally the cause of most cases of genital herpes.

Chapter 24

Food and Industrial Microbiology

OBJECTIVES

Microorganisms are important in the food industry not only as producers of certain foods, but as contaminants of others. These applications of microbiology occur not only at the level of the homemaker, but also at the industrial level, where foods and other products are produced in massive quantities. This chapter will survey how the microbial influence is felt. Its objectives are to:

1. Understand why milk products and other foods are easily contaminated by microorganisms.

2. Survey a number of dairy products, such as yogurt and cheese, that are produced through the fermentation action of microorganisms.

3. Explain how various foods such as meats, fish, and canned foods become spoiled, and explore the effect of that spoilage.

4. Specify a number of methods used for food preservation in the home and in the industrial setting.

5. Note the various types of diseases that can be caused by foodborne microorganisms and outline methods by which laboratory technologists can detect contaminating microorganisms in foods.

6. Survey the standard plate count, reduction, phosphatase, and other techniques used to ensure the safety of milk and milk products.

7. Discuss some of the industrial applications in which microorganisms are cultivated in mass quantities for use.

8. Summarize the fermentation processes that lead to alcoholic beverages, and show how microorganisms can be used to produce nutritional supplements.

9. Recognize the importance of microorganisms as insecticides and in the processes of genetic engineering.

THEORY AND PROBLEMS

MICROORGANISMS AND FOODS

24.1 What is the difference between natural and artificial contamination of foods?

When microorganisms are present in food, it is considered contaminated. The contamination can be **natural contamination** if the microorganisms are attached to foods growing in the field (e.g., yeasts on the skins of grapes), but it is **artificial contamination** if it occurs as a consequence of handling or processing the food (e.g., acid-producing bacteria in grape juice cause it to become sour).

24.2 Is contamination of food always considered harmful?

Although contamination is generally considered in a negative light, since pathogens may cause disease, there are several instances in which contamination is considered helpful. For example, the controlled contamination of cucumbers yields pickles, and the controlled contamination of cabbage gives us sauerkraut. These contaminations are desirable. To control undesired contamination, chemical compounds can be added to processed foods to discourage microbial growth.

24.3 Why are milk and milk products easily contaminated with microorganisms?

Milk is an excellent growth medium for microorganisms, and it supports the proliferation of a variety of microorganisms. Production, processing, and marketing of dairy products therefore requires special attention to the possibilities of contamination. Most milk and dairy products are **pasteurized** for 30 minutes at 62°C (holding method) or for 15 to 17 seconds at 72°C (flash method). These methods kill all pathogens in milk or dairy products, with particular attention to the organisms that cause tuberculosis and Q fever, since these are among the most heat-resistant bacteria known .

24.4 What are some dairy products produced by the natural fermentation of milk?

The natural fermentation of milk yields fermented milk products that fall into two classes: fermented milks and cheeses. **Fermented milks** include buttermilk, in which strains of *Streptococcus lactis* have produce acid from the carbohydrate. Another fermented milk product is sour cream, in which streptococci produce acid to make the cream sour. A third fermented milk product is yogurt. Here, species of *Lactobacillus* and *Streptococcus* ferment the components of milk and produce a custardlike sour milk product rich in bacteria.

24.5 What sort of cheeses can be produced by the fermentation of milk?

Cheese products are produced from casein, the protein of milk. The casein is precipitated either by bacterial action or by the enzyme rennet added to the milk. The precipitated protein is unripened cheese, such as cottage cheese or pot cheese. Further growth of microorganisms in the milk curds produces ripened cheeses. Some examples are hard cheeses such as cheddar, American, and Swiss (the holes in Swiss cheese are gas holes produced by growing microorganisms). Soft ripened cheeses include Camembert and bleu cheese. Bleu cheese is produced by the growth of *Penicillium* species in the milk curd.

24.6 Which microorganisms are involved in the pickling processes in the food production?

In the pickling process microorganisms grow and ferment the food in a salty environment while producing large amounts of acid from available carbohydrates. Species of *Leuconostoc* and *Lactobacillus* are commonly used in the pickling process. As Figure 24-1 illustrates, to produce **sauerkraut**, cabbage is grated and salted and allowed to ferment naturally (since the bacteria are already present among the leaves). For **pickled cucumbers**, salt is added to fresh cucumbers, and the naturally occurring bacteria ferment the vegetable carbohydrates over a period of weeks. Different spices are added to prepare various forms of sauerkraut and pickled cucumbers.

FOOD CONTAMINATION AND PRESERVATION

24.7 What is the source of food contamination in spoiled foods?

There are many sources of microbial contamination in foods. The general sources are air, soil, water, sewage, and animal waste. The natural flora clinging to foods grown in the ground (such as beets) are also potential spoilers. Food is considered spoiled when it is aesthetically unfit for consumption. Fungi and bacteria are among the chief microbial forms associated with spoilage.

24.8 What are the sources of contamination in meats and fish products?

Meats and **fish** products are often contaminated by bacteria during the processing procedures. Microorganisms from the animal's waste, skin, feet, and internal organs contain potential contaminants. Contaminations usually affect only the surface of the meat , but fish tissues may be affected on their skin and in their deeper tissues (e.g., muscle tissues) because the tissues are relatively loose. Meat is easily contaminated when it is ground for hamburger or sausage because the chopped meat has many air pockets contaminated with bacteria formerly on the surface.

24.9 Which types of spoilage are particularly important in canned food?

Because **canned food** is held for long periods of time, the potential for food spoilage is high if the food contents have not been sterilized. Gas spoilage is produced primarily by species of the genus *Clostridium*. Flat-sour spoilage is due to members of the genus *Bacillus*. This kind of spoilage occurs when microorganisms metabolize a food component and produce acid but no gas. Acid spoilage may be produced by acid-producing organisms such as lactobacilli, or it may be due to the food's acid reacting with the metal of the can.

24.10 Which criteria must be observed when considering prevention of food spoilage?

The three important criteria that determine the type of food spoilage prevention techniques are: the nature of the food to be preserved, the length of time it is to be kept before consumption, and the methods necessary for handling and processing the food. These three criteria determine which preservation method is to be used.

24.11 How is temperature regulation used for food preservation techniques?

Because microorganisms are sensitive to temperature, the **refrigerator** and **freezer** can be used to preserve the food by reducing the metabolism and enzyme activity going on in the microorganisms. At refrigerator temperatures of 4 to 8°C, fresh vegetables can be preserved, and leftovers and dairy products can be held free of excessive spoilage. However, there are limits to the time that refrigeration can be used, and these limits vary with the type of food, how it was prepared, the amount of exposure to the air, and so on. Freezer temperatures can be used for long-term preservation for foods such as meats, but dehydration ("freezer-burn") is a potential problem. It should be noted that refrigeration and freezing do not kill microorganisms as such, but they reduce their metabolism. Spores usually remain viable during refrigeration and freezing.

24.12 Which high-temperature methods are used for food preservation?

To preserve foods high temperatures can be used in the processes of boiling, pressurized steam, or pasteurization. In **boiling**, the food is immersed in water held at 100°C for 30 to 60 minutes to destroy bacterial vegetative cells, but not bacterial spores. Boiling also inactivates toxins possibly deposited by bacteria in leftovers. It inactivates many enzymes in foods, which is why plant foods are often blanched before freezing. Although the enzymes are inactivated, the flavor and texture of the food are preserved. Boiling is more effective against microorganisms in high-acid foods, such as tomatoes, than low-acid foods.

24.13 What are the TDP and the TDT as they are utilized in food preservation?

The TDP and the TDT refer to the **thermal death point** and the **thermal death time** for microorganisms associated with food. Each microorganism has a TDP, which is the temperature required for its destruction at a given time. It also has a TDT, which is the amount of time necessary for its destruction at a given temperature. TDP and TDT values differ considerably for a particular food and its processing method, as well as the size of the container, the acidity of the food, and other factors. The values are established for canning procedures that employ pressurized steam. Important temperature considerations related to foods are displayed in Figure 24-2.

24.14 Can pasteurization be applied to foods other than dairy products?

Although **pasteurization** is most widely used in the dairy industry, it can be applied to any food that cannot be boiled or sterilized under pressurized steam. For example, fruit juices are usually pasteurized to eliminate wild yeasts before fermentation, and the wine is pasteurized after fermentation to destroy the fermenting yeast. Pasteurization is also used to preserve honey.

24.15 Which principles underlie food preservation by drying?

During the **drying** and **dehydration** processes, food is subjected to procedures that remove most of its water. This water removal inactivates any microorganisms present because they need water for their chemical activities and growth. However, spores survive the drying process and remain in a dormant condition until the water is restored. Drying is used for raisins, dates, fruits, dried meats, and sun-dried foods.

24.16 Which chemicals are used in the food preservation process?

Over the centuries a variety of **chemicals** have been used for food preservation. Salt, for example, has been widely used as a preservative because it draws water out of foods by osmosis. Sugaring and smoking meats are other time-honored practices. Smoking deposits chemical preservatives in foods and dehydrates them. Among the chemicals approved by the United States Food and Drug Administration are sodium propionate for breads and bakery products, benzoic acid for soft drinks, sulfur dioxide gas for certain fruits and vegetables, and sorbic acid for jams, jellies, and a number of other foods.

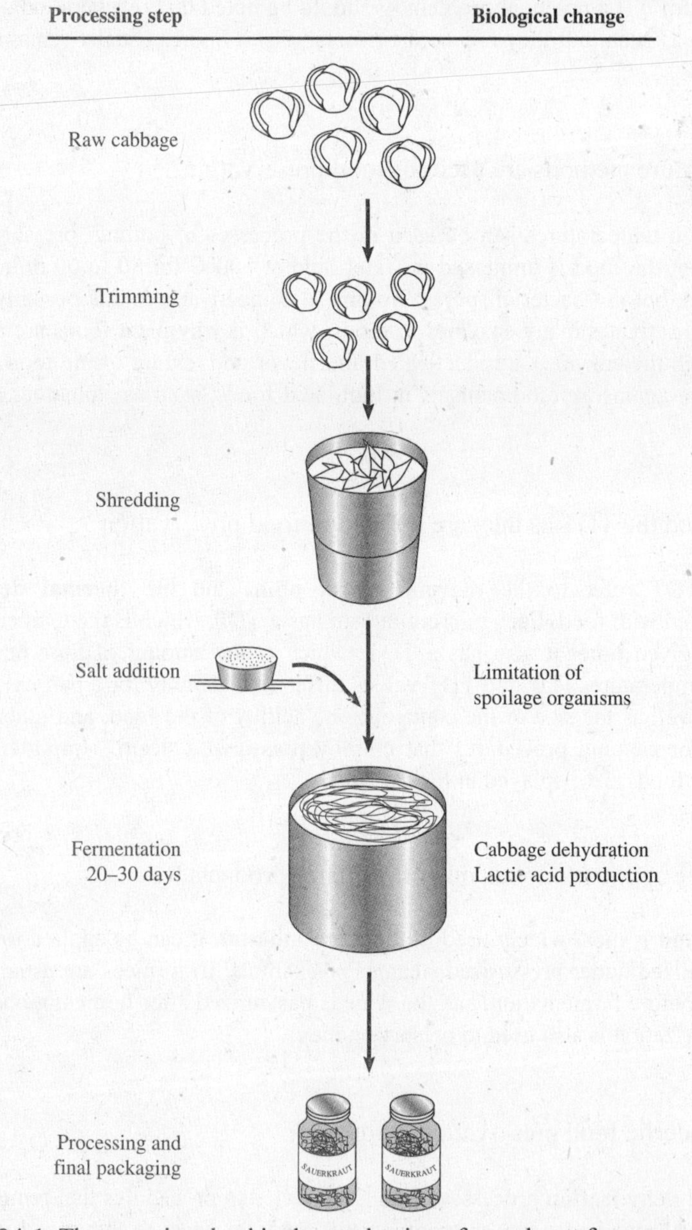

Fig. 24-1 The steps involved in the production of sauerkraut from raw cabbage.

24.17 Can physical methods be used for preserving foods?

There are several physical methods employed for preserving foods. For instance, microorganisms are **filtered** out of beverages such as wine, cider, and beer. **Ultrasonic vibrations** (high-frequency sound waves) can also be used in liquid foods because the vibrations produce shock waves that crash into microorganisms and obliterate them. **Ultraviolet light** has been used in certain cooked foods and cheeses, and **gamma irradiation** is now used for several types of preserved fruits and vegetables.

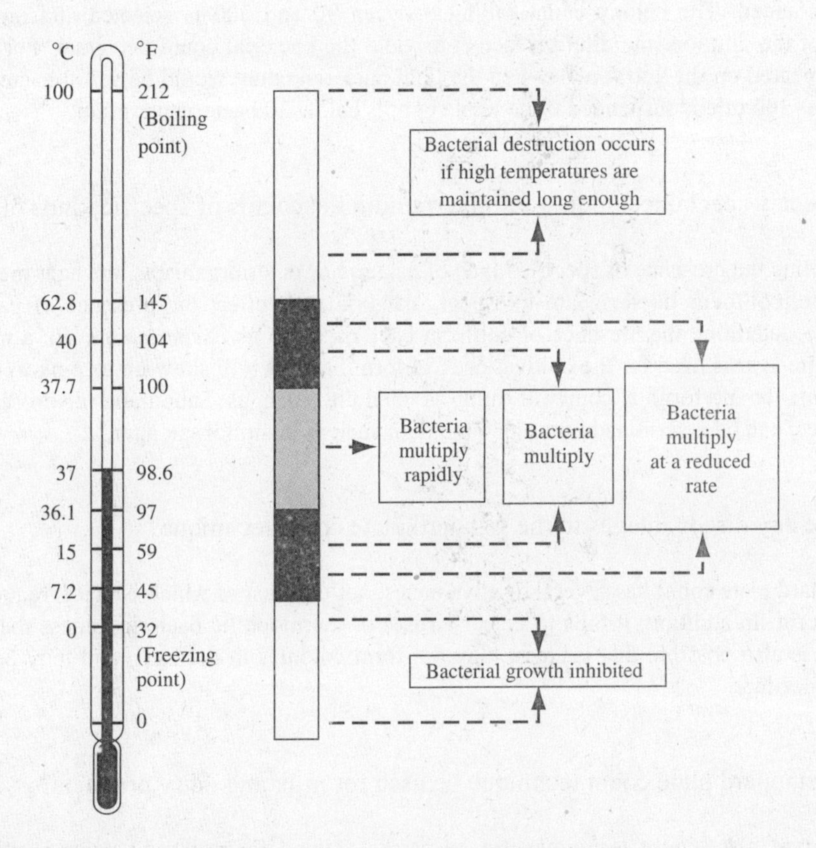

Fig. 24-2 Some important temperature considerations related to spoilage and preservation.

24.18 Which types of human diseases can be caused by foodborne microorganisms?

Foodborne microorganisms can be involved in food poisonings or food infections. Food poisonings are caused by species of staphylococci and clostridia. The toxins produced by these microorganisms interfere with body processes and disease results. Food infections are due to contaminating microorganisms that grow and multiply in the body and cause such diseases as salmonellosis, cholera, shigellosis, typhoid fever, gastroenteritis, amoebiasis, and numerous others (Chapter 21).

LABORATORY TESTING

24.19 How can food be tested for the presence of microorganisms?

The bacterial content in a sample of food can be determined by the **standard plate count.** This procedure determines the total number of bacteria in a gram of food. The procedure is performed by aseptically adding one gram of food to 99 ml of sterile distilled water or buffer. A technician then transfers 1-ml and 0.1-ml samples to sterile Petri dishes, thereby representing 1:100 and 1:1000 dilutions of the food sample. Other dilutions are prepared, and a growth medium such as plate-count agar are added. The plates are incubated at 37°C for 24-48 hours and the number of colonies appearing on the plates are counted. The colony count falling between 30 and 300 is selected and multiplied by the reciprocal of the dilution (the dilution factor) to yield the bacterial count per gram. For example, if 97 colonies appeared on the 1:100 plate, then the total bacterial count would be 9,700 bacteria per gram of food. The 30-300 rule is suspended if the total count is below 30 bacteria per gram.

24.20 Are there any special techniques for determining the counts of specific kinds of bacteria?

To determine the presence of specific kinds of bacteria or other organisms, the agar medium is varied. For example, **coliform bacteria** are intestinal bacteria that reflect the presence of pathogens of the intestine. To determine the presence of coliform bacteria (such as *Escherichia coli*), a medium such as violet red bile agar is used for the cultivation. Coliform bacteria will show up as pink to red colonies on this medium. To perform a count of fungi a medium such as Sabouraud dextrose agar is used. Staphylococci can be determined by using a medium such as mannitol salt agar.

24.21 Are there any disadvantages to the standard plate count technique?

The standard plate count has several disadvantages, not the least of which is that it requires at least one day to perform. In addition, it fails to detect viruses or thermophilic bacteria, unless the temperature is adjusted. It is also possible that bacteria may not form colonies in the agar, and may be missed in the counting procedure.

24.22 Can the standard plate count technique be used for milk and dairy products?

The standard plate count technique can be easily adapted for milk and dairy products by simply substituting 1 ml for the 1 gram of food. The remainder of the test is performed as for food.

24.23 Are there any other specialized tests available for milk and dairy products?

For a rapid determination of the bacterial content of milk, the **phosphatase test** is available. Phosphatase is an enzyme normally found in cow's milk. It has a heat tolerance similar to that of the most heat-resistant forms of bacteria (*Mycobacterium* and *Coxiella* species). Therefore a rapid test for phosphatase is performed by incubating a sample of the milk with a substance containing phosphate groups. If phosphatase is present in the milk, it will release the phosphate groups and bring about a color change. However, if the phosphatase has been destroyed during pasteurization, no phosphate groups will be released and no color change will happen. The test is performed in an automated device and the results are available quickly.

24.24 What is the reduction test for milk?

Another test available to determine the bacterial quality of milk is the **reduction test**, presented in Figure 24-3. A sample of milk is added to a solution of methylene blue, and the mixture is incubated. If bacteria are present, they will grow in the milk and liberate electrons. The electrons cause the methylene blue to change from its normal blue color to clear, and the milk becomes white. If no bacteria are present, the milk will remain blue for several hours. The quality of the milk can be determined by how rapidly the blue color disappears.

24.25 Are there any tests available for the direct counting of microorganisms in food and milk samples?

In order to get a direct count of microorganisms, it is possible to use a special slide containing a grid. A sample of milk is spread on the grid using a special Breed pipette, and a special stain is used. The bacteria in a selected section of the grid are counted and multiplied by a dilution factor representing the original amount of milk placed on the slide.

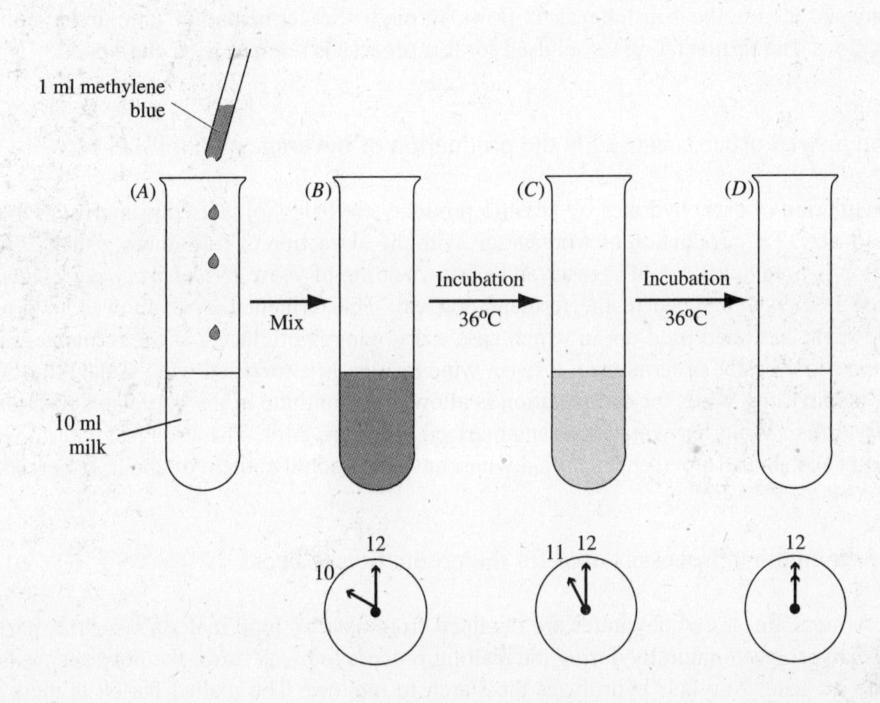

Fig. 24-3 A reduction test used to determine the bacterial quality of milk. Poor quality milk causes a rapid loss of color, while good quality milk retains the blue color for several hours.

MICROORGANISMS AND INDUSTRY

24.26 Are there any industrial processes that relate to microorganisms and foods?

In many cases, microorganisms themselves are used as foods, and industrial processes are designed to produce these microorganisms in large quantities. For instance, **mushrooms** are an important type of produce, and "mushroom farming" is a major industry. Yeast cells are sometimes mixed with livestock feeds, and **yeast** is also used as active dry yeast in baking processes. To produce dry yeast, yeast cells are suspended in a starchy paste and dried slowly in a vacuum. Microorganisms are also used in the form of **single-cell protein (SCP)**. Single-cell protein consists of algae and edible molds, which have plentiful supplies of inexpensive carbohydrate and rich amounts of vitamins and protein. Bacteria can also be used in SCP.

24.27 How are microorganisms grown in industrial situations where huge masses of microorganisms are necessary?

For the production of microorganisms in industrial quantities, two types of fermentations are used: the batch technique and the continuous-flow technique. In the **batch technique**, a fermenting vat or industrial tank is filled with a nutritive substrate consisting of numerous materials including industrial waste products such as corn steep. The fermentation vat may hold up to 100,000 gallons of substrate. A single strain of microorganism is added to the substrate and permitted to ferment it over a long period of time under conditions that are relatively still. For the **continuous-flow technique**, substrate is added at a regular interval, and the fermenting material is withdrawn to recover the end product. The microorganisms act on the substrate as it flows through the fermentation unit under conditions of continuous flow. The fermentation vessel used for this process is referred to as **chemostat**.

24.28 Which industrial process is used in the production of beverage alcohol?

The fermentation of carbohydrates by yeast to produce ethyl alcohol is used primarily in the alcoholic beverage industry. The production of **wine** begins with the extraction of fruit juices, usually grape juices. Pasteurization is used to kill "wild" yeasts, and a known strain of yeast (*Saccharomyces cerevisiae* strain *ellipsoideus*) is then inoculated to the fermentation vat. The fermentation is allowed to proceed until most of the sugar has been used up, in which case a dry wine is produced. If the fermentation is halted before all the sugar has been fermented, a sweet wine results. In a **fortified wine**, a quantity of brandy is added. For a **sparkling wine**, the fermentation is allowed to continue in the bottle and additional carbon dioxide gas forms. Wine fermentation is summarized in Figure 24-4. The strongest natural wine has an alcohol content of about 16 percent (fortified wines have an alcohol content of about 22 percent).

24.29 Which fermentation process is used for the production of beer?

In **beer** fermentations, carbohydrates are obtained from starchy food materials such as barley grains. The starch is hydrolyzed naturally during the malting process to break down the polysaccharides into the disaccharide maltose. Amylase hydrolyzes the starch to maltose. The malted barley is then allowed to remain for further digestion to produce wort, which is pasteurized. Hops are added, and the yeast *Saccharomyces cerevisiae* is added. The fermentation takes about 15 days, then the beer is aged for approximately 6 months. Beer has a final alcohol content of about 4 percent.

24.30 How are distilled liquors produced in comparison to wines?

Wines are the product of alcohol fermentation on fruit sugars by the yeast *Saccharomyces*. To prepare a **distilled liquor**, the alcohol is heated and drawn off as fumes, which are then condensed to yield the liquor. Brandy is prepared from numerous kinds of fruit juices (e.g., apple, peach, blackberry, cherry), while whiskey is produced from corn, rye, or barley. Rum is prepared from molasses, and neutral spirits such as gin and vodka result from the fermentation of potatoes. The proof number of a liquor is twice the percentage of the alcoholic content.

24.31 In what ways does sake compare to beer and wine?

Although **sake** is often considered a wine, it is more correctly a beer because it is produced from a grain (rice) rather than from a fruit. The mold *Aspergillus oryzae* is used in sake production.

24.32 How are antibiotics produced on industrial scales by microorganisms?

Antibiotics were originally produced by the metabolism of microorganisms. Over the decades, the processes were adapted to industrial scales, and *Penicillium* species and other molds were grown in huge industrial tanks. Irradiations were used to produce mutant species with enhanced antibiotic yields. The antibiotics were isolated from the liquor in the aerated fermentation vats then purified for medicinal use. These industrial processes are still used, but many antibiotics are produced by purely chemical means.

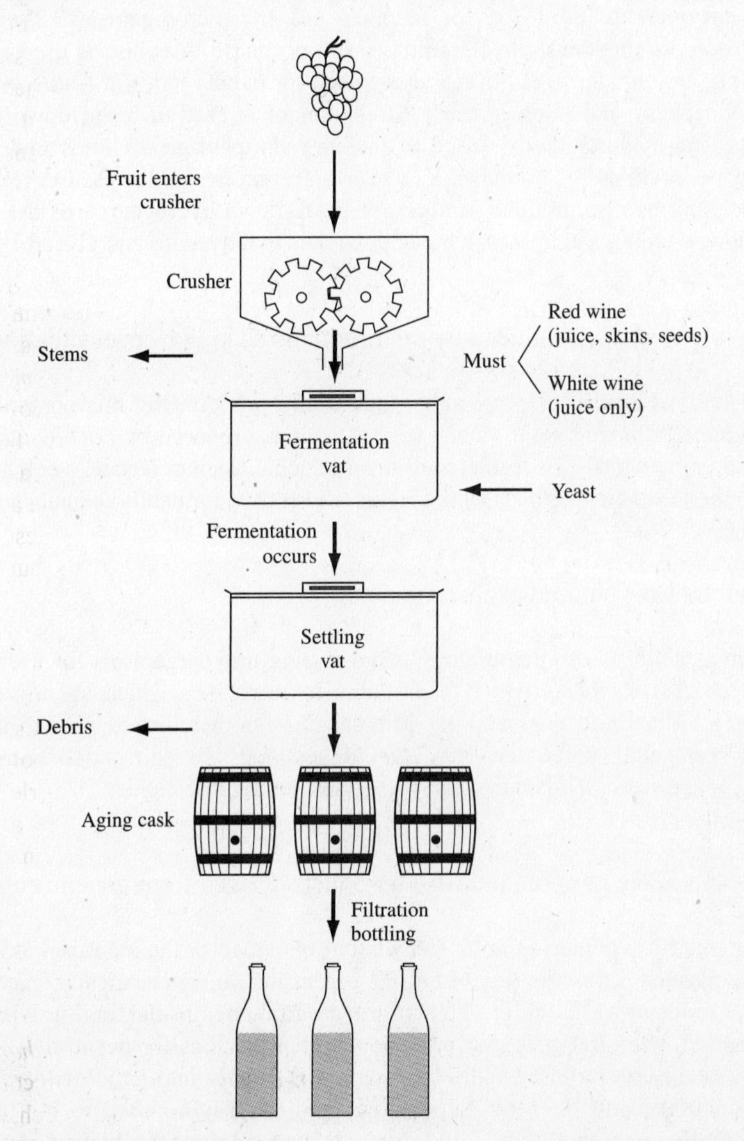

Fig. 24-4 Some of the major steps involved in the fermentation of grape juice to wine.

24.33 Can microorganisms be used to produce vitamins and amino acids?

Numerous **vitamins** are currently produced by industrial fermentations. Vitamin B_{12}, for example, is used to prevent certain types of anemia in the body and riboflavin is used to encourage metabolic processes in the body. Both are produced by bacteria cultivated in industrial plants. Yeasts are rich in B vitamins and are used as a food supplement. Among the **amino acids** produced by microorganisms in industrial processes are glutamic acid and lysine. Both are used as dietary growth supplements.

24.34 Which enzymes can be produced industrially by the activity of microorganisms?

Numerous **enzymes** are produced for practical use by microorganisms. These enzymes include amylase, proteases, streptokinase, hyaluronidase, and pectinase. Amylase is used as a spot remover in laundry presoaks and in adhesives. Proteases are used for baiting hides in leather manufacturing and in glues, meat tenderizers, and drain openers. Streptokinase is used to break down blood clots in heart attack patients, and hyaluronidase is used to facilitate the passage of fluids under the skin. Another industrial enzyme, pectinase, is produced by *Clostridium* species and is used to ret flax and turn it into linen. Microorganisms also produce catalase (to break down hydrogen peroxide), cellulase (to digest cellulose to glucose units), and lipase (to break down fats to fatty acids and glycerol).

24.35 Which organic acids are produced by microorganisms in industrial situations?

Among the most important **organic acids** industrially produced by microorganisms are citric acid, which is used in soft drinks, candies, and inks. Lactic acid, produced by *Lactobacilli,* is used as a food preservative, to prepare hides for leather, and for producing calcium lactate, a calcium source. Gluconic acid is used in the feed of laying hens, and itaconic acid is used for paints and adhesives.

24.36 Can microorganisms be used as insecticides?

One of the major thrusts of microbiology is developing microorganisms for use as **insecticides**. The sporeformer *Bacillus thuringiensis* (BT) produces toxic crystals which are ingested by caterpillars, which develop intestinal infections and die. The bacillus can therefore be used to inhibit the growth of Gypsy moths, hornworms, and other pests. *Bacillus popillae* is used against Japanese beetles where it causes milky spore disease. Viruses can also be used as natural insecticides.

24.37 What are some of the possible industrial products derived from genetic engineering?

The field of **genetic engineering** holds tremendous potential for the industrial use of microorganisms. By altering the plasmids of carrier organisms, the organisms can be chemically changed to produce such pharmaceutical products as interferon, human growth hormone, insulin, and urokinase. Yeasts are also being reengineered with viral genes to produce vaccines, such as for hepatitis B. The genes for toxin production are being used as insecticides by inserting BT genes into harmless organisms that live with common agricultural plants. Genetic engineering may one day be used to increase the capacity for nitrogen fixation in numerous bacteria and in plants, and cellulase production may be introduced to a variety of microorganisms. Bacteria are being reengineered with the ability to produce Factor VIII for hemophiliacs, and viruses are being used for carriers for brining helpful genes into individuals who have abnormal genes. For example, it may be possible to correct the deficiency in cystic fibrosis patients by introducing genes for the proteins the patients lack. Discoveries and insights of genetic engineering are far-reaching and extraordinary. They typify the industrial uses of microbiology in future years.

Review Questions

True/False. For each of the following statements, mark the letter "T" next to the statement if the statement is true. If the statement is false, change the underlined word to make the statement true.

___ **1.** The pasteurization process kills all microorganisms present in milk or dairy products.

___ **2.** The fermented milk products produced by the activity of microorganisms include buttermilk, sour cream, and cottage cheese.

___ **3.** The soft, ripened cheeses such as bleu cheese are produced by the growth of *Penicillium* species in the milk curd.

___ **4.** To produce pickled cucumbers, naturally occurring bacteria ferment the protein in cucumbers over a period of weeks.

___ **5.** Bacteria and protozoa are among the most important forms of microorganisms associated with food spoilage.

___ **6.** Among the important genera of bacteria used in the pickling process are *Leuconostoc* and *Corynebacterium*.

___ **7.** Flat-sour spoilage of foods is generally due to microorganisms that produce no gas.

___ **8.** During refrigeration and freezing, bacterial spores may remain viable and later be transmitted in contaminated food.

___ **9.** When food is immersed in boiling water for 30 to 60 minutes, the spores are usually destroyed.

___ **10.** The thermal death point is the amount of time necessary for the destruction of a microorganism at a given temperature.

___ **11.** In the preparation of fruit juices for market, it is usually practice to pasteurize the product in order to eliminate any bacteria that may be present.

___ **12.** The usual method for preserving many fruits is drying, a process that removes most of its ions.

___ **13.** The chemical used as a preservative in many soft drinks is sorbic acid.

___ **14.** Ultrasonic vibrations are high-frequency light waves that obliterate microorganisms by crashing into them.

___ **15.** Among the important bacteria that can cause food poisoning are species of streptococci.

___ **16.** To assess the effectiveness of pasteurization of milk and dairy products, microbiologists test for the presence of the enzyme glucuronidase.

___ **17.** Among the microorganisms used as food are two types of fungi, the yeasts and the mushrooms.

___ **18.** A fermentation vessel used for a continuous flow fermentation is the biostat.

___ **19.** The letters SCP stand for single-cell polysaccharide, a form of microorganisms used as food.

__ **20.** Among the many uses of bacteria is their application to the fermentation industry where they produce a variety of alcoholic beverages.

__ **21.** For beer fermentations, the carbohydrates are obtained from starchy materials such as grapes.

__ **22.** The original source of antibiotics for medical use was a variety of microorganisms.

__ **23.** Among the many enzymes industrially produced by microorganisms are amylase, catalase, cellulase, and lipase.

__ **24.** A valuable use of *Bacillus thuringiensis* in a practical situation is as a herbicide.

__ **25.** Genetic engineering is commonly performed by altering the cell membranes of carrier organisms with new genes.

Multiple Choice. Select the letter of the item that correctly completes each of the following statements.

1. Microorganisms are an important factor in the production of all the following foods except (*a*) sauerkraut (*b*) milk (*c*) pickled cucumbers (*d*) Swiss cheese.

2. Which of the following foods or food products would be likely to spoil the quickest: (*a*) a box of pasta (*b*) a sack of flour (*c*) a jar of salt (*d*) a pound of hamburger meat.

3. Species of *Clostridium* are primarily responsible for the contamination of canned food with (*a*) gas (*b*) acid (*c*) alkaline substances (*d*) alkene substances.

4. Cheese and cheese products are produced from the milk component (*a*) lactose, a carbohydrate (*b*) lactose, a protein (*c*) butterfat (*d*) casein, a protein.

5. All the following are recognized sources of microbial contamination in food with the exception of (*a*) animal waste (*b*) ultraviolet light (*c*) soil (*d*) sewage.

6. One of the drawbacks of preservation using refrigeration is that (*a*) refrigeration does not kill microorganisms (*b*) refrigeration does not reduce the metabolism of microorganisms (*c*) refrigeration kills spores that may be present (*d*) refrigeration lowers the enzyme activity of microorganisms.

7. Boiling is an effective way of preserving foods, especially if the foods are high in (*a*) fungi (*b*) acidity (*c*) calcium ions (*d*) vitamins and minerals.

8. Drying as a preservation method has a negligible effect on (*a*) viruses (*b*) the hyphae of fungi (*c*) protozoal cells (*d*) bacterial spores.

9. All the following are chemicals approved for preservation purposes in foods except (*a*) sulfur dioxide (*b*) sorbic acid (*c*) nitric acid (*d*) benzoic acid.

10. Filtration can be used as a physical method for preserving foods in such things as (*a*) meats (*b*) dairy products (*c*) cider and beer (*d*) canned vegetables.

11. Species of staphylococci and clostridia are widely acknowledged as causes of (*a*) food infections (*b*) food poisonings (*c*) food toxemias (*d*) food septicemias.

12. All the following would be required to perform a standard plate count procedure on a sample of food with the exception of (*a*) a bacteriological growth medium (*b*) a number of Petri dishes (*c*) a device to measure 1 ml samples of diluted food (*d*) a light microscope.

13. To determine whether food has been contaminated with fecal material, it is convenient to test for the presence of (*a*) protozoal cysts (*b*) adenoviruses (*c*) coliform bacteria (*d*) *Bacillus* spores.

14. When milk has been pasteurized successfully, the milk will no longer contain the enzyme (*a*) polymerase (*b*) phosphatase (*c*) peroxidase (*d*) purinase.

15. Two types of fungi commonly used as foods are (*a*) *Aspergillus* and *Cladosporium* species (*b*) *Candida* and *Cryptococcus* species (*c*) yeasts and mushrooms (*d*) ergot and aflatoxin.

16. The continuous-flow and batch techniques are both used to (*a*) pasteurize large quantities of milk (*b*) produce microorganisms in industrial quantities (*c*) perform the standard plate count technique (*d*) perform the membrane filter technique.

17. A dry wine is produced by yeast when they ferment grapes and (*a*) produce large number of aromatic compounds (*b*) produce large quantities of protein in the wine (*c*) use up all the minerals in the wine (*d*) use up all the sugar in the wine.

18. The organism *Saccharomyces cerevisiae* is best known for its importance in (*a*) antibiotic production (*b*) alcohol fermentation (*c*) sauerkraut production.

19. All the following enzymes are produced by microorganisms for practical uses except (*a*) protease (*b*) streptokinase (*c*) pectinase (*d*) hydrolase.

20. The bacterium *Bacillus thuringiensis* is widely used in contemporary biology as (*a*) a source of fermentation enzymes (*b*) a producer of cheese and cheese products (*c*) an insecticide (*d*) a purifier of water systems.

Completion. For each of the following, add the word or words that best completes the thought .

1. Milk and dairy products are pasteurized for 30 minutes in the ____________ method, or from 15 to17 seconds in the flash method.

2. In the production of buttermilk and sour cream, a strain of *Streptococcus lactis* produces ____________ from the carbohydrate in milk.

3. Soft, ripened cheeses such as ____________ cheese are produced by the growth of *Penicillium* in the milk curd.

4. Cabbage is grated, salted, and allowed to undergo fermentation by microorganisms in the production of ____________.

5. The contamination usually affects only the surface of meat, but ____________ tissues may also be affected because they are relatively loose.

6. Meats can be preserved for long periods of time, but freezer burn due to ___________ is a potential problem.

7. Boiling at 100°C for 30 to 60 minutes destroys the ___________ cells of bacteria but has no effect on spores.

8. The TDP is the thermal death point, which is the ___________ required for the destruction of a microorganisms at a given time.

9. Foods are dehydrated and chemical preservatives are deposited in foods when the process of ___________ is used.

10. Several types of fruits and vegetables are now preserved by using ___________ irradiation.

11. Such bacterial diseases as salmonellosis, cholera, ___________, and typhoid fever may be due to contamination of foods by microorganisms.

12. To produce a ___________ wine, the fermentation is allowed to continue in the bottle and gas is formed by the fermentation.

13. Although sake is often considered a wine, it is more correctly a ___________ because it is produced from a type of grain.

14. Glutamic acid and lysine are among the ___________ produced for human consumption by microorganisms.

15. *Bacillus popillae* is used as a bacterial insecticide against Japanese beetles where it causes ___________.

Chapter 25

Environmental Microbiology

OBJECTIVES

Microorganisms occupy an important niche in the environment, because of the significance of their metabolic activities. These activities are explored in this chapter, whose objective are to:

1. Define and ecosystem and show how microorganisms fit into aquatic and soil ecosystems.

2. Outline in detail the critical activities played by microorganisms in the cycles of oxygen and carbon on Earth.

3. Delineate where microorganisms make key contributions to the cycle of nitrogen, and show how life on Earth would probably be impossible without the microbial activity of the nitrogen cycle.

4. Discuss the breakdown of sewage in wastewater with an emphasis on the steps involving microorganisms.

5. Summarize the processes in the purification of water used by consumers.

6. Explain some of the laboratory tests used to determine whether waters are safe for drinking.

THEORY AND PROBLEMS

25.1 Why is it important to study microorganisms living in the natural environment?

Preservation of the natural environment depends heavily on the roles played by microorganisms in maintaining the chemical balance of the available nutrients and the waste products of metabolism. In addition, microorganisms are key elements in the cycles occurring in soils, and they are often responsible for breaking down the pollutants that enter the environment.

MICROBIAL ECOLOGY

25.2 What is an ecosystem and how do microorganisms fit into ecosystems?

An **ecosystem** is the community of organisms found in a physically defined space. Ecosystems may be as small as a swamp or pond or as huge as one of the Great Lakes. The sum total of living organisms within the ecosystem is the **biota**. The dynamic interactions of the microorganisms of the biota with the physical and chemical makeup of the ecosystem comprises the subject matter of **environmental microbiology**.

25.3 What is the relationship of microbial ecology to the ecosystem?

Microbial ecology is the study of the relationship occurring between different microbial populations and their environments. An ecosystem consists of numerous microbial communities each influenced by local physical and chemical conditions such as temperature, pH, and nutrient and oxygen availability.

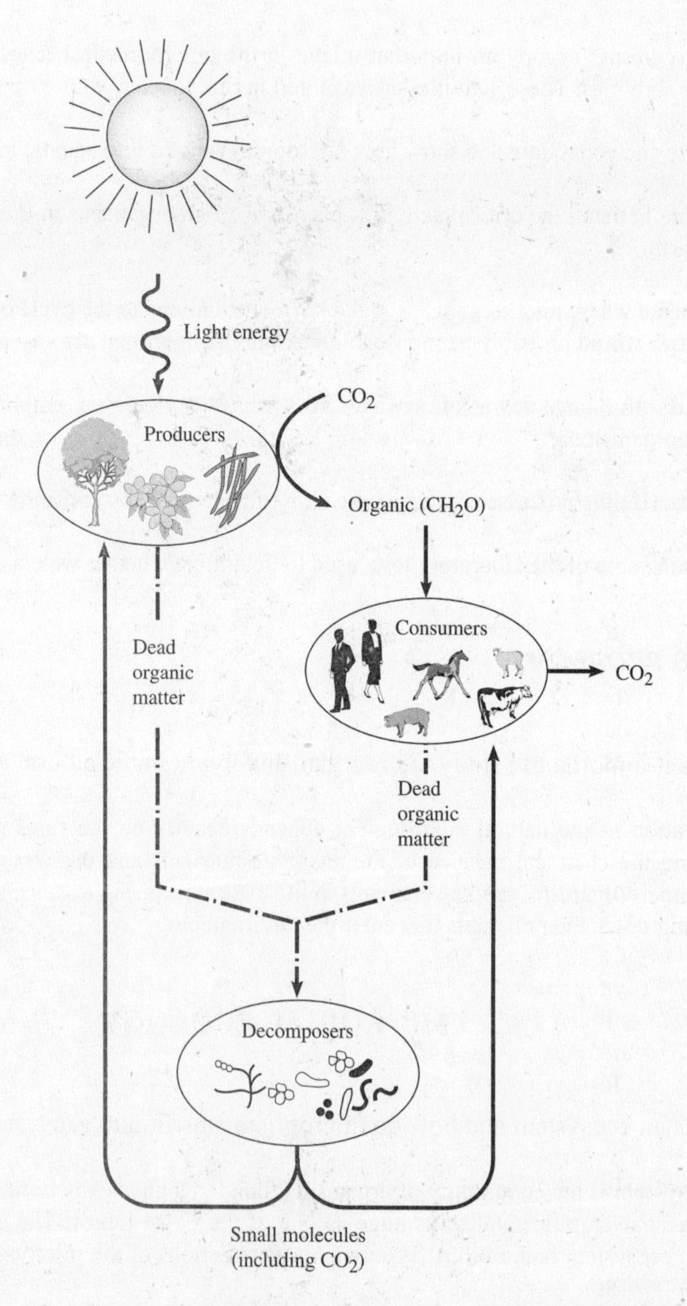

Fig. 25-1 The cycle of organic matter as performed by producers, consumers, and decomposers in the ecosystem.

25.4 Describe the roles played by microorganisms in aquatic ecosystems such as freshwater lakes.

In **aquatic ecosystems** such as freshwater lakes, microorganisms are **producers**. On sunny days light saturates the surface water and cyanobacteria and oxygenic green algae perform photosynthesis to generate carbohydrates. The photosynthetic microorganisms derive their minerals and carbon dioxide from the metabolic activities performed by other bacteria and zooplankton, which break down organic matter into their simple elements. Organisms as these are **decomposers**. The photosynthetic microorganisms thus generate new biomass. **Consumers** such as animals and humans use the biomass produced in photosynthesis as food. When they die, the decomposers take over once again. Figure 25-1 depicts these transformations.

25.5 Are anaerobic bacteria important in aquatic ecosystems?

Many types of photosynthetic bacteria exist without oxygen in an anaerobic environment. In the lower regions of an aquatic ecosystem, some light penetrates and the anaerobic photosynthetic organisms carry out the synthesis of carbohydrates. These organisms use as a substrate for photosynthesis a variety of organic compounds or reduced sulfur compounds such as H_2S.

25.6 Which types of bacteria are found in shallow aquatic environments?

Bacteria as well as brown and green algae cover the rocks in stream beds and colonize the surfaces of particulate matter in shallow aquatic environments. Examples of aquatic bacteria in this region are species of *Caulobacteria* and *Gallionella*, both of which have stalks for adhering to objects in a stream. *Gallionella* commonly clogs drains, water pipes, and wells, as it deposits iron compounds such as iron oxides.

25.7 Which is the largest contributor of organic matter to the soil?

In the **soil ecosystem**, microorganisms are the largest contributors of organic matter. The organic matter is derived through the metabolism of animal and plant waste. The material not recycled combines with mineral particles to form the dark-colored material of soil called **humus**. Humus is composed primarily of decay-resistant organic matter. It increases the soil's ability to retain air and water.

25.8 Which types of microorganisms are present in the soil?

Most soils contain a rich and heterogeneous population of protozoa, primarily flagellates, amoebas, and ciliates. Photosynthetic microorganisms include cyanobacteria, green algae, and diatoms. Most bacteria are found in the upper 25 cm of soil, and representative genera include *Arthrobacter, Bacillus, Actinomycetes, Pseudomonas,* and *Agrobacterium.* Fungi present in the soil generally exist in their spore form, but the hyphal forms grow actively in acidic soil.

25.9 What are renewable resources and what is the importance of microorganisms in their recycling?

Renewable resources are resources that can be recycled through the interactions of natural processes to be used over and over again. Such things as trees, animal foods, plant foods, and cotton are renewable resources. Microorganisms are absolutely essential in the web of metabolic activities that renew resources. Microorganisms thus maintain an ecological balance on Earth.

BIOGEOCHEMICAL CYCLES

25.10 What chemical activities are performed by microorganisms in the oxygen cycle?

In the **oxygen cycle**, oxygen is a key element for the chemical reactions of cellular respiration (glycolysis, Krebs cycle, electron transport, chemiosmosis). The atmosphere is the chief reservoir of oxygen available for these processes. Oxygen is returned to the atmosphere for use in metabolism by photosynthetic green plants and photosynthetic microorganisms such as cyanobacteria. During the process of photosynthesis, these organisms produce oxygen from water and liberate the oxygen to the atmosphere, as Figure 25-2 illustrates. The oxygen is then made available for the reactions of cellular respiration by heterotrophic organisms including plants, animals, and microorganisms.

25.11 How do microorganisms influence the carbon cycle on Earth?

Most of the organic matter present in soil originates in plant material from leaves, rotting trees, decaying roots, and other tissues. In the **carbon cycle**, soil bacteria and fungi recycle this carbon by using the organic matter in their metabolism. Without the recycling action of these organisms, life would suffer an irreversible decline as the nutrients essential for life became tied up in complex molecules.

25.12 What happens to dead vegetable matter when it is deposited in the soil?

Dead vegetable matter is digested by extracellular microbial enzymes into soluble products available to microbes. The fungi and bacteria then metabolize the soluble organic products to simpler products such as carbon dioxide, small acids, and other materials available for plant growth. These elements are made available to the root systems of plants. Undigested plant and animal matter becomes part of the humus.

25.13 Why is nitrogen an essential chemical element in organic compounds?

Nitrogen is an essential cellular element of amino acids, purines, pyrimidines, and certain coenzymes. Nitrogen accounts for about 9 to 15 percent of the dry weight of a cell. The organic compounds of life could not be formed without nitrogen.

25.14 In the nitrogen cycle what are the sources of nitrogen for plants and animals?

In the **nitrogen cycle**, many organisms obtain their nitrogen from organic sources such as amino acids or purines, while other organisms obtain their nitrogen from inorganic compounds such as nitrogen gas (N_2), ammonia (NH_3), or nitrate (NO_3^{1-}). Before nitrate or nitrogen gas can be used, however, the nitrogen in the compounds must be reduced to the nitrogen in ammonia. The ammonia is then assimilated by an enzyme-catalyzed pathway in which glutamic acid and glutamine are formed. These amino acids are then used to form other nitrogen compounds in the cell.

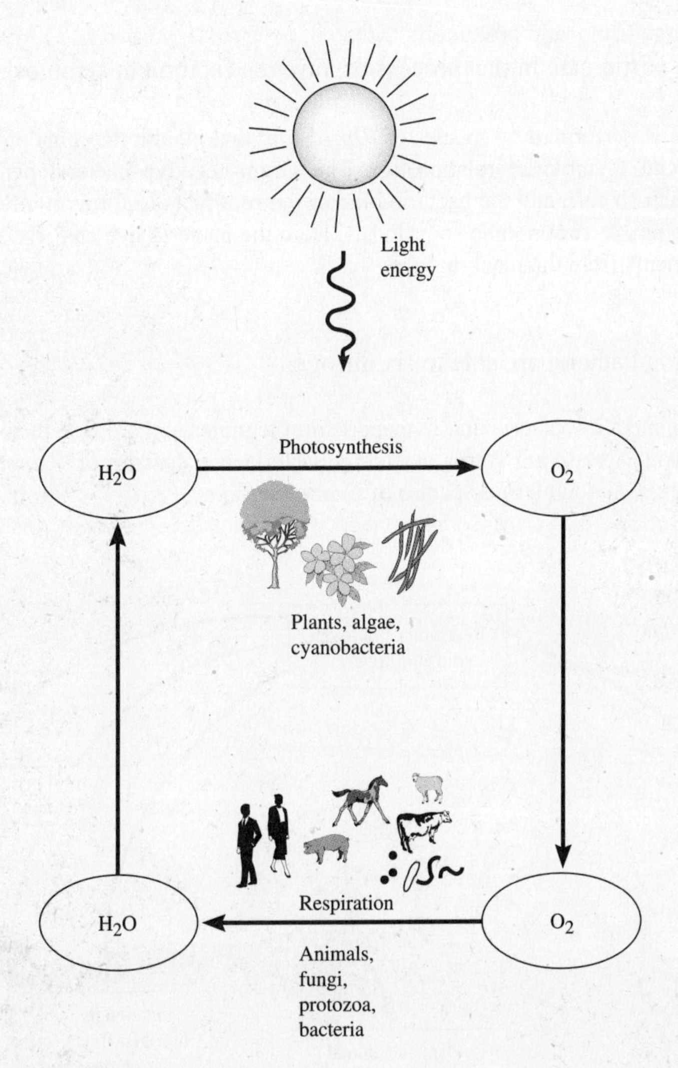

Fig. 25-2 The oxygen cycle occurring in the Earth ecosystem and showing the importance of microorganisms in the two major phases.

25.15 What are the main reservoirs of nitrogen on the Earth?

The principal reservoir of nitrogen on Earth is the atmosphere, which contains 80 percent nitrogen. In the marine environment, nitrate supplies approximately 60 percent of the nitrogen for life forms. The interconversions of various forms of nitrogen are essential aspects of the nitrogen cycle in an ecosystem.

25.16 By what process is nitrogen fixed on Earth?

In the process of **nitrogen fixation**, nitrogen gas from the atmosphere is used to form ammonia by the chemical process of reduction. Nitrogen fixation is performed by free-living bacteria as well as by bacteria growing in symbiosis with leguminous plants (plants that bear their seeds in pods , such as peas, beans, alfalfa, clover, and soybeans). The nitrogen-fixing bacteria use the enzyme nitrogenase to reduce nitrogen to ammonia, which is then used for the production of amino acids.

25.17 Which bacteria participate in the process of nitrogen fixation in symbiosis?

Nitrogen fixation is performed by species of *Rhizobium* that inhabit the roots of leguminous plants in a mutually beneficial (symbiotic) relationship. The Gram-negative bacteria penetrate the root hair, stimulate the root hair to curl, and the bacteria bind to the root hairs to form an infection thread growing toward the root proper. A root nodule soon forms. Here the bacteria live and fix atmospheric nitrogen, while deriving nutrients from the plant, in turn.

25.18 Which free-living bacteria are able to fix nitrogen?

There are many genera of bacteria that live apart from legumes and are able to fix nitrogen in the soil. Among the important free-living nitrogen fixing bacteria are species of *Azotobacter, Azospirillum, Bacillus, Beijereinckia,* and numerous species of *Cyanobacteria.*

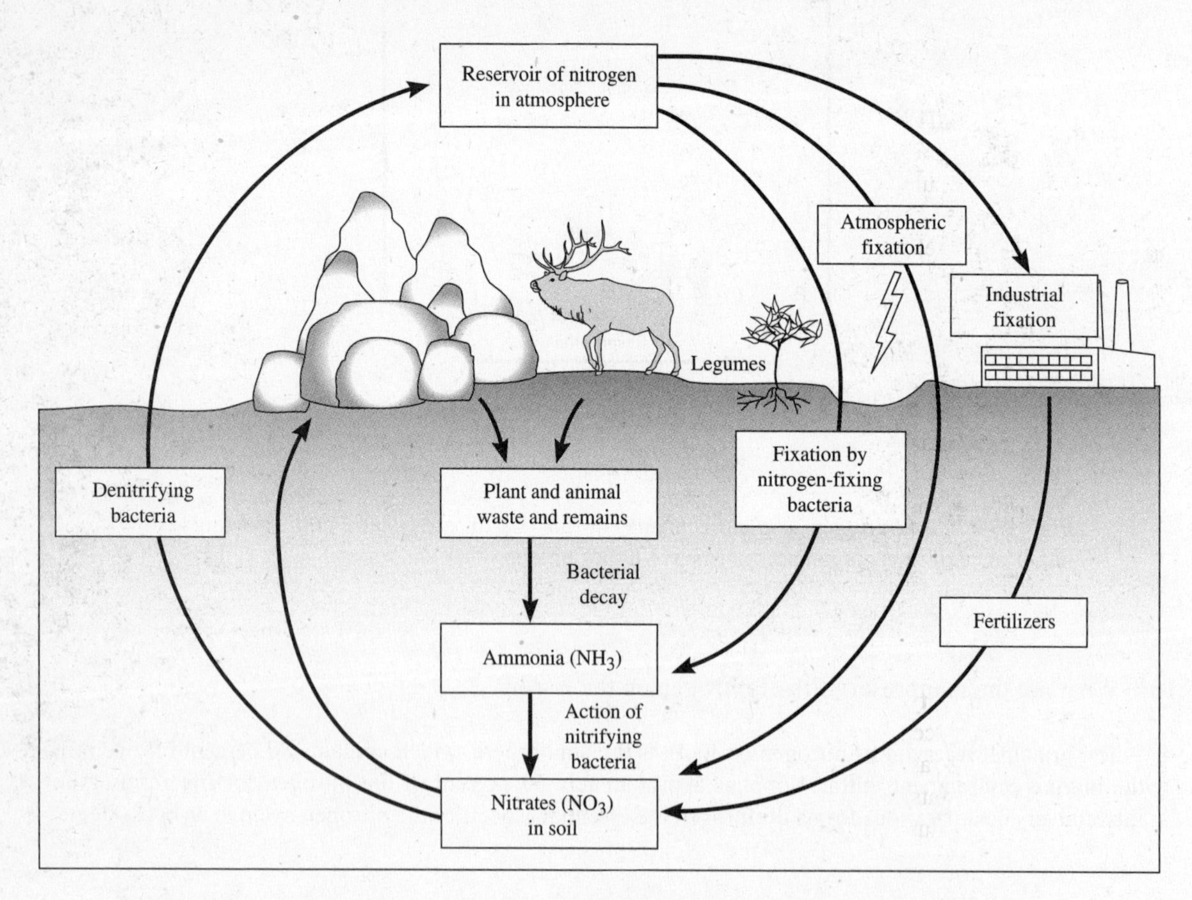

Fig. 25-3 The nitrogen cycle and its various aspects. Bacteria use their metabolic activities to influence numerous parts of the cycle.

25.19 What does the process of ammonification entail?

Once nitrogen has been incorporated into ammonia, the ammonia is utilized for various organic substances, using glutamic acid and glutamine as intermediaries. Later, when plants, animals, and microorganisms die, the nitrogen is recycled when the nitrogen from various compounds is released and used to form ammonia. This is the process of **ammonification**. For example, proteins and nucleic acids are broken down first to amino acids and purines, and then to acids, gases, and ammonia. The ammonia is released to the environment where it is used as a nitrogen source for animals, plants, and other organisms. Ammonification also occurs from the excretory products of animals such as urea, the major component of the urine. The urea is broken down by urea-digesting bacteria, and ammonia is produced through this ammonification process.

25.20 Which chemical processes occur in the nitrification process?

The conversion of ammonia to nitrate (NO_3^{1-}) is the process of **nitrification**. Nitrifying bacteria, such as species of *Nitrosomonas* and *Nitrosococcus*, are involved. *Nitrosomonas* species convert ammonia to nitrite (NO_2^{1-}), then *Nitrosococcus* species convert the nitrite to nitrate (NO_3^{1-}). Nitrification occurs in soils, water, and marine environments where the bacteria are found. The nitrate that results serves as an important nitrogen source for plants. The nitrogen cycle is illustrated in Figure 25-3.

25.21 What is the advantage of denitrification, and what takes place in this process?

Denitrification is the process in which the nitrogen of nitrate is converted into gaseous nitrogen. The process makes nitrogen available to bacteria that use it in the process of nitrogen fixation. Denitrification is accomplished by numerous bacteria that reduce nitrite (NO_2^{1-}) to nitrous oxide (N_2O), and then to atmospheric nitrogen (N_2).

25.22 Explain the importance of sulfur in the metabolism of living cells.

Sulfur makes up a small percentage of the dry weight of a cell (approximately 1 percent), but it is an important element in the formation of certain amino acids such as methionine and glutathione, and it is used in the formation of many enzymes.

25.23 How do bacteria express their importance in the sulfur cycle?

Many bacteria have an important place in the **sulfur cycle** in the soil. Sulfate-reducing bacteria grow in mud and anaerobic water environments where they reduce sulfate compounds to hydrogen sulfide (H_2S). Photosynthetic sulfur bacteria then grow anaerobically and oxidize the H_2S, thereby releasing the sulfur as elemental sulfur (S). Species of colorless sulfur bacteria, including members of the genus *Thiobacillus*, *Beggiatoa,* and *Thiothrix*, also grow in hydrogen sulfide, but they convert the hydrogen sulfide to sulfate, which is then made available to plants.

25.24 What is the importance of phosphorus in cellular metabolism?

Living things use organic and inorganic phosphorus sources in the synthesis of nucleotides, phospholipids, and phosphorylated proteins. Phosphorus enters the soil and water as phosphate during the breakdown of crops, decaying garbage, leaf litter, and other sources.

25.25 How do microorganisms enter into the phosphorus cycle in the soil?

In the **phosphorus cycle**, microorganisms use phosphorus in the form of calcium phosphate, magnesium phosphate, and iron phosphate. They release the phosphorus from these complexes and assimilate the phosphorus as the phosphate ion (PO_4). This ion is incorporated into DNA, RNA, and other organic compounds utilizing phosphate, including phospholipids. When the organisms are used as foods by larger organisms, the phosphorus is concentrated in the food chain.

WASTEWATER MICROBIOLOGY

25.26 What is the importance of wastewater in environmental microbiology?

Natural waterways can consume and dispose of wastes up to a point. However, human activities generate a tremendous volume of wastewater that requires treatment before discharge into waterways. Often this wastewater contains excessive amounts of nitrogen, phosphorus, and metal compounds, as well as organic pollutants that would overwhelm waterways with an unreasonable burden. Wastewater also contains chemical wastes that are not biodegradable as well as pathogenic microorganisms that can cause infectious disease. Wastewater treatment facilities are therefore an important aspect of environmental microbiology.

25.27 Which principle is used to measure the pollution level of wastewater?

The extent of pollution in wastewater can be determined by measuring the **biochemical oxygen demand (BOD)**. The BOD is the amount of oxygen required by the microorganisms during their growth in wastewater. The BOD is determined by measuring the oxygen concentration in a sample of water before incubation and after incubation in an airtight stoppered bottle for a period of about five days at a temperature between 5 and 20°C. The difference in the dissolved oxygen is the BOD. A higher BOD indicates a higher amount of organic matter in the wastewater. High BOD values are found in wastewater from agricultural communities, food processing plants, and certain industries.

25.28 What are the primary steps in the treatment of wastewater?

The first or **primary treatment** of wastewater involves the removal of particulate matter in settling tanks. The solids that sediment are drained off and the sludge is collected to be burned or buried in landfills. Alternately, it can be treated in an anaerobic sludge-digesting tank.

25.29 Explain the steps in the secondary treatment of wastewater and sewage?

During the **secondary treatment** of wastewater and sewage, the BOD of liquid and sludge waste is reduced. In the anaerobic sludge digester, microorganisms break down the organic matter of proteins, lipids, and cellulose into smaller substances that can be metabolized by other organisms. Organic acids, alcohols, and simple compounds results. Methane is a common gas produced in the sludge tank, and it can be burned as a fuel to operate the waste treatment facility. The remaining sludge is burned or buried in a landfill and its fluid is recycled and purified.

25.30 What occurs in aerobic treatment of sewage in the secondary stage?

In aerobic secondary sewage treatment, the fluid waste is aerated then placed through a **trickling filter** such as pictured in Figure 25-4. In this process, the liquid waste is sprayed over a bed of crushed rocks, bark, or other filtering materials. Colonies of bacteria, fungi, and protozoa grow in this bed and act as filters to remove organic materials passing through. The microorganisms metabolize the organic compounds and convert them to carbon dioxide, sulfate, phosphates, nitrates, and other ions. The material that comes through the filter has been 99 percent cleansed of microorganisms.

25.31 Which other method is available for treating liquid waste?

Liquid waste can also be treated in an activated digester after it has been vigorously aerated. Slime-forming bacteria form on accumulating flocs and trap other microorganisms to remove them from the water. Treatment for several hours reduces the BOD significantly, and the clear fluid is removed for purification. The sludge is disposed of in a landfill.

25.32 What is the purpose of tertiary sewage treatment?

In the **tertiary treatment** of sewage, the fluid from the secondary treatment process is cleansed of phosphate and nitrate products that otherwise might cause pollution. The ions are precipitated as solids, often by combining them with calcium or iron, and ammonia is released by oxidizing it to nitrate in the nitrification process. Adsorption to activated charcoal removes many organic compounds such as polychlorinated biphenyls (PCBs) that might otherwise work as pollutants.

Fig. 25-4 A trickling filter used to spray water over filtering materials during the process of water purification.

25.33 What processes are used in water purification when water is to be used for drinking?

Essentially the same processes are used for **drinking water** as used for wastewater. Initially, the solid matter is allowed to settle out, then the water is filtered either through a slow sand or a rapid sand filter, and these processes remove 99 percent of the microorganisms. Many communities then add chlorine to water to maintain the low microbial count and ensure that the water is safe for drinking purposes. Chlorine gas or hypochlorite (NaOCl) is used for these purposes.

25.34 How is environmental microbiology practiced in the home septic system?

The home septic system is a miniature waste treatment facility. In a **septic tank**, household sewage is digested by anaerobic bacteria, and the solids settle to the bottom of the tank. Solid waste is carried out of the outflow apparatus into the septic field composed of tile beneath the ground. The water seeps out through holes in the tiles and enters the soil where bacteria complete the breakdown processes. A similar process occurs in **cesspools**, except that the sludge enters the ground at the bottom of the pool and liquids flow out through the sides of the pool into the ground.

25.35 What methods are available for detecting the bacterial contamination of water and water environments?

It is generally impossible to test for all pathogenic organisms, but indicator bacteria can be detected with several established tests. One test is the **standard plate count**, such is used for food and milk samples (Chapter 24). There are other tests available depending upon the needs of the situation.

25.36 Which indicator organisms are used in water bacteriology?

The most common indicator organisms in water bacteriology are the **coliform bacteria**. These are a group of Gram-negative intestinal rods normally found in the intestine and typified by *Escherichia coli*. If the indicator organisms are present, then it is likely that the water has been contaminated with human feces.

25.37 How is the membrane filter technique used to detect bacteria in water?

The **membrane filter technique** uses a filtration apparatus and a cellulose filter known as a membrane filter. A 100-ml sample of water is passed through the filter and the filter pad is then transferred to a medium that will support the growth of bacteria. Bacteria trapped in the filter grow on the medium and form colonies. By counting the colonies, an estimation can be made of the number of bacteria in the original 100-ml sample.

25.38 Can the membrane filter technique be used to detect coliform bacteria?

The membrane filter technique can easily be adapted for coliform bacteria by using a medium such as **violet red bile agar**. Coliform bacteria such as *E. coli* will form red colonies on this medium.

25.39 How is the most probable number (MPN) test performed in water bacteriology?

In the **most probable number (MPN) test**, tubes of lactose broth are inoculated with water samples measuring 10 ml, 1 ml, and 0.1 ml. During incubation, coliform organisms will produce gas. Depending upon which tubes display gas from which water samples, an MPN table is consulted and a statistical range of the number of coliform bacteria is determined. The MPN test is very easy to perform and interpret, but it does not determine the exact number of bacteria, nor does it test for bacteria other than coliform bacteria.

25.40 Which tests are used to determine the presence of *E. coli* in water?

In order to test for the presence of *E. coli* in water, a medium called **eosin methylene blue (EMB) agar** is used. *E. coli* colonies grow on this medium and become green with a metallic fluorescent sheen. Thus, if *E. coli* is present in the water, the characteristic colonies will form, and the water is designated polluted.

25.41 Are there any methods for detecting bacteria in water that employ biotechnology procedures?

Among the most sophisticated tests for water bacteriology are those that employ **gene probes**. Gene probes are fragments of DNA that seek out and combine with complementary DNA fragments. For example, if a test for *E. coli* were to be performed, the water would be treated to disrupt any bacteria present and release its nucleic acid. Then a specific *E. coli* probe would be added to the water. Like a left hand seeking a right hand, the probe would search among all the nucleic acid in the water and unite with the *E. coli* DNA if present. A radioactive signal indicates that a match has been made and *E. coli* has been detected. If no radioactivity is emitted, then the gene probe has been unable to locate its matching DNA, and *E. coli* is probably absent from the water.

Review Questions

Completion. Add the word or words that correctly complete each of the following statements.

1. A community of organisms such as microorganisms found in a physically defined space is an ____________.

2. The oxygenic green algae in a freshwater lake generate carbohydrates by the process of ____________.

3. Iron compounds such as iron oxides may be deposited in drains, water pipes, and wells by bacteria belonging to the genus ____________.

4. Most fungi found in the soil environment exist as ____________.

5. Microorganisms influence the carbon cycle on Earth by decaying and using the organic matter found in cells primarily from the ____________.

6. Nitrogen is an essential cellular element because it is found in such compounds as purines, pyrimidines, certain coenzymes, and the building blocks of proteins known as ____________.

7. The roots of leguminous plants are inhabited by species of nitrogen-fixing bacteria belonging to the genus ____________.

8. A major source of nitrogen deposited in the soil is the major component of urine known as ____________.

9. The process in which the nitrogen from nitrate ions is converted into gaseous nitrogen is correctly known as ____________.

10. The element sulfur must constantly be recycled in nature because it is an important aspect of certain amino acids such as glutathione and ____________.

11. Sulfate-reducing bacteria grow in the anaerobic mud of a water environment and reduce sulfate compounds to ____________.

12. Phosphorus, which is used for nucleotides and phospholipids, enters the soil during the breakdown of crops as ____________.

13. The extent of pollution in wastewater can be determined by measuring the ____________.

14. In the primary steps of wastewater treatment, particulate matter is siphoned off into ____________.

15. One of the gases produced during the secondary treatment of sewage can be burned as a fuel and is known as ____________.

16. In the tertiary treatment of sewage, the fluid from secondary treatment is cleansed of phosphate and nitrate products that might cause ____________.

17. At least 99 percent of the microorganisms are removed from water when water is placed through a ____________.

18. In water bacteriology the common indicator organism is *Escherichia coli*, which belongs to a group called ____________.

19. Bacteria can be trapped for colony growth when a sample of water is trapped in a ____________.

20. Tubes of lactose broth are inoculated with samples of different amounts of water in the laboratory test known as the ____________.

True/False. Enter the letter "T" if the statement is true, enter the letter "F" if the statement or any part of the statement is false.

___ **1.** The biota is the name given to the sum total of living organisms found within an ecosystem.

___ **2.** Photosynthetic microorganisms are responsible for generating new biomass in an aquatic ecosystem.

___ **3.** Animals and plants are the largest contributors of organic matter in the soil ecosystem.

___ **4.** Most soils contain a rich and heterogeneous population of microorganisms, including protozoa, cyanobacteria, algae, bacteria, and fungi.

___ **5.** Photosynthetic green plants and photosynthetic microorganisms are responsible for returning sulfur to the atmosphere for use by other organisms.

___ **6.** Without the recycling action of soil bacteria and fungi, nutrients essential for life would remain tied up in complex molecules.

___ **7.** At least 50 percent of the dry weight of a cell is nitrogen, an essential element of amino acids.

___ **8.** Nitrogen fixation can be accomplished only by bacteria living in symbiosis with leguminous green plants.

___ **9.** Species of bacteria that inhabit the roots of leguminous plants are all acid-fast.

___ **10.** In the process of ammonification, proteins and nucleic acid are first broken down into their constituent parts.

___ **11.** In the process of denitrification, gaseous nitrogen is converted into nitrate, which is then made available to bacteria.

___ **12.** Such amino acids as valine, alanine, and histidine contain sulfur as an essential part of their molecular makeup.

___ **13.** Phosphorus is important in cellular metabolism because it is used in the synthesis of nucleotides and phospholipids.

___ **14.** The BOD is determined by measuring the nitrogen concentration in a sample of water before and after incubation.

___ **15.** To maintain low microbial counts, many communities add chlorine gas to water.

___ **16.** In a home septic tank, sludge enters the ground at the bottom of the tank and liquid flows out through the sides of the tank.

___ **17.** The coliform bacteria used as indicators of pollution are Gram-negative intestinal rods that include *Escherichia coli.*

___ **18.** In the membrane filter technique, bacteria grow in a liquid medium, and the density of the medium is used as an indication of how many bacteria are present.

___ **19.** A medium such as eosin methylene blue agar can be used to cultivate *Escherichia coli* from water, and form green colonies with a metallic fluorescent sheen.

___ **20.** Gene probes are useful in water bacteriology because they will detect the presence of DNA from a certain organism if that organism is present in the water.

Multiple Choice. Select the letter of the item that correctly completes each of the following statements.

1. The study of the relationship occurring between different microbial populations and their environments is (*a*) microbial evolution (*b*) microbial physiology (*c*) microbial ecology (*d*) microbial biochemistry.

2. An essential factor that must be present for cyanobacteria and green algae to perform photosynthesis is (*a*) nitrate ions (*b*) sunlight (*c*) vitamins A, C, and E (*d*) other microorganisms such as fungi.

3. At the bottom of aquatic ecosystems where there is not oxygen, photosynthesis is carried out using as substrates (*a*) phosphorus and ammonia compounds (*b*) reduced sulfur and organic compounds (*c*) carbon dioxide and oxygen as substrates (*d*) minerals such as nitrate and nitrite.

4. The soil material known as humus is composed primarily of (*a*) phosphates and nitrates (*b*) inorganic substances such as iron oxides (*c*) fermented acids and bases (*d*) organic matter that resists decay.

5. In the soil environment, most bacteria are found in (*a*) the deepest part of the soil (*b*) within the first few millimeters (*c*) only where there are large quantities of sand (*d*) in the upper 25 centimeters.

6. Fungi are expected to be found in soil that is (*a*) rich in nitrogen compounds (*b*) acidic (*c*) neutral in character (*d*) also filled with bacteria and protozoa.

7. Renewable resources are those resources (*a*) that can be recycled (*b*) that are used for the increase in biomass (*c*) that participate in the sulfur cycle of life (*d*) that are transported in natural waterways.

8. Photosynthetic green plants and photosynthetic microorganisms such as cyanobacteria are responsible for returning to the atmosphere large quantities of (*a*) nitrogen (*b*) oxygen (*c*) potassium (*d*) zinc.

9. All the following are essential components of the nitrogen cycle on Earth except (*a*) calcium phosphate (*b*) ammonia (*c*) nitrate ions (*d*) amino acids.

10. Nitrogen-fixing bacteria may be located (*a*) on the roots of leguminous plants (*b*) in the outer fringes of space (*c*) in the human intestine (*d*) associated with most beef and fish products.

11. The major product of the process of ammonification occurring in the soil is (*a*) urea (*b*) amino acids (*c*) protein (*d*) ammonia.

12. Bacteria species of *Thiobacillus* and *Beggiatoa* play an important role in the (*a*) water cycle on Earth (*b*) phosphorus cycle (*c*) sulfur cycle in the soil (*d*) breakdown of sewage.

13. The BOD helps determine the (*a*) extend of pollution in wastewater (*b*) filtering capacity of the soil (*c*) number of bacteria in a 100 ml sample of water (*d*) types of biota in the ecosystem.

14. One of the first steps in wastewater treatment is the (*a*) addition of chlorine (*b*) addition of hydrogen sulfide (*c*) removal of particulate matter (*d*) addition of phosphorus to the water.

15. One of the purposes of secondary treatment of wastewater and sewage is to (*a*) increase the chlorine content (*b*) reduce the BOD (*c*) encourage the formation of PCBs (*d*) discourage ammonification.

16. Sludge that accumulates during the secondary treatment of wastewater and sewage can be (*a*) placed in a landfill (*b*) used as fertilizer for crops (*c*) placed in a local stream or river (*d*) used as a source of nitrogen-fixing bacteria.

17. Chlorine gas is used to maintain (*a*) a low microbial count in water (*b*) the development of particulate matter in primary sewage treatment (*c*) the development of humus in the soil (*d*) the purity of a slow sand filter.

18. In the analysis of wastewater, *Escherichia coli* is used as (*a*) an indicator of fecal contamination of the water (*b*) a standard organism for performing a plate count (*c*) an indicator of the number of nitrogen-fixing bacteria in the water (*d*) a measure of the amino acid content of the water.

19. The membrane filter is useful for the detection of bacteria in the water because (*a*) it kills unwanted bacteria (*b*) it provides nitrogen for the development of bacteria (*c*) bacterial filters perform on it (*d*) it can be used to remove chlorine from the water.

20. In the processes of water bacteriology, gene probes can be used to locate (*a*) PCBs in water (*b*) nitrate ions in water (*c*) decaying trees in water (*d*) DNA fragments of microorganisms in water.

Answers to Review Questions

CHAPTER 1

Multiple Choice

1. (*b*) **2.** (*c*) **3.** (*c*) **4.** (*a*) **5.** (*a*) **6.** (*d*) **7.** (*d*) **8.** (*b*) **9.** (*a*) **10.** (*b*) **11.** (*d*) **12.** (*b*) **13.** (*c*) **14.** (*c*) **15.** (*c*) **16.** (*d*) **17.** (*b*) **18.** (*a*) **19.** (*d*) **20.** (*c*)

Matching

1. (*c*) **2.** (*d*) **3.** (*e*) **4.** (*c*) **5.** (*b*) **6.** (*a*) 7. (*b*) **8.** (*e*) **9.** (*c*) **10.** (*a*)

True/False

1. diatoms **2.** true **3.** infectious **4.** fungi **5.** true **6.** spontaneous generation **7.** wine **8.** Louis Pasteur **9.** true **10.** 1940s **11.** bacteria **12.** cytoplasm **13.** eukaryotes **14.** protein **15.** true **16.** Fungi **17.** true **18.** acidic **19.** true **20.** plants **21.** hepatitis **22.** van Leeuwenhoek **23.** true **24.** true **25.** World War II **26.** immunizations **27.** Carolus Linnaeus **28.** underlined **29.** five **30.** spirals

CHAPTER 2

Completion

1. elements **2.** nitrogen **3.** organic compounds **4.** molecules **5.** atomic weights **6.** hydrogen bonds **7.** water **8.** photosynthesis **9.** glucose **10.** lactose **11.** cellulose **12.** lipids **13.** dehydration synthesis **14.** amino acids **15.** peptide bond **16.** enzymes **17.** deoxyribonucleic acid **18.** ribose **19.** uracil **20.** semiconservative

True/False

1. F **2.** T **3.** F **4.** F **5.** F **6.** T **7.** F **8.** T **9.** T **10.** F **11.** T **12.** T **13.** F **14.** F **15.** F **16.** T **17.** F **18.** T **19.** T **20.** T

Multiple Choice

1. (*c*) **2.** (*c*) **3.** (*d*) **4.** (*c*) **5.** (*d*) **6.** (*c*) **7.** (*a*) **8.** (*c*) **9.** (*c*) **10.** (*b*) **11.** (*b*) **12.** (*c*) **13.** (*b*) **14.** (*a*) **15.** (*c*) **16.** (*c*) **17.** (*a*) **18.** (*d*) **19.** (*c*) **20.** (*d*)

CHAPTER 3

Completion

1. light microscope **2.** nanometer **3.** mycoplasmas **4.** one million **5.** *Balantidium coli* **6.** two lenses **7.** resolution **8.** refractive index **9.** lenses **10.** diaphragm **11.** spirochetes **12.** yeasts **13.** ultraviolet light **14.** wavelength **15.** scanning electron microscope **16.** transparent **17.** acid-fast procedure **18.** basic dyes **19.** cell wall **20.** *Mycobacterium*

Microscope Matching

1. (*e*) **2.** (*d*) **3.** (*e*) **4.** (*d*) **5.** (*c*) **6.** (*b*) **7.** (*a*) **8.** (*d*) **9.** (*a*) **10.** (*b*)

Multiple Choice

1. (*c*) **2.** (*c*) **3.** (*b*) **4.** (*d*) **5.** (*b*) **6.** (*a*) **7.** (*c*) **8.** (*c*) **9.** (*c*) **10.** (*c*) **11.** (*a*) **12.** (*b*) **13.** (*d*) **14.** (*c*) **15.** (*d*) **16.** (*c*) **17.** (*b*) **18.** (*d*) **19.** (*a*) **20.** (*b*)

Stain Technique Matching

1. (*d*) **2.** (*a*) **3.** (*a*) **4.** (*b*) **5.** (*a*) **6.** (*d*) **7.** (*c*) **8.** (*b*) **9.** (*a*) **10.** (*b*)

CHAPTER 4

Multiple Choice

1. (*b*) **2.** (*c*) **3.** (*c*) **4.** (*c*) **5.** (*a*) **6.** (*d*) **7.** (*b*) **8.** (*d*) **9.** (*b*) **10.** (*d*) **11.** (*b*) **12.** (*c*) **13.** (*b*) **14.** (*d*) **15.** (*c*) **16.** (*a*) **17.** (*b*) **18.** (*c*) **19.** (*d*) **20.** (*a*)

True/False

1. F **2.** T **3.** F **4.** T **5.** T **6.** T **7.** F **8.** F **9.** T **10.** F **11.** T **12.** T **13.** F **14.** F **15.** T **16.** F **17.** T **18.** F **19.** T **20.** T

Completion

1. fluid mosaic model **2.** water **3.** spore **4.** chitin **5.** staphylococcus **6.** glycocalyx **7.** phospholipid **8.** enzymes **9.** cytoplasm **10.** tetanus **11.** cilia **12.** endoplasmic reticulum **13.** RNA **14.** phagocytosis **15.** protozoa **16.** lophotrichous **17.** fimbriae **18.** viruses **19.** bacillus **20.** cell wall

CHAPTER 5

Multiple Choice

1. (*c*) **2.** (*a*) **3.** (*d*) **4.** (*b*) **5.** (*b*) **6.** (*c*) **7.** (*a*) **8.** (*d*) **9.** (*b*) **10.** (*c*) **11.** (*b*) **12.** (*a*) **13.** (*d*) **14.** (*b*) **15.** (*a*) **16.** (*d*) **17.** (*b*) **18.** (*c*) **19.** (*c*) **20.** (*c*)

True/False

1. logarithmic **2.** true **3.** generation **4.** true. **5.** cysts **6.** salt **7.** protozoa **8.** true **9.** true **10.** synthetic **11.** red blood **12.** true **13.** mitosis **14.** yeasts **15.** true **16.** true **17.** liquid **18.** true **19.** acidic **20.** true

Matching

1. (*a*) **2.** (*d*) **3.** (*e*) **4.** (*d*) **5.** (*c*) **6.** (*a*) **7.** (*b*) **8.** (*e*) **9.** (*g*) **10.** (*f*)

Selection

1. (*d*) **2.** (*b*) **3.** (*b*) **4.** (*a*) **5.** (*b*) **6.** (*c*) **7.** (*d*) **8.** (*b*) **9.** (*a*) **10.** (*a*)

CHAPTER 6

Multiple Choice

1. (*c*) **2.** (*a*) **3.** (*b*) **4.** (*a*) **5.** (*d*) **6.** (*c*) **7.** (*d*) **8.** (*a*) **9.** (*a*) **10.** (*c*) **11.** (*d*) **12.** (*a*) **13.** (*c*) **14.** (*c*) **15.** (*c*) **16.** (*c*) **17.** (*b*) **18.** (*d*) **19.** (*b*) **20.** (*d*) **21.** (*b*) **22.** (*a*) **23.** (*c*) **24.** (*a*) **25.** (*b*)

Completion

1. catabolism **2.** protein **3.** -ase **4.** adenosine triphosphate **5.** oxidation reaction **6.** glucose **7.** Embden-Meyerhoff pathway **8.** yeasts **9.** acetyl-CoA **10.** NAD **11.** oxygen **12.** cytochromes **13.** chemiosmosis **14.** 38 **15.** anaerobic **16.** galactose **17.** amino acid **18.** beta oxidation **19.** oxaloacetic acid **20.** light energy **21.** ATP **22.** oxygen **23.** carbon dioxide **24.** glucose **25.** starch

Matching

1. (*d*) **2.** (*a*) **3.** (*d*) **4.** (*c*) **5.**(*b*) **6.** (*d*) **7.** (*a*) **8.** (*b*) **9.** (*d*) **10.** (*a*)

CHAPTER 7

True/False

1. uracil **2.** true **3.** thymine **4.** true **5.** amino acids **6.** transcription **7.** semiconservative **8.** messenger **9.** nucleoid **10.** true **11.** introns **12.** ribosomes **13.** transfer **14.** true **15.** CUG **16.** peptides **17.** endoplasmic reticulum **18.** true **19.** true **20.** operon

Multiple Choice

1. (*c*) **2.** (*b*) **3.** (*c*) **4.** (*b*) **5.** (*d*) **6.** (*c*) **7.** (*a*) **8.** (*b*) **9.** (*b*) **10.** (*d*) **11.** (*c*) **12.** (*c*) **13.** (*d*) **14.** (*a*) **15.** (*b*)

Completion

1. DNA **2.** ribose **3.** genes **4.** DNA polymerase **5.** amino acids **6.** three **7.** transcription **8.** nucleus **9.** mRNA **10.** amino acids **11.** 20 **12.** binding site **13.** secondary **14.** nitrogenous bases **15.** *Escherichia coli*

CHAPTER 8

Completion

1. point mutation **2.** induced mutation **3.** visible light **4.** recombination **5.** transformation **6.** competent **7.** plasmids **8.** *Streptococcus pneumoniae* **9.** F^+ **10.** recombined **11.** jumping genes **12.** virus **13.** generalized transduction **14.** *Salmonella* **15.** restriction enzymes **16.** DNA ligase **17.** chimera **18.** insulin **19.** interferon **20.** biotechnology

Multiple Choice

1. (*c*) **2.** (*a*) **3.** (*c*) **4.** (*d*) **5.** (*c*) **6.** (*c*) **7.** (*a*) **8.** (*d*) **9.** (*c*) **10.** (*c*) **11.** (*c*) **12.** (*b*) **13.** (*a*) **14.** (*d*) **15.** (*c*) **16.** (*b*) **17.** (*b*) **18.**(*d*) **19.** (*b*) **20.** (*d*)

True/False

1. F **2.** T **3.** F **4.** F **5.** T **6.** F **7.** T **8.** F **9.** F **10.** F **11.** T **12.** F **13.** T **14.** F **15.** T **16.** T **17.** F **18.** T **19.** F **20.** F

CHAPTER 9

Multiple Choice

1. (*c*) **2.** (*d*) **3.** (*a*) **4.** (*b*) **5.** (*c*) **6.** (*c*) **7.** (*b*) **8.** (*b*) **9.** (*a*) **10.** (*d*) **11.** (*a*) **12.** (*d*) **13.** (*d*) **14.** (*b*) **15.** (*c*) **16.** (*d*) **17.** (*c*) **18.** (*c*) **19.** (*a*) **20.** (*c*)

Completion

1. organic matter **2.** sanitizing agents **3.** sanitizing agents **4.** spores **5.** oxidation **6.** autoclave **7.** tyndallization **8.** milk **9.** osmosis **10.** formaldehyde **11.** 70 percent **12.** silver **13.** fungi **14.** hydrogen peroxide **15.** broad spectrum antibiotic **16.** cell wall **17.** *Streptomyces* **18.** *Pneumocystis carinii* **19.** acyclovir **20.** folic acid

True/False

1. T **2.** F **3.** F **4.** T **5.** F **6.** T **7.** F **8.** T **9.** F **10.** T **11.** T **12.** T **13.** F **14.** F **15.** T **16.** F **17.** F **18.** T **19.** F **20.** T

CHAPTER 10

Multiple Choice

1. (*c*) **2.** (*b*) **3.** (*a*) **4.** (*d*) **5.** (*b*) **6.** (*c*) **7.** (*b*) **8.** (*d*) **9.** (*c*) **10.** (*a*) **11.** (*c*) **12.** (*c*) **13.** (*c*) **14.** (*d*) **15.** (*b*) **16.** (*a*) **17.** (*c*) **18.** (*c*) **19.** (*b*) **20.** (*d*)

True/False

1. tuberculosis **2.** pleomorphic **3.** photosynthesis **4.** *Neisseria* **5.** true **6.** true **7.** *Yersinia pestis* **8.** true **9.** pertussis **10.** *Veillonella* **11.** enteric **12.** true **13.** axial filament **14.** animals **15.** true **16.** *Moraxella lacunata* **17.** true **18.** peptidoglycan **19.** actinomycetes **20.** true

Matching

1. (*l*) **2.** (*n*) **3.** (*h*) **4.** (*j*) **5.** (*r*) **6.** (*m*) **7.** (*c*) **8.** (*k*) **9.** (*p*) **10.** (*e*) **11.** (*t*) **12.** (*a*) **13.** (*o*) **14.** (*s*) **15.** (*g*) **16.** (*b*) **17.** (*c*) **18.** (*q*) **19.** (*f*) **20.** (*i*)

CHAPTER 11

Multiple Choice

1. (*d*) **2.** (*b*) **3.** (*c*) **4.** (*c*) **5.** (*d*) **6.** (*b*) **7.** (*b*) **8.** (*d*) **9.** (*c*) **10.** (*a*) **11.** (*b*) **12.** (*c*) **13.** (*d*) **14.** (*b*) **15.** (*c*) **16.** (*d*) **17.** (*c*) **18.** (*c*) **19.** (*b*) **20.** (*a*) **21.** (*d*) **22.** (*c*) **23.** (*b*) **24.** (*a*) **25.** (*c*)

Completion

1. molds **2.** acid **3.** oxygen **4.** reproduction **5.** sporangiospores **6.** heterotrophic **7.** glucans **8.** aerial hyphae **9.** oval **10.** ethyl alcohol **11.** "plus" and "minus" **12.** Eumycophyta **13.** Oomycetes **14.** *Rhizopus* **15.** *Amanita*

Matching

1. (*d*) **2.** (*e*) **3.** (*f*) **4.** (*c*) **5.** (*a*) **6.** (*d*) **7.** (*f*) **8.** (*e*) **9.** (*f*) **10.** (*d*) **11.** (*c*) **12.** (*b*) **13.** (*c*) **14.** (*f*) **15.** (*e*)

CHAPTER 12

True/False

1. fluid **2.** cysts **3.** true **4.** trophozoite **5.** pseudopodia **6.** chlorophyll **7.** intestinal **8.** true **9.** AIDS **10.** eukaryotic **11.** motility **12.** true **13.** Apicomplexa **14.** true **15.** Sexually transmitted **16.** *Paramecium* **17.** Sporozoa **18.** true **19.** fungus **20.** mosquitoes

Matching

1. (*c*) **2.** (*e*) **3.** (*d*) **4.** (*i*) **5.** (*b*) **6.** (*g*) **7.** (*a*) **8.** (*c*) **9.** (*e*) **10.** (*f*) **11.** (*j*) **12.** (*a*) **13.** (*b*) **14.** (*d*) **15.** (*f*) **16.** (*j*) **17.** (*a*)**18.** (*c*) **19.** (*b*) **20.** (*d*)

Multiple Choice

1. (*d*) **2.** (*b*) **3.** (*a*) **4.** (*a*) **5.** (*d*) **6** . (*b*) **7.** (*d*) **8.** (*c*) **9.**(*b*) **10.** (*b*) **11.** (*d*) **12.** (*c*) **13.** (*c*) **14.** (*b*) **15.** (*a*) **16.** (*d*) **17.** (*d*) **18.** (*d*) **19.** (*c*) **20.** (*b*)

CHAPTER 13

Multiple Choice

1. (*c*) **2.** (*b*) **3.** (*a*) **4.** (*a*) **5.** (*d*) **6.** (*c*) **7.** (*b*) **8.** (*d*) **9.** (*a*) **10.** (*b*) **11.** (*c*) **12.** (*c*) **13.** (*d*) **14.** (*c*) **15.** (*a*)

True/False.
1. F **2.** F **3.** T **4.** T **5.** T **6.** F **7.** T **8.** F **9.** F **10.** T **11.** F **12.** F **13.** T **14.** F **15.** T

Completion

1. oxygen **2.** bacteria **3.** multicellular **4.** asexual **5.** two **6.** light **7.** diatoms **8.** cellulose **9.** red **10.** sexual

CHAPTER 14

True/False

1. protein **2.** true **3.** rabies **4.** true **5.** receptor **6.** true **7.** larger **8.** capsomeres **9.** envelope **10.** formaldehyde **11.** true **12.** lysogenic **13.** true **14.** liver **15.** true **16.** Negri bodies **17.** cell walls **18.** interferon **19.** antibodies **20.** true

Multiple Choice

1. (*b*) **2.** (*d*) **3.** (*c*) **4.** (*d*) **5.** (*d*) **6.** (*b*) **7.** (*c*) **8.** (*b*) **9.** (*d*) **10.** (*b*) **11.** (*a*) **12.** (*c*) **13.** (*c*) **14.** (*b*) **15.** (*b*) **16.** (*c*) **17.** (*a*) **18.** (*c*) **19.** (*c*) **20.** (*a*)

Completion

1. living cells **2.** DNA **3.** bacteria **4.** capsomeres **5.** envelopes **6.** bacteria **7.** sphere **8.** phenol **9.** detergent **10.** poxviruses **11.** antibodies **12.** lysosome **13.** genome **14.** ribosome **15.** reverse transcriptase **16.** lysozyme **17.** disease **18.** grown **19.** tumors **20.** azidothymidine

CHAPTER 15

Multiple Choice

1. (*b*) **2.** (*c*) **3.** (*c*) **4.** (*a*) **5.** (*d*) **6.** (*c*) **7.** (*b*) **8.** (*c*) **9.** (*a*) **10.** (*c*) **11.** (*d*) **12.** (*d*) **13.** (*d*) **14.** (*c*) **15.** (*c*) **16.** (*a*) **17.** (*b*) **18.** (*c*) **19.** (*a*) **20.** (*c*)

Completion

1. infection **2.** *Staphylococcus* **3.** infectious dose **4.** respiratory tract **5.** chronic infections **6.** asymptomatic **7.** pneumococcal pneumonia **8.** mosquito **9.** high **10.** contaminated **11.** blood **12.** disease **13.** collagenase **14.** intoxications **15.** Centers for Disease Control and Prevention **16.** reservoir **17.** contagious **18.** red blood cells **19.** lactobacilli **20.** opportunistic **21.** hepatitis B **22.** toxemia **23.** botulism **24.** systemic **25.** Grarr negative **26.** symptom **27.** endemic **28.** food **29.** fomites **30.** exotoxins

True/False

1. F **2.** F **3.** T **4.** T **5.** T **6.** F **7.** T **8.** F **9.** F **10.** T **11.** F **12.** T **13.** T **14.** T **15.** T **16.** F **17.** F **18.** T **19.** F **20.** T

CHAPTER 16

True/False

1. mechanical **2.** lysozyme **3.** protein **4.** true **5.** helper **6.** plasma **7.** artificially acquired active immunity **8.** true **9.** true **10.** true **11.** lymph nodes **12.** neutrophils **13.** artificial passive **14.** true **15.** haptens **16.** B-lymphocytes **17.** antibody **18.** IgM **19.** perforin **20.** B cells **21.** true **22.** days **23.** six months **24.** T cells **25.** IgD **26.** plasma cells **27.** thymus **28.** true **29.** pathogen **30.** true

Multiple Choice

1. (*c*) **2.** (*d*) **3.** (*b*) **4.** (*a*) **5.** (*b*) **6.** (*d*) **7.** (*c*) **8.** (*b*) **9.** (*b*) **10.** (*a*) **11.** (*b*) **12.** (*d*) **13.** (*b*) **14.** (*b*) **15.** (*a*) **16.** (*b*) **17.** (*c*) **18.** (*c*) **19.** (*c*) **20.** (*c*)

Completion

1. T cells **2.** mononuclear phagocytic system **3.** naturally acquired passive immunity **4.** MHC molecules **5.** proteins **6.** variable region **7.** IgA **8.** membranes **9.** viruses **10.** complement system **11.** memory cells

CHAPTER 17

Multiple Choice

1. (*a*) **2.** (*c*) **3.** (*b*) **4.** (*c*) **5.** (*a*) **6.** (*c*) **7.** (*d*) **8.** (*d*) **9.** (*c*) **10.** (*b*) **11.** (*c*) **12.** (*a*) **13.** (*b*) **14.** (*c*) **15.** (*a*) **16.** (*d*) **17.** (*b*) **18.** (*d*) **19.** (*d*) **20.** (*c*)

Matching

1. (*c*) **2.** (*d*) **3.** (*b*) **4.** (*a*) **5.** (*c*) **6.** (*d*) **7.** (*e*) **8.** (*c*) **9.** (*b*) **10.** (*a*)

True/False

1. F **2.** T **3.** F **4.** F **5.** T **6.** F **7.** T **8.** F **9.** T **10.** T **11.** F **12.** T **13.** F **14.** F **15.** F **16.** T **17.** F **18.** T **19.** T **20.** F

CHAPTER 18

Completion

1. *Staphylococcus* **2.** alpha-hemolytic **3.** skin **4.** acyclovir **5.** *Pseudomonas aeruginosa* **6.** poxvirus **7.** smallpox **8.** Koplik spots **9.** rubella **10.** A **11.** Herpesviridae **12.** pregnant **13.** griseofulvin **14.** gonorrhea **15.** eye **16.** measles **17.** genital herpes **18.** exfoliative toxin **19.** scarlet fever **20.** papilloma viruses **21.** 14 days **22.** MMR vaccine **23.** tinea diseases **24.** fungi **25.** *Candida albicans* **26.** shingles **27.** fungi **28.** conjunctiva **29.** eye **30.** *Nocardia*

Multiple Choice

1. (*d*) **2.** (*d*) **3.** (*b*) **4.** (*c*) **5.** (*b*) **6.** (*c*) **7.** (*d*) **8.** (*d*) **9.** (*c*) **10.** (*a*) **11.** (*b*) **12.** (*a*) **13.** (*d*) **14.** (*d*) **15.** (*b*) **16.** (*a*) **17.** (*a*) **18.** (*d*) **19.** (*a*) **20.** (*d*)

True/False

1. T **2.** T **3.** F **4.** T **5.** F **6.** T **7.** F **8.** T **9.** F **10.** F **11.** T **12.** F **13.** F **14.** T **15.** F **16.** F **17.** T **18.** T **19.** T **20.** F

CHAPTER 19

Multiple Choice

1. (*b*) **2.** (*d*) **3.** (*c*) **4.** (*d*) **5.** (*b*) **6.** (*d*) **7.** (*a*) **8.** (*c*) **9.** (*d*) **10.** (*c*) **11.** (*d*) **12.** (*b*) **13.** (*a*) **14.** (*d*) **15.** (*d*) **16.** (*b*) **17.** (*c*) **18.** (*c*) **19.** (*a*) **20.** (*d*) **21.** (*d*) **22.** (*d*) **23.** (*a*) **24.** (*b*) **25.** (*c*)

Matching

1. (*j*) **2.** (*e*) **3.** (*n*) **4.** (*g*) **5.** (*b*) **6.** (*l*) **7.** (*a*) **8.** (*o*) **9.** (*d*) **10.** (*m*) **11.** (*h*) **12.** (*c*) **13.** (*k*) **14.** (*f*) **15.** (*i*)

True/False

1. T **2.** T **3.** F **4.** F **5.** T **6.** F **7.** F **8.** T **9.** F **10.** T **11.** F **12.** T **13.** T **14.** F **15.** T **16.** F **17.** F **18.** F **19.** T **20.**T

CHAPTER 20

Completion

1. acid-fast **2.** cell wall **3.** amphotericin B **4.** blood cells **5.** pertussis **6.** tuberculosis **7.** *Paragonimus westermani* **8.** protozoan **9.** rodents **10.** antigenic shift **11.** vaccine **12.** lung **13.** *Streptococcus* **14.** exotoxins **15.** *Coxiella burnetii* **16.** ornithosis **17.** *Mycobacterium* **18.** droplets **19.** amantadine **20.** valley fever **21.** Darling's disease **22.** *Pneumocystis* **23.** scarlet fever **24.** Mantoux **25.** diphtheria

Multiple Choice

1. (*c*) **2.** (*a*) **3.** (*c*) **4.** (*d*) **5.** (*b*) **6.** (*d*) **7.** (*b*) **8.** (*c*) **9.** (*b*) **10.** (*b*) **11.** (*a*) **12.** (*c*) **13.** (*a*) **14.** (*d*) **15.** (*b*) **16.** (*d*) **17.** (*c*) **18.** (*b*) **19.** (*b*) **20.** (*d*)

True/False

1. gamma **2.** valves **3.** scarlet fever **4.** true **5.** no **6.** BCG **7.** diphtheria **8.** pneumococcal **9.** true **10.** chlamydiae **11.** Legionnaires' disease **12.** influenza **13.** fungus **14.** amphotericin B **15.** AIDS

CHAPTER 21

True/False

1. F **2.** T **3.** F **4.** T **5.** T **6.** F **7.** T **8.** T **9.** F **10.** F **11.** T **12.** F **13.** T **14.** T **15.** F **16.** T **17.** T **18.** T **19.** F **20.** F

Multiple Choice

1. (*d*) **2.** (*b*) **3.** (*a*) **4.** (*c*) **5.** (*d*) **6.** (*a*) **7.** (*a*) **8.** (*b*) **9.** (*c*) **10.** (*b*) **11.** (*c*) **12.** (*b*) **13.** (*a*) **14.** (*d*) **15.** (*b*) **16.** (*c*) **17.** (*b*) **18.** (*a*) **19.** (*c*) **20.** (*c*)

Completion

1. trench mouth **2.** *Fusobacterium* **3.** orchitis **4.** staphylococcal food poisoning **5.** serological types **6.** feces **7.** carrier **8.** water **9.** negative **10.** shellfish **11.** bipolar staining **12.** peptic ulcers **13.** gastroenteritis **14.** myocarditis **15.** attenuated viruses **16.** flagella **17.** *Amanita* **18.** cyst **19.** tapeworms **20.** roundworm

CHAPTER 22

Multiple Choice

1. (*c*) **2.** (*b*) **3.** (*c*) **4.** (*c*) **5.** (*d*) **6.** (*d*) **7.** (*a*) **8.** (*d*) **9.** (*b*) **10.** (*c*) **11.** (*a*) **12.** (*d*) **13.** (*d*) **14.** (*c*) **15.** (*b*) **16.** (*a*) **17.** (*d*) **18.** (*b*) **19.** (*c*) **20.** (*d*) **21.** (*d*) **22.** (*c*) **23.** (*a*) **24.** (*b*) **25.** (*b*) **26.** (*c*) **27.** (*a*) **28.** (*c*) **29.** (*b*) **30.** (*c*)

True/False

1. valves **2.** rod **3.** bipolar **4.** true **5.** undulant **6.** spirochete **7.** maculopapular **8.** tick **9.** virus **10.** true **11.** true **12.** true **13.** fifth disease **14.** protozoan **15.** ticks

Matching

1. (*l*) **2.** (*k*) **3.** (*o*) **4.** (*i*) **5.** (*a*) **6.** (*d*) **7.** (*m*) **8.** (*n*) **9.** (*f*) **10.** (*d*) **11.** (*h*) **12.** (*b*) **13.** (*c*) **14.** (*g*) **15.** (*j*)

CHAPTER 23

Multiple Choice

1. (*c*) **2.** (*b*) **3.** (*d*) **4.** (*a*) **5.** (*a*) **6.** (*b*) **7.** (*d*) **8.** (*b*) **9.** (*d*) **10.** (*c*) **11.** (*d*) **12.** (*c*) **13.** (*c*) **14.** (*c*) **15.** (*a*) **16.** (*d*) **17.** (*b*) **18.** (*d*) **19.** (*a*) **20.** (*b*) **21.** (*a*) **22.** (*c*) **23.** (*b*) **24.** (*a*) **25.** (*c*)

Completion

1. cystitis **2.** Weil's disease **3.** animals **4.** *Staphylococcus aureus* **5.** flagella **6.** negative **7.** urethra **8.** males **9.** gonococcal ophthalmia **10.** spirochete **11.** secondary stage **12.** penicillin **13.** *Haemophilus ducreyi* **14.** gonorrhea **15.** erythromycin **16.** chlamydia **17.** mononuclear cells **18.** herpes simplex **19.** genital warts **20.** TORCH **21.** *Candida albicans* **22.** nerves **23.** herpes whitlow **24.** lactobacilli **25.** acyclovir

True/False

1. T **2.** T **3.** F **4.** T **5.** T **6.** F **7.** F **8.** T **9.** T **10.** F

CHAPTER 24

True/False

1. pathogens **2.** yogurt **3.** true **4.** carbohydrates **5.** bacteria **6.** *Lactobacillus* **7.** true **8.** true **9.** vegetative cells **10.** time **11.** yeasts **12.** water **13.** benzoic **14.** sound **15.** staphylococci **16.** phosphatase **17.** true **18.** chemostat **19.** protein **20.** yeasts **21.** barley grains **22.** true **23.** true **24.** insecticide **25.** plasmids

Multiple Choice

1. (*b*) **2.** (*d*) **3.** (*a*) **4.** (*d*) **5.** (*b*) **6.** (*a*) **7.** (*b*) **8.** (*d*) **9.** (*c*) **10.** (*c*) **11.** (*b*) **12.** (*d*) **13.** (*c*) **14.** (*b*) **15.** (*c*) **16.** (*b*) **17.** (*d*) **18.** (*c*) **19.** (*d*) **20.** (*c*)

Completion

1. holding **2.** acid **3.** bleu **4.** sauerkraut **5.** fish **6.** dehydration **7.** vegetative **8.** yeasts **9.** smoking **10.** gamma **11.** shigellosis **12.** sparkling **13.** beer **14.** amino acid **15.** milky spore disease

CHAPTER 25

Completion

1. ecosystem **2.** photosynthesis **3.** *Gallionella* **4.** spores **5.** plants **6.** amino acids **7.** *Rhizobium* **8.** urea **9.** denitrification **10.** methionine **11.** hydrogen sulfide **12.** phosphate **13.** biochemical oxygen demand **14.** settling tanks **15.** methane **16.** pollution **17.** filter **18.** coliform bacteria **19.** membrane filter **20.** most probable number test

True/False

1. T **2.** T **3.** F **4.** T **5.** F **6.** T **7.** F **8.** F **9.** F **10.** T **11.** F **12.** F **13.** T **14.** F **15.** T **16.** F **17.** T **18.** F **19.** T **20.** T

Multiple Choice

1. (*c*) **2.** (*b*) **3.** (*b*) **4.** (*d*) **5.** (*d*) **6.** (*b*) **7.** (*a*) **8.** (*b*) **9.** (*a*) **10.** (*a*) **11.** (*d*) **12.** (*c*) **13.** (*a*) **14.** (*c*) **15.** (*b*) **16.** (*a*) **17.** (*a*) **18.** (*a*) **19.** (*c*) **20.** (*d*)

Index

To: Dept 9410
Creative Homeowner Press®
24 Park Way
Upper Saddle River, New Jersey 07458

Enclose $3.75 (includes 1st class postage & handling) for each pattern or plan ordered.
Add $2.95 for catalog and allow 6 to 8 weeks for delivery.
Place an X in the box beneath the pattern number for your order.

9410

103 630 666 599 113
632 411 313 392 56
Catalog

Enclose check or money order for total amount of order (California residents please add 6% sales tax) and make payable to U-Bild Enterprises.

Name (Print) ____
Address ____
City ____
State ____ Zip ____